Garden Plants Taxonomy

Bijan Dehgan

Garden Plants Taxonomy

Volume 1: Ferns, Gymnosperms, and Angiosperms (Monocots)

Springer

Bijan Dehgan
Emeritus Professor
Environmental Horticulture
University of Florida
Gainesville, FL, USA

ISBN 978-3-031-11560-8   ISBN 978-3-031-11561-5   (eBook)
https://doi.org/10.1007/978-3-031-11561-5

This Springer imprint is published by the registered company Springer Nature Switzerland AG
The registered company address is: Gewerbestrasse 11, 6330 Cham, Switzerland

Foreword

This textbook/reference book on the taxonomy of ornamental garden plants presents the results of a lifetime of study of the horticultural systematics and cultivation of plants grown in the United States, from Minnesota to southern Florida, and from the Atlantic coast to California, although the focus is clearly on ornamental herbs, shrubs, and trees of Florida and adjacent regions. The book covers 3 families of lycophytes, 19 families of monilophytes (ferns and allies), 11 families of gymnosperms (including 2 cycad and 6 conifer families), and 177 families of flowering plants (angiosperms; including 2 early divergent familial clades, 6 families of magnoliids, 32 families of monocots, and 137 families of eudicots). The circumscription of these families follows that of the Angiosperm Phylogeny Group IV classification (in the case of the angiosperms), while the circumscription of families in other major clades follows Peter Stevens' Angiosperm Phylogeny website or other recent phylogenies. The monophyly of nearly all these families is well supported, but Calophyllaceae are included within an expanded Clusiaceae. Subfamilies are usually mentioned, and economically important genera for each family are listed. The book provides detailed taxonomic treatments (with numerous photographs indicating diagnostic features, morphological variation, landscape uses, and ornamental value) for 970 genera and/or species, including 48 gymnosperms, 18 magnoliids, 214 monocots, and 657 eudicots; in addition, many additional genera and/or species are given brief coverage (with included photos). For each taxon receiving full coverage, the scientific and common names are provided, along with etymological information (and sometimes synonyms), including a description of the plant's morphology, geographical distribution, economic uses, culture and propagation, hardiness zone, and interesting, miscellaneous comments. Cultivars are included and illustrated as appropriate. Detailed introductions to morphological variation and associated terminology are provided for gymnosperms and angiosperms (with numerous helpful photos and illustrations). The author also briefly introduces the International Code of Nomenclature for Cultivated Plants and explains how it works together with the botanical code (i.e., International Code of Nomenclature for algae, fungi, and plants). Descriptive botanical terminology is rather complex; thus, an extensive glossary is provided.

This book is sure to be very useful to horticulture and botany students, those working in horticultural occupations, individuals seeking to improve the aesthetic quality of the landscapes around them through the cultivation of vascular plants, or even to those who merely wish to better enjoy and appreciate private or public botanical gardens (in the United States). The book will place the wonderful diversity of vascular plants within a phylogenetic framework, especially for those who have not spent much time thinking about evolutionary relationships. Finally, the book is sure to be a cure for plant blindness, i.e., the inability to see or notice plants in one's own environment, a condition that is so negatively affecting our popular American culture.

Distinguished Professor Emeritus, Department of Biology　　　　　　　　　Walter S. Judd
University of Florida,
Gainesville, FL, USA

Preface

"Plant Blindness" was first described by academic botanists James Wandersee and Elisabeth Schussler in 1998. In short, the name invokes the inability or unwillingness among modern folk "to see or notice the plants in their environments."

It can well be argued that horticulturists are among the least "plant blind" people that one can encounter. Couple the love of growing plants with doctoral training in plant taxonomy, and the antithesis of plant blindness is formed. Such has been the professional career of Bijan Dehgan.

Dr. Dehgan has blended the fine art of growing plants with the rigor of the botanist's enterprise of knowing, identifying, and classifying plants. His career has embraced the systematics of *Jatropha*, a large and difficult genus of the family Euphorbiaceae, along with efforts to improve the growth and conservation of cycads, a worldwide group of ancient gymnosperms. Individuals like Bijan are bridge-builders between the worlds of plant growers and plant knowers.

In the last few decades, DNA technology has led to a revolution in our understanding of plant evolution and the relationships among many plant groups The speed with which this revolution has occurred has often left horticulturists grumbling, as names changed, and genera leapfrogged from one long-established family to another. Dr. Dehgan's comprehensive book places the cornucopia of valued ornamental plants into the latest system of plant classification as realized by the genomic evidence that has accumulated about the evolution of the plant kingdom.

"Garden Plants Taxonomy" begins by first introducing the reader to the rudiments of plant taxonomy, the how and why of plant classification, both botanical and horticultural, including the concept of cultivars. Without rules for nomenclature, any attempt at classification soon becomes chaos. The lion's share of the volume is devoted to a systematic exploration of horticultural biodiversity, beginning with the ferns and fern allies, followed by cycads and conifers (gymnosperms), and culminating with flowering plants, angiosperms. Salient characteristics of these large subdivisions are provided with excellent expository illustrations. Within these broad categories, each a major division of the plant kingdom, horticulturally significant plant families and their constituent genera are detailed, and lavishly illustrated with representative species valued in cultivation. Cultural notes, as well as scientific information, are often provided.

Professional horticulturists, gardeners, and students seeking a deeper understanding of the relationships among garden plants will find the book a meaningful addition to their reference libraries. It will become an essential tool for teachers and students of plant identification courses.

On a personal note, Dr. Dehgan was the graduate committee chair for my own doctoral degree in horticultural taxonomy 40 years ago, and like him, my professional career followed a combined path of horticultural and botanical research. Thank you, Bijan, both for that and this magnificent contribution to alleviating incipient plant blindness!

Retired geneticist and systematist   Alan W. Meerow
U.S. Department of Agriculture – Agricultural Research Service
Washington, DC, USA

References

Wandersee, James H., and Elisabeth E. Schussler, "Toward a Theory of Plant Blindness," Plant Science Bulletin 47, no. 1 (March 2001): 2–9.

Wandersee, James H., and Elisabeth E. Schussler, "Preventing Plant Blindness," The American Biology Teacher 61, no. 2 (February 1999): 82–86.

Introduction

Botanists have generally neglected cultivated varieties as beneath their notice. (Darwin 1888)

All too often among orchid growers, the mention of taxonomy and taxonomists elicits a cacophony of hisses and boos. Actually, most orchid taxonomists are also avid orchid growers, and this article attempts to bridge the communication gap between what would appear to be two distinct and antagonistic categories of people. (Kiat Tan 1980)

The residual failures of harmony between botanical and cultivated plant systematists is related to review of their methodologies. Because some cultivated plant groups overlap the distinction, it is argued that special classification systems are not appropriate generally to cultivated plants. (John Lewis 1988)

The two classification systems (taxonomic and cultonomic) should be seen as separate but linked at some point. (R. G. Van Den Berg 1999)

Are taxonomists an endangered species, wrote Panczyk (1996) in American Nurseryman; "The future of taxonomists, that rare breed responsible for classifying and identifying plants, is uncertain – not because anyone is out to get them, but because the world is changing and their profession may be left behind." She was correct in one respect; the world was changing and taxonomy right along with it. Classifications that were simply based on visual characteristics (morphology) were changing to anatomical, chemical, numerical, and ultimately to systematic DNA-based phylogeny grounded on common ancestry (monophyly). That most botanical gardens and arboreta could no longer afford maintaining the essential facilities to conduct such studies has effectively been transferred to large well-financed gardens and university botany (biology) departments. So, where does horticulture fit in the scheme of things? After all, horticulturists are the ones that directly deal with growing the fruits, vegetables, and flowers necessary for human survival.

Heiser (1969) noted that with a few notable exceptions, notably L.H. Bailey, taxonomists seem to have avoided cultivated plants. Edgar Anderson (1949), in his famous book, *Pants, Man, and Life*, called it "the greater paradox." He noted that plant taxonomists pass up cultivated plants and rush to cloud forests to throw plants into their press. The fact that cultivated plants originated from the wild seems to have been overlooked, particularly by horticultural taxonomists, the rare entities that do not simply list individual species without any explanation of their origin or relationships. In some respects, this is understandable since most crop and landscape plants are selections of choice, or more appropriately, cultigens. As a general rule, perhaps subconsciously, we recognize (identify) plants based on their relationship or comparison with other plants. This is best explained by Hui-Lin Li (1974), "In utilizing and cultivating plants, it was imperative that primitive man learn to recognize and differentiate the various types. This identification of plants gradually led to appreciating their resemblances and differences. This rudimentary knowledge of grouping and classifying plants, accumulating slowly through the ages, finally led to development of taxonomy." Recognition and grouping of plants probably began before man learned to cultivate.

The first man was nomadic and gatherer of food, vegetables, and fruit from the wild. To do that he/she would have to recognize which fruits and vegetables were edible and which inedible or toxic. Thus, it was necessary to be familiar with the characteristics of useful plants and their similarities and differences with one another. They were the first botanists and essentially the first plant taxonomists. Once land ownership became the norm, plants were brought from the wild and the art of gardening and horticultural practices had to be learned. Thus, recognition of plant relationships was necessary to expand the diversity of their plants to meet their nutritional needs. Gradual progress from growing plants for food to gardening for pleasure and visual gratification was only the next logical step.

"Cultivated Plant Taxonomy" is defined as "the study of theory and practice of the science that identifies, describes, classifies, and names cultigens—those plants whose origin or selection is primarily due to intentional human activity." Although there is some merit to this definition, the fact remains that cultivated plants originate from wild natural plants and therefore cannot be completely divorced from botanical taxonomy. There is no conceivable means of assembling cultigens in a specific classification system without using some botanical principles, at least a genus, and, when appropriate, the specific epithet of the parental taxon. Indeed, the Cultivated Code is for proper naming and application of Cultivar, Group, and Grex, not for their classification. Several interesting and thought-provoking articles in Andrews et al. (1999) have suggested the classification of cultigens. Cultigen is a term strictly used for plants that originated in cultivation by humans and cannot be accepted or organized as a substitute for cultivated plants that originated from wild species. Citing a cultigen without a binomial or at least a generic name is meaningless. Plant lists that are created for and by United States Department of Agriculture (USDA) are arranged alphabetically by a genus name followed by species (a binomial) and then cultigens (usually cultivars, nothospecies, or nothogenus). If there is to be a classification for cultivated plants, it must be based on more logical concepts with botanical names as the root of cultigens. The term "Culton" has been introduced presumably, as a substitute for species, to facilitate the classification of cultivated plants. An appropriate application of this term as a substitute for species, a term that has been discussed and debated for centuries, is questionable.

Horticultural taxonomy is defined as "study of the identification, naming, and classification of horticulturally important plants." The very names of plants are in continuous flux since publication of the *Species Plantarum* by Linnaeus in 1753. The International Code of Nomenclature in 1867, by Alphonse de Candolle (Lois de la Nomenclature Botanique), set the standards for naming organisms. The current Botanical Code, in its 17th edition, can make valid today a name that was obsolete yesterday, and a name may be totally discarded, replaced, or transferred to one published earlier, and even by a new one (several cases have been shown in this book). Things are not as complicated as it seems. An understanding has developed between horticulturists and botanists. The botanical nomenclature and its rules are accepted and complied with by horticulturists. But it became necessary to create the International Code of Nomenclature for Cultivated Plants (ICNCP) in 1953 to publish rules and regulations for naming and application of cultigens: Cultivars, Groups, and Grexes; plants that are created or selected by humans. Hence, cultigens cannot be written without accurate applicable botanical nomenclature, as has been used throughout this book. A good example of cultivated plant taxonomy (systematics) is those of Pickersgill and Karamura (1999) of banana, Moller and Cronk (1999) of *Saintpaulia* and *Streptocarpus*, and Zhang et al. (1999) for *Cephataxus*, among others. These systematic approaches form the foundation of proper classification of species and cultigens (see, for example, several publications by Meerow and Meerow et al., in references for Amaryllidaceae). There are, however, some reasonably well-organized groupings without background systematics, such as that of *Taxus* (Hoffman 2004).

If interpreted as practical botany, horticulture has remained far behind in understanding its botanical principles. Recent phylogenetic (DNA-based) reorganization of higher plants has revolutionized taxonomic treatments of all biological entities, even when morphology does not completely agree with their organization, as noted several times in this book. This book is an example of applying principles of botanical phylogenetic taxonomy to assemble genera and species of 200 vascular plant families of ferns, gymnosperms, and angiosperms that are cultivated for enhancement of human living spaces: homes, gardens, and parks. The emphases are on cultivated species but examples of same plants are often shown in the wild and in landscapes. In providing descriptions, it is assumed that students and other interested individuals have no background in general botany (plant characteristics), or nomenclature. Fundamental features of all plant groups discussed are fully illustrated by original watercolor drawings or photographs. Orders, higher classes, and authority of names are usually not introduced because they are of little utility in horticultural practices (except at least once at the outset of scientific publications). Anyone interested in detailed discussions of taxa above families should refer to recent publications by Judd et al. (1915), Christenhusz et al. (2017), Soltis et al. (2018), and Angiosperm phylogeny website – Missouri Botanical Garden by P. F. Stevens (2016 et seq.) www.mobot.org/MOBOT/Research/APweb.

Discussion of the families is grounded on recent botanical phylogenetic treatments, which is based on common ancestry (monophyly). There is nothing new about the concept of similarity or dissimilarity of plants within botanically designated families and genera relevant in horticulture. There is no conceivable means of creating a special classification for cultivated plants without employing fundamental botanical principals, as noted above. This has been long reflected in older as well as newer relevant horticultural publications, beginning with Theophrastus, the so-called father of botany, in *Historia Plantarum* (c. 350 BC), and Linnaeus, the father of taxonomy, in *Species Plantarum* (1753), to modern works such as Hortus III (1976), *The European Garden Flora* (1986, 2011), *The New York Botanical Garden Illustrated Encyclopedia of Horticulture* (Everett 1980), *The New Royal Horticultural Society Dictionary of Gardening* (1999), and numerous encyclopedic publications (*cf.* references), as well as the more recent excellent comprehensive works such as *Tropical Flowering Plants* (Llamas 2003), *Tropical Garden Flora* (Staples and Herbst 2005), *Plants of the World* (Christenhusz et al. 2017), and *Phylogeny and Evolution of the Angiosperms* (Soltis et al. 2018). The chief audiences of these publications are botanists and horticulturists. Students of botanical taxonomy may take field trips to wild natural habitats, but essentially, they learn much of their plant relationships from garden plants, as do students of plant identification in horticulture and landscape architecture classes. In reality, a vast majority of cultivated landscape plants originated from the wild. The difference lies in the topics covered in the botany class and use of such textbooks as Heywood et al. (2007), Simpson (2010), and particularly that of Judd et al. (2015). The latter book is the most comprehensive and authoritative treatment of plant systematics (research-based taxonomy) by five well-known plant taxonomists (systematists). It includes the latest phylogenetically based taxonomic information as opposed to the more common horticulture treatments that include descriptions of individual taxa, often presented without providing the basic knowledge of familial or generic relationships. It is of notable interest that in earlier years, before the advent of chemotaxonomy, numerical taxonomy, and the current DNA-based phylogenetic systematics, the most common textbook in both botany and horticulture identification classes was Bailey's *Manual of Cultivated Plants* (1949). Of course, phylogenetic taxonomy is not a new concept and was originally based on morphological characteristics; it is the DNA-based phylogeny that has revolutionized modern biological classifications. In practical terms, this book represents the horticultural treatment that corresponds to phylogenetic-based botanical taxonomy, to which is added a list of cultigens and related cultivated genera and species. Hence, the harmony between horticultural and botanical taxonomy.

The distinction between botanical and horticultural treatments of plant families and lower taxa is a bewildering issue. Should horticultural concerns (cultigens) override the phylogenetic bases of botanical classifications? As this question frequently surfaced while writing this book, it became clearly evident that the answer is not as simple as it may seem. Recent phylogenetic plant classifications have led to profound rearrangement, reorganization, and in some cases, essentially dismantling and reassembling of some families. Take the case of Malvaceae, which traditionally was known because of familiarity with such commonly cultivated plants as *Hibiscus*, okra, cotton, and related genera. The current phylogenetic classification system combines nine previously familiar and, for the most part, readily recognizable families (Berryaceae, Bittneriaceae, Bombacaceae, Dombeyaceae, Grewiaceae, Helictraceae, Malvaceae, Sterculiaceae, and Tiliaceae) and are treated as subfamilies with tribes. The same may be said of Fabaceae that was readily recognizable in seemingly three coherent subfamilies but are now in six monophyletic subfamilies. While it is difficult to argue with DNA-based phylogeny, the issue becomes one of proper description, which is often based on shared DNA and chemical and anatomical features (systematics), and only secondarily by vegetative and floral morphology (at least by Angiosperm Phylogeny Group: 1998–2016 et seq.). The morphological features, however, are fundamental in horticulture and landscape architecture. It is, therefore, not surprising that many horticulture plant identification classes and books simply rely on individual species and cultivar characteristics and avoid any discussion of higher taxa (e.g., families, subfamilies, tribes) than genera. In some cases, families are mentioned without description or clarification of shared generic features. To remain current and accurate, in cases where phylogenetic classification demanded the inclusion of more than one family, such as Malvaceae, Fabaceae, Apocynaceae, and several others, subfamilies are listed within the given family, with descriptions that exclude visually unobservable chemical and anatomical features, followed by generic and specific examples. To reiterate, in presenting family descriptions (and generic examples), in this book a distinction is made between what J. Lewis (1989) noted as "methodologies," by disregarding the more basic technical visually non-recognizable features (chemical, anatomical, etc.), used in botanical phylogenetic treatments and relying solely on morphological characteristics, but treated based on the current phylogenetic system. However, credit should be given to Judd et al. (1999–2015), now in its 4th edition, and other more recently published books, for providing detailed visually observable characteristics, as well as other relevant shared features, despite some horticulturally unfamiliar treatments. Of course, the same may be said of a number of other less current books. In this book, commercially available cultivars in the trade and list of related cultivated genera and species are added, and many are fully illustrated with color photographs to make this book horticulturally more useful and utilitarian.

The taxa presented in this book are not exclusively examples of plants hardy in warmer subtropical climatic zones, as did *Landscape Plants for Subtropical Climates* (1998), by this author, or those strictly on tropical plants. Moreover, in many respects, this book has little similarity with the previous publication. Not only there are greater details and many more plants, but black and white drawings have been replaced with color illustrations and photographs. Families and representative taxa have been significantly increased, and agree with and in general are based on the recent phylogenetic classifications with brief comments on relationships (subfamilies, etc.). Every attempt has been made to photographically present major characteristics with more than a single picture. Vegetative and floral characteristics have also been presented in color as are general features of major groups. For centuries black and white drawings have been, and continue to be, the primary source of images in plant books. Indeed, most, if not all, revisionary, monographic, and floristic publications (see for example Flora of North America, Dehgan 2012; Judd 2015), either do not allow color photographs or only in limited numbers. This policy mostly follows many centuries of tradition. There are, however, many examples of earlier centuries that have been masterly illustrated with color paintings (e.g., Köhler's Medizinal-Pflanzen, 1887–1898, Otto Wilhelm Thomé, 1885 for Flora of Germany, Austria, etc.), as well as newer publications such as Heywood et al. (2007) and Simpson (2010). There are several recently published books with color photographs. One of the two earlier mentioned books, Llamas (2003), includes a photograph in color for each tropical taxon discussed while that of Staples and Herbst's (2005) voluminous treatment of Hawaii's cultivated plants includes only 12 collated color plates but a large number of black and white drawings that illustrate major features. Two more recently published books by well-known botanists, Christenhusz et al. (2017) and Soltis et al. (2018), are superb examples of using color photographs, but drawings only where necessary. It would be a difficult task to replace drawings that readily represent most morphological features of a given species with one or more color photographs. Such undertaking requires several years of photography and significant time and travel, not to mention difficult publication and cost issues. Moreover, representing all major features of any taxon in color may be notable but overwhelming.

Writing this book has involved many years of travel to various countries, such as the United States, and numerous botanical gardens and photography as far back as 50 years ago. With very few exceptions, every attempt has been made to include the plant's use in designed landscape situations or in the wild. In some instances, as a matter of interest, the country or the garden where the photographs were taken is noted in the captions. This was done in part to show climatic ranges of where plants may be grown. Nearly all listed species and cultivars, however, are in cultivation in the United States, from south Florida to Minnesota, and western Mediterranean climate regions and beyond. To avoid making the book too voluminous, additional photographs of related taxa are added with a brief description to better illustrate members of families and genera. Relevant citations are included only where most helpful, otherwise noted in the reference list.

Last but not least, as author of this book I take full responsibility for any errors or lapses in updating the accuracy and organization of the taxa. Changes in phylogeny and nomenclature are practically a daily affair, and only a botanical genius such as Professor Walter Judd can keep up and recollect such changes. Although this book is intended as a reference or textbook for horticulture scientists and students, it is no less a useful tool for botanists. To reiterate, I am wholeheartedly grateful to my friend and colleague for editing this book and writing the Foreword.

Acknowledgments

Although majority of the photographs in this book are those of the author taken over many years, and indeed some scanned from photographic slides as old as 50 years old, I wish to acknowledge my sincerest gratitude to the well-known and truly talented botanical illustrator Ms. Susan Trammell for her excellent watercolor illustrations of angiosperm characteristics. I also wish to express my long-held appreciation to my valued friend and coworker of nearly 30 years, Ms. Fe Almira, for allowing me to use some of her excellent photographs. Unfortunately, as the author and Ms. Almira have often been on same US and international locations and photographed the same subjects, it has ultimately proven difficult to distinguish the source of each photograph with certainty. I have been most fortunate to have a well-known excellent illustrator and a superb photographer and scientist on my team. Permission to use 4000 superb photographs by professor Kurt Stüber of Max Planck Institute has been a particular blessing and an important contribution to the presentation of some families. It is of notable interest that some of professor Stüber's photographs appear on Wikipedia. Approval by Dr. Edward Gilman, emeritus professor of Environmental Horticulture, of possible inadvertent use of his tree photographs is also greatly appreciated. I also wish to thank Dr. Gerald Carr, professor emeritus at the University of Hawaii, for permission to use his photograph of *Araucaria columnaris* seed. Ernesto Sandoval, the curator of University of California, Davis Conservatory, provided a close-up photograph of *Dorstenia gigas*. Dr. Bart Schutzman provided photographs of blueberry (*Vaccinium* cv.). My interminable gratitude goes to Professor Walter S. Judd for his advice, corrections, criticisms, and editing this book, as well as contribution of a few photographs. His infinite knowledge of plant taxonomy and up-to-date phylogeny of plants, natural and cultivated, is in no small measure that of a true botanical scientist. Most assuredly he deserves to be the coauthor of this book. He declined on the grounds that he did not make significant contributions to justify his coauthor. That is attributable only to his modesty. I am privileged to have his association. I am equally pleased and gratified by association with and thoughtful forward of Dr. Alan W. Meerow, USDA-ARS-SHRS, renowned horticultural plant taxonomist's Preface contribution. The large and technical research and his contributions to phylogeny of Amaryllidaceae are nothing short of a perfect example of how horticultural systematics research should be conducted. Last but not least, Nancy Dehgan, my dear wife of 55 years, encouraged me to write and publish this book and in fact earlier spent much time checking spellings, contents, and scanning photographic slides. With much regret and profound sadness, she did not live long enough to see the result. This book is dedicated to her memory.

Contents

About the Author

Bijan Dehgan received degrees in agricultural engineering from Shiraz University (in Iran), certificate in American language from Columbia University, BS and MS in landscape and environmental horticulture, and PhD in botany (plant taxonomy, under Professor Grady L. Webster) at the University of California, Davis. While working on systematics of the genus *Jatropha* (Euphorbiaceae), he was a lecturer in Environmental Horticulture and curator of the Botanical Conservatory. He was the principal researcher on selection and propagation of drought tolerant plants for California highways. He transferred to the Department of Environmental Horticulture at the University of Florida in 1978 and retired as emeritus professor in 2006. Although the principal investigator in several research projects, his primary focus remained on taxonomy of *Jatropha* and successful conservation of the endangered cycads through their propagation and growth, so as to prevent the collection of wild plants. Abundance of *Zamia pumila* (=Z. *floridana*) in landscapes is testimony to successful conservation of an endangered plant. He was also the principal investigator in revegetation of phosphate mines in Florida. His publications include *Landscape Plants for Subtropical Climates*, *Public Garden Management: A Global Perspective*, and *Jatropha* (Euporbiaceae) by University of California and New York Botanical Gardens, among others. He has been a recipient of awards for teaching and for contribution to the advancement of botanical science by the Botanical Society of America.

Alphabetical List of Families

A

ACANTHACEAE	III
ACERACEAE (see Sapindaceae)	II
ACORACEAE	I
ADOXACEAE (see Viburnaceae)	II
AIZOACEAE	II
ALISMATACEAE	I
ALSTROEMERIACEAE	I
ALTINGIACEAE	II
AMARANTHACEAE	II
AMARYLLIDACEAE	I
ANACARDIACEAE	II
ANNONACEAE	II
APIACEAE	II
APOCYNACEAE	III
AQUIFOLIACEAE (see Schisandraceae)	III
ARACEAE	I
ARALIACEAE	III
ARAUCARIACEAE	I
ARECACEAE	I
ARISTOLOCHIACEAE	II
ASCLEPIADACEAE (see Apocynaceae)	II
ASPARAGACEAE	I
ASPHODELACEAE	I
ASPLENIACEAE	I
ASTERACEAE	III

B

BALSAMINACEAE	III
BEGONIACEAE	II
BERBERIDACEAE	II
BETULACEAE	II
BIGNONIACEAE	II

BISCHOFIACEAE (see Phyllanthaceae)	II
BIXACEAE	II
BLECHNACEAE	I
BORAGINACEAE	III
BRASSICACEAE	II
BROMELIACEAE	I
BURSERACEAE	II
BUXACEAE	II

C

CACTACEAE	III
CALCEOLARIACEAE	III
CALYCANTHACEAE	II
CAMPANULACEAE	III
CANNACEAE	I
CAPPARACEAE	II
CAPRIFOLIACEAE	III
CARICACEAE	II
CARYOPHYLLACEAE	II
CASUARINACEAE	II
CHRYSOBALANACEAE	II
CISTACEAE	II
CLEOMACEAE	II
CLUSIACEAE	II
COLCHICACEAE	I
COMBRETACEAE	II
COMMELINACEAE	I
CONVOLVULACEAE	III
CORNACEAE	III
COSTACEAE	I
CRASSULACEAE	II
CUNONIACEAE	II
CUPRESSACEAE	I
CYATHEACEAE	I
CYCADACEAE	I
CYPERACEAE	I

D

DAVALLIACEAE	I
DICKSONIACEAE	I
DIDIEREACEAE	III
DILLENIACEAE	II
DIOSCOREACEAE	I
DIPSACACEAE	II
DROSERACEAE	II
DRYOPTERIDACEAE	I

E

EBENACEAE	III
ELAEAGNACEAE	II
EQUISETACEAE	I
ERICACEAE	III
ERYTHROXYLACEAE	II
EUPHORBIACEAE	II

F

FABACEAE	II
FAGACEAE	II
FOUQUIERIACEAE	III

G

GELSEMIACEAE	III
GENTIANACEAE	III
GERANIACEAE	II
GESNERIACEAE	III
GINKGOACEAE	I
GNETACEAE	I
GOODENIACEAE	III
GUNNERACEAE	II

H

HAEMODORACEAE	I
HAMAMELIDACEAE	II
HELICONIACEAE	I
HELWINGIACEAE	III
HIPPOCASTANACEAE	II
HYDRANGEACEAE	III
HYPERICACEAE	II

I

IRIDACEAE	I
ISOETACEAE	I

J

JUGLANDACEAE	II

L

LAMARIOPSIDACEAE	II
LAMIACEAE	III
LAURACEAE	II
LECYTHIDACEAE	III
LENTIBULARIACEAE	III
LILIACEAE	I
LINACEAE	II
LINDERNIACEAE	III
LOASACEAE	III
LORANTHACEAE	II
LYCOPODIACEAE	I
LYGODIACEAE	I
LYTHRACEAE	II

M

MAGNOLIACEAE	II
MALPIGHIACEAE	II
MALVACEAE	II
MARANTACEAE	I
MARATTIACEAE	I
MARSILIACEAE	I
MELANTHIACEAE	I
MELASTOMATACEAE	II
MELIACEAE	II
MENYANTHACEAE	III
MONTIACEAE	III
MORACEAE	II

MUSACEAE	I
MYRICACEAE	II
MYRTACEAE	II

N

NANDINACEAE (see Berberidaceae)	II
NELUMBONACEAE	II
NEPENTHACEAE	II
NYCTAGINACEAE	III
NYMPHAEACEAE	II
NYSSACEAE	III

O

OCHNACEAE	II
OPHIOGLOSSACEAE	I
OLEACEAE	III
ONAGRACEAE	II
ONOCLEACEAE	I
ORCHIDACEAE	I
OSMUNDACEAE	I
OXALIDACEAE	II

P

PAEONIACEAE	II
PANDANACEAE	I
PAPAVERACEAE	II
PASSIFLORACEAE	II
PAULOWNIACEAE	III
PEDALIACEAE	III
PENTAPHYLACEAE	III
PHILESIACEAE	I
PHYLLANTHACEAE	II
PHYTOLACCACEAE	II
PINACEAE	I
PIPERACEAE	II
PITTOSPORACEAE	III
PLANTAGINACEAE	II
PLATANACEAE	II
PLUMBAGINACEAE	II
POACEAE	I
PODOCARPACEAE	I
POLEMONIACEAE	III
POLYGALACEAE	II
POLYGONACEAE	II
POLYPODIACEAE	I
PORTULACACEAE	III

PRIMULACEAE	III
PROTEACEAE	II
PSILOTACEAE	I
PTERIDACEAE	I

R

RANUNCULACEAE	II
RHAMNACEAE	II
ROSACEAE	II
RUBIACEAE	III
RUTACEAE	II

S

SALICACEAE	II
SALVINIACEAE	I
SAPINDACEAE	II
SAPOTACEAE	III
SARRACENIACEAE	II
SAXIFRAGACEAE	II
SCHISANDRACEAE	II
SCIADOPITYACEAE	I
SCROPHULARIACEAE	III
SELAGINELLACEAE	I
SIMONDSIACEAE	II
SOLANACEAE	III
STRELITZIACEAE	I

T

TACCACEAE	I
TAXACEAE	I
THEACEAE	III
THYMELAEACEAE	II
TROCHODENDRACEAE	II
TROPAEOLACEAE	II
TURNERACEAE	II

U

ULMACEAE	II
URTICACEAE	II

V

VELLOZIACEAE	I
VERBENACEAE	III

NOMENCLATURE FOR CULTIVATED PLANTS

INTRODUCTION

The purpose of giving a name to a taxon is to supply a means of referring to it and to indicate to which category it is assigned, rather than to indicate its characters or history. (International Code of Nomenclature for Cultivated Plants, 9th edition, 2016)

The importance of applying specific names to all objects in nature is clearly evident. Names are indispensable to the order of things and life would be much too chaotic if names were not used for identification of everything we see, handle, make, or communicate about. Plants as well as all other natural objects or biological entities are known by a geographically and linguistically restricted common name and by a scientific name, the latter internationally recognized by the academic (scientific) community.

The number of higher plants is estimated to be about 620 plant families, 16,167 genera, and about 321,000 species. However, for centuries, horticulturists have deliberately selected plants to meet a certain demand for food, fiber, or a visual desire for beauty. These selected forms are referred to as cultivars (cultivated varieties), thus, a rough estimate of the number of horticultural plants, including ornamentals, is perhaps in excess of 500,000. To get a clearer mental picture of what we are discussing here, imagine yourself in a grocery store purchasing some common fruits and vegetables. It is highly unlikely that even one of the items you are observing, in long rows of such commodities, is exactly the same as was originally found in the wild. In addition to the names applied to their wild parental taxa, these now have been given names of their own, so that they can be distinguished on the basis of one or more characteristics (size, shape, color, taste, origin, etc.) and are known as cultigens (produced by humans).

COMMON NAMES

These various examples testify to the inadequacy of great numbers of English or 'common' names of plants. Many of them as those just cited, are erroneous and misleading. Some of them are duplicates and a few of them designate the same plant the world around. Perhaps it is time to start a reformation in vernacular names, or at least to drop many of them from catalogues and books. (Bailey 1933)

Common, vernacular, or folk names are names with which we identify plants used or observed frequently in our daily lives. For example, we all grow up knowing such names as lettuce, carrot, coffee, tea, etc., although in some cases we may not recognize the plant itself. Familiarity with the name, however, facilitates recognition and recall of such plants. This occurs because of the mental process of association. If we consume coffee on a daily basis and become aware that it is the product of a roasted and ground seed from a shrub with shiny medium size leaves and red berries, the picture remains in the mind's eye. The same may be said of coconuts which are born on palm trees growing on or near the beach, or tea plants which are related to the common landscape shrubs known as camellia. Thus, we are able to recognize certain commonly used plants by associating their name with one or more of their conspicuous features or uses. Such features may be morphological, such as leaf shape (arrowhead), flower shape (sunflower), fruit color (blueberry), and so on; based on one of the human senses, such as taste (bittersweet), smell (skunk cabbage), or touch (thornbush); places or situations in which they occur (mountain laurel, water hyacinth); or blooming time (Christmas cactus, Easter lily, four o'clock, morning glory). Similarity of a plant part with organs of the human body (liverworts, gravel root) in the old herbals signified a cure for afflictions of that organ by the plant or plant part which resembled it. Liverworts, for example, were thought useful for cure of liver diseases and gravel root was used for kidney stone. This is known as the "Doctrine of Signatures."

In as much as use of botanical names is the preferred option, because of their popular usage, common names should not be discarded, despite numerous difficulties that arise from their usage, as noted by L. H. Bailey (1933).

With all due respect to Shakespeare, that which we call a rose by any other name may not necessarily be a rose, let alone smell as sweet. It is notable that "Rose" is one of the

B. Dehgan, *Garden Plants Taxonomy*, https://doi.org/10.1007/978-3-031-11561-5_1

exceptionally few names used internationally in practically all languages. Stated unequivocally, linguistic limitation is the primary reason for use of botanical Latin names in scientific publications. Among other disadvantages of common names is application of a single name to several unrelated plants. For example, in English texts the name "jasmine" is applied to 10 species of different genera, and "cedar" to at least five unrelated taxa. Conversely, a single plant may be known by several names, as is frequently noted in this book. Moreover, relatively few naturally occurring plant species are used by man in their wild state, and there are no appropriately known common names for all. New food and medicinal plants are being discovered each day and hundreds of ornamental plants are added every year, often resulting in duplicity of common names and further confusion.

Perhaps the greatest disadvantage of common names is the absence of laws or regulations that govern their application. A common name cannot be right or wrong, and even the most extensive discussions may not resolve the problem of accuracy. This is a particularly challenging problem with landscape and ornamental plants, where common names are used alone or in combination with cultivar names. In the United States, such practice is commonplace. Current checklists of woody ornamental plants published by several states are specifically intended to encourage use of botanical names in order to maintain some degree of accuracy and uniformity, as required by the Cultivated Code of Nomenclature.

NOTE It is unrealistic to discuss all articles and recommendations of botanical and horticultural codes in the brief presentation that follows. It is highly recommended that the reader who wishes to publish a new taxon consults the Botanical Code for the botanical ranks and the Cultivated Code for designation of cultivar names and graft chimeras (cultigens).

BOTANICAL NAMES

Scientific or botanical plant names are nearly always derived from Latin or Greek, but regardless of their origin, they must be in Latinized form. The use of botanical names is governed and regulated by **The International Code for Botanical Nomenclature (ICN)** and is referred to as the "**Botanical Code" [2018 International Code of Nomenclature for Algae, fungi, and Plants (Shenzhen Code)].**
https://www.iapt-taxon.org/nomen/pages/main/art_3.html.

The principles and rules of the Botanical Code are uniformly applicable to cultivated as well as the wild plants. The articles and recommendations of the code are intended to promote "uniformity, fixity, and accuracy" in naming of plants and, at least in principle, have the force of international law.

Scientific names are often descriptive, frequently more so than the common names. Learning a few simple Latin or Latinized words simplifies the learning process considerably. At the rank of species ("the specific epithet"), morphological characteristics such as shape, number, color, etc.; sensual features, such as taste, smell, or touch; geography and ecological conditions; or the name of a person to be honored, have all been used for designation. The reference list provides a number of resources that give the meaning and/or origin of the frequently used botanical names. In contrast to common names, the two primary advantages of botanical names far outweigh the disadvantages in terms of their importance:

1. A plant can have only one valid (legitimate) scientific name. Any other name illegitimately applied is invalid, and usually is relegated to synonymy. A synonym is regarded as less adequate or essentially unacceptable than the name considered legitimate.
2. A name may not apply to more than a single type of plant and is used for that plant only, regardless of its geographical distribution and origin.

These two basic rules eliminate any confusion with regard to identity of a given type (a species) of plant and make it possible for all people to recognize it irrespective of their spoken language. Annoying as it may be, rules of priority may dictate use of a previously applied name, or removal of an illegitimate name and its replacement with a valid name, hence occasional name changes. This is a particularly difficult problem for growers of cultivated plants, because once a botanical name is established in general usage, it may take a few years for users to adopt a new one.

Use of Botanical Names in Horticulture

The same body of botanical organization that sets the rules and amends or changes the Botanical Code is also responsible for the "**International Code of Nomenclature for Cultivated Plants" or the "Cultivated Plant Code" (ICNCP).** Article 2 of the Cultivated Code clearly states that *"The International Code of Botanical Nomenclature (Botanical Code) governs the use of botanical names in Latin form for both cultivated and wild plants, except for graft chimeras...."* Since horticulturists are obligated to accept the Botanical Code, it is appropriate to begin this section with a review of the taxonomic or hierarchical arrangements of ranks, with the understanding that only those below the rank of 'order' (family, genus, species, etc.) are of significance in classification of cultivated plants. Of course, the word "classification" is irrelevant and inappropriate in horticulture since botanical classification is equally applicable to horticulture as are botanical names. Horticultural plants can be grouped by growth habit, flower color, and any other feature, but cannot legitimately be classified, unless based on research (systematics).

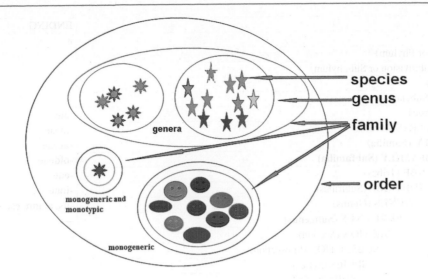

Illustrated plant hierarchies with **species** as the basic unit, one or more of which constitute a **genus**, one or more genera establish a **family**, and one or more families make up an **order**. A family with a single genus is said to be **monogeneric** and a genus with a single species is **monotypic**. The term **"nested sets"** is used to describe this hierarchical arrangement, which is essentially analogous to smaller tables or baskets fitted below or inside larger ones.

The principal ranks of taxa in descending sequence are illustrated below. The anglicized form is considered the most useful for practical purposes. The endings for appropriate ranks are noted in red. [for detail specific applications of ranks see: "Stevens, P. F. (2001 onwards)]". Angiosperm Phylogeny Website. Version 14, July 2017 and more or less continuously updated since. http://www.mobot.org/MOBOT/research/APweb/.

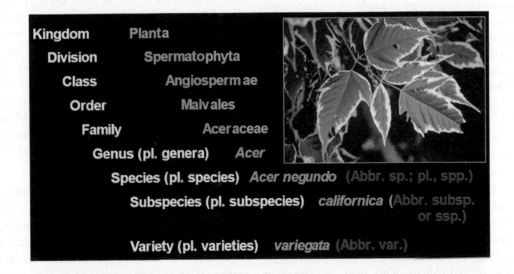

The principal ranks of hybrid taxa (nothotaxa) are nothogenus and nothospecies. These ranks are the same as genus and species. The prefix "notho-" (from ancient Greek, nŏthos, "bastard") indicates the hybrid character.

The most important categories in horticulture are given in CAPITALIZED **BOLD** letters and are discussed following the outline:

RANK	ENDING
KINGDOM (Regnum)	a
DIVISION (Division or Phylum)	
SUBDIVISION (Subdivision or Subphylum)	
CLASS (Classis)	
SUBCLASS (Subclassis)	
ORDER (Ordo)	-ales
SUBORDER (Subordo)	-ineae
FAMILY (Familia)	**-aceae**
SUBFAMILY (Subfamilia)	**-oideae**
TRIBE (Tribus)	**-eae**
SUBTRIBE (Subtribus)	**-inae**
GENUS (Genus)	**-us, a, um, es, on, etc.**
SUBGENUS (Subgenus)	**—**
SECTION (Section)	**—**
SUBSECTION (Subsection)	
SERIES (Series)	
SUBSERIES	
SPECIES (Species)	**-us, a, um, es, on, etc.**
SUBSPECIES (Subspecies)	"
VARIETY (Varietas)	"
FORM (Forma)	"
SUBFORM (Subforma)	"

Concepts of Ranks

It is nearly impossible to give an example of a plant which would fit all of the ranks outlined above. It should be clear, however, that we are dealing with the kingdom of plants, of which only two Divisions: Spermatophyta (the seed-bearing plants) and Pterophyta (the spore-bearing plants), are of primary interest. With this understanding, we can now proceed to define each of the major categories, beginning with the species.

SPECIES (plural, Species). As a basic taxonomic unit, species has been the subject of much discussion and the primary topic of many books since publication of Species Plantarum by Linnaeus in 1753. Species represent a lineage (descent from an ancestor), and the least inclusive of monophyletic (same ancestor) group. A detailed discussion of the species concept is beyond the scope of this book. However, the simplest definition of species is one that combines morphological resemblance of individuals with their interbreeding capability. Thus, in a homogeneous population, species may be defined as (biological species concept) **"a group of interbreeding individuals that are morphologically similar but not necessarily identical."** In this definition, some degree of variation (heteromorphy) among individuals is assumed, but it is also clear that each plant can usually interbreed with another of the same kind. In other words, no genetic barrier exists between two individuals of a given species and their progenies are also fertile. As an analogy, races of humans may differ in color or certain morphological fea-

tures but are fully capable of interbreeding. By the same token, cultivars of plants may differ in growth habit or any other characteristic, but in most cases retain the ability to cross with any other cultivar of the same species. For example, the yaupon hollies *Ilex vomitoria*, *I. vomitoria* 'Nana', and *I. vomitoria* 'Pendula', despite their dissimilar growth habit, interbreed and have fertile progeny when grown in proximity to one another and resemble the original species. [**NOTE**: Definition and delimitation of "species" is a complicated and controversial topic; it has and continues to be debated. The statement provided here is merely intended as a simplification to facilitate understanding of the concept]

Based on the binomial system (Article 23.1 of the Botanical Code and Appendix VII of the Cultivated Code), the name of each species is a binary combination, consisting of a genus name and a specific epithet, such as *Cycas revoluta*, where *Cycas* is the genus and *revoluta* is the specific epithet. If an epithet consists of two or more words, these should be used in united or hyphenated form, as in *Adiantum capillus-veneris*, *Macfadyena unguis-cati*, etc. The specific epithet alone is meaningless without reference to the name of the genus. In the way of analogy, the name John is a very common name and may refer to thousands of persons. However, the name John Smith refers to a specific individual about whom one may have considerable knowledge. The name "rubrum" simply means red, and as such, it does not refer to any particular plant or it may refer to any and all red plants. *Acer rubrum* specifically refers to the red maple.

The specific epithet may be taken from any source or may be taken arbitrarily. There is no particular rule for selecting names of species, although the most common practice is to use the Latinized form of the name of a locality (*Pistacia chinensis*), characteristic (*Ilex crenata*), or a person to be honored (*Rhododendron simsii*). The specific epithet is written in lower case, regardless of its derivation. In some older literature, such words are capitalized if used as noun, named after a person or a specific locality.

For writing purposes, if the specific name of a species is not known but the generic name is recognized, the word "species" is abbreviated as **sp.** if singular or **spp.** if plural. For example, if one or more cactus plants are recognized as belonging to genus *Mammillaria* but without certainty as to which species, then it is written as *Mammillaria* sp. (singular). If several plants representing different species of the genus *Mammillaria* are present, then they are written as *Mammillaria* spp. (plural).

VARIETY AND SUBSPECIES of a species differ with respect to their geographical occurrence. A **variety** is considered to be a wild species population in a given locality whose individuals possess consistent major difference in characteristic(s), whereas **subspecies** are geographically or attitudinally isolated (disjunct) populations of wild species which differ in one or more characteristics. The two terms, however, are probably erroneously used interchangeably in taxonomic literature.

Names of Infraspecific Taxa (taxa below the rank of species), excluding **Form**, are most commonly Latin or Latinized words for locality (*Acer negundo* subsp. *californica*), or a characteristic (*Acer negundo* var. *variegatum*). The endings must agree in gender and number with that of the specific epithet. Articles 24–27 of the Botanical Code specifically deal with naming of taxa below the rank of species (except cultigens). In writings, the word "variety" is abbreviated as **var.** and the word "subspecies" preferably as **subsp.** (plural, **subspp.**) but also as **ssp.** (plural, **sspp.**).

FORMS AND SUBFORMS are plants varying in a single characteristic, such as flower color, and are usually treated as cultivars in horticulture and discussed in detail below. In botanical writings, these are used as "**forma**" and "**subforma**" or abbreviated as **f.** and **subf.**, respectively. For example, *Trifolium stellatum* forma *nanum* or *Trifolium stellatum* f. *nanum*. Naming of forms is no longer common in botanical literature.

GENUS (plural, Genera). A genus may consist of a single species or a group of species which have one or more major characteristics in common. The term **monotypic** is used for genera with a single species, for example, the genus *Ginkgo* which consists of only *G. biloba*. Although subjective, genera are recognized by taxonomist and layman alike because of their shared characteristics. Lime, lemon, grape fruit, orange, etc., for example, are all in the genus *Citrus* because they share the characteristic fruit (hesperidium), leaves with winged petioles and typical citrus fragrance, etc. In this case, a rose is a rose, if they all belong to the genus *Rosa* and share all characteristics by which we recognize a rose.

The name of the genus is a noun in singular form and is always capitalized. It may be taken from any source or composed in any arbitrary manner. For example, old Latin names of plants (*Quercus*, *Fagus*, etc.), names honoring individuals (*Adansonia*, *Camellia*, etc.), modern names formed from vernacular names (*Bambusa*, *Musa*, etc.), names formed from geographical names (*Taiwaniana*, *Utahia*, etc.), anagrams of names (*Lobivia* for *Bolivia*, *Docynia* for *Cydonia*, etc.), or simply meaningless words (*Liatris*, *Ipsea*, etc.). The exception in the naming of genera is in Article 20 of the Botanical Code which expressly forbids the formation of generic names based on currently used technical terms in plant morphology (e.g., Tuber, Radicula, Lanceolatus, etc.), names consisting of two words (unless joined by a hyphen), and words not intended for use as names (e.g., Anonymous, Scripioides, etc.). The names of genera are always capitalized and italicized (or underlined). The reader is referred to Article 20–22 of the Botanical code for details of rules and recommendations for naming of genera and subgeneric ranks.

FAMILY (plural, families). The highest taxonomic rank used in horticulture, a family consists of one or more related genera. A family that consists of only one genus is referred to as **Monogeneric**, as in Gingkoaceae (*Gingko*), Stangeriaceae (*Stangeria*), Cycadaceae (*Cycas*), etc. Unlike generic and specific names, family names are not italicized (underlined) but they are always capitalized.

The name of a family is a plural adjective used as a substantive (a noun), formed by adding the suffix -aceae to the stem of a generic name which is included in that family. Families are almost always named after a genus, for example, *Solanum* + aceae (Solanaceae), *Juglans* + aceae (Juglandaceae), *Orchis* + aceae (Orchidaceae), etc. Because of long usage, the following eight families which do not terminate with -aceae are considered validly published and accepted by the Botanical Code, but with alternative names ending with -aceae equally acceptable. These include:

Compositae = Asteraceae, based on the genus *Aster*
Cruciferae = Brassicaceae, based on the genus *Brassica*
Gramineae = Poaceae, based on the genus *Poa*
Guttiferae = Clusiaceae, based on the genus *Clusia*
Labiatae = Lamiaceae, based on the genus *Lamium*
Leguminosae = Fabaceae, based on the genus *Faba* (= *Vicia*)
Palmae = Arecaceae, based on the genus *Areca*

Umbelliferae = Apiaceae, Based on the genus *Apium*

For additional information on rules pertaining to formation and usage of family and subfamilial ranks (subfamily, section, etc.) the reader is referred to Articles 18 and 19 of the Botanical Code. It is notable that APG IV (Angiosperm Phylogeny 2016–21) gives preference to the use of the original names, as does The Plant List.

ORDER Only rarely used in horticultural taxonomic treatments, this category is quite significant in botanical taxonomy. An order may include one or more families, based on some characteristics in common. Delimitation of orders and suborders is often controversial and may differ according to the authority consulted. Generally, names of orders are based on generic limits, as are those of the family. For example, Malpighiales is based on the family Malpighiaceae, which in turn is based on the genus *Malpighia*, or the order Malvales which is based on the Malvaceae and in turn based on the genus *Malva*. Addition of the suffix -ales to the stem of the genus name makes up the name of the order, for example, *Cycas* + ales (Cycadales) and *Myrtus* + ales (Myrtales). For further information on naming and usage of orders and higher ranks, refer to Articles 16 and 17 of the Botanical Code.

AUTHORITY OF PLANT NAMES Although authority names are not used in the examples cited in various discussions in this book or chapters, the name of the individual or individuals who described and named a given plant must be included with the plant's name, when in print. For example, *Cycas revoluta* Thunb. (abbreviation for Carl Peter Thunberg, 1743–1822, who wrote Flora Japonica in 1784), *Brassica oleracea* L. (abbreviation for Carolus Linnaeus or Carl von Linne, 1707–1778, author of the binomial nomenclature), etc. It has become customary in recent times not to abbreviate the name of the authorities, as in *Jatropha moranii* Dehgan and Webster. It should be kept in mind that the discoverer of a plant is not necessarily the person who assigns a name to it. Frequently plants are sent to herbaria or to individuals who are engaged in study or monographic treatments of various taxa for identification and naming. To avoid redundancy and to save on publication costs, in most scientific publications authorities of plants are cited only the first time a name appears. Guidelines of accuracy, however, require that the authority be mentioned at least once at the outset of published articles, but for all names cited in any monographic work.

Jaropha moranii
B. Dehgan & G. L.
Webster

Jatropha mutabilis
(Pohl) Baillon

Jatropha platyphylla
Müll. Arg.

There are a few complexities in authority citations, some of which should be noted:

1. When a name has been published jointly by two authors, their names are cited by addition of "&" or "et" (= and). For example, *Eucalyptus rhodantha* Blakely & Steedman or *E. rhodantha* Blakely et Steedman. If more than two authors are responsible for description of a taxon, however, only the name of the first author is cited followed by "et al." (= and others), a rule often ignored by taxonomists. Moreover, in recent years, the names of all authorities are commonly cited.

2. When a name is proposed by one author but validly published by another, the connecting word "ex" (= described by) is cited rather than & or et. For example, *Euphorbia pulcherima* Willd. ex Klotsch. In this case, Klosch was the first to propose or publish the name, but Willd made later changes or corrections.

3. Any alteration (including transfer) in the rank of the genus or below is indicated by placing the name of the original author in parenthesis, followed by that of the individual which affected the change. *Penstemon laetus* subsp. *roezlii* (Regel) Keck., for example, was originally described by Regel as *Penstemon roezlii* and subsequently lowered to the rank of subspecies by Keck. Citation of the first author is not necessary if the taxon is given a new specific epithet, or a name is relegated to synonymy.

4. It is not necessary to cite the author's name when naming cultivated plants (cultivars, grexes, and groups), even if the given plant was originally published as a botanical name, unless there is a valid reason for doing so.

HORTICULTURAL NAMES

It is implicit in horticulture that all botanical names, regardless of taxonomic rank, are accepted and all rules that govern their usage must be followed by the horticultural taxonomist. Articles 1–9 (9th edition) of the Cultivated Code specifically refer to this fact (except graft chimeras which are governed by the Cultivated Code). For specific details, recommendations, notes, and examples, see Principle 2 in ICNCP (the International Code of Nomenclature for Cultivated Plants, 9th ed., 2016)

Cultigens: Cultivars (cultivated varieties), Grexes, and Groups consist of uniform artificial populations of economically useful plants which are selected, propagated, and maintained by man and must be named in accordance to the Cultivated Code. For this reason, the botanical hierarchy of infraspecific categories is not applicable to cultivated plants (ICNCP, Recommendation 1A). Only in vernacular usage are the words "variety" and "form" may be used to designate a particular cultivated plant (a cultivar) under discussion. To avoid confusion, it is far preferable to use cultivar when discussing cultivated plants, except when the plant in question is one collected from the wild.

The Botanical Code states that "Plants brought from the wild into cultivation retain the names that are applied to the same taxa growing in nature." Article 2 of the Cultivated Code, however, defines cultivar as: An assemblage of cultivated plants clearly distinguished by a particular character or combination of characters (morphological, physiological, cytological, chemical, or others), and which, when produced (sexually or asexually), retains its distinguishing character(s). Thus, the distinction lies in whether the plant is collected from the wild or selected under cultivation (not to be confused with the term **Cultigen**, which refers to a plant that originated in cultivation and is known only from cultivation). Although not clearly expressed in either of the codes, any plant found in the wild that differs in minor morphological or physiological characteristics, such as flower or fruit color or cold hardiness, not considered of sufficient importance by botanical taxonomists to have been given an infraspecific (subspecies, variety, or form) name, should be regarded as a cultivar when brought into cultivation. If such characteristic is considered botanically significant, then the botanical name must be accepted. For example, *Ilex opaca* (American holly) with normally red fruit has individuals throughout its range which bear yellow fruit. This has been named *Ilex opaca* f. *xanthocarpa* and refers to any yellow-fruited American

holly. *Ilex opaca* f. *xanthocarpa* 'Canary', however, is a vegetatively propagated cultivar.

As of 1953 when the term cultivar was first introduced, it was determined that they should be fancy names (except when botanical taxa are reduced to cultivars), consisting of three words or fewer, and which are always capitalized but never italicized. Latinized names given to cultivated plants prior to January 1, 1959, are retained but should be capitalized and not italicized. In printed form the name of the cultivar was abbreviated as cv. and single quotes. Articles 14 of the Cultivated Code (2016), however, discourage use of cv., var., double quotes, or any other designation and only single quote is acceptable for cultivar designation, and never italicized. For example,

Lonicera japonica 'Purpurea'; *Cephalotaxus harringtonia* 'Fastigiata'; *Stenotaphrum secundatum* 'Floratam' etc.

FORMATION OF CULTIVAR NAMES In the same manner that a species name consists of a generic name and a specific epithet, the name of a cultivar also becomes a part of the species name. For example, 'Golden Delicious' or 'McIntosh' would be meaningless if not associated with *Malus pumila*, which would then specifically designate cultivars of the common apple. In another example, the Latin word "nana" means dwarf or small and is often used as a cultivar name. This word has no significance unless we know specifically to which plant it refers. The correct designation would then be *Ilex vomitoria* 'Nana' (dwarf yaupon holly) or *Punica granatum* 'Nana' (dwarf pomegranate). For this reason, only one cultivar name may be given in a cultivar class (recognized cultivars within a taxonomic group), since application of the same cultivar name to two or more taxa in the same genus or species will lead to confusion. For example, it is not appropriate to have an *Ilex rotunda* 'Nana' since there is already an *Ilex vomitoria* 'Nana'.

To be valid, cultivar names, as well as descriptions, must be published and distributed to the horticultural community. The primary medium for publication of new cultivars are such journals as HortScience, Scientia Horticulturae, Kew Bulletin, Baileya, and other recognized publications of plant societies. Cultivar or grex names may not be translated when they appear in a different language publication but may be transliterated (e.g. written in the alphabet of that language, character by character). Such translations are considered a different form of the original name for trade (marketing) designation (Chapter VIII, article 32 of the code).

A cultivar name is invalidly published if it is the same as a botanical or common name of a species or a genus, since it may lead to confusion. For example, *Citrus reticulata* 'Lime' or *Citrus paradisi* 'Apple' is not acceptable. However, names that include the words variety or form may not be rejected, but they have to be written in full. For example, *Citrus reticulata* 'Variegated Valencia' is acceptable.

In short, a cultivar name is legitimate only if it is in accordance with the articles of the code, and it is illegitimate if it does not conform. Each cultivar has one correct name, but may have one or more synonyms. These may be used only when a correct name is not commercially acceptable.

The Cultivated Code, same as the Botanical Code, does not have the force of national or international law. Compliance with the codes is therefore entirely optional but highly encouraged. Both codes are accepted by the community of world horticulturists and botanists. To protect the legal rights of the plant breeders and patent seekers, and to ascertain the accuracy and public comprehension of published materials, articles of the codes are fully enforced.

TYPES AND ORIGINS OF CULTIVARS

As was indicated earlier, cultivars (cultivated varieties) are selections from plants in cultivation, differing from their parental taxa in one or more desirable morphological, physiological, and/or chemical characteristics. However, it is precisely because such plants originate in cultivation that they do not merit a specific Latin epithet (except hybrids). Specific epithets are reserved for plants from wild populations, where uniformity is neither desired nor expected. Maintenance of uniformity in cultivars requires a precise method of reproduction. The determining factor in cultivar selection, therefore, is their mode of propagation. Thus, two distinct categories of cultivars may be recognized:

Asexually Reproduced Cultivars

CLONAL CULTIVAR The horticultural term clone (from the Greek word clon, meaning a twig) introduced in 1903 by Herbert John Weber (1865–1946) refers to the many fruit trees and field crops as well as urban and timber plants which are reproduced by vegetative means (cutting, grafting, budding, division, layering, or micropropagation). Commercial varieties propagated in such manner (topophysic clones) are uniform heterozygous genotypes (except where mutations occur), which usually originate from a single plant, or even a single cell. The word clone can thus be used to refer collectively to all apple trees produced by grafting of the same stock, or all the strawberry plants produced by runners from a single plant. A clone derived from a male member of a dioecious species will consist of only male plants (e.g., *Ilex cornuta* 'Rotunda'), whereas a female clone will consist of only female plants (e.g., *Ilex cornuta* 'Dazzler').

The most common type of clonally propagated cultivars in woody plants arises by bud mutation. For example, *Euonymus japonica*, which normally has green leaves, produces branches with leaves that have some degree of yellow, white, or silver variegation. Plants propagated from such branches have been given cultivar names. These mutations are commonly referred to as **sports**.

Although the term clone in the context of this book is used to denote vegetatively propagated cultivars, it should be kept in mind that such phenomenon also occurs in the wild. Populations of clonal plants are well-known in nature (e.g., *Elodea canadensis*) and remain clonal in cultivation. When such a plant is brought into cultivation, its botanical name is retained. However, if a particular selection is made from it while under cultivation, then it should be given a fancy cultivar name.

Apomicts The term Apomixis was first employed by Hans Günter Winkler (1926–2018) in 1908, to refer to reproduction by seed which originated without the union of the gametes (i.e., without pollination and/or fertilization). In this case, the genotype is unchanged, and the resulting plant(s) is therefore genetically identical to the mother plant. In flowering plants this is accomplished by (1) elimination of the reduction division, so that the gametophyte remains diploid, or (2) development of the egg without fertilization (Parthenogenesis). Expressed in simple terms, apomicts are plants that are produced from unfertilized ovules and are thus genetically identical to the mother plant. In ferns, apomixis occurs when chromosomes fail to separate in the last mitotic division, retaining the original diploid chromosome numbers in the spores, once again giving rise to genetically identical plants. Apomixis is known in more than 80 genera of vascular plants, including such common fruit crops as *Citrus* and landscape plants such as *Crateagus*. For all practical intents and purposes, apomicts are principally clones, despite their reproduction by seed, and as such, they are treated as cultivars when specific forms are in cultivation.

Sexually Produced Cultivars

Sexual reproduction, i.e., reproduction by seeds, is the primary means by which new and improved cultivars of horticultural crops are artificially created by man. Fertilization in agricultural crops, as in all angiosperms and gymnosperms, occurs either because of self-pollination (**autogamy**) or cross-pollination (**xenogamy**). Recurrent self-pollination (autogamy) results in inbreeding, wherein seeds are produced by union of gametes from one individual, therefore eventually giving rise to plants that are homozygous (genetically more or less uniform). By contrast, cross-pollination or outcrossing (xenogamy) results in seeds that are produced by the union of gametes from two genetically different individuals (genotypes), therefore giving rise to heterozygous plants (genetically different). Thus, the outcome of specific breeding programs geared toward selection of plants that possess certain characteristics would be determined by whether the parental taxa are auto- or xenogamous. In other words, the desired degree of uniformity is based on selfing or outcrossing; selfing results in one or more identical characters, whereas outcrossing may result in a combination of uniform characteristics. On this basis, several groups of cultivars which are produced by seed are often a result of artificial pollination.

PURE LINE CULTIVARS A line is a group of phenotypically (observable physical or biochemical characteristics) uniform plants whose genetic identity is maintained by self-pollination. Development of homozygous lines generated by self-pollination is accomplished by selection of specific strains (pedigree selection). These are hybridized until a desired homozygous individual is created. In most cases, pedigree selection is a long process from the time when parents are hybridized to when a new cultivar is released. Certain crop plants such as tomato breed true only after six generations of selfing because these lines become homozygous. Seeds produced through repeated self-pollination of homozygous or pure lines invariably breed true and appear uniform in growth habit, color of flower, etc. Many annual crops and bedding plants are produced by this method.

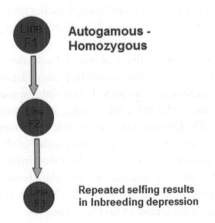

Autogamous - Homozygous

Repeated selfing results in Inbreeding depression

INBRED LINE CULTIVARS An inbred line is a group of cross-pollinating plants maintained as self-pollinating lines through artificial restraints on cross-pollination. These are then generally crossed to produce F1 hybrid cultivars. Crops in which F1 hybrids are common include bedding plants, vegetable crops (cabbage, onion, carrot, sugar beet, etc.), and two major agronomic crops (sorghum and maize). The greatest advantage of inbred lines is the vigor and uniformity of their F1 hybrids. However, unlike pure line cultivars that come true from seed in successive generations because of their homozygosity, F1 inbred line hybrids, when selfed,

segregate into various nonuniform entities because they are heterozygous. When this process is carried past one generation (F2 or beyond), the cultivars are known as Synthetic Cultivars.

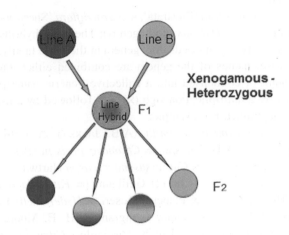

Xenogamous - Heterozygous

HYBRIDS These are the offspring of two plants of different species (or lower ranks) of the same genus or different genera in the same family or subfamily, which are either sterile or do not breed true from seed because of heterozygosity. Nomenclature of species (Interspecific) and generic (Intergeneric) hybrids and their derivatives are governed by the Botanical Code, irrespective of wild origin or artificial generation. These are discussed under horticultural names here because cultivars of hybrids (selections) are very common in cultivation and are treated in detail by the Cultivated Code (Article 2 and Division III: in Cultivated Code, 2016). The binary names of hybrids are subject to the same rules as are species, and authorities are cited for them in the same manner as well. Hybridity may be indicated by × (multiplication sign) or prefix "notho" (e.g. nothogenus or nothospecies)

Interspecific Hybrids (nothospecies): are hybrids between two species of the same genus. They may be designated by names or a formula and expressed by a multiplication sign (×). The order of names may be alphabetical, or if known, the name of the female parent is written first. For example,
Berberis × *mentorensis* L. M. Ames. [= *Berberis thunbergii* DC. × *Berberis julianae* Schneid.]
Abelia × *grandiflora* (Andre) Rehd. [= *Abelia chinensis* R. Br. × *Abelia uniflora* R. Br. ex Wallich.]
Magnolia × *soulangeana* Soul.-Bod.[= *Magnolia heptapeta* (Buc'hoz) Dandy × *Magnolia quinquepeta* (Buc'hoz) Dandy]
In these examples, specific epithets have been assigned for the interspecific hybrids according to the rules of the Botanical Code, with the obvious intent of brevity and sim-

plicity. Selections from interspecific hybrids, however, are given cultivar names. For example,

Magnolia × *soulangiana* 'Alba', *M.* × *soulangiana* 'Burgundy', *M.* × *soulangiana* 'San Jose', etc.

And

Abelia × *grandiflora* 'Prostrata', *A.* × *grandiflora* 'Sherwoodii'

Intergeneric Hybrids (nothogenus): These are hybrids of species in two or more distinct genera in the same family. In this case, names of the genera are combined either into a formula, or condensed into a collective generic name preceded by a multiplication sign (×), and followed by a Latin specific epithet. For example;

× *Cupressocyparis leylandii* (A.B. Jackson & Dallim.) Dallim & A.B. Jackson [= *Chamaecyparis nootkatensis* (D. Don) Spach. × *Cupressus macrocarpa* Hartw.]

× *Fatshedera lizei* (Cochet) Guillaum [= *Fatsia japonica* (Thunb.) Decne. & Planch. 'Moseri' × *Hedera helix* L.]

× *Citrofortunella floridana* J. Ingram & H. E. Moore [= *Fortunella japonica* (Thunbg.) Swingle × *Citrus aurantiifolia* (Christm.) Swingle]

NOTE: the code essentially gives the option of placing the multiplication sign (×) immediately next to or spaced away from the name.

As in the case of interspecific hybrids, the intent of designating a collective generic name is brevity and simplicity. For valid publication of the collective generic name, statement of parentage is sufficient. No diagnosis or description is necessary because the parental taxa are presumably known. Hybrids of different species of the same two genera may be quite dissimilar in their morphological characteristics. It would lead to confusion if such hybrid genera were to be described each time one was created. However, as of 1 January 1935, assignment of a specific epithet to a collective generic name requires a diagnosis (short statement) and a description. Selections of intergeneric hybrids in cultivation are given cultivar names. For example;

× *Cupressocyparis leylandii* 'Green Spire', × *C. Leylandii* 'Naylor's Blue', etc.

× *Fatshedera lizei* 'Variegata'

× *Citrofortunella floridana* 'Eustis'

GROUPS Species that have two or more cultivars, interspecific or intergeneric hybrids which result from crossing three or more taxa (hybrid complexes), and assemblages of similar cultivars (nonuniform assemblages), are designated as groups ["All members of a group must share characteristic(s) by which that group is defined." (Article 3 of Cultivated Code: The Group)]. This is analogous to "species complexes" in the Botanical Code, where, because of hybridization or ecogeographical reasons, considerable variation may exist in a single species. No nomenclatural provisions, other than recognition of lower ranks (subsp., var., f., etc.), are made in the Botanical Code for treatment of such complexes. Where parental taxa become known through biosystematic studies, the specific epithets are combined to form a normal

binomial (e.g., *Camellia saluensis-pitardii-reticulata* complex). In horticulture, however, a phrase used as a collective epithet and which may contain a word such as Group, Hybrid, Hybrids, Cross, Crosses, Grex, etc. may be used to indicate collective nature of the group (e.g., *Brassica oleracea* Sabauda Group). If a collective epithet in a modern language is followed by a cultivar name, it should be placed in parentheses. For example,

Rosa (Hybrid Tea) 'Richmond'

Lilium (Bellingham Hybrids) 'Shunksan'

Tulipa (Darwin group) 'Bartigon'

Beta vulgaris (Sugar Beet group) 'Klein Wanzleben E'

Lolium perenne (Early group) 'Devon Eaver'

Cattleya (Fabia g.) 'Prince of Wales'

Rhododendron cinnabarinum subsp. *xanthocodon* (Concotenans group) 'Copper'

Such situations usually arise in large groups of commonly cultivated plants, as in Orchidaceae, *Rhododendron, Begonia, Camellia, Lilium*, etc. In these cases, the plants are sufficiently well-known to permit omission of the parentheses and, if the status of the collective epithet is also clear, the word grex or its abbreviation (g.) – see below. For example,

Cattleya Fabia 'Prince of Wales'

Paphiopedilum Leyburnense 'Magnificum Dragonstone'

Saphrolaeliocattleya Falcon 'Westonbirt'

Begonia Tuberhybrida, Crispa group

Begonia Semperflorence-cultorum, Gracilis group

Rhododendron cinnabarinum Blandfordiiflorum group

GREX Article 4 of the Cultivated Code defines grex as "The formal category for assembling plants based solely on specified parentage is the grex. It may only be used in orchid nomenclature." The rules for forming grex names are noted in Article 23. Certain rules applicable to nomenclature of orchids are somewhat different from those of the Botanical or the Cultivated Codes. For additional information, the interested reader is referred to the Handbook on Orchid Nomenclature and Registration, 4th Edition (Cribb et al. 1994). See Orchidaceae in text for examples.

GRAFT-CHIMERAS Treated in Articles 20–24 of the Cultivated Code, Graft-Chimeras or graft hybrids are plants that are composed of two different tissues. These are named in exactly the same fashion as sexual hybrids, except that a plus sign (+) rather than a multiplication sign (×) is used to designate the chimera. The formula of a graft-chimera consists of the name of the two components, an addition sign, and an epithet. For example,

Syringa + *correlata* [= *Syringa* × *chinensis* + *S. vulgaris*]

+ *Laburnocytisus adamii* [= *Cytisus purpureus* + *Laburnum anagyroides*]

The specific epithet of a chimera must not be the same as that of a sexual hybrid when the same species are components of both. For example,

+ *Crataegomespilus dardarii* [= *Crataegus monogyna* + *Mespilus germanica*], but

× *Crataegomespilus gillotii* [= *Crataegus monogyna* × *Mespilus germanica*]

So as to avoid any confusion, "chimeras" need not necessarily be grafted and may arise naturally, such as different colors side by side (see *Dioon edule* in text)

PRONUNCIATION OF NAMES

The scientific names of plants are in Latin or Latinized derivation from Greek or modern languages. There are no rules, however, that govern their pronunciation. The best one can hope for is sufficient clarity and proper enunciation by correctly accenting the syllables. There is general agreement among taxonomists that words should be phonetically divided into syllables by using vowels, as in the English language, but accented according to the rules of classical Latin. The long (soft) sound of the vowel is indicated by a grave (/) accent (as in fate, scene, bite, note, tube) and the short (hard) sound by an acute (\) accent (as in fat, set, bit, not, tub). Thus, three steps are necessary for pronunciation of plant names:

1. Determine where the syllables are.
2. Determine which vowels are long and which are short.
3. Determine proper placement of primary and secondary stress on the syllables.

The following table which summarizes the sounds of vowels, consonants, and diphthongs is adopted from Stearn (1966, 1972) and Radford et al. (1974).

REFORMED ACADEMIC	TRADITIONAL LATIN	EXAMPLE ENGLISH
Vowels		
a as in father	fate	*Acer* (Ay-ser)
a as in apart	fat	*Fatsia* (Fat-see-a)
e as in they	me	*Melia* (Me-lee-a)
e as in pet	pet	*Petunia* (Pe-tew-nee-a)
i as in machine	ice	*Picea* (Pie-see-a)
i as in pit	pit	*Pittosporum* (Pi-toe-spaw-rum)
o as in note	note	*Opuntia* (Oe-pun-tee-a)
o as in not	not	*Nolina* (No-lie-na)
u as in brute	brute	*Brunfelsia* (Brun-fell-see-a)
u as in full	tub	*Turricula* (Ter-ri-cue-la)
Diphthongs		
ae as ai in aisle	as ea in meat	*Caesalpinia* (See-sal-pi-nee-a)
au as ou in house	as aw in bawl	*Aucuba* (Aw-cue-ba)
ei as in rein	as in height	*Eichornia* (Eye-cor-nee-a)
eu as u in cute	cute	*Euphorbia* (You-for-bee-a)
oe as in toil	as ee in bee	*Oenothera* (Ee-no-thee-ra)
ui as in oui (French),	we, ruin	*Luisia* (Loo-is-ee-a)
Consonants		
c always as in cat	before a, o, u as in cat	*Cassia* (Ka-see-a)
	before e, i, y as in center	*Cedrus* (See-drus)
cc	followed by a, o, u as k	*Baccharis* (Ba-ka-ris)
	followed by i or y as k-si	*Vaccinium* (Vak-si-nee-um)
ch (of Greek words) as k	as k or ch	*Pachypodium* (Pa-kee-po-dee-um)
ci	followed by a vowel as sh	*Senecio* (Se-nee-see-oh)
cn	as n	*Cnidoscolus* (Ni-dah-sko-lus)
ct	as t	*Ctenanthe* (Te-nan-thee)
g always as in go	before a, o, u as in go	*Gardenia* (Gar-dee-nee-a)
	before e, i, y as in gem	*Genista* (Je-nis-ta)
gn	as n	*Gnidia* (Ni-dee-a)
j (= i)	as y in yellow	*jalapa* (ha-la-pa),
	as j as in jam	*Jatropha* (Ja-troh-fa)
ph as p	like f	*Photinia* (Foh-ti-nee-a)
pn	as n	pneumatophore (New-ma-toh-for)
ps	as s	*Psilotum* (Sie-loh-tum)

REFORMED ACADEMIC	TRADITIONAL LATIN	EXAMPLE ENGLISH
pt	as t	*Pterocarya* (Te-roh-ka-ree-a)
s as in sit, gas	sit, gas	*Simmondsia* (Si-mond-see-a)
si	followed by a vowel as zh	*Ardisia* (Ar-dee-zhee-a)
t	as in table, native, Table	*Taxus* (Tak-sus)
	but ti in nation	*Thevetia* (Te-vee-shee-a)
v (consonant u)	as w as in van	*Viburnum* (Vie-bur-num)
x	initial letter as z	*Xylosma* (Zie-lohs-ma)
	middle letter as ks	*Taxus* (Tax-sus)
y as u in French pur	as in cypher	*Cydonia* (Sie-doh-nee-a)
y as in French du	as in cynical	*Euonymus* (You-ah-ni-mus)

A Comment on Use of Double Names

Commercial usage of cultivar synonyms (double names), as is currently in vogue by nurseries and plant producers, has not been mentioned in the International Code and no specific recommendation is provided. The double naming of cultivars has created significant confusion in producers and horticultural community. Some have referred to use of cultivar double names as "nonsense variety denomination"; it has led to problems in correct cultivar recognition. It also has contributed to difficulty and confusion by students in horticulture plant identification classes. If the primary objective of using double names is to protect a registered or patented cultivar, why merely marking the cultivar with a trademark symbol ™, or registered ® is not sufficient, as has been done in some cases? This can be done on every plant label and easily written in every publication. In practical terms, this system actually avoids confusion and clearly informs everyone, from large propagation houses to backyard nurseries, that propagation and sale of a given cultivar designated with the proper symbols are actually illegal. This system has worked well with copyright laws and there is no reason it should not work equally well with plants. Copyrights designated with the symbol © are protected for 75 years, as will plants designated with ™ or ® symbols. Using genus *Scaevola* as example, if any propagation nursery attempts propagating *S. emulata* 'Bombay'® or 'Santasfic'® they would be breaking the law. But is it more legal or absolutely necessary to write *Scaevola emulata* 'Newton' (New Wonder® pp 10,584) or Scaevola 'Wescaetopi' (Top Pot Pink® pp 19,729)? Which of the two names will the plant be marketed after the patent is issued? Protecting one's property is understandable, but causing confusion and difficulty in the process is not warranted.

There are situations where double cultivar or other designations may be justified and perhaps even necessary and helpful. To avoid cultivar name, name duplication in *Rose, Rhododendron, Hemerocallis, Hosta, Camellia,* and other genera with large numbers of cultivars, consistent with respective society registrations (e.g., American Rose Society), letters, or other designation, could be useful. Such a system as is already in use for 20,000 rose cultivars provides a good example. A Grex or Group name with or without a cultivar name, as has been done for nearly 100,000 orchids, is also a reasonable practice. Of course, this would not necessarily preclude use of ™ or ®, as may be appropriate.

For a further detailed discussion see: https://www.plant-delights.com/blogs/articles/name-that-plant and similar articles on the web.

LYCOPHYES AND MONILOPHYTES (FERNS)

Ferns in a conservatory at Montreal Botanic Garden (above) and in tropical Amazon Forest (below)

INTRODUCTION

There have been significant changes (updates) in classification and grouping of ferns and their phylogenies in recent years. Fern classifications have historically been subjected to controversies. With the new molecular phylogeny, much of the familiar terminology of higher groupings applied to ferns has been replaced as well. It may appear as an error by anyone familiar with earlier classifications when Equisitaceae and Ophioglossacace are actually included with the common familiar ferns. As noted in the introductory remarks in this book, horticulture, if interpreted as practical botany, has remained far behind in understanding fern classifications. Although there have also been numerous changes in gymnosperms and angiosperms, none has been as sweeping as that of ferns. The most recent treatment of ferns is that of Chritenhusz et al. (2017). Regretfully, its complex organization, arrangement, and recognition of many new taxa are far above and beyond the scope of this book, which is primarily intended as a practical teaching and learning tool. The somewhat unfamiliar treatments that follow are primarily based on APG IV (2016–2021), the most important source of plant classifications based on DNA phylogeny (the evolutionary history of organisms). Thus, for all intents and purposes, taxonomy of ferns remains unsettled. The following is primarily based on Judd et al. (2015).

LYCOPODIOPSIDA/LYCOPHYTES

LYCOPODIACEAE

CLUBMOSSES, GROUND PINES, WOLF FOOT

15 GENERA AND 390 SPECIES
GEOGRAPHY Worldwide in temperate and tropical regions, particularly in South America.

GROWH HABIT Erect or prostrate terrestrial or epiphytic herbaceous, often pendulous plants; about 200 species are epiphytic. **STEM**: roots are adventitious, dichotomously branched; root-bearing stems from angles of branches, dichotomous, growing at an oblique angle; shoots arise by bifurcation of apical meristem. **LEAVES**: microphylls with a single vein; spiral or alternate, to about ¾ in. (2 cm), not ligulate, often densely cover the stem. **REPRODUCTIVE STRUCTURES**: sporangia lateral, one per leaf; often heart-shaped or kidney-shaped; spore cases occur on the upper side of the microphylls and contain only one kind of spore (homosporous).

Three subfamilies have been recognized: Subfamily **LYCOPODIELLOIDEAE**: *Lateristachys*, *Lycopodiella*, *Pseudolycopdiella*, and *Palhinhaea*; Subfamily **LYCOPODIODEAE**: *Austrolycopodium*, *Dendrolycopodium*, *Diphasiastrum*, *Diphasium*, *Lycopodiastrum*, *Lycopodium*, *Pseudodiphasium*, *Pseudolycopodium*, and *Spinulum*; and subfamily **HUPERZIODEAE**: *Huperzia*, *Phlegmariurus,* and *Phylloglossum*. Terrestrial Neotropical species are found in Alpine grasslands. *Lycopodium* and *Phylloglossum* powder which is from their spores is used for medicinal purposes. *Huprezia*, in a broad sense, was thought not to be monophyletic (species with common ancestry), hence, divided into two genera: *Huprezia* (terrestrial and mostly temperate) and *Phlegmariurus* (tropical and usually epiphytic).

CULTIVATED GENERA Some species are probably commercially produced in limited numbers by fern growers, but *Huperzia* spp. and *Lycopodium* spp. are often seen in botanical conservatories. Considering the unsettled uncertainty of Lycopodiaceae classification, however, it would be unwise to suggest members of the genera noted above are or are not in cultivation. Suffice it to note that some species may be commercially produced.

COMMENTS Intergeneric hybridization has been reported in Lycophytes. Apparently, the spores are flammable and used in early photography and may have been used as industrial lubricants for rubber cohesion in condoms and rubber gloves. Some lycopsids of the past may have reached 150 ft. (50 m) tall with branching along their trunk. The name *Lycopodium* is from Greek *lukos*, wolf, and *podion*, diminutive of *pous*, foot, hence the common name wolf's foot. It is of interest to note that if *Huperzia* is shown not to be a monophyletic genus (not of common ancestry), then it has to be split up to *Huperzia* s.s. (in a restricted sense) and *Phlegmariurus*. Although recognized by Pteridophyte Phylogeny Group (PPG), it has not been accepted by other sources.

Huperzia phlegmaria (= *Lycopodium phlegmaria*,
Phlegmariurus phlegmaria; tassel fern)

Huperzia carinata (keeled   Tassefern) Fertile ***Lycopodium annotinum***,
(= *Lycopodium carinatum;* interrupted
clubmoss, stiff clubmoss)

SELAGINELLACEAE

SPIKE MOSSES, LESSER CLUBMOSSES

±750 SPECIES IN 1 GENUS

GEOGRAAPHY Worldwide but predominantly in tropical rainforests, although some species in the arctics and a few in deserts.

GROWH HABIT Usually perennial herbaceous, terrestrial, rarely epiphytic, some species are poikilohydric (= resurrection). **STEM**: dichotomously branched; erect or prostrate, spread by creeping stems (e.g., *S. kraussiana*), suberect (e. g., *S. trachyphylla*), or erect (e.g., *S. erythropus*); roots (rhizophores) are produced at branch forks. **LEAVES**: scale-like monomorphic sterile microphylls with a ligule, a single vein, and spirally arranged, often 4-ranked; sometimes iridescent; there is considerable variation in leaves; to $\frac{1}{5}$ in. (1 cm). **REPRODUCTIVE STRUCTURES**: unlike *Lycopodiaceae* they produce two kinds of spores [heterosporous: megaspores and microspores (female or male in gametophyte generation)], in terminal usually 4-sided cone-like structures; sporangia are stalked and on upper side of the ligule.

CULTIVATED GENERRA *Selaginella* is known from cultivation.

COMMENTS *Selaginella* is from the New Latin, diminutive of *selago,* based on the species name *Lycopodium selago* (northern fir moss; probably the same as *L. clavatum*). Selaginellaceae is monogeneric, includes only *Selaginella*. Species grow in various habitats, from very humid with considerable shade to open sunny areas. Some species are commercially produced and used as ground cover in landscapes or as hanging baskets. *Selaginella lepidophylla* (rose of Jericho), is sold in dry condition but resurrected when placed in water; *S. kraussiana* (golden clubmoss), and *S. moellendorffii* (gemmiferous spikemoss), are also commercially produced. The detailed discussion of *Selaginella* morphology see APG IV (2017 and later updates), Judd et al. (2015), and Christenhusz et al. (2017).

Selaginella uncinata 'Rainbow'

Selaginella kraussiana   *Selaginella willdenowii*

Selagineella pilifera (terrestrial resurrection plant), native to Texas and northern Mexico.

ISOETACEAE

QUILLWORT FAMILY

Isoëtes (from Greek *Iso*, equal and *etos*, yearly) is widely distributed member of the Lycophytes with worldwide distribution. Roots arise from beneath the corm/rhizome. Stems arise from the corm, unbranched. They are predominantly herbaceous evergreen or deciduous aquatics or occur in marshes. They have narrow tuft of quilt-shaped hollow leaves that arise from a swollen base that also contains the female and male (heterosporous) sporangia in leaf bases. Some lycopsids of the past may have reached 150 ft. (50 m) tall with branching along their trunk. For details see APG IV, 2021.

Isoetes echinospora and I. *lacustris*, both temperate species of Europe, Asia, and North America, with awl-shaped, olive-green leaves, are often submerged and may be used in well-lighted aquaria.

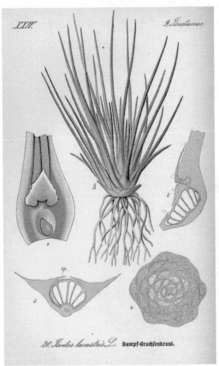

Isoetes lacustris (quillwort).
Illustrations from Otto Wilhelm Thomé (1885).

MONILOPHYTES (FERNS)

The vascular nonflowering **Monilophytes**, the more familiar ferns (plus Equisetaceae, Ophioglossaceae, and Psilotaceae), constitute an ancient plant assemblage of worldwide distribution. The delicate lacy and often much divided frond characteristic of most ferns explains the word **"fern"** that originated from the old English **"fearn"** meaning feather

and is related to the Old High German **"farn,"** and Sanskrit **"parna,"** leaf. Of course, ferns do not always have feathery fronds as illustrated in such species as bird's nest fern (**Asplenium nidus**), staghorn ferns (**Platycerium spp**.), among others. Some ferns have very small and diminutive fronds while others are tall as trees with large fronds. Most ferns are tropical and subtropical, others occur in moist temperate forests, and some can be found in deserts (resurrection ferns). They may be terrestrial, epiphytic and grow on tree trunks, rupestral and grow on rock crevices, while others are climbers (e.g., **Lygodium spp**.), and a few, such as **Salvinia spp**., are aquatics.

It is noteworthy that while some ferns are readily identifiable, many are seemingly difficult because of similarity of their growth habit and **frond** (leaf) characteristics. Thus, classification of ferns has traditionally been based on such characteristics as location of **sori** (singular **sorus,** group of **sporangia**), presence or absence and shape of **indusia** (indusium, the hard cover of sori, when present), shape of the **spores**, and anatomical features of the **stipe** (the petiole), such as number and arrangement of vascular bundles, and of course the fronds, when possible. The complexity of fern classifications is best illustrated by older literature where the largest number of ferns was pigeon-holed in the Polypodiaceae. Recent treatments, particularly that of APG IV (2016 and updates), Judd et al. (2015), Christenhusz et al. (2017), have recognized Equistales, Psilotales, and Ophioglossales as true ferns under Monilophytes.

Ferns can be readily distinguished from other plants by their spore-producing cases (**sporangia**) that are clustered in brownish spots usually at terminus of veins or along midveins or midribs and are usually located on the underside of the fronds. In most cases, sporangia may be covered by an **indusium** (a hard cover; see figure for reproductive structure of ferns). The rather unique process of emerging fronds is recognized as **"circinate vernation,"** is known as **"fiddlehead"** or **"crozier,"** and is another distinguishing feature of ferns, although this feature is also present in expanding leaves of some cycads. Considering the distinct features of ferns, as opposed to gymnosperms or angiosperms, it may be more appropriate to restrict the term frond to ferns (the term is sometimes used in palms and cycads).

While any detailed discussion of fern life cycle and propagation is beyond the scope of this book (see figure 7.4 in Judd et al. 2015), it should be noted that in contrast to seedlings of angiosperms and gymnosperms, that of germinated fern spores after the early **gametophyte** stage are known as **sporlings**. A fully grown fern is often referred to as **sporophyte**. Ferns may be propagated by spores, division of rhizomes, or as currently commercially practiced, by means of tissue culture (a few examples are presented under species discussions). Students interested in more comprehensive information on ferns should refer to the listed references, particularly to the Encyclopedia of Ferns by Jones (1987a, b) and detailed explanations of their phylogeny in APG IV (2016 and updates) and Judd et al. (2015).

Ferns are not as common in the landscapes as are seed plants. They are well-represented in temperate botanical garden conservatory collections around the world (see Dehgan 2014). They are, however, a regular feature of gardens and landscapes in the tropics. For example, Staples and Herbst (2005) have mentioned about 260 species of ferns and fern allies representing 31 families and 80 genera grown in the Hawaiian Islands, including some that have become invasive (as are some in Florida and elsewhere). The relatively few species that are discussed in this book are intended to familiarize students with diversity of ferns. Examples of the latter are illustrated but noted only briefly because they are uncommon in cultivation. Nevertheless, general familiarity with them is essential for students of plant science. General fern characteristics are noted below:

FERN FROND TERMINOLOGY

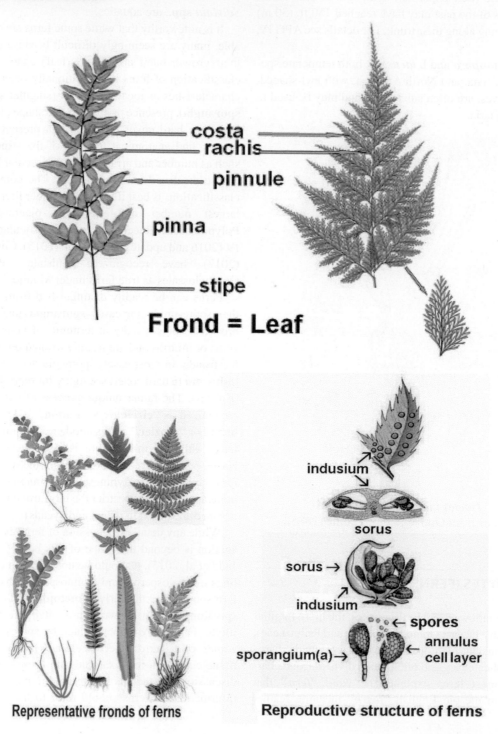

Representative fronds of ferns

Reproductive structure of ferns

FAMILIES OF EXTANT FERNS

ANEMIACEAE	HYMENOPHYLLACEAE	PLAGIOGYRACEAE
ASPLENIACEAE	LINDSAEACEAE	POLYPODIACEAE
BLECHNACEAE	LOMARIOPSIDACEAE	PSILOTACEAE
CIBOTIACEAE	LOXOMATACEAE	PTERIDACEAE
CULCITACEAE	LYGODIACEAE	SACCOLOMATACEAE
CYATHEACEAE	MARATTIACEAE	SALVINIACEAE
DAVALLIACEAE	MARSILEACEAE	SCHIZAEACEAE
DENNSTAEDTIACEAE	MATONIACEAE	TECTARIACEAE
DICKSONIACEAE	METAXYACEAE	THELYPTERIDACEAE
DIPTERIDACEAE	OLEANDRACEAE	THYRSOPTERIDACEAE
DRYOPTERIDACEAE	ONOCLEACEAE	WOODSIACEAE
EQUISETACEAE	OPHIOGLOSSACEAE	
GLEICHENIACEAE	OSMUNDACEAE	

PSILOTACEAE

WISK FERN FAMILY

2 GENERA AND ±12 SPECIES

GEOGROPHAY Warm temperate to pantropical; in southern US, southern Mediterranean Europe, Japan, and Southeast Asia.

GROWTH HABIT Small to medium size terrestrial or epiphytic plants without roots, but with fleshy dichotomously branched rhizomes and rhizoids. **STEM**: dichotomously branched, erect or hanging; green; scale-like or awl-shaped; simple, maybe once-forked; leaves may or may not have central vein and resemble microphylls (likely reduced leaves); spirally arranged. **REPRODUCTIVE STRUCTURES**: The large sporangia are often fused to form synangium (synangia), usually on the lower sides of branches (2 in *Tmesipteris* and 3 in *Psilotum*); spores large and numerous. Psilotaceae are eusporangiate (as are all vascular plants, except for leptosporangiate ferns), having sporangia that arise from a group of epidermal cells (as opposed to leptosporangiate, where sporangium is formed from a single epidermal cell).

CULTIVATED GENERA *Psilotum* and *Tmesipteris* are infrequently sold by fern growers, but are often seen in conservatories.

COMMENTS Based on molecular studies, Psilotaceae have been shown to be ferns and sister group to Ophioglossaceae. ***Psilotum nudum*** growth is of interest. It often spreads to greenhouse containers, but appears to have a preference for plants with which it grows: species of Euphorbiaceae and Begoniaceae (personal observations). *Psilotum nudum* is regarded as a sacred plant in Hawaii, Japan, and Philippines. *Tmesipteris* is not commonly seen, only in conservatories and personal fern collections.

Psilotum nudum (whisk fern) in a conservatory (left), dichotomous branching habit with **synangia** (3-chambered (fused) spore bearing structure; middle), and synangium (fused spore cases) close up (right). The structures beneath the synangia are referred to as **enations**. *Psilotum* is from the Greek *Psilos*, naked, in reference to leafless branches.

Tmesipteris elongata (hanging fork fern); from Greek *tmesis*, cutting and *pteris*, fern is in reference to the slight split on the leaf tips. Species of the genus are native to Australasia. Photographed in Montreal Botanic Garden Conservatory.

OPHIOGLOSSACEAE

ADDER'S-TONGUE FAMILY

4 GENERA AND 80 SPECIES
GEOGRAPHY Cosmopolitan, except in dry areas.

GROWTH HABIT Small terrestrial and epiphytic perennials; roots without root hairs and unbranched; dimorphic (separate sterile leaf-like part and fertile part – sporophore). **STEMS**: rhizomes succulent and with secondary growth (woody), unbranched. **LEAVES**: vernation nodding (as opposed to circinate); singly or few at growing season; common stalk divided into sterile laminate (photosynthetic), and fertile portions; fronds simple (in *Ophioglossum*) or compound (in *Botrychium*). **SORI**: sporangia globose, marginal or terminal. **SPORES**: globose-tetrahedral (four triangular faces). The family is traditionally known as eusporangiate ferns; only one kind of spore produced (homosperous), not green.

CULTIVATED GENREA *Botrychium* and *Ophioglossum* are well-known and produced by some fern nurseries.

COMMENTS Flora of North America considers Ophioglossaceae to be distantly related to ferns and more closely related to Marattiales and certain seed plants, especially cycads, on anatomical bases.

Ophioglossum pendulum

Ophioglossum vulgatum (left) and *Botrychium virginianum* (right). Both taxa are dimorphic, having distinct sterile and fertile fronds. The name *Ophioglossum* (Addler's tongue fern), is from Greek *ŏphis*, a snake; *glossa*, a tongue, referring to the fertile spike of *O. vulgatum*. *Botrychium* is from Greek *botrys*, a bunch of grapes, in reference to the bunch-like formation of the spore-bearing organs. Members of this family have a cosmopolitan distribution.

EQUISETACEAE

HORSETAIL FAMILY

GEOGRAPHY Mostly in temperate northern hemisphere of Eurasia, Canada, US, and South America, but with only one species in the Old World.

GROWTH HABIT Rhizomatous evergreen perennial plants with adventitious roots. **STEM**: ribbed jointed hollow stems, except at the nodes where branches may arise. **LEAVES**: whorled, fused into sheets at base surrounding the nodes. **REPRODUCTIVE STRUCTURE**: sporangia elongate, beneath the expanded base of the sporangiophores;

only one kind of spore (homosporous) is produced on terminal compact cone-like structures, but in some species on separate strobilus-bearing, none green shoots.

CULTIVATED GENERA *Equisetum* is the only genus in the family and only infrequently produced and sold.

COMMENTS *Equisetum,* from Latin *equus,* horse and *saeta,* bristle, hair, hence the name horsetail, is the only living genus of about 20 herbaceous terrestrial species, in the family.

Equisetum species are not common in cultivation, but are colonizers of unforested areas, lake margins, and wetlands. Best-known species are *E. arvense* (field horsetail) and *E. humale* (scouring rush), which are often seen in containers in conservatories, but are sometimes noxious weeds. Hybridization is reported, as in *Equisetum × dycei,* *Equisetum × ferrissii* (*E. humale* × laevigatum). Additional species reported for sale include *E. variegatum* and *E. fluviatile*; *Equisetum* is sometimes referred to as living fossils. Used for scouring pots and pans (hence the common name), and for medicinal purposes.

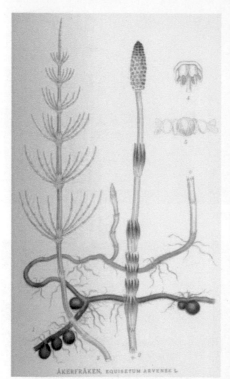 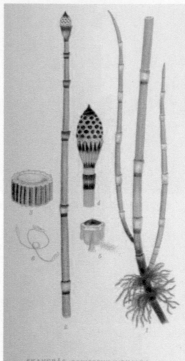 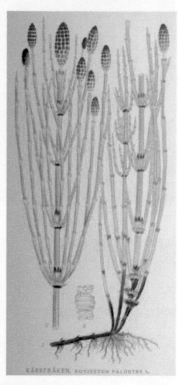

ÅKERFRÄKEN, EQUISETUM ARVENSE L.   SKAVGRÄS, EQUISETUM HIEMALE L.   KÄRRFRÄKEN, EQUISETUM PALUSTRE L.

Illustrations from Otto Wilhelm Thomé (1885).

Equisetum hyemale (photographed at Kanapaha Botanical Garden, Gainesville, Florida)

OSMUNDACEAE

OSMUNDA FAMILY

6 GENERA AND 18 SPECIES

GEOGRAPHY Nearly Cosmopolitan, missing from Middle East, Siberia or polar

GROWTH HABIT Terrestrials with erect or decumbent, often dichotomously branched stems, or with a single arborescent trunk and wiry roots. **FRONDS:** fronds dimorphic, with the sterile ones once pinnate but fertile ones bipinnate; glabrous at maturity; stipe with an expanded stipular base; pinnae with free veins. **SORI:** sporangia separate or in loose clusters or in wholly fertile bipinnate fronds; indusia absent. **SPORES:** tetrahedral-globose, green.

ECONOMIC USES Ornamentals or cultivated specifically for trunks that are extensively used for growing orchids, bromeliads, and other epiphytic plants.

CULTIVATED GENERA *Leptopteris, Osmunda,* and *Todea.*

COMMENTS Although except for *Osmunda*, species of other genera in this family are uncommon in the trade but much sought after for tropical gardens and landscaping. Classification of Osmundaceae is surprisingly complex, despite its small size. Listed genera include, according to APG IV (2016): *Claytosmunda, Leptopteris, Osmunda, Osmundastrum, Plenasium,* and *Todea*. Hybrids between *Osmundastrum cinnamomeum* and *Osmunda* have been reported.

Osmunda regalis

ROYAL FERN

ETYMOLOGY [Derivation unknown but may be named for *Osmunder*, Scandinavian God also known as Thor] [Regal, royal].

GROWTH HABIT Deciduous or evergreen (in warmer regions) terrestrial fern; to about 6 ft. (2 m) in height with equal spread; rhizomatous, acaulescent (the fertile frond). **FRONDS:** distinct fertile and sterile: fertile fronds often shorter and narrower, with brown sporangial clusters near the apex; the sterile fronds triangular to ovate, usually longer than the fertile fronds, with pinnae divided into 3–4 in. (8–10 cm), oblong-lanceolate pinnules with oblique bases and obtuse tips; stipes long and attached to a rhizome covered with root fibers. **SORI:** sporangia large, not in sori. **SPORES:** tetrahedral-globose, green. **NATIVE HABITAT:** Europe, Africa, and from Canada to South America. It is often found in moist woods, swamps, and lake or stream banks. Also native to Florida, including *O. regalis* var. *spectabilis.*

CULTURE Best grown in wet, acidic soils in shady locations. Requires a night temperature of about 45° F to remain green through winter months; may be unsightly during winter months when dormant; caterpillars may also be a problem. **PROPAGATION:** division of the crown, spores, and tissue culture. **LANDSCAPE USES:** hedge, specimen, and especially good for tropical effect in cooler climates. **HARDINESS ZONE:** 3A–10A.

COMMENTS Related species, ***Osmunda cinnamomea, O. claytoniana,*** and ***O. spectabilis*** are also in cultivation. Young fiddleheads and winter buds are cooked and eaten. See Christenhusz et al. (218) for additional uses.

Osmuda cinnamomea (cinnamon fern), photographed at U.S. National Arboretum, Washington
D. C.

Osmunda regalis (royal fern) in the landscape (left, photographed in Italy), fertile and sterile
fronds (right, in Florida)

Todea barbara

KING FERN
ETYMOLOGY [In honor of Heinrich Julius Tode (1733–1797), German mycologist] [Freign]

GROWTH HABIT Large fern said to grow to 10 ft. (3 ⅓ m), but much shorter in cultivation, with multiple crowns and potential for trunk development which are covered with wiry black fibrous roots. **FRONDS**: monomorphic, erect, leathery, bipinnate. **SORI**: sporangia arc produced on the pinnules. **NATIVE HABITAT:** New Guinea, Australia, New Zealand, and South Africa, predominantly in moist areas but occasionally in drier sites and sometimes in rock crevices; often on organic substrates.

CULTURE Best in a shady location; no reported problems. **PROPAGATION:** division of the crown, spores, and tissue culture. Unlike trunks of tree ferns that can be removed and replanted, those of *Todea* will not transplant by this method. **LANDSCAPE USES:** accent plant for its attractive foliage but is known to grow in containers and on rocks. **HARDINESS ZONE:** 10B–13B. Unlikely to grow in cooler climates.

COMMENTS Apparently the only other known *Todea* species, *T. papuana*, is endemic to Papua, New Guinea, but not reported from cultivation.

Todea Barbara (king fern), grown on rocks (both photographed in Bogor Botanical Gardens, Indonesia

MARSILIACEAE

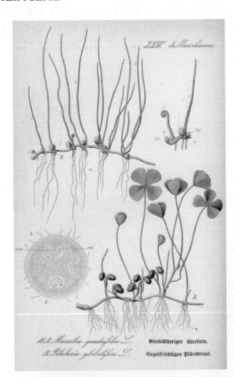

Pilularia globulifera (pillwort, above)
Marsilea quadrifolia (water clover, below)

Illustrations from Otto Wilhelm Thomé (1885).

Pilularia (from Latin *Pilula*, diminutive of *pila*, ball) is a small relatively widespread genus with creeping rhizomes, threadlike leaves, and spherical sporangia (= sporocarps) in the axils of leaves. *Marsillea* [named for Count Luigi Fernando Marsigli (1656–1730), Italian botanist] *Marsilea quadrifolia*, water clover (in reference to the four wedge-shaped pinnae) is said to be occasionally grown as ornamental, a good luck plant: four-leaf clover.

CYATHEACEAE

SCALY TREE FERN FAMILY

5 GENERRA AND 500+ SPECIES
GEOGRAPHY New and Old World tropical wet montane and cloud forests and some in south temperate forests (New Zealand and southern South America), and northern India, China, and Japan.

GROWTH HABIT Terrestrial, large to very large, or sometimes medium sized, with short or long, erect, slender to robust, apically scaly stems; covered with large or small scales or hairs. **FRONDS**: pinnately or bipinnately compound; usually large, up to 15 ft. (5 m); stipe usually with pneumathodes (**ae**rating tissue), in two lines, straw color or dark or black; lamina usually 1–3 pinnate; veins simple or forked, free. **SORI**: on lower side or terminal on veins and marginal, orbicular, without indusia, or indusia saucer or cup-shaped, or globose and completely surrounding sporangia; homosporous; **SPORES**: tetrahedral, variously ornamented.

ECONOMIC USES Ornamentals and fibrous rhizomes are used for growing epiphytic taxa, such as orchids and bromeliads.

CULTIVATED GENREA *Alsophila*, *Cyathea*, and *Sphaeropteris*

COMMENTS According to Christenhusz and Chase (2014), the family has 13–14 genera. According to IUCN, species of Cyatheaceae are threatened or endangered, but some are invasive. Full description of the species of Cyatheaceae is difficult primarily because of their nomenclatural uncertainty. In fact, descriptions vary from one source to another.

Sphaeropteris medullaris (= *Alsophila extensa*, *Cyathea medullaris*; mamaku, black tree fern, sago fern), homosporous fern with black trunk that is covered with stipe bases; scaley leaves, native to wetter coastal areas of New Zealand, Fiji, and Polynesia; the tallest tree fern, to 60 ft. (20 m) and fronds said to be up to 15 ft. (5 m) long. Photographed in Bogor, Indonesia. It is cultivated in San Francisco, California.

Cyathea cooperi (= *Alsophila cooperi*, *Sphaeropteris cooperi*; Cooper's tree fern, lacy tree fern, Australian tree fern). Note the finer texture of the fronds. Native to New South Wales and Queensland, Australia. Evergreen, grows up to 13 ft. (4 m), with gracefully arching fronds. New fronds and fiddleheads said to be particularly attractive. Listed as invasive.

PTERIDACEAE

(INCLUDING ADIANTACEAE)

MAIDENHAIR FERN FAMILY

53 GENERA AND ±1210 SPECIES

GEOGRAPHY Nearly cosmopolitan, but widespread and diversified in Central and South America; in damp cool locations as well as arid regions.

GROWTH HABIT Mostly terrestrial; small to medium; rhizomes variable, long or short, erect or ascending; bearing scales. **FRONDS:** monomorphic or dimorphic in a few genera; clustered or spaced along creeping rhizomes; pinnately or palmately compound, pinnae variously shaped, from oblong rhomboid to often flabellate (entire, semicircular, or fan-shaped); veins free and dichotomously branched or reticulate. **SORI:** at terminus of veins at the margins of pinnae and covered by reflexed false indusia along the veins. **SPORES:** tetrahedral or globose.

ECONOMIC USES Ornamentals: in rock gardens, along brooks, and as potted plants.

CULTIVATED GENERA *Acrostichum, Adiantum, Antrophyum, Bommeria, Ceratopteris* (?)*, Cheilanthes, Doryopteris, Hemionitis, Llavea, Pellaea, Pityrogramma, Pteris*, and *Vittaria*.

COMMENTS Any detailed discussion of the genera and species in Pteridaceae in a broad sense is beyond the scope of this book. Suffice it to note that in the current classification of the family, according to Smith et al. (2006), 15 previously recognized families of which, if distinct families were to be recognized, only five could be accepted: **Parkeriaceae, Adiantaceae, Cryptogrammaceae, Sinopteridaceae,** and **Pteridaceae.** Apparently in APG IV (2016–2021) and in Flora of North America (vol. 2), only Pteridaceae is accepted. The uncertainty and confusion in this group preclude any discussion of family delimitation or enumeration of genera. Nevertheless, a few of the more common ones reported in the trade are noted above and enumerated by

APG IV (2016–2021): *Pteris* (250), *Adiantum* (225), *Cheilanthes* (100), *Jamesonia* (50), *Myriopteris* (45), *Aleuritopteris* (40), *Antrophyum* (40), *Haplopteris* (40), *Pellaea* (40). *Adiantum* is quite diverse in the West Indies. Cheilianthoid ferns (subfamily **Cheilanthoideae**) (±400 species) can grow in very dry conditions and are essentially resurrection ferns. Vittaroid (subfamily **Vittaroideae**) ferns are commonly epiphytic and are one of the major monilophyte groups. Some species of *Pteris* and *Pityrogramma* accumulate arsenic.

Acrostichum danaeifolium

GIANT LEATHER FERN

SYNONYMS *Acrostichum excelsum, A. leptophlebium, Chrysodium lomarioides*.

ETYMOLOGY [From Greek *akros*, terminal and *stichos*, a row, in reference to arrangement of the spores that completely cover back of the pinnules (leaflets)] [with leaves like *Danaë*, Alexandrian laurel, *Danae racemosa*]

GROWTH HABIT Large fern, to 12 ft. (3½ m) long and 5 ft. (1 $\frac{2}{3}$ m) wide. **FRONDS:** pinnately compound; pinna opposite, oblique pointing mostly downward. **SORI:** spread across the entire back of the fronds (= acrostichoid). **NATIVE HABITAT:** Central and South America, Caribbean, Florida Keys to Dixie and St. Johns Counties. It occurs in brackish as well as freshwater marshes.

CULTURE May be planted in diverse soils and can tolerate sun and salt, as long as it is not allowed to dry out; dead and unattractive fronds may be removed once each year; no drought tolerance but does not withstand prolong submersion; scales and slugs may be problems. **PROPAGATION:** rhizomes or spores. **LANDSCAPE USES:** as specimen plants or mass background or hedge planting; appropriate for wet sites but inappropriate as groundcover because of its large size. **HARDINESS ZONE:** 9B–13B.

COMMENTS The genus includes three species: *Acrostichum danaeifolium, A. aureum,* and *A. speciosum,* all known from cultivation.

Acrostichum danaeifolium (giant leather fern)

Adiantum spp.

MAIDENHAIR FERN

ETYMOLOGY [From Greek *adiantos,* unwetted or dry, in reference to the pinnules which repel water (remain dry) if plunged in water] [pl. species]

GROWTH HABIT Evergreen ferns; terrestrial or rarely epiphytic; ground cover; short rhizomes; generally seen smaller. **FRONDS:** delicate, airy, foliage erect or drooping, graceful; fine texture; size variable with species to 3 ft. (1 m) tall, varies with species. from once to 4 times pinnate, generally 2–3 times pinnate; triangular to ovate to nearly orbicular; from 6 to 36 in. (15–90 cm) long; pinnules 4-sided, oblong or wedge-shaped; without midribs; veins evenly forked. size varies with species, ranging from ½ to 2½ in. (1½–7 cm); margins more or less notched; light green; stipe and rachis slender, wiry, shiny black or purplish. **SORI:** usually round or oblong; borne at pinnule upper margins; covered by recurved pinnule margin (false indusium); **SPORES:** minute.

CULTURE Partial to full shade outdoors; inside in bright, indirect sunlight; moist but well-drained soils with high organic matter content. The southern maidenhair (*A. capillus-veneris*) and brittle maidenhair (*A. tenerum*) grow best in somewhat alkaline soils; the other species grow best in acid soils. Plants do best in humid locations, requiring 60% or more humidity indoors to thrive. **:** varies with species: the commercially grown *A. raddianum* (delta maidenhair) is restricted to tropical zones (10B–13B). *Adiantum tenerum* (brittle maidenhair) can be grown in zones 9A–10B. No salt tolerance for any species. Scale, mites, mealy bugs, snails, slugs, and root rots may be problems. **PROPAGATION:** division, spores, but primarily by tissue culture. **LANDSCAPE USES:** outdoors as rock garden or border plants in shady, moist locations. The southern and brittle maidenhairs do best in high-pH areas among limestone rocks, masonry walls, or along walks. Inside in containers or hanging baskets in high-humidity rooms or in terrariums.

COMMENTS About 200 or more species of *Adiantum,* and several named cultivars are available in the trade. Some examples include: *A. aethiopicum, A. aleuticum, A. capillus-veneris, A. capillus-veneris* 'Imbricatum', *A. capillus-veneris* 'Mainsii', *A. caudatum* (trailing maidenhair), *A. formosum, A. fulvum, A. hispidulum, A. latifo-lium, A. macrophyllum, A. mairisii, A. pedatum, A. peruvianum, A. polyphyllum, A. pulverulentum, A. raddia-num, A. raddiatum* 'Californicum', *A. raddiatum* 'Microphylum', *A. tenerum, A. trapeziforme, A. venustum,* among others.

Adiantum capillus-veneris (Venus maidenhair)

Adiantum polyphyllum (giant maidenhair), from Venezuela to Peru, it is the largest maidenhair fern, with its divided fronds and long black stipes it illustrates diversity of the genus (left), ***Adiantum pedatum*** (northern maidenhair fern), from northern and western U.S., as the name implies has palmately branched fronds.

Adiantum polyphyllum (giant maidenhair), from Venezuela to Peru, is the largest maidenhair fern; with its divided fronds and long black stipes, it illustrates diversity of the geus (left), *Adiantum pedatum* (northern maidenhair fern), from northern and western U.S., as the name implies, has palmately branched fronds.

Pteris cretica

CRETAN BRAKE FERN
SYNONYMS *Pteris pentaphyla, P. serrata, P. treacheriana, P. trifoliata, Picnodoria cretica*

ETYMOLOGY [Classical Greek name for a fern, *pterôn*, feather or wing, in allusion to shape of the fronds] [Greece's Crete Island]

GROWTH HABIT Terrestrial evergreen fern with slender short creeping, sparingly brown scaly rhizome. **FRONDS:** erect or slightly arching, pinnate but often appears pedately divided to nearly palmate in younger specimens; pinnae triangular, linear to linear-lanceolate, mostly opposite; 1–5 pairs but often with a terminal pinna; fertile pinna longer and narrower than the sterile ones. **SORI:** narrow, in continuous lines at frond margins and covered by them creating false indusia. **NATIVE HABITAT:** uncertain but widely culti-vated and may have spread in Europe and Africa, but according to Staples and Herbst (2005), it is perhaps native to the Old-World tropics and possibly southern Mexico and is considered native to Hawaii.

CULTURE Easy plant to grow in relatively humid warm conditions; tolerate most media, although shows preference for well-drained organic soils. Leaf spots, scale and other typical insects, leaf burns as a result of dryness or too much sun, although they can tolerate moderate periods. **LANDSCAPE USES:** ornamentals, for landscape as well as indoor use. Should be used with caution in warmer climates as it has a tendency to become invasive, as for example in Florida and even in moderately cooler climates. **HARDINESS ZONE:** 9A–13B.

COMMENTS *Pteris cretica* is a widespread variable species with an apparent tendency to allow cultivar selections, including variegated and crested examples: **'Albo-lineata'**, **'Alexandreae'**, **'Childsii'**, **'Distinction'**, **'Gautheri'**, **'Major'**, **'Mayi'**, **'Maxii'**, **'Ovardii'**, **'Parkeri'**, **'Rivertoniana'**, **'Roweii'**, **'Wimsettii'**, and **'Wilsonii.'** In addition, there are a number of related species also in cultivation: *P. ensiformis* (including **'Arguta'**, **'Evergemiensis'** and **'Victoriae'**), *P. excela*, *P. multifida*, *P. nipponica*, *P. tremula*, *P. tricolor*, *P. umbrosa*, *P. wallichiana*, and *P. vittata*.

***Pteris cretica* 'Albo Lineata'** (silver ribbon fern)

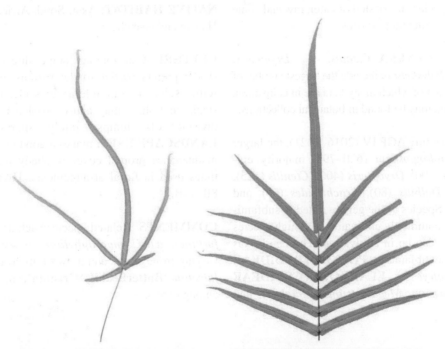

Pteris multifida (Chinese brake)   ***Pteris vittata*** (Chinese ladder brake)
Both species are of Old-World origin and escaped from cultivation in Florida and elsewhere.

DRYOPTERIDACEAE

THE SHIELD FERN OR WOOD FERN FAMILY

26 GENERA AND ABOUT ±2235 SPECIES

GEOGRAPHY Cosmopolitan; pantropical with many temperate relatives.

GROWTH HABIT Perennial, terrestrial, or epiphytic; rhizomes creeping, ascending or erect, sometimes climbing. **FRONDS:** pinnately compound, stipes with round vascular bundles; pinna monomorphic or infrequently dimorphic, sometimes with scales, glands, or uncommonly pubescent; veins pinnate or dichotomous, and sometimes anastomosing. **SORI:** round, indusia round-reniform or peltate, or sori exindusiate (without indusium); sporangia with 3-rowed, short or long stalks; **SPORES:** monolete (there is a single line on the spore indicating the axis).

ECONOMIC USES Container and landscape plants, used in shady, moist areas, or extensively in floral arrangements. Rhizomes of some species are roasted or eaten raw and some species are used for medicinal purposes.

CULTIVATED GENERA *Ctenitis,* *Dryopteris,* *Elaphoglossum*, and *Polystichum* include the largest number of species in the family and of which many taxa are in cultivation. Species of other genera may be found in botanical collections.

COMMENTS According AGP IV (2016–2021), the larger genera include: *Elaphoglossum* (620–795, majority epiphytic), *Polystichum* (500), *Dryopteris* (400), *Ctenitis* (125), *Megalastrum* (90), *Dolbitis* (80), *Arachnoides* (60), and *Stigmatopteris* (40). Species of the genera noted in subfamilies are more or less commonly cultivated, although species of other genera may also be in cultivation. The family has been divided into three subfamilies: **POLYBOTRYOIDEAE** (*Maxonia,* *Polybotrya*), **ELAPHOGLOSSOIDEAE** (*Bulbitis,* *Elaphoglossum, Megalastrum, Rumohra*), and

DRYOPTERIDOIDEAE (*Ctenitis,* *Cyrtomium, Dryopteris, Phenorephlebia*, and *Polystichum*).

Cyrtomium falcatum

HOLLY FERN

ETYMOLOGY [From Greek *kyrtos,* arched; from the habit of growth of these ferns] [Sickle-shaped].

GROWTH HABIT Evergreen fern; terrestrial or rarely epiphytic; ground cover; rhizomes stout, densely covered with brown scales; clumping with a dense, spiral crown; arching; medium texture; rhizomes densely scaly, brown; reaches a height of 3 ft. (90 cm) with a 2–3 ft. (60–90 cm) spread; moderate growth rate. **FRONDS:** odd-pinnately compound; spirally arranged; to 2½ ft. (66 cm) long; pinnae ovate to elliptic, to 5 in. (13 cm) long; dark green above, lighter beneath; glossy, waxy, leathery; netted veins; margins entire or weakly wavy or toothed; rachis base densely covered with brown scales. **SORI:** large, round, brown, scattered on the undersides of mature pinna; **SPORES**: minute. **NATIVE HABITAT:** Asia, South Africa but naturalized in Hawaii and elsewhere.

CULTURE Can tolerate some direct sunlight and full shade; prefers partial shade; various, moist, well-drained, fertile soils; no salt tolerance; scale, mites, mealy bugs, snails, and slugs may cause problems. **PROPAGATION:** division of clumps; rarely spores, tissue culture. **LANDSCAPE USES:** makes a good interior plant or a low-maintenance ground cover in shady locations, and sometimes used in floral arrangements. **HARDINESS ZONE:** 8B–13B.

COMMENTS Related species such as *C. caryotidium*, *C. fortunei*, and *C. macrophyllum*, as well as cultivars with varying frond forms and a dwarf are available, including *C. falcatum* 'Butterfieldii', 'Cristata', and 'Rockfordianum', among others.

Cyrtomium falcatum, growth habit (above), tissue cultured (below left), portion of a fertile frond

Polystichum setiferum

SOFT SHIELD FERN

SYNONYMS *Polypodium setiferum, Polysticum angulare*

ETYMOLOGY [From Greek, *pŏlys*, many; *strichos*, a row, in reference to the row of sori] [Bearing bristles].

GROWTH HABIT Terrestrial semievergreen fern with an erect crown basal rhizome. **FRONDS:** long, leathery, pale green, erect but arching, bipinnate, with pale golden-brown scales; pinnae opposite, pinnule margins and tips with soft bristle-like teeth; stipe bases scaly. **SORI:** round, in two rows on both sides of the midrib, covered with shield-shaped brown indusia. **SPORES**: and sporangia are light yellow. **NATIVE HABITAT:** species of southern and western Europe, which is essentially warm Mediterranean climate.

CULTURE Filtered light or mostly shade and well-drained organic soil which should not be allowed to dry out. **PROPAGATION**: spores, tissue culture. **LANDSCAPE USES:** suitable for woodland gardens singly or in mass planting; in shadier locations; cannot tolerate cold dry conditions or direct sun for long periods. **HARDINESS ZONE:** 6A–10A.

COMMENTS There is considerable variation in this species resulting in several cultivars of which the crested forms include: *Polystichum setiferum* 'Capitatum', 'Cristatum', 'Congestum', 'Divisilobum', 'Dahlem', 'Grandiceps Fox', 'Herrenhausen', 'Perscristatum', 'Polydactylum', and 'Ramosum', A number of other *Poysticum* species are also in cultivation such as: *P. acrostichoides, P. aculeatum, P. braunii, P. californicum, P. dycei, P. makinoi, P. munitum, P. neolobatum, P. rigens, P. stigerum*, and *P. tsussimense*, among others.

Polystichum setiferum growth habit (right) and close up of the frond (left)

Rumohra adiantiformis

LEATHERLEAF FERN

SYNONYMS *Aspedium clypeolarium, A. coriaceum, A. politum, Dryopteris adiantiformis, Lastrea adantiformis, Nephrodium duriusculum, Polypodium adiantiforme, P. calypteratum, P. coriaceum, P. politum, Polystichum adiantiforme, P. capense, P. coriaceum, Rumohra aspidoides, R. capuroni*

ETYMOLOGY [In honor of Karl Friedrich von Rumohr (1785–1843), German art expert] [Resembling *Adiantum* or Maidenhair Fern].

GROWTH HABIT Evergreen fern; terrestrial or sometimes epiphytic; rhizomes stout, clumping; symmetrical, compact, dense ground cover; grows to 3 ft. (1 m) tall with a 4–5 ft. (1½–1 $\frac{2}{3}$ m) spread; moderate growth rate; spreads by a stout creeping rhizome densely covered with brown, papery scales. **FRONDS:** broadly triangular, to 3 ft. (1 m) long and 2½ ft. ($\frac{2}{3}$ m) wide; 3–4 times pinnate at the base, less compound up the rachis; stiff, leathery, glabrous, waxy; lower pinnae triangular; pinnules oblong to narrowly ovate

with a cuneate base and acute to acuminate apex; coarsely toothed margins; dark green; medium-fine texture. **SORI:** minute spores produced in large sori on the undersides of mature leaves. **NATIVE HABITAT:** tropical and subtropical regions of the southern hemisphere.

CULTURE Partial or full shade on a wide range of well-drained soils; no salt tolerance; fern borers, scale, mites, mealy bugs, snails, slugs, and fungus diseases, all are potential problems. **PROPAGATION:** rhizome division, spores, tissue culture. **LANDSCAPE USES:** makes a durable, dependable ground cover. The stiff, harsh feeling fronds are long-lasting when cut and popular in flower arrangements. **HARDINESS ZONE:** 9B–12B; normally grown outdoors in south and central Florida. Can be grown in protected locations in lower zones.

COMMENTS Cut leatherleaf fern production is a major industry in central Florida. In the past, Florida produced nearly all the cut leatherleaf fern in the world. Labor costs are forcing production offshore. The more compact darker green frond cultivar **'Iberia'** is more commonly used for commercial reproduction.

Rumohra adiantiformis

Dryopteris filix-mas (Dryopteridaceae; male fern),
deciduous fern with erect, stout rhizomes;
native to Europe and North America

POLYPODIACEAE

POLYPODIUM FAMILY

63 GENERA AND ±1650 SPECIES
GEOGRAPHY Cosmopolitan.

GROWTH HABIT Predominantly epiphytic, small to massive, slender to filiform plants, with erect or short- or long-creeping stems, and often with many fibrous roots; rhizomes are monomorphic, with fronds, little-branched, their morphology varies according to the species; some taxa rhizomes are monomorphic, but some taxa are dimorphic, much-branched, and hollow. **FRONDS:** mostly monomorphic or sometimes dimorphic, in clusters or individually spaced, simple and entire or variously pinnatifid or forked; stipe jointed near the base; veins free or anastomosing. **SORI:** round to elongated, on an entire vein or only the vein terminus; exindusiate (not covered with indusia); **SPORES:** ellipsoidal, globose, tetrahedral, trilete-grammitids (gametophytes strap-like).

ECONOMIC USES Commonly used as landscape or potted plants, with some of the epiphytic species, such as ***Platycerium***, extensively grown for decorative purposes on walls and trees, or a piece of wood.

CULTIVATED GENERA *Aglaomorpha, Campyloneurum, Drynaria, Lemmaphyllum, Microgramma, Microsorum, Nephidium, Phlebodium, Platycerium, Polypodium,* and *Pyrrosia,* among several others.

COMMENTS This family in the past treatments included many taxa now recognized as distinct genera and species. In the years past, this family functioned as damping ground for any taxon that would not fit elsewhere. The Polypodiaceae (in a strict sense) is still the largest and most diverse family of ferns and includes the genera cited above and others cited by APG IV (2016–2021), where the number of species in some genera is noted. *Lecanopteris* is an epiphytic myrmecophyte (ant plant).

Platycerium spp.

STAGHORN FERNS

ETYMOLOGY [From Greek *platys,* broad; and *keras,* a horn, in allusion to the form of the fronds].

GROWTH HABIT Evergreen ferns; epiphytic; large, spreading, or drooping, clasping; some forming large clumps; coarse texture; variable with species, 1–4 ft. (30–120 cm) long, but some fronds to 6 ft. (2 m) long; slow growth rate; stemless. **FRONDS:** dimorphic; sterile clasping fronds shield like, rounded to oblong, turning brown and papery; green fertile fronds entire or lobed, typically dichotomously forked into antler-like lobes, covered with stellate hairs, upright or drooping. **SORI:** none; spores formed in dense clusters on tips of lower surface of fertile leaves; no indusia. **SPORES**: minute. **NATIVE HABITAT:** Old World Tropics, Subtropics (1 in New World), usually in wet or cloud forests.

CULTURE Partial to deep shade on organic well-drained fibrous materials; no salt tolerance; scale, ants, mites, snails, and fungus diseases may be problems. **PROPAGATION:** spores, offsets, or 'pups' from some spp., and tissue culture. **LANDSCAPE USES:** as a specimen or accent on a patio wall, palm trunk, or hanging from the branches of a tree. Caution should be exercised with regard to size and weight which in time may become a problem. **HARDINESS ZONE:** variable with species, generally zones 10A–13B; A few species can be grown outdoors in warmer climates.

COMMENTS About 8 or more related species; some with selected cultivars, including: *Platycerium andinum* (South American staghorn); *P. angolense*, *P. bifurcatum* (common staghorn fern); *P. coronarium*; *P. elephantotis* (cabbage fern); *P. elisii*, *P. grande* (giant staghorn fern); *P. stemaria* (triangular staghorn fern); *P. superbum* (silver elkhorn fern); and *P. willinckii* (Java staghorn fern), among a few others.

Platycerium superbum (= *P. grande*?), giant staghorn fern (left) and tissue cultured *P. bifurcatum* (right)

Platycerium bifurcatum (common staghorn fern)
Photographed at Selby Botanical Garden, Florida

Pleopeltis michauxiana

RESURRECTION FERN

SYNONYMS *Acrostichum feruginosum, A. polypodioides, Marginaria polypodioides, Polypodium chrysoconia, P. incarnoides, P. mesetae, P. polypodioides*

ETYMOLOGY [From the Greek name *Polypodion: polys,* many; and *pons,* a foot, in reference to the much-branched, spreading rhizomes] [Resembling *Polypodium*].

GROWTH HABIT Evergreen fern; epiphytic; with creeping rhizomes; to 7 in. (18 cm) tall; spread variable; rhizomes; creeping, thin, wiry, much-branched. A drought-resistant fern, the fronds fold and curl up during dry periods and revive again when moist (resurrection Fern). **FRONDS:** oblong; to 7in. (18 cm) long and 2 in. (5 cm) wide; pinnatifid, leathery, gray-scaly beneath; borne on stipes attached to the rhizomes with a distinct joint; pinnae oblong, entire; medium green. **SORI:** round; in rows along each side of the midvein; indusia absent. **SPORES:** minute. **NATIVE HABITAT:** found throughout the southeastern U.S. as far north as Delaware and west as Texas; it is also native to tropical America

CULTURE Partial to deep shade, naturally grows primarily on the bark of oaks, but infrequently on magnolias, and some other trees, and on rock surfaces; can be grown in an organic, very well-drained medium; no salt tolerance; generally, pest free. **PROPAGATION:** division or spores. **LANDSCAPE USES:** can be used in naturalistic landscapes on the bark of trees, rock crevices, and in terrariums. **HARDINESS ZONE:** 7A–13B.

COMMENTS Many other *Polypodium* species are in culti vation, though primarily in private or botanical collections.

Pleopeltis michauxiana

Drynaria quercifolia

BASKET FERN
SYNONYM *Polypodium quercifolium*

ETYMOLOGY [Origin of the word obscure, but perhaps from Greek *druas* (dryad), mythological tree nymphs inhabiting the oaks, in reference to epiphytic nature of the species] [Resembling oak leaves]

GROWTH HABIT Predominantly epiphytic or epipetric (= grow on rocks); fronds grow from rhizomes that are creeping and covered with brown scales. **FRONDS**: dimorphic, sterile (nest fronds) overlap and are basal (often under green leaves), sessile, cordate-oblong and sinuate (wavy margins) or lacinate (divided into deep narrow irregular segments), with little to no chlorophyll and noticeably elevated venation, these do not shed and form a characteristic basket that covers and essentially protects the rhizomes; fertile fronds large, pinnate-pinnatifid, dark green. **SORI:** round, without indusia, often run together to form a continuous line, located at the angle of the anastomosing veinlets. **NATIVE HABITAT:** Tropical Africa, Southeast Asia, Australia, and Oceania, in tropical rainforests.

CULTURE May be grown on tree trunks, large rocks, or brick walls by attaching a piece of the rhizome. **PROPAGATION**: from spores, sporophytic buds on fronds, or by tissue culture. It can tolerate short periods of sun; may house snakes, ants, and other creatures. **LANDSCAPE USES:** ornamentals, but important as medicinal plant and commercially sold for this purpose. **HARDINESS ZONE:** 9B–13B.

COMMENTS Fifteen species of *Drynaria* are recognized, of which in addition to *D. quercifolia*, *D. fortunei*, *D. rigidula* (basket fern), *D. roosii* (Gu-Sui-Bu) and its cultivar **'Whitei',** and perhaps others are sometimes offered for sale. Also powdered rhizomes sold as "bone building" medicine.

Drynaria quercifolia (?)
Green fertile fronds

Drynaria rigidula
Dried sterile ("nest") fronds

ADDITIONAL GENERA

Niphidium crassifolium
crassifolium, graceful fern,
C. and S. America)

Microgramma percussa
(=_Polypodium percussum_, Indonesia)

ASPLENIACEAE

SPLEENWORT FAMILY

25 GENERA AND 515 SPECIES
GEOGRAPHY Nearly cosmopolitan.

GROWTH HABIT Small- to medium-size plants, terrestrial or mostly epiphytes or rupestral (rock inhabitants), with rhizomes erect, uncommonly long-creeping, stout to slender, occasionally branched, and usually covered with scales and/or hairs. Some species reproduce plantlets on the fronds. **FRONDS:** simple, monomorphic, forming a rosette, and variously pinnately compound; stipe rigid; stipules absent. **SORI:** solitary, short or oblong-linear, along the upper portion (one side) of branched veins; indusia oblong to linear; **SPORES:** spheroid to ellipsoid.

ECONOMIC USES Ornamentals, in moist locations or on trees or walls.

CULTIVATED GENERA *Asplenium, Camptosorus, Ceterach, Phyllitis, Scolopendrium,* and perhaps others depending on taxonomic treatments.

COMMENTS As in other fern families and genera, there is much disagreement on delimitation of taxa or their placement. Interspecific hybridization in *Asplenium* is common and of interest. Students should consult appropriate references for additional information. As in most other fern families, Aspleniaceae has been the subject of much controversy, with some authors accepting only one genus while others up to 10 genera. For example, flora of North America (vol. 2) recognizes a single large genus, *Asplenium,* with mostly epiphytic species, with 700 species. Significantly, APG IV (2016–2021) also recognizes only a single species, *Asplenium,* with mostly epiphytic species. Different genera are recognized in this book to familiarize students with diversity of the family, not to argue with taxonomy of the family.

Asplenium nidus

BIRD'S NEST FERN
SYNONYMS *Asplenium australasicum, A. ficifolium, Neottopteris australasicum, N. mauritiana, N. musaefolia, N. nidus, N. rigida*

ETYMOLOGY [From Greek *a,* not; *spleen,* the spleen, in reference to this fern's traditional virtues in afflictions of the spleen and liver] [Nest].

GROWTH HABIT Epiphytic; evergreen fern with erect stiff fronds arising from the crown giving a bird's nest-like appearance to the center; grows 2–4 ft. (24–48 cm) tall, spread variable. **FRONDS:** simple; leathery, arise from a crown on short stipes; 2–4 ft. (24–48 cm) long, up to 8 in. (20 cm) wide; bright green; irregularly undulate; margins entire, sinuate, or irregularly lobed; light green; veins numerous, penniparallel. **SORI:** spores produced in linear, elongate, black sori alongside secondary veins on frond undersides; lacking indusia or covered with false indusia; **SPORES**: minute. **NATIVE HABITAT:** Australia, Tropical Asia, Polynesia.

CULTURE Partial or full shade on a wide range of moist but well-drained organic soils; high humidity; no salt tolerance. Foliar nematodes and scales may become a problem. **PROPAGATION:** spore or tissue culture. **LANDSCAPE USES:** as epiphytes on trees, in containers for patios, or as foliage plant. **HARDINESS ZONE:** 10B–13B; outdoors only in warmer climates; indoors in any location.

COMMENTS Cultivars *Asplenium nidus* 'Aves' and *Asplenium nidus* 'Osaka' have been introduced. Fronds of other species in the genus may be entire, pinnate, or lobed. Some are native to US, Europe, and temperate Asia and are hardier than the tropical species, which require greenhouse conditions. The genus includes both evergreen and deciduous species. Examples in the trade include species, selected cultivars, and hybrids: *A. antiquum, A. bulbiferum* (plantlets arise from pinnules), *A. dimorphum × difforme, A. ebenium, A. platyneuron, A. rhizophyllum, A. scolopendrium* **Cristatum Group**, and *A. trichomanes,* among others.

Asplenium nidus. Tissue cultured plantlets (above left), circinate vernation (fiddleheads) of developing fronds (above right), plants in tropical forest in Indonesia (middle left), fertile frond (middle right), and plants in a landscape (below)

Asplenium thunbergii [after Carl Peter Thunberg (1743-1828), professor of botany at Uppsala University]. Photograph included here to illustrate diversity of the genus with its pinnate fronds. It is native to Malaysia and Indonesia. It superficially resembles *Belechnum* but its fronds are finely divided.

Asplenium scolopendrium (= *Phyllitis scolopendrium*)

Asplenium trichomanes

WOODSIACEAE

(INCLUDING ATHYRIACEAE)

LADY FERNS

1 or 2 GENERA AND 65 SPECIES

GEOGRAPHY Nearly cosmopolitan but diverse in temperate regions of northern hemisphere and tropical south American mountains, and a few in Africa and Madagascar.

GROWTH HABIT Predominantly terrestrial or sometimes on rocks (epipetric) or cliffs; rhizomes creeping, ascending, or erect, scales at the tips usually not clathrate (= latticed). **FRONDS**: variable but usually monomorphic, pinnately compound; veins pinnate or forked. **SORI**: round, j-shaped, or linear with kidney-shaped to linear indusia or exindusiate. **SPORES**: kidney-shaped.

ECONOMIC USES Ornamentals; *Diplazium esculentum* (vegetable fern) is used as vegetable.

CULTIVATED GENERA *Athyrium, Cystopteris, Deparia, Diplazium, Gymnocarpium,* and *Woodsia,* among others.

COMMENTS Woodsiaceae, according to Smith et al. (2006), also includes **Athyriaceae** and **Cystopteridaceae**. Nearly 85% of the species are included in the two main genera: *Athyrium* and *Diplazium,* that apparently hybridize, and divided into five subgenera. Only a single genus is recognized by Pteridopyte working Group (2016).

Athyrium niponicum

JAPANESE PAINTED FERN

ETYMOLOGY [Derivation unknown, possibly from Greek *anthoros*, breeding well, alluding to the diverse forms of sori] [Japanese]

GROWTH HABIT Compact terrestrial deciduous fern with a basal rootstock. **FRONDS**: taller than broad; pinnate, with 6–10 pairs of pinnae that are also longer than broad; stipes and rachises dark red; foliar colors vary with cultivars. **SORI**: on underside of all pinnae; j-shaped with kidney-shaped indusia. **SPORES**: kidney-shaped. **NATIVE HABITAT**: Eastern Asia, Japan, northern China, Korea, and Taiwan.

CULTURE One of the easiest ferns to grow in moist organic-rich soils; partial to full shade but colors intensify with some early morning sun exposure; a small amount of slow-release fertilizer is usually sufficient; essentially no serious problems noted other than the deciduous habit and necessity of removing dead fronds. **PROPOGATION**: tissue culture, division in early spring or fall, or by spores; **LANDSCAPE USES**: shady borders, alpine and rock gardens, or in containers; may also be used as groundcover or mass or mixed plantings. **HARDINESS ZONE**: 5B–11A. Although more suited to cooler climates, *A. niponicum* can be grown in cooler shady subtropical regions.

COMMENTS A popular fern with several cultivars in the trade, including **'Apple Court'** (crested with purple and silver markings), **'Metalicum'** (silver green and red fronds), **'Pictum'** (Japanese painted fern which has metallic silver-gray with red and blue hues), **'Pictum Cristatum'** (same color but pinna crested), **'Burgundy Lace'**, **'Red Beauty'** (yellow fronds and bright red stems), **'Regal Red'** (triangular silver fronds and red rachis), **'Silver Falls'** (silver fronds with red veins), and others.

Athyrium niponicum (above) and *A. niponicum* **'Pictum'** (below)

Diplazium proliferum

MOTHER FERN
SYNONYMS *Asplenium proliferum, Athyrium proliferum, Callipteris prolifera, Diplazium incisum, D. repandum, D. serratum, D. spinulosum*

ETYMOLOGY [From Greek, *diplasios*, double, in reference to paired sori] [prolific, spreading readily]

GROWTH HABIT Evergreen terrestrial fern with a short erect rhizome that is covered with scales at the apex and that can resemble a short trunk. **FRONDS:** monomorphic, pinnately compound, clustered at the apex of the rhizome 3–6 ft. (1–2 m) long; bulbils (!) and eventually plantlets are produced on the rachis (hence, the common name); pinna many, alternate, lanceolate, narrowed at the apex to a tip, reduced near the base, sometimes shallowly lobed; veins anastomosing. **SORI:** arranged in v-shape along and on both sides of veinlets, indusia linear, attached along the veins. **NATIVE HABITAT:** widespread in tropical Africa and Madagascar, in thickets and marshes, usually as understory plant.

CULTURE In bright full or partial shade in organic well-drained soil. **PROPAGATION**: easily grown by spores or by plantlets, and probably by tissue culture; no specific problems reported, but probably dryness and insects may cause some damage. **LANDSCAPE USES:** grown in the landscape or in containers, but there is no report of it being grown specifically for food. **HARDINESS ZONE:** 9A–13B.

COMMENTS Widely used as an ornamental but, similar to *D. esculentum* young fronds are also utilized as food. Two genera, *Callipteris* and *Monomelangium,* have been included in *Diplazium* by Smith et al. (2006). The genus is pantropical and includes about 400 species referred to as "twinsorus ferns." *Diplazium esculentum*, as the specific epithet implies, is more commonly used as vegetable. Other cultivated species include *D. hymenodes* (peacock fern), *D. molokaiens* (the Molokai twinsorus fern), and *D. lanchophyllum* (the lance-leaved glade fern).

Diplazium proliferum

BLECHNACEAE

BLECHNOIDS FAMILY, CHAIN FERN FAMILY

24 GENERA AND ±265 SPECIES
GEOGRAPHY Native to the Old and New worlds, especially in tropical and subtropical Africa, Hawaii, South America, Australia, and New Zealand, and *woodwardia* in North America.

GROWTH HABIT Predominantly terrestrial, with erect, decumbent (lying along a surface with tips curving upwards), or creeping rhizome, often bearing stolons, and sometimes with trunk-like stem. **FRONDS:** usually monomorphic but sometimes dimorphic, pinnately compound, in a cluster or spaced individually, new fronds often pinkish (anthocyanic); pinnae entire or serrate, glabrous or pubescent; veins free, pinnate, forked, or sometimes anastomosing. **SORI:** roundish or slightly in chains or linear (elongate), often parallel and adjacent to midrib or secondary veins; indusia linear, opening towards the midrib or absent; sporangia with short or long stalks with more than one row of cells; **SPORS:** reniform (kidney-shaped). Indusia, according to Judd et al. (2015), is the distinctive feature of this family.

ECONOMIC USES Ornamentals; often used in the landscape and some species as ground covers, although others are tree-like and suitable for container growing.

CULTIVATED GENERA *Blechnum, Brainea, Doodia, Sadleria, Stenochlaena, Woodwardia,* and probably others

COMMENTS Smith et al. (2006) recognize nine genera in the family, but The Plant List notes 21 genera and 219 species. This is indicative of difficulties in fern classification. Genera other than those noted above are probably grown in public garden and private collections as well. As in other families, Blechnaceae classification is also controversial. Occurrence of sori in chains or parallel rows and indusia are distinctive features. The number of genera noted here by APG IV (2016–2021) is a significant increase from the nine genera included by Smith et al. (2006).

Blechnum brasiliense

ANGELFIRE, BRAZILEAN TREE FERN
SYNONYMS *Blechnum cocovadense, B. fluminense, B. nigrosquamatum, B. nitidum*

ETYMOLOGY [From the classical Greek name *blcknŏn* for a fern used by Dioscorides, although probably not specifically in reference to this genus] [Brazilian]

GROWTH HABIT Although not a true tree fern, as the common name implies, with age it develops a short prominent trunk with a crown of spreading fronds. **FRONDS:** monomorphic, once-pinnate; young developing ones pinkish-red, on short stipes. **SORI:** linear, in bands on both sides of the midrib; indusia linear and open towards the midrib. **NATIVE HABITAT:** in moist to wet soils often in shaded habitats, along banks of streams, In South America (Brazil and Peru).

CULTURE Full bright shade, prefers acid loamy well-drained moist soils; dryness causes necrotic spots and frond death; sunburn. **LANDSCAPE USES:** often grown in landscapes and large tubs in well-lighted situations, but not in extended sunny location. **HARDINESS ZONE:** 10B–13B.

COMMENTS There are related species including the more common *B. gibbum* and its cultivar **'Silver Lady'**, *B. indicum* (swamp fern), *B. occidentale* (hammock fern), and *B. penna-marina* (alpine hard fern). Also listed are a number of more warm temperate species such as *B. capense, B. niponicum, B. spicant,* and its cultivars, as well as others. *Blechnum brasiliense* is not specifically noted in APG IV (2016–2021), but it is the most common species in the trade for its reddish new leaves and its small tree-like growth habit.

Blechnum brasilense (Photographed at Longwood Gardens, Pennsylvania)

Blechnum brasilense (?) in Amazonian Forest, photographed in Bolivia

Blechnum gibbum (dwarf tree fern) [name translates to swollen in a lopsided way]. Note the difference between specimens grown in a shadier conservatory (above) as opposed to outdoor landscape (below).

Blechnum spicant

ADDITIONAL GENUS

Lomaria spannagelii (= *Blechnun spannagelii*).   Genus *Lomaria* is actually a member of
Lomariopsidaceae, but resembles *Blechnum* spp.

DAVALLIACEAE

DAVALLIA FAMILY

1 GENUS AND 65 SPECIES

GEOGRAPHY Native to the Old World, especially in tropical and subtropical Southeast Asia, with a single species in western Mediterranean and tropical Africa.

GROWTH HABIT Mostly epiphytic, with erect, decumbent, or creeping densely scaly rhizomes, sometimes with stolons. **FRONDS:** usually monomorphic; 3–5 times pinnate, entire or variously pinnately compound, in a cluster or often spaced individually; stipe jointed near the base; pinnae entire or serrate, glabrous or pubescent; veins free, pinnate or forked. **SORI:** roundish or slightly elongate, indusia roundish, lunar (half-moonshaped), reniform, or orbicular; on terminus of veins near the margin and usually opening towards the margin; **SPORES:** ellipsoidal.

ECONOMIC USES Ornamentals; used in hanging baskets and as groundcovers.

CULTIVATED GENERA *Davallia, Humata, Nephrolepis, Rumohra, Scyphularia,* and perhaps others in private collections.

COMMENTS Smith et al. (2006) have included *Humata, Parasorus,* and *Scyphularia* in *Davallia.* Number of genera is debatable. Status of Davalliaceae in APG IV (2016–2021) is not clearly defined.

Davallia fejeensis

RABBIT'S FOOT FERN
ETYMOLOGY [After Edmund Davall (1713–1798), English botanist of Swiss origin] [From Fiji Islands].

GROWTH HABIT Fine textured evergreen fern; terrestrial or sometimes epiphytic; rhizomes stout, clumping, with long pubescence, appear furry (hence, the common name); to ½ in. (1½ cm) thick; grows 1½–2 ft. (45–60 cm) tall with variable spread. **FRONDS:** triangular; 4 times pinnate; pinnules linear, terminal, and single-veined; **SORI:** minute spores produced in sporangia along the pinnule margins; indusia cylindrical or cup-shaped. **NATIVE HABITAT:** Fiji Islands.

CULTURE Partial or full shade on a wide range of well-drained but moist organic soils; high humidity; no salt tolerance; leaf drop due to dry, hot air, or cold; scales may be problems. **PROPAGATION:** division of old plants and tissue culture. **LANDSCAPE USES:** often grown in sphagnum moss as hanging baskets or "fern balls." **HARDINESS ZONE:** 10B–13B. Not sufficiently cold hardy to be grown permanently outdoors in most of the continental US; hardy to 60–65°F.

COMMENTS According to APG IV (2016–2021), the genus consists of 65 species, tropical and subtropical plants of the Old World, which are mostly evergreen. One cultivar ***Davallia fejeensis* 'Plumosa'** has gracefully drooping, more feathery foliage. Other species such as ***Davallia denticulata, D. trichomanoides, D. solida*** (giant hair's foot fern), ***D. tryphylla, D. tyermanii*** (= *Humata tyermanii*), and other species are also in the trade.0

Davallia fejeensis (photographed at Longwood Botanical Garden, Pennsylvania)

Scyphularia pentaphylla (= *Davallia pentaphylla*, black caterpillar fern). Photographed at University
of California Davis Conservatory.

DICKSONIACEAE

DICKSONIA FAMILY

3 GENERA AND ±35 SPECIES

GEOGRAPHY Tropical America, and subtropical Asia, Australia, New Caledonia, and the Pacific Islands (including Hawaii).

GROWTH HABIT Terrestrial; usually with an arborescent, unbranched, erect, or creeping stem, covered with dense matted trichomes and/or fibrous roots at base. **FRONDS:** usually monomorphic, in a loose cluster, to 12 ft. (4 m) long, variously pinnately compound or pinnatifid; pinnae usually much dissected; stipe covered with dense trichomes near the base; veins free. **SORI:** at the margins of pinnules on a somewhat elevated receptacle, on terminus of veins, and enclosed by the reflexed margins; **SPORES**: tetrahedral-globose.

ECONOMIC USES Ornamentals; often used as garden or indoor plants; trunks of some species used for planting epiphytic orchids and small bromeliads.

CULTIVATRED GENERA *Balantium, Calochlaena, Dicksonia,* and *Lophosoria*

COMMENTS *Calochlaena, Dicksonia,* and *Lophosoria* are the three genera recognized by Smith et al. (2006) and by APG IV (2016–2021). *Balantium* is the fourth genus included by the Plant List, but not accepted by APG IV. Other genera such as *Cibotium* (all endemic to Hawaiian Islands) that were previously placed in Dicksoniaceae are recognized as distinct Cibotiaceae. In general, "the core tree ferns" are currently recognized as members of four distinct families primarily based on characteristics of the sori: Cibotiaceae, Cyatheaceae, Dicksoniaceae, and Metaxyaceae. *Cyathea* and *Dicksonia* are the most common in cultivation.

Balantium antarcticum

AUSTRALIAN TREE FERN, TASMANIAN TREE FERN; TASMANIAN DICKSONIA

SYNONYM *Dicksonia antarctica*

ETYMOLOGY [New Latin, from Ancient Greek, diminutive of *balantion,* a bag or pouch; *Dicksonia,* after James Dickson (1738–1822), a British botanist and nurseryman] [Of the South Polar Region].

GROWTH HABIT Terrestrial ferns with erect arborescent stems, to 50 ft. (17 m), but usually considerably shorter in cultivation; single, unbranched, densely covered with trichrome and fibrous roots. **FRONDS:** to 4 ft. (1¼ m) long, more or less leathery, monomorphic or dimorphic, 2–4-pinnate-pinnatifid, with long, dense trichrome at the base of the stipe; pinnules to 2 in. (5 cm) long, with 5–7 teeth at margins, veins free. **SORI:** indusium clam-shaped, consisting of an actual indusium attached to a segment tissue; sori marginal or rarely terminal. **NATIVE HABITAT:** Australia, Tasmania, as understory plants in high rainfall areas of southeastern Australia.

CULTURE Partly shaded sites, in moist, well-drained, preferably under large, evergreen trees; does not tolerate direct sun (sun scorch) or prolonged freezing temperatures; no salt tolerance. **PROPAGATION:** spores on screened peat moss or in sterile culture, tissue culture. **LANDSCAPE USES:** often used in landscape in warm, moist regions, or as indoor foliage plants, attractive in mass plantings or as an accent plant. **HARDINESS ZONE:** 9A–13B. It does tolerate frost but may lose its leaves.

COMMENTS *Balanitium* (*Dicksonia*) *fibrosa* wooly tree fern; golden tree fern, with which *Balantium antarcticum* is sometimes confused; *Balantium swllowianum* (= *Dicksonia sellowiana*), *D. squarosa* (slender tree fern), and others are sometimes cultivated. Species of tree ferns are often confused and misidentified. For example, *Cyathea cooperi* (Cyatheaceae) with finer frond texture is sometimes sold as *Dicksonia antarctica* (Dicksoniaceae). The generic name *Balantium* does not appear to have been accepted by APG IV (2016–2021), but it is commonly known in the trade, but more often sold as *Dicksonia*.

Balantium antarcticum (= *Dicksonia antarctica*) Photographed in Golden Gate Park, San Francisco, California)

LOMARIOPSIDACEAE

LOMARIOPSID FAMILY

5 GENERA AND 70 SPECIES

GEOGRAPHY Tropical regions of the Old and the New Worlds, few in temperate zones.

GROWTH HABIT Rhizomes creeping or sometimes climbing hemiepiphytic (epiphytic early in life cycle), but mostly terrestrial, often producing long slender stolons (runners). **FRONDS:** simple pinnate, pinnae linear to linear oblong, with entire or crenate margins, sometimes divided or auriculate; veins free or penniparallel. **SORI:** in a single row on both sides of the veins, roundish or slightly elongate, indusia roundish, lunar (half-moon-shaped), reniform, or orbicular or exindusiate; on terminus of veins near the margin. **SPORES:** ellipsoidal.

ECONOMIC USES Ornamentals.

CULTIVATED GENERA *Cyclopeltis*, *Lomariopsis* (=*Thysanosoria*), *Nephrolepis*.

COMMENTS Although in its original classification Nephrolepidaceae was considered monogeneric (*Nephrolepis*), in the current treatment of the family by Smith et al. (2006), four genera are recognized. Additional information is not provided in APG IV (2016–2021).

Nephrolepis exaltata

BOSTON FERN, SWORD FERN
ETYMOLOGY [From Greek *nephros,* a kidney; and *lepis,* a scale, in reference to the kidney-shaped indusium] [Very tall].

GROWTH HABIT Evergreen, terrestrial, or epiphytic fern; ground cover or often on palm trees, with particular preference for *Phoenix canariensis* (Canary Island date palm); short rhizomes; size variable with cultivar, range 6 in. (15 cm) to 3 ft. (1 m) tall, commonly to 2 ft. (50 cm) tall; variable spread; rapid growth rate; spreads by thin, green runners (stolons) from the crowns that root to form new plants. **FRONDS:** highly variable; foliage erect or drooping; medium, fine texture; 1–5 times pinnate in some cultivars; arching, flat, narrow, sword-shaped; pinnae alternate, narrowly deltoid; to 3 in. (8 cm) long; numerous, close together; various shades of light green; not waxy, papery texture; entire or crenate to serrate margins. **SORI:** kidney-shaped brown, submarginal on the underside of mature pinnae, terminal on veins. **SPORES:** minute. **NATIVE HABITAT:** Tropics and subtropics of the world.

CULTURE Partial to deep shade; moist, but well-drained soil of some fertility; no salt tolerance scale, mites, mealy bugs, snails, slugs, and fungus disease may become problems. **PROPAGATION:** division of rooted runners, spores, tissue culture. **LANDSCAPE USES:** makes a good, fast-spreading ground cover; requires frequent thinning once established; gives tropical, lush effect to **Sabal, Butia, Phoenix,** and other palms when growing epiphytically on their trunks. Several cultivars are used as pot or hanging basket plants. **HARDINESS ZONE:** 8A–13B.

COMMENTS Historically, the most common cultivar was **'Bostoniensis'** which is more graceful, drooping, and compact than the species and has given rise to numerous sports. Today, many cultivars for leaf form have been selected, a few

of the common ones are **'Compacta', 'Emerald Vase', 'Fluffy Ruffles', 'Hillii', 'Roosveltii',** and **'Whitmanii'.** However, several other *Nephrolepis* species and cultivars are also available in the trade, including *N. acutifolia, N. biserrata, N. cordifolia, N. duffii, N. falcata, N. hirsutula, N. obliterata, N. pendula*, and probably others including the inappropriately named *N. multiflora. Nephrolepis exaltata* in particular is considered invasive in many areas, including Florida.

Nephrolepis exaltata growing on trunk of *Phoenix canariensis* (above, in Florida) and *N. duffii* (below, photographed in Bogor Botanical Garden, Indonesia))

LYGODIACEAE

CLIMBING FERNS

1 GENUS AND ±25 SPECIES

GEOGRAPHY Pantropical but with temperate species in Asia.

GROWTH HABIT Terrestrial, with slender creeping hairy rhizomes. **FRONDS**: pinnate, with indeterminate (consistent) growth, the rachis of a single frond climbing up to 10 ft. (2–3 m), forming an individual plant; pinnae pseudo-dichotomously branched with a bud in angle of branch; veins free or anastomosing. **SORI**: solitary, on lobes of the ultimate segments with a single sporangium per sorus; indusium antrorse (= facing upwards); **SPORES**: tetrahedral (= four-sided).

ECONOMIC USES Ornamentals; potential medicinal uses; leaves of some species are used for weaving.

COMMENTS *Lygodiuim* had been placed in Schizaeaceae in some earlier treatments.

Lygodium spp.

JAPANESE CLIMBING FERN
ETYMOLOGY [From Greek, *lygōdēs*, like a willow, twining, in reference to the climbing habit of the species]

GROWTH HABIT Terrestrial evergreen climbing ferns; the horizontal shallow subterranean short rhizome constituting the stem. **FRONDS:** deciduous or evergreen, dimorphic, said to be 6–20 ft. (2–6½ m) long, with thin stipe and rachis; pinna alternate, pinnules divided or palmate; veins free or anastomosing. **SORI:** resemble finger-like projections on pinnule margins; sporangia in two rows. **NATIVE HABITAT:** predominantly in tropical and subtropical forest margins of eastern Asia, Japan, northern China, Korea, and Taiwan, but exceptionally in cold temperate Eastern US (*L. palmatum*).

CULTURE Require vertical support on which they can climb, readily adjust to most soils, and become unwieldy, otherwise no major problems. **PROPAGATION**: spore germination, at about 85°F (30°C) **LANDSCAPE USES:** primarily as ornamentals. **HARDINESS ZONE:** 4A–13B. Species dependent.

COMMENTS *Lygodium japanicum* (Japanese climbing fern), *L. microphyllum* (Old World climbing fern), and *L. flexuosum* have been reported as invasive exotics in Florida. As in vast majority of invasive exotics, the species were originally introduced as ornamentals and in fact at least *L. japanicum* is currently produced and sold as potted plants. Other species occasionally offered in the trade include *L. circinatum*, *L. palmatum* (American climbing fern), *L. scandens*, and *L. venustum*. *Lygodium flexuosum* extract is sold for therapeutic purposes.

Lygodium japanicum   *Lygodium circinatum*

MARATTIACEAE

MARATTIA FAMILY

6 GENERA AND 110 (135?) SPECIES

GEOGRAPHY Pantropical, from Mexico to Argentina, with *Marattia* extending to Hawaii from Southeast Asia.

GROWTH HABIT Terrestrial or rarely epiphytic; roots large, rhizome fleshy, short, upright or creeping, about 0.08 in. (2 mm) across, root hairs few; stems, and fronds with mucilage canals. **FRONDS:** large, fleshy, mostly 1-3-pinnate, with enlarged, fleshy stipules at the base and swollen pulvinae along petioles and rachises; stems and pinna-bearing scales. **SORI:** sporangia free or in round or elongate synangia (fused sporangia, *cf. Psilotum*), annulus absent. **SPORES:** bilateral or ellipsoid.

ECONOMIC USES Ornamentals and minor consumption of *Angiopteris* and *Marattia* rhizomes.

CULTIVATED GENERA *Angiopteris, Christensenia, Danaea*, and *Marattia*.

COMMENTS Smith et al. (2006) included four genera in Marattiaceae, all of which are reported from cultivation: predominantly in botanical conservatories and tropical climate regions. The Plant list, however, lists 9 genera and 148 species. According to Christenhusz et al. (2018), the genera include: *Angiopteris* (±30), *Christensenia* (1), *Danaea* (55), *Eupodium* (3), *Marattia* (10), and *Ptisana* (±35). Marattiaceae is considered to be one of the most primitive eusporangiate (sporangium develop from a group of cells and with many spores), fern families.

Angiopteris evecta

ELEPHANT FERN, GIANT FERN, KING FERN
SYNONYM *Polypodium evectum*

ETYMOLOGY [From Greek *angeion*, a vessel; *pteris*, a fern]; [Expanded, springing forward, in reference to the large upright fronds].

GROWTH HABIT Large terrestrial fern with comparatively bulky stem (not a tree fern) covered with persistent swollen stipe bases of old fronds; rhizomes up to 3 ft. (1 m). **FRONDS:** large, 16–26 ft. (5–8 m) long, twice pinnate (bipinnate) with thick fleshy stipe attached to the brown woody stem with swollen bases (pulvinae); pinnules are born on secondary branches (rachilla). **SORI:** round sporangia in synangia (fused sporangia) along pinnule margins on terminus of the penniparallel veins; annulus absent. **SPORES:** ellipsoid. **NATIVE HABITAT:** lowland moist humid tropical forest of Madagascar, Malaysia, southern India, New Guinea, Australia, and Pacific Islands.

CULTURE May be grown only in wetter warmer climates in rich acidic well-drained soils, and primarily in shadier locations. It may be propagated from leaf bases, spores, or tissue culture. Large leaves may collapse as a result of water loss from the fleshy water-filled bases; sunburn; it does not tolerate dryness. **LANDSCAPE USES:** this is one of the most attractive ferns suitable for tropical landscapes or in conservatory collections; either as accent or container specimen. **HARDINESS ZONE:** 11B–13B.

COMMENTS *Angiopteris* is naturalized in Hawaii, Jamaica, Cuba, and parts of Central America (Costa Rica), and is considered invasive. It is known in the trade as *A. opaca, A. robusta,* and *A. fokiensis*, among others. Genus *Macroglossum* is now included in *Angiopteris*. *Archangiopteris* is considered siter to *Angiopteris*.

Angiopteris evecta   *Angiopteris fabiensis*

ADDITIONAL FAMILIES

Salvinia natans (Salviniaceae), is a widespread aquatic fern (floating fern, floating moss).
Originally native to Southern Europe, northern Africa, and Asia but now cosmopolitan in
standing water. It is a deciduous, free-floating perennial aquatic with a pair of opposite leaves on
the surface and a submerged one that functions as root. Plant is considered an aggressive aquatic
weed. Genus name honors Antonio Maria Salvini (1633-1729), professor of Greek at Florence
who helped Micheli in his botanical studies.

Illustration from Otto Wilhelm Thomé (1885).

Introduction

They are a morphologically and geographically diverse assemblage of ancient cone-bearing woody seed plants of considerable ornamental and economic importance. In some cases, they may be threatened or endangered. These include four interesting subclasses (see table of extant gymnosperms): **Cycadidae** (cycads), **Ginkgoidae** (gingko), **Pinidae** (conifers), and **Gnetidae** (gnetums, etc.), Vast majority of conifers are common landscape plants in temperate as well as tropical climate regions. Gingko and most conifers, however, have also proven to be ideal economic and landscape plants, as well as subjects for bonsai specimens, and are used as such by many amateurs and in most conservatories. Species of the three genera, *Gnetum* spp., *Welwitschia mirabilis*, and *Ephedra* spp., are also seen primarily in botanical conservatories, but the first two are briefly mentioned and illustrated in this book to familiarize readers with their unusual and remarkable characteristics. Although extensive recent DNA studies have shown these genera to be related to other gymnosperms (Pinaceae), they share some features that are characteristics of angiosperms, hence they were considered by some to be the link between the two classes.

Gymnosperms do not possess a flower with an ovary that would produce fruit, but naked ovules that once fertilized ultimately become seeds. These are born directly on **megasporophylls** and *not* enclosed within a fruit. Hence, gymnosperm (from Greek, *gymnos* "naked" and *spermus* "seed": "naked seed"); typical **strobili** (in cycads) and **cones** (in conifers) that are usually an aggregation of the megasporophylls (= seed leaves) and cone scales. Other gymnosperm characteristics are illustrated in the following photographs and noted in the glossary. In common (and sometimes in published works) usage, strobili of cycads are referred to as cones. Use of "cone" should be restricted to conifers, but reference to cycads is understandable.

All ten genera and ±250 species of cycads are in cultivation and used extensively in landscapes and conservatories. Despite the superficial similarity of their leaves to those of palms and common reference to them as "sago palms", they have no relationship with palms. They are strobilus-bearing plants belonging to two currently recognized families: Cycadaceae (*Cycas*) and Zamiaceae (Bowenia, *Stangeria, Dioon, Ceratozamia, Lepidozamia, Macrozamia, Microcycas, Encephalartos,* and *Zamia*). These are natives of tropical, subtropical, and warm temperate regions of the world. Only *Zamia integrifolia* (=*Z. floridana*) occurs in Florida and southern Georgia, in the United States. Cycads are considered to be the most primitive living seed plants, remnants of a more widespread Mesozoic group of plants. The 1000 living gymnosperms, according to Christenhusz et al. (2017), are arranged in 12 families (see below), while Judd et al. (2015) recognized 15 families. Individuals interested in more details should consult APG IV (2016–2021), Judd et al. (2015), Christenhusz et al. (2017), and their extensive cited literature.

Cycads are dioecious (sexes occur on separate plants). An interesting feature of cycads is their root system that in addition to normal roots, have **coralloid roots** that are apogeotropic (grow upwards to the soil surface and their nodules contain the cyanobacteria *Nostoc* that fix atmospheric nitrogen, similar to that of the legumes). It is a common misconception that cycads are only green. Some cycads have bright red leaves (particularly *Ceratozamia* spp.) in early development and some plants are bluish or more or less silvery gray. Although somewhat slow growing, cycads are easy to cultivate. A majority inhabit open drier sites in their natural habitat, but a few are understory (e.g., *Bowenia* and some *Zamia*) plants that require some shade and adequate moisture. Although they prefer warm conditions, most tolerate cooler temperatures above freezing and a few (e.g., *Cycas revoluta, C. taitungensis, C. panzhihuaensis,* some *Ceratozamia* spp., *Zamia integrifolia,* and other species) even withstand temperatures of a few degrees below freezing for short time periods. Generally, they are propagated by seeds and in a few exceptional cases by 'pups' along the trunk in female *Cycas* and *Encephalartos* and by division of the subterranean stems in *Zamia.* They can be chemically manipulated to branch (Dehgan and Almira, unpublished, Dehgan 1999). Consult the listed references for additional information on propagation, growth, and development of cycads. As a whole, cycads are beetle pollinated, but in some cases, such as *Cycas* spp. that have open megastrobilus (female cones), are also wind pollinated. Nearly all cycads are under CITES (Convention on International Trade and Endangered Species) protection and are illegal to collect without proper permits and often allowed only for research purposes. A significant number of new species have been described, as well as many published botanical and horticultural studies in recent decades.

Conifers constitute the largest group of gymnosperms and are nearly cosmopolitan in their distribution. A vast majority of conifers are native to temperate regions and not uncommonly cover vast areas, but many are also found in the tropics. Unlike cycads that are entirely dioecious, a majority of conifers are monoecious [produce separate male (pollen) and female (ovulate) structures on the same plant]. All conifers are woody, ranging from groundcovers to shrubs, and the tallest trees (e.g., *Sequoia sempervirens*, redwood). A majority of temperate conifers are excurrent (possess a central leader and lateral branches). While cycads are used primarily as ornamentals, conifers provide much of the world's lumber (softwood) and other wood products. Moreover, they are of a much greater ecological significance.

GROWTH HABITS

Juniperus horizontalis
Prostrate

Chamaecyparis lawsoniana
Arborescent

Cycas rumphii
Palm-like

LEAVES

Except for cycads, which have pinnately compound leaves, all other gymnosperms have simple leaves.

Acicular (Needle-like)

Subulate (Scale-like)

Pinnately Compound

Laminate (Blade-like)

DURATION OF GROWTH

Most gymnosperms are evergreen, only a few conifers are deciduous and a few *Cycas* lose their leaves. Deciduous conifers include:

Larix decidua   *Taxodium distichum*   *Metasequoia glyptostroboides*

MICROSTROBILUS

Pollen-bearing (= male) strobilus in cycads

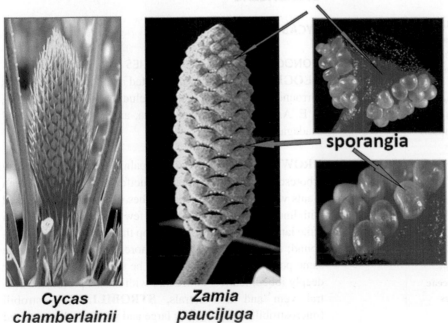

Cycas chamberlainii   *Zamia paucijuga*

MEGASTROBILUS

seed-bearing (= female) strobilus in cycads

megasporophyll

seed

Zamia loddigesii   *Encephalartos* spp.   **megasporophyll**

FAMILIES AND GENRA OF EXTANT GYMNOSPERMS (Based on Christenhusz et al. 2011)

Subclass CYCADIDAE:	Subclass PINIDAE:[a]
Cycadaceae	Cupressaceae
Cycas	*Cunninghamia*
Zamiaceae	*Metasequoia*
Bowenia	*Sequoia*
Ceratozamia	*Cryptomeria*
Dioon	*Taxodium*
Encephalartos	*Chamaecyparis*
Lepidozamia	*Cupressus*
Macrozamia	*Juniperus*
Microcycas	*Platycladus*
Stangeria	Pinaceae
Zamia	*Cedrus*
	Pinus
Subclass GINKGOIDAE:	Podocarpaceae
Ginkgoaceae	*Nageia*
Ginkgo	*Podocarpus*
	Sciadopityaceae
	Sciadopitys
Subclass GNETIDAE:	Taxaceae
Welwitschiaceae	*Taxus*
Welwitschia	*Cephalotaxus*
Gnetaceae	*Torreya*
Gnetum	Araucariaceae
Ephedraceae	*Araucaria*
Ephedra	*Wollemia*
	Agathis

[a]See conifers for example of cone structure

CYCADIDAE

CYCADACEAE

CYCAS FAMILY

MONOGENERIC WITH ±92 SPECIES

GEOGRAPHY Widely distributed in mainland and islands surrounding the Indian Ocean, including northern Australia, S. E. Asian Islands, Japan, China, and eastern Africa, and Madagascar.

GROWTH HABIT Dioecious palm-like shrubs or nearly arborescent and sometimes caudiciform, mostly evergreen plants with primarily annual flushes of growth interspersed with linear cataphylls. **LEAVES:** few to numerous and often quite large, clustered in spirals atop the stem; pinnately compound; in a few species two or more leaflets arise from the same point on the rachis or may be seemingly branched or deeply lobed, leaflets entire and with only a prominent central vein and no laterals, **STROBILI:** male strobili (microstrobili) are generally large and with spirally arranged numerous microsporophylls which are covered with many pollen sacs (sporangia); female strobili (= megastrobili) also often large but in rounded, imbricate clusters of leaflike megasporophylls, each bearing 1–9 ovules, dependent on the species. **SEEDS:** large, resembling drupes, orange to reddish-brown **fleshy integument** (outer seed coat) and a hard bony **inner integument** (inner stony seedcoat) beneath;

with or without spongy **integument** (tough outer protective layer).

ECONOMIC USES Ornamentals; seeds of some species eaten, but not recommended.

CULTIVATED *Cycas* The vernacular name "sago palm" is problematic since sago palm is the true palm *Metroxylon sagu*, from trunk of which starchy material is harvested. Although nearly all known species are offered for sale and are found cultivated in various botanical collections and con-servatories, few are actually common in public landscapes or in the trade: *Cycas angulata, C. apoa, C. armstrongii, C. balansae, C. basaltica, C. cairnsiana, C. calcicola, C. circinalis, C. furfuracea, C. hainanensis, C. macrocarpa, C. media, C. megacarpa, C. multipinnata, C. ophiolitica, C. panzhihuaensis, C. papuana, C. pruinosa, C. revoluta, C. rumphii, C. siamensis, C. taitungensis, C. tansachana, C. thouarsii, C. wadei, C. zeylanica*, and others. Only the common species are discussed here in detail. Notable collections are maintained by several botanical gardens and private individuals.

CYCAS SEEDS

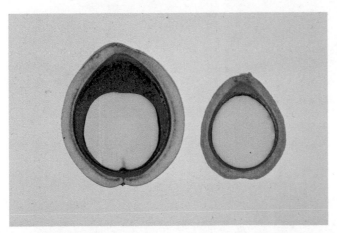

Cycas rumphii (left) with spongy tissue (fleshy inner integument) and *Cycas revoluta* (right) without the spongy integument. Note the fleshy outer layer integument (testa) and the bony inner integument in both species. Typical for all cycads is the presence of testa and the stony layer that enclose the **female gametophyte tissue** (megagametophyte, analogous to endosperms of angiosperms).

Cycas multipinnata

ROYAL SAGO
SYNONYMS *Cycas   longipetiolula,   Epicycas multipinnata,*

ETYMOLOGY [From Greek *koikos*, classical name for a kind of palm which *Cycas* resembles in habit and leaf] [From Latin, *multi*, many, *pinnatus*, pinnate, in reference to branched leaflets]

GROWTH HABIT Although it shares the basic characteristics, with its slight trunk, few leaves, and divided leaflets, it has little similarity with other cycads. This and the few related species are unique among all cycads. **STEM/BARK**: trunk almost subterranean, to about 16 in. (40 cm) above ground; bark brown-gray, scaley. **LEAVES**: few, 1–3; grow 9–21 ft. (3–7 m) tall and 24–39 in. (60–100 cm) wide; pinnate, pinnae ±trullate (similar to brick layers trowel, narrow at tip, wide at base) in outline, flat, petiole subterete (nearly round); spines 30–50 along each side; primary leaflet 12–22 pairs per leaflet, longitudinally inserted to primary rachis, glossy dark green, lanceolate, middle leaflets largest, dichotomously 3–5 forked; midvein slightly elevated on both surfaces; margins entire to somewhat wavy; glabrous. **STROBILI**: male strobili fusiform-cylindric, smaller than other *Cycas* spp. microsporophylls tomentose, apex acute; megasporophylls brown tomentose, glabrescent; ovules 3–5 on each side of the stalk; seeds 6–10, greenish to yellow, obovoid, 1 in. (2.5 cm). **NATIVE HABITAT**: Yunnan and Guangxi in southwestern China and Yen Bai Province in northern Vietnam. Grows as an understory plant.

CULTURE Partial to full shade, grows healthy in well-drained regular moist soil mixes; no specific problem other than sunburn reported. **PROPAGATION**: seeds. **LANDSCPE USES**: specimen plant in shady location. **HARDINESS ZONE**: not reported for this and related species, but they appear to tolerate lower temperatures if not directly exposed to frost and probably under evergreen trees. [the cultural needs expressed are based on authors' personal experience]

COMMENTS *Cycas multipinnata* and related species such as *C. debaoensis* and *C. micholitzii* (endemic to China), with branched leaflets are on the IUCN Red List of Threatened Species (2010), primarily as a result of overcollection by commercial dealers and habitat destruction by farmlands.

Cycas multipinnata (above left) and *C. debaoensis* (above right), species native to China, are distinct from the common species in having bipinnate or multipinnate as compared to simple pinnate leaves; *C. debaoensis* female strobilus and seeds (middle); *C. micholitzii* (forked leaf cycas, also endemic to China), with bifurcating leaflets (left) and the relatively small male strobilus (right). Photographed at Nong Nooch Botanical Garden, Thailand (above), and authors personal collection (middle and below)

Cycas revoluta

SAGO PALM, KING SAGO PALM
SYNONYMS *Cycas miquelii, Epicycas miquelii*

ETYMOLOGY [Greek *koikos*, classical name for a kind of palm to which *Cycas* resemble in habit and leaf] [Rolled backward, in reference to leaflet margins].

GROWTH HABIT Dioecious, evergreen palm-like plants, upright, often suckering, female plants from the base and males sometimes branch from the apex after coning; to 15 ft. (5 m) tall, usually shorter; slow growth. **STEM/BARK:** leaf scars and persistent leaf bases. **LEAVES:** pinnately compound, 3–4 ft. (1–1½ m) long, in rosettes, leaflets to 7 in. (18 cm) long, ¼ in. (7 mm) wide, glossy green, sharp apex, margins revolute, stiff and leathery, leaflets reduced to prickles at base of rachis; petiole spinose. **STROBILI**: male strobilus cylindrical, to 24 in. (60 cm) tall; female with modified scale-like leaves, cream-brown felt-covered, grouped into a globose cone-like mass. **SEEDS:** ovate, orange-red to dark red, to 2 in. (5 cm), in diameter, somewhat flattened. **NATIVE HABITAT:** Southern Japan, as understory plants in woodland or savanna.

CULTURE Full sun to partial shade, various well-drained soils; moderately salt-tolerant. **PROPAGATION**: seeds should be allowed to mature and the embryo fully developed before planting; rooting of "side shoots". (Cycads lack lateral buds; branching initiates from callus in leaf axils in female plants or at the base, almost never from the apex). **PROBLEMS:** various scales, mealybugs, leaf spot, micronutrient deficiencies, especially manganese. **LANDSCAPE USES**: specimen, accent, urn, used to create a tropical effect. Perhaps the most commonly cultivated cycad in warmer climates. **HARDINESS ZONE:** 8B–13B.

COMMENTS *Cycas revoluta* is perhaps the most common cycad in the trade and in landscapes. No variation, hence no cultivar, is known, although some specimens appear to have intermediate characteristics of *C. revoluta* and *C. taitungensis*, and are probably hybrids between the two, as often they are planted together. Species of *Cycas* are the only wind-pollinated cycads (Dehgan and Dehgan 1988), all other cycads are beetle-pollinated.

Cycas revoluta older female specimens, one with lateral growth developed from the "pups" along its trunk (above left) and the other from suckers at its base (above right), female strobilus opened and ready for pollination (middle left) and after pollination and seed development (middle right), close up of female strobilus with seed showing megasporophylls (below left), and a mature male strobilus (below right). NOTE: apical branching in female Cycas is rare.

Cycas rumphii

QUEEN SAGO PALM

SYNONYMS *Cycas celebica, C. corsoniana, C. speciosa, Zamia corsoniana*

ETYMOLOGY [Greek *koikos*, classical name for a kind of palm to which *Cycas* resembles in habit and leaf] [Dutch explorer Georg Eberhard Rumphius (1627–1702)].

GROWTH HABIT Dioecious evergreen palm-like plant, upright, often suckering, occasionally branching from apex on older male plants; to 20 ft. (7 m), commonly 9–15 ft. (3–5 m); slow growth. **LEAVES**: glossy, pinnately compound, 6–8 ft. (2–3 m) long in rosettes, leaflets to 12 in. (30 cm) long and ¾ in. (2 cm) wide, acuminate tip, margins flat, pliable, petiole partly to completely spinose or mostly entire; basal leaflets not reduced to prickles; apical leaflets in pairs. **STEM/BARK**: pithy trunk, leaf scars, and persistent leaf bases. **STROBILI**: male strobilus cylindrical to 2 ft. (36 cm) long, 8 in. (20 cm) or more in diameter; females with a cluster of modified narrow triangular seed-bearing leaves (megasporophylls), brown felt-covered, bearing 4–9 seeds. **SEEDS**: ovate, initially green but reddish-brown when mature, to 3 in. (7½ cm) in diameter, somewhat flattened, red at maturity; possess a spongy inner integument that results in flotation. **NATIVE HABITAT**: widespread species around Indian Ocean islands and mainland (see comments below, and Dehgan and Yuen 1983).

CULTURE Full sun to partial shade, various well-drained soils; moderately salt-tolerant. **PROBLEMS**: various scales, mealybug, (these serious problems), leaf spot, micronutrient deficiency. **PROPAGATION**: seeds, rooting of "side shoots" (cycads do not possess lateral buds and therefore do not possess lateral branches). **LANDSCAPE USES**: specimen, accent, urn, used to create a tropical effect. **HARDINESS ZONE:** 10A–13B.

COMMENTS Several related species (including *C. rumphii,* that until recently erroneously cultivated as *C. circinalis*. *Cycas circinalis* is otherwise native to India and rare in cultivation. However, several other species are often mistakenly referred to as *Cycas rumphii*. For a detailed discussion of historical complexity of this species, see the cited references (e.g., Dehgan and Yuen 1983), but particularly: http://plantnet.rbgsyd.nsw.gov.au/cgi-bin/cycadpg?taxname =Cycas+rumphii

Cycas rumphii branched male specimen at Montgomery Botanical Center (above); a container grown specimen (middle left), trunk of female plant with "pups" which can be separated and grown independently (middle right), developing female megasporophylls (below left), megasporophylls with unripe seeds (below middle), and multiple male strobili on a single plant (below right). There are no lateral buds in cycads.

Cycas taitungensis

PRINCE SAGO OR EMPEROR SAGO

SYNONYMS *Cycas taiwaniana, Cycas revoluta* var. *taiwaniana.*

ETYMOLOGY [From Greek *koikos*, classical name for a kind of palm to which *Cycas* resembles in habit and leaf] [After Taitung, Hsien region of Taiwan].

GROWTH HABIT Dioecious, palm-like gymnosperm, sometimes branching in male plants; to about 15 ft. (5 m) tall. **STEM/BARK:** trunk with permanent leaf bases, to 20 in. (50 cm) or more in diameter; male plants branch frequently, apex usually covered densely with yellow tomentum. **LEAVES:** pinnately compound, mostly upright at first then pendent; leaflets sequentially reduced from top to bottom so as to become spine-like at the base, 5–10 in. (12–25 cm) long and about ¼ in. (1 cm) wide, lanceolate-linear, straight, ending in a rigid, pungent point; margins thickened but not as revolute as that of *C. revoluta*. **STROBILI:** male strobili (microstrobili) upright, to about 20 in. (50 cm) long, ovoid-cylindric, with triangular microsporophylls; female strobili (megastrobili) with imbricate leaf-like, yellowish tomentose megasporophylls tightly grouped before and after pollination, gradually becoming distinct as seeds mature. **SEEDS:** usually 2–3 per sporophyll, but often up to six, narrowly obovate, to about 2 in. (5 cm) long, becomes purplish-red to nearly black at maturity. **NATIVE HABITAT:** Taitung, Taiwan, in full sun on rocky and steep slopes.

CULTURE A faster growing plant than *C. revoluta*, this species should be grown in a well-drained, fertile soil preferably in full sun or partial bright shade; mealy bugs and *Aulacaspis* scales may become troublesome although not as devastating as *C. revolute* or *C. rumphii*; leaves become somewhat long and bent when grown in shady locations. **PROPAGATION:** seed when allowed to fully mature or when possible, by "side shoots" of female plans. **LANDSCAPE USES:** specimen plants or in groups of three or more with a minimum of 6–8 ft spacing. **HARDINESS ZONE:** 9A–13B.

COMMENTS This plant may be superior to *C. revoluta* because of its rapid growth, initially upright but soon gracefully pendent leaves, and trouble-free aspect, although both species possess positive features for cooler landscapes. The Plant List does not record any synonyms for *C. taitungensis* and recognizes *C. taiwaniana* as a distinct species.

Megasporophyll of *Cycas taitungensis*

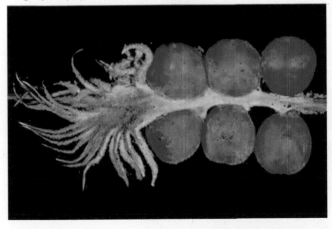

Cycas taitungensis flush of spring growth (above left), old leaves removed to show chemically induced multi branching (above right), growth of multi- and single-trunk specimens (middle), female strobilus after pollination (below left), and developing male strobilus (below right). Notice the dense yellow pubescence at base of the male strobilus which is characteristic of this species.

ADDITIONAL *CYCAS* SPECIES

Cycas schumanniana represents an excellent example of a cycad with attractive seeds.
Photographed at Montgomery Botanical Center, Florida

Cycas siamensis illustrates caudiciform (enlarged base) habit of the species as well as its deciduous habit.

Cycas wadei, endemic to the Philippines, is reportedly nearly disappeared from the wild, but is available in the trade.

Cycas calcicola new flush of growth following fire (above), the silvery foliage color is natural and eventually turns to normal green (middle), and seeds from previous year (below). Photographed in vicinity of Darwin, Australia

Containerized **Cycas sp.** (probably *Cycas micronesica*) showing unjust treatment of field collected plants and branching patterns of female plants under stress. (Photographed at a nursery in Indonesia)

ZAMIACEAE

ZAMIA FAMILY

8 GENERA AND ±200 SPECIES

GEOGRAPHY Central & South America, the Caribbean extending into Florida, Australia, and Africa.

GROWTH HABIT Dioecious palm-like erect shrubs or caudiciform, sometimes with subterranean branches, but above-ground branches uncommon; usually with nitrogen fixing nodules (through symbiotic relationship with blue-green algae), in **apogeotropic** (= vertically upward, often above soil level), dichotomously branched roots. **STEM-BARK**: taxa with cylindrical stems usually covered with leaf bases, those with caudiciform stems usually smooth. **LEAVES**: once-pinnately compound; leaflets spirally arranged, lacking a midrib, with parallel, dichotomously branched longitudinal veins or penniparallel (in *Stangeria*); entire or variously serrate or margins and apices spinose. **STROBILI**: plants dioecious; both male and female strobili with well-organized, spirally arranged sporophylls that may be distinct or imbricate; microsporophylls each with numerous sporangia (= pollen sacs); megasporophylls individually peltate, each generally with 2 ovules (rarely 3 in *Ceratozamia* spp.). **SEEDS**: drupe-like, predominantly orange or red but may be cream or yellowish; similar to *Cycas* they possess a fleshy coat (outer integument) that envelopes the bony layer (inner integument); usually produce two cotyledons.

ECONOMIC USES Ornamentals; otherwise of no significance. Starch is extracted from the stems and seeds of few taxa, which when detoxified were made into bread. It should be noted that all members of Zamiaceae and Cycadaceae have toxic glycoside properties known as cycasins and are not recommended as food source.

CULTIVATED GENERA Species of all 9 currently recognized genera, *Bowenia, Ceratozamia, Dioon, Encephalartos, Lepidozamia, Macrozamia, Microcycas*, *Stangeria*, and *Zamia,* are in the trade, with extensive collections in some tropical public gardens (e.g., Nong Nooch Botanical Gardens, Montgomery Botanical Center, and Fairchild Tropical gardens, among others), and common in conservatories of temperate gardens, and with extensive private collections, that over time have resulted in placing many species on the endangered list. Only the most common representatives are presented in this book. Interested students should consult the listed references for more detailed information.

COMMENTS Recent classification of cycads by Christenhusz et al. (2011), which was based on DNA and morphological studies, has placed nine traditionally recog-nized genera into two subfamilies (**ENCEPHALARTOIDEAE** and **ZAMIOIDEAE**). In all previous classifications, the genera *Bowenia* and *Stangeria* were considered as separate families, **Boweniaceae** and **Stangericaceae**, respectively, primarily based on their distinct foliar characteristics, although more recently the two genera were combined into a single family. Despite some reservations, there is no agreement or disagreement with any arrangement of the group as no taxonomic classification decisions have been made in this book. The current classification, however, does not appear to be logical on morphological grounds. Of course, this issue appears again and again in ferns and flowering plants as well.

BOWENIA

2 Extant Species

ETYMOLOGY [In honor of Sir George F. Bowen (1821–1899), first governor of Queensland]

GROWTH HABIT Fern-like caudiciform shrubs with short subterranean stems. **STEM/BARK:** caudex subspherical with 5–20 leaf-bearing branches (*B. serrulata*) or ovoid with 1–7 leaf-bearing branches (*B. spectabilis*). **LEAVES:** bipinnate but with a terminal leaflet, emerging singly and, similar to ferns, with circinate vernation; leaflets mostly ovate entire (*B. spectabilis*) or ovate-lanceolate serrate (*B. serrulata*); with numerous parallel longitudinal veins; petioles without prickles. **STROBILI:** microsporophylls with numerous sporangia and spirally arranged and vertically aggregated into an ovoid male strobilus; megasporophylls usually with two ovules, peltate, also spirally arranged into ovoid-globose strolilus. **SEEDS:** with whitish bony integument, becoming purple as they mature, germination cryptocotylar (upon germination seed remains attached to the cotyledons, as do all cycads). **NATIVE HABITAT:** endemic in lowland tropical eastern Queensland, Australia, mostly as understory plants.

CULTURE Should be planted in well-drained somewhat organic soils but not allowed to dry out. They perform best in partial shade preferably under tree canopies. **PROPAGATION:** seed. **LANDSCAPE USES:** container or specimen plants. **HARDINESS ZONE:** 9B–13B. May be planted in zone 9B, but should be in protected area or under tree canopy.

COMMENTS Although fern-like, any references to leaves as fronds or to leaflets as pinnules make what is already a confused state even more so. Such terminology should be used solely in fern descriptions, not in cycads or palms. The known species, *Bowenia serrulata* and *B. spectabilis*, are available in the trade and often seen in public garden conservatories or in landscapes in warmer climates. The status of *Bowenia* **'Tinaroo'** is yet to be clarified. The genus has been variously placed in Zamiaceae, in its own family (Boweniaceae), and sometimes together with *Stangeria* in Stangeriaceae. In the current treatment, it is placed in Zamiaceae.

Bowenia serrulata, fern-like growth habit without prickles or spines (above left), close up of leaves with serrate margins (above right), female strobili ready for pollination (below left), and peltate microsporophyll with sporangia that have already released their pollen.

Bowenia spectabilis with its bipinnate leaves and entire leaflets without a midrib exemplifies this Australian native cycad

CERATOZAMIA

About 25 Species

ETYMOLOGY [From Greek *ceratos*, a horn: in reference to the paired horns of sporophylls; *zamiaie*, a false rendering of Plinny for *azaniae*, a reference to pine cones]

GROWTH HABIT Palm-like shrubs with cylindrical or globose, usually unbranched caudex. **STEM/BARK:** caudex smooth with age and without persistent leaf bases; basal offsets (in trade referred to as side shoots, suckers, or pups) are produced particularly if plants are disturbed. **LEAVES:** pinnate, spirally arranged; petioles and in some cases the rachis are variously covered with prickles; develop straight, without circinus; leaflets simple, entire, coriaceous, opposite or nearly so, without a distinct midrib but with dichotomously branched parallel veins; inserted directly onto the petiole with a slight swelling at point of attachment (articulate); evenly spaced or clustered in groups of 2–6 along the rachis (in **C. hildae**); developing leaves usually in flushes and few at a time, often light to dark red. **STROBILI:** microsporophylls with two distinct spines, aggregated spirally into a stalked elongated male strobilus with numerous micrsporangia; megasporophylls peltate, also with two stout spines, aggregated spirally into a sessile or subsessile strobilus; predominantly with two ovules (seeds) but rarely with three. As in all other cycads, the female strobili are significantly wider and often shorter than the male. **SEEDS:** subglobular to oblong with a creamy white to yellowish coat, cotyledons two, which remain within the seed coat (cryptocotylar germination). **NATIVE HABITAT:** primarily in mountainous seasonally wet tropical forest at relatively higher elevations in Mexico, extending into Guatemala, Honduras, and Belize. Species with wider leaflets usually occur in wetter areas.

CULTURE Most *Ceratoaamia* species prefer partial shade, although their developing leaves often present brighter pinkish to red colors when in more direct sun. **PROPAGATION:** generally, by seed; rooting of lateral shoots is possible but not commercially viable. Earlier tissue culture was reported as having been successful, although results did not show significant positive results. **LANDSCAPE USES:** excellent choices as landscape plants in warmer climates. **HARDINESS ZONE:** 9A–13B (species-dependent).

COMMENTS Species of *Ceratozamia* are for the most part considered threatened or endangered by IUCN and are under CITES Appendix 1 protection, making it illegal to collect and trade. However, several species are in cultivation and available in nurseries, for example: ***Ceratozamia hildae, C. kuesteriana, C. latifolia, C. mexicana*** (some confusion as to the true identity of the species), ***C. microstrobila, C. norstogii*** (***C. plumosa***?), ***C. miquelliana, C. robusta, C. sabatoi, C. whitelockiana, C. zaragozae***, and few others. Some of the species such as *C. hildae, C. kuestriana*, and few others are surprisingly tolerant of cooler weather. The following *Ceratozamia* species key may facilitate correct species determinations.
Key to species of *Ceratozamia*
http://www.plantapalm.com/vce/taxonomy/ceratozamia.htm

Ceratozamia hildae
BAMBOO CYCAD

ETYMOLOGY [From Greek *ceratos*, a horn: in reference to the paired horns of sporophylls; *zamiaie*, a false rendering of Plinny for *azaniae*, a reference to pine cones] [Honoring Hilda Guerra Walker, original collector's daughter].

GROWTH HABIT A distinct species in the genus and essentially for all cycads based on the unique arrangement of the leaflets on the rachis. **STEM/BARK:** short and subterranean, plants without an aerial stem. **LEAVES:** several long petiolate leaves arise directly from the underground caudex; petioles densely pinkish pubescent early but glossy green and with or without prickles when mature; leaflets folded together in developing leaves and in groups of usually 2–3 in opposite or subopposite arrangement on the rachis; margins entire, venation vertically mostly parallel; unlike other *Certoazamia* species, emerging leaves are green. **STROBILI:** microsporophylls initially green then become grayish-brown, aggregate to a stalked cylindrical male strobilus, one from each branch of the subterranean stems; megasporophylls also aggregate into stalked female strobilus that are initially green but become grayish-brown as they mature; strobili of both sexes have a pair of small spiny projections. **SEEDS:** comparatively smaller than other species, ovoid, with yellowish-white fleshy coat that becomes soft and brown and easily removable as the seeds mature. **NATIVE HABITAT**: endemic to Mexico, in somewhat higher elevations of moist coastal deciduous oak woodlands of Queretaro and San Luis Potasi states.

CULTURE Surprisingly cold-tolerant species that thrives in partial shade or full sun, in well-drained soils and, with some irrigation and fertilization, develops an attractive crown of leaves that somewhat resemble a dwarf clumping bamboo plant. Female strobili are particularly susceptible to scales. **PROPAGATION:** as in other *Ceratozamia* spp., seeds germinate readily (within a month or so) and seedlings are for the most part trouble-free. **LANDSCAPE USES**: make a most attractive specimen in either landscape or appropriate size containers. **HARDINESS ZONE:** 8B–13B.

COMMENTS Wild populations are listed by IUCN as critically endangered. However, there is no justification for this predicament since seeds germinate readily and there could be in situ and/or ex situ plantings. The species is well-known in the trade and readily available.

Ceratozamia hildae, with its bamboo-like growth habit (above left), densely pubescent developing leaves that resemble fern fiddleheads (above middle), and long leaves with clustered soft leaflets (above right) on the rachis, is unique among all cycads. The female (below left) and male (below middle) strobili, however, bear a pair of rather sharp thorn-like projections which are shown with unripe white and ripe brown seeds embedded in the megasporophylls (below right).

Ceratozamia kuesteriana

ETYMOLOGY [From Greek *ceratos*, a horn: in reference to the paired horns of sporophylls; *zamiaie*, a false rendering of Plinny for *azaniae*, a reference to pine cones] [Honoring Baron K. von Kuester (–1894), nineteenth-century plant collector for Kew Gardens]

GROWTH HABIT Low growing, comparatively small but spreading, more or less caudiciform species with leaves arising from the base. **STEM/BARK**: stems mostly subterranean, but in time develop a short caudex; sucker from the base. **LEAVES:** in spirally arranged crown, long, arching, and spreading leaves with mostly spineless petiole; developing foliage appear reddish to light copper brown; leaflets are linear lanceolate, somewhat falcate, coriaceous, glabrous; margins entire; venation straight, more noticeable on the underside. **STROBILI**: microsporophylls aggregate into solitary stalked elongated conical brown male strobilus; megasporophylls also aggregate into a solitary stalked ovoid, but much wider than the male, cylindrical, initially bright green but brown when mature, female strobilus with potential to produce a few hundred seeds if hand-pollinated. Sporophylls with a pair of horns and peduncles in both sexes tomentose. **SEEDS:** ovoid-obovoid, fleshy coat initially cream-colored but turn brown, soft, and easily removable soon after the strobilus falls apart. **NATIVE HABITAT**: endemic to higher elevations at pine-oak cloud forests of the southern Mexican state of Tamaulipas, along the Sierra Madre Oriental.

CULTURE Prefer shadier locations but tolerate full sun, leaves may become discolored; in well-drained but moist sandy loam soil. **PROPAGATION:** fleshy coat can be removed without difficulty and seeds germinate readily in less than 2–3 months; suckers may also be rooted. **LANDSCAPE USES:** when properly grown, this species makes a superb specimen in landscapes, containers, and even indoors in a bright spot but becomes more susceptible to scales. **HARDINESS ZONE:** 8B–13B. May require some winter protection in lower zones.

COMMENTS Listed as endangered by IUCN Red List Appendix 1, as a result of overcollecting and land clearing. Once again, there is no justification for lack of establishments of in situ or ex situ populations. Seeds germinate readily and plants are easy to grow.

Ceratozamia kuesteriana with its new narrow leaflets that appear reddish-brown and mostly lack any prickles or spines (above left), fully developed but still young leaves (above right), an unusually large female strobilus in a unique prostrate specimen (middle left), a pollinated strobilus with seeds (middle, right), young female strobilus (below left), and male strobilus (below right)

Ceratozamia microstrobila

ETYMOLOGY [From Greek *ceratos*, a horn: in reference to the paired horns of sporophylls; *zamiaie*, a false rendering of Plinny for *azaniae*, a reference to pine cones] [From Latin *micro*, small, and *strobilus*].

GROWTH HABIT Dioecious palm-like shrub with a caudex. **STEM/BARK:** nearly completely subterranean, ovoid to sub-cylindrical, light brown in color, covered with persistent leaf bases. **LEAVES:** usually few, 2–4, but often more in cultivated than reported in wild plants; petiole with numerous prickles; rachis with few prickles; leaflets ±broadly lanceolate, alternate on the lower parts of the rachis, but subopposite and opposite on the upper portion, slightly falcate, coriaceous, apex acute, margins entire, glaucous on the upper surface and lighter green on the lower; veins straight and noticeable. **STROBiLI:** microstrobili aggregate into a solitary brown, tomentose stalked male strobili, each with a pair of short horns; sporangia cover the underside; megasporophylls peltate, aggregating into greenish-brown densely tomentose, solitary stalked female strobilus, each with a short pair of horns. **SEEDS:** oval, with a yellowish-white fleshy coat that turns brown with age. **NATIVE HABITAT:** Mexico, San Luis Potosi, Municipio of Ciudad del Maiz, at higher elevations of 2800 ft (850 m), in shallow reddish clay soil, rich in humus, on limestone outcrops, in transition zone between low deciduous forest and mixed oak woodland.

CULTURE An easy-to-grow and rewarding plant in partial shade and in well-drained somewhat organic soil. The plant is sufficiently cold hardy to tolerate frost in zone 8B without protection, except for rare hard freezes for long periods; scales, particularly on female strobili, may be troublesome. **PROPAGATION:** female strobili that are produced in alternate years can be hand-pollinated by simply shaking the shedding pollen from the male strobili on them at an appropriate developmental stage. Pollen may also be collected and used sporadically over a period of time to assure pollination. Seeds germinate within the first two months. **LANDSCAPE USES:** this taxon is an excellent candidate for landscapes or in containers, that may also be used indoors (with some sharp spine caution). **HARDINESS ZONE:** 8B–13B.

COMMENTS One of the three smaller *Ceratozamia* spp. and perhaps the most attractive in the genus because of its dark reddish-brown new leaves that last a month or longer and the coarse texture of its somewhat variable (wider or narrower) leaflets. As in related taxa, according to IUCN Red List, this species is also endangered as a result of overcollection and habitat destruction. However, to reiterate earlier notations, there is no justification for not establishing in situ or ex situ populations.

Ceratozamia microstrobila with its developing metallic red-brown leaves would be easily recognized as one of the most attractive cycads. From early development (above left) to fully developed leaves (above right and below left), it remains red for several weeks. Male strobili (below middle) and female strobili (below right) have the pair of spines typical for the genus.

ADDITIONAL *CERATOZAMIA* SPECIES

Ceratozamia zaragozae has the narrowest leaflets in the genus.

Ceratozamia euryphyllidia has the largest and broadest leaflet in the genus. Photographed at
Fairchild Botanical Gardens, Florida

DIOON

About 12 Species

ETYMOLOGY [From Greek *dis* (= twice) and *oon* (= an egg), in reference to the paired seeds on the megasporophyll]

GROWTH HABIT Palm-like shrubs or more or less arborescent, species initially caudiciform but in some cases develop into a distinct cylindrical trunk. STEM/BARK: some species with persistent leaf bases, but others ultimately with smooth bark; aerial branching is not known (except when chemically treated), but offsets from base are common in some species. LEAVES: pinnate, spirally arranged, erect when young, lower leaflets often reduced to spines; petioles swollen at base, without spines; leaflets simple, margins often spinose; no distinct midrib, veins parallel, dichotomously divided; some species pubescent when young. STROBILI: macrosporophylls spirally aggregated into a stalked male strobilus, each with numerous microsporangia on the lower side; megasporophylls also spirally aggregated into a female strobilus in some species (e.g., *D. spinulosum*), but in others widened and flattened (e.g., *D. edule*), somewhat similar to that of *Cycas* species. Ovules usually two per megasporophyll. SEEDS: nearly spherical or ellipsoidal, with whitish or yellowish fleshy coat; cotyledons 2, united at the tip and remain within the bony integuments for a significantly longer period than after first leaves develop (cryptocotylar germination). NATIVE HABITAT: Mexico, Honduras, and Nicaragua, in tropical pine-oak forests, dry hillsides, and coastal dunes.

CULTURE Full sun, in well-drained soil and in dryer sites. Although occasional irrigation is helpful, wet soils are detrimental to most species. PROPAGATION: seed or uncommonly by division of offsets. LANDSCAPE USES: specimen or container plants, but located at a distance from the public. Leaflets are sharp and particularly painful when their margins are spinose. Nearly all specimens are useful in xeric landscapes. HARDINESS ZONE: 8A–13B, species-dependent.

COMMENTS Most known species are in cultivation and include: *Dioon angustifolium, D. califanoi, D. caputoi, D. edule, D. holmgrenii, D. mejiae, D. merolae, D. purpussii, D. rzedowskii, D. sonorensis, D. spinulosum,* and *D. tomasellii. Dioon edule* in particular is offered in the trade with uncertain nomenclatural designation, but with several names of localities where it occurs.

Dioon edule
CHAMAL, CHESTNUT DIOON, MEXICAN SAGO

SYNONYMS *Dioon aculeatum, D. imbricatum, D. strobilaceum, Macrozamia littoralis, M. pectinata, Platyzamia rigida, Zamia maelenii*

ETYMOLOGY [From Greek *dis,* twice and *oon,* an egg, in reference to the paired seeds on the megasporophyll] [Edible, in reference to the seeds].

GROWTH HABIT Dioecious, palm-like arborescent (with age), plants, evergreen, caudiciform for many years before actual upright trunks develop; a large specimen with approximately 10 ft. (2½ m) of trunk length dated to more than 2600 years old. Specimens very seldom branch above ground, but frequently produce multiple stems at ground level; similar to palms, many leaves are borne at the top of each stem, to 12 ft. (4 m); LEAVES: pinnately compound, glaucous or glabrous, green, bluish or reddish-brown when emerging, bluish to bright green at maturity, with sharp-pointed but smooth-margined leaflets. lower leaflets reduced to spines. STROBILI: microsporophylls aggregated spirally to male strobili, which are relatively slender and have bearing pollen in many small sporangia; megasporangia also aggregated spirally to female strobili which are larger in diameter, possessing wide flattened spade-shaped tomentose sporophylls, each bearing two ovules. Both strobili are densely tomentose and are cream to light gray-brown in color. SEEDS: globular and covered with a cream to yellow easily removable outer integument at maturity. NATIVE HABITAT: eastern Mexico, from the state of Tamaulipas southward to Veracruz.

CULTURE Full sun and excellent drainage are the two primary requirements, otherwise they grow well in a variety of soils; plants survive adverse nutritional, soil, and light conditions for long periods; for the most part trouble-free, but occasionally scale, mealybug, and thrips during leaf emergence may become problems. PROPAGATION: seeds germinate readily after the fleshy coat is removed. LANDSCAPE USES: grows into a magnificent specimen plant where under proper growing conditions it could potentially require a large space; should not be planted close to walkways or areas where pedestrians and particularly children frequent because leaflets are dangerously sharp. HARDINES ZONE: 8A–13B. A subtropical and tropical plant limited to the warmer regions of the U.S. Provenance is important in establishing hardiness of individual plants; plants from southernmost populations in Mexico do not tolerate frost but those from northern populations, such as those in Tamaulipas, can stand some freezing.

COMMENTS Two subspecies are recognized: the typical southern *Dioon edule* subsp. *edule* and the northern *Dioon edule* subsp. *angustifolium,* from drier sites. Plant characteristics of specific localities are discussed by J. Chemnick in Cycad Pages on line: [http://plantnet.rbgsyd.nsw.gov.au/cgi-bin/cycadpg?taxname=Dioon+edule].

Despite the specific epithet, "edule" emerging leaves and seeds contain carcinogens and may be poisonous. Seeds should be handled carefully during attempts at propagation. Leaves are commonly used in religious holidays in Central America as sub-stitute for palms to adorn crosses; hence, the religious common name "Palma de la Virgen." The related species such as **D. mero-lae** are also in cultivation, but grow outdoors in warmer climates. Plant populations in the wild are considered threatened.

Dioon edule subsp. **Edule,** despite its sharp-pointed leaflets, is an attractive cold hardy and drought-tolerant plant suitable for landscape and containers (above), with female (middle left) and male (middle right) strobili; detailed structure of the female strobilus with mature seeds (below left) and particularly the megasporophylls with paired seeds (below right) are useful in understanding of cycads.

   As one of the ***Dioon edule*** variations, "Palma Sola", although listed on the web as variety, a subspecies, and a cultivar, currently does not appear to have a formal botanical or horticultural name designation, but it is one of the most attractive, cold-hardy, and drought-tolerant cycads.
   As shown above, even within its own cultivated specimens it displays variation in color intensity. Specimen in the middle right is a unique individual and those below with deep blue and yellow leaves (left) and orange (right) are chimeras that unfortunately outgrow in time.

Dioon spinulosum
GIANT DIOON

ETYMOLOGY [From Greek *dis* (= twice) and *oon* (= an egg), in reference to the paired seeds on the megasporophyll] [Spiny, in reference to leaflet margins].

GROWTH HABIT Dioecious arborescent plants to about 30–40 ft. (9½–12¼ m) tall, reputed to be the tallest American cycad (*cf. Microcycas calocoma*). **STEM/BARK:** trunk wide, to 16 in. (40½ cm) in diameter. **LEAVES:** pinnate, grow 5–7 ft. (1½–2¼ m); leaflets numerous, flat, leathery, lanceolate, glossy bright green, basal ones reduced to sharp spines; margins spiny; apexes pointed. **STROBILI:** microsporophylls spirally aggregated into a stalked male strobilus $1\frac{1}{3} - 1\frac{3}{4}$ ft. (40–55 cm) long and $\frac{1}{3} - \frac{1}{2}$ ft. (7–10 cm) in diameter, each with numerous microsporangia on the lower side; megasporophylls also spirally aggregated into a large female strobilus 1–3 ft. (30–90 cm) long and ¾–1 ft. (25–30 cm) in diameter; both strobili densely covered with whitish pubescence; ovules usually two per megasporophyll. **SEEDS:** ovoid, with cream to white easily removable sar-

cotesta. **NATIVE HABITAT**: in tropical evergreen forest, to 1500 ft (457.2 m) on rocky limestone cliffs in states of Oaxaca, Veracruz, and Yucatan in Mexico.

CULTURE Prefers partial sunlight although it can tolerate some shade; too much shade, however, can result in droopy appearance, in well-drained soil, and with occasional irrigation. Application of slow-release fertilizer with minor elements enhances growth rate as well as appearance. **PROPAGATION:** seeds germinate readily; can also be accomplished by separation of the offsets (suckers). **LANDSCAPE USES:** it makes an attractive container as well as accent landscape plant, but as the specific epithet implies it should be used with caution. **HARDINESS ZONE:** 9B–13B. May suffer cold damage in lower zones.

COMMENTS One of the most common *Dioon* species grown in the tropics and a favorite of botanical conservatories. Plant populations in the wild are listed as endangered. **Caution—** regardless of how the plant is used, it is highly recommended that it be located at a distance from public walkways. The spiny leaflet apexes and margins can be quite painful.

Dioon spinulosum is one of the more common cycads in warmer climates. It is known to reach great age and produce an interesting large heavy female strobilus. Photographed at Montgomery Botanical Center, Coral Gables, Florida (above) and male (left below) and female (right below), at Fairchild Tropical Botanical Garden, Coral Gables, Florida.

Dioon merolae is another arborescent species similar to *D. edule*, but not as cold-tolerant. Photographed at Fairchild Tropical Botanical Garden, Coral Gables, Florida.

Dioon mejiae

ENCEPHALARTOS

About 60 Species

ETYMOLOGY [From Greek *ĕn-*, within, *cĕphale*, the head, and *artos*, bread, in reference to the starchy bread made from the pith, top center part of the trunk, "bread in head"]

GROWTH HABIT Dioecious palm-like small or large shrubs with subterranean, caudiciform, or stout erect stems to about 10–15 ft. (3–5 m). **STEM/BARK:** trunks often stout, covered with persistent leaf bases; offsets from the base in several species often result in multi-stemmed plants, but no aerial branching. **LEAVES:** pinnate, spirally arranged, lower leaflets reduced to spines; petioles swollen at base and without prickles; leaflets simple, often with sharp stiff spiny, dentate or lobed margins; no discrete midrib, veins parallel, dichotomously branched, inserted towards the lower part of the petiole. **STROBILI:** microsporophylls spirally aggregated into stalked or sessile elongated male strobili, each with numerous sporangia on the lower surfaces, male strobili often emit a strong scent when pollen begin to shed; megasporophylls appearing peltate, thickened, also spirally aggregated into much larger strobili than the male, each with two ovules on the inner side of the megasporophylls (see illustration at the beginning of the chapter). **SEEDS:** oblong or ellipsoidal, with variously colored (yellow, orange, red, or brown) outer integument; cotyledons two, joined at the tip and remain within the seed at germination (cryptocotylar germination). **NATIVE HABITAT:** Africa, wide ranging, in open shrublands on steep rocky slopes to evergreen and coastal dune forests.

CULTURE Direct sun and well-drained soil in subtropical and tropical climates. Removal of lower leaves is recommended for safety reasons if larger species are grown as specimen or particularly container plants. **PROPAGATION:** seed; it is important in cultivated plants that female strobili be hand-pollinated so as to assure viable embryo development. Often inviable seeds develop without pollination and do not germinate. Use of offsets (suckers) is recommended only if species are rare or one or the other sexes not available. A perfect example is that of *E. woodii*, which apparently is no longer found in the wild and is known only as male plants. As in other cycads, seeds of *Encephalartos* are toxic and should be handled with gloves. **LANDSCAPE USES**: in many cases (but with few exceptions) attractive and equally dangerous plants for landscaping. Most species require large open areas to grow but as the specific name *E. horridus* is explicitly expressive, the spinose nature of many of the species limits their use in public gardens and parks, unless planted in inaccessible far-reaching locations. Nevertheless, they are valued popular landscape plants in drier warmer regions and sought after by private collectors. **HARDINESS ZONE:** 10A–13B.

COMMENTS Despite the potential of most (not all) species to be physically harmful, *Encephalartos* spp. are among the most sought-after cycads perhaps for their large attractive strobili. Hence, there is a long history of poaching, not only from the wild but also from cultivation. All species of the genus are listed by CITES (Appendix 1) as endangered. Conservation of the species has included insertion of microchips into trunks of wild plants. Fortunately, many of the 60+ known species are currently available in the trade. Examples include: ***Encephalartos aemulans, E. altensteinii, E. aplanatus, E. arenarius, E. barteri, E. bubalinus, E. caffer, E. chimanimaniensis, E. concinnus, E. cycadifolius*** (few spines)***, E. dyerianus, E. eugene-maraisii, E. fridericiguilielmi, E. ferox, E, gratus, E. hildebrandtii, E. horridus, E. inopinus, E. lanatus, E. latifrons, E. laurentianus, E. lebomboensis, E. lehmannii, E. manikensis, E. middelburgensis, E. natalensis, E. sclavoi, E. transvenosus, E. trispinosus, E. villosus, E. whitelockii***, as well as others including several hybrids, particularly involving *E. woodii* (e.g., *E. natalensis* × *woodii*). It is of notable interest that apparently at present live nursery-propagated plants can be sold only in the United States. Photographs of several species are shown to illustrate diversity of this genus.

Encephalartos gratus mature plant (above left), younger plant with male strobili (above right), immature female cone (below left), megasporophyll, and female sporophylls with mature seeds (below right). Native to NW Mozambique and SE Malawi, Africa. Photographed at Fairchild Tropical Botanical Garden, Coral Gables, Florida.

Encephalartos laurentianus female specimen (left) and plant with female strobilus (right). Native to Angola and Zaire in Africa. Photographed at Tenerife Botanical Garden, Canary Islands.

Encephalartos altensteinii mature plant in conservatory (left, photographed at Amsterdam Botanical Garden, Holland) and plant with male strobili in landscape (right, at Fairchild Tropical Garden, Coral Gables, Florida). Native to eastern Cape and Natal Provinces of South Africa. Photographed at Hortus Botanicus Leiden, Amsterdam, Netherlands (left) and at Nong Nooch Botanical Garden, Thailand (right).

Encephalartos woodii is known only as male plant and not known to exist in the wild. Suckers have been distributed to several botanical gardens (e. g., Longwood and Kew Gardens). Native to Natal, South Africa. Photographed at Kew Gardens, London.

Encephalartos villosus is among the species with subterranean stem, long arching leaves with prickles and spines, and attractive large yellow female stroboli. Native to southeastern Africa. Photographed at Fairchild Tropical Botanical Garden, Coral Gables, Florida.

Encephalartos trispinosus is blueish similar to *E. horridus* but with flat (not twisted) leaflets. Native to eastern Cape Province, South Africa. Photographed at Fairchild Tropical Botanical Garden, Coral Gables, Florida.

Encephalartos ferox is an arobescent species with prickled petiole and broad spinose leaflets, and attractive large bright orange–red female strobili. Native to northern Natal Province. Photographed at Fairchild Tropical Botanical Garden, Coral Gables, Florida.

Encephalartos arenarius (left), is about 3 ft. (1 m) tall at maturity and with lobed leaflets. *Encephalartos lehmannii* (right), is a caudiciform species with entire leaflets. Both species are conservatory favorites. Native to eastern Cape Province of South Africa. Photographed at US National Botanical Garden conservatory, Washington D.C.

Encephalartos horridus has blueish, often twisted, lobed leaflets. Its painfully spinose horrid characteristic is best expressed in its specific epithet. Obviously not recommended for landscaping unless far away from public. Photographed at US National Botanical Garden conservatory, Washington D.C.

Encephalartus princeps (left) is characterized by its arborescent growth, crown of leaves, and entire overlapping leaflets. It is native to eastern Cape Province of South Africa. ***Encephalartus pterogonus*** (right) has long leaves with short petioles and nonoverlaping spinose leaflets. It is native to Mozambique.

LEPIDOZAMIA

2 Species

ETYMOLOGY [From Greek *lepis*, scale and *zamia*, the genus, in reference to scale-like leaf bases which clothe the stem]

GROWTH HABIT Dioecious generally unbranched palm-like plants of two columnar arborescent species: *Lepidozamia peroffskyana* said to reach to about 23 ft. (7 m), and *L. hopei* to about 57 ft. (17 m), making the latter tallest living cycad in nature (*cf. Microcycas*). Neither species, however, attain these stated maximum heights in cultivation. **STEM/BARK:** erect cylindrical trunks, covered with more or less persistent leaf bases, shedding in the lower parts of the trunk. **LEAVES:** pinnate, spirally arranged leaves 6–9 ft. (2–3 m) long, appearing in annual flushes; leaflets simple, alternate, glossy, petioles, rachis, as well as leaflets all without spines or prickles and those of the lower ones not reduced to spines; no callus present at the point of attachment to the rachis; veins numerous and parallel. **STROBILI:** microsporophylls spirally aggregated into large sessile male strobili, each bearing numerous microspongia on its lower surface; megasporangia also spirally aggregated to sessile large female strobili, sporophylls appear peltate and with a triangular apex; ovules two or rarely three or one, sessile, attached on the inner face and directed inward on the megasporophyll. **SEEDS:** large, somewhat rounded or oblong, with red or rarely yellow integument; embryo with two cotyledons that remain inside the seed at their tips upon germination (cryptocotylar germination). **NATIVE HABITAT:** endemic to steep slopes of mountain ranges in tropical rainforests of northern and southern Queensland in northeastern New South Wales, Australia. Both species grow in high annual rainfall areas. According to Kennedy (2013), at least *L. peroffskyana* tolerates some frost days (not in this author's experience).

CULTURE They prefer filtered sun, but tolerate full sun as well, in well-drained but moist soils and need occasional irrigation. They will do well in warmer Mediterranean climates. As in most other cycads, they are subject to attacks by scales and borers, particularly on indoor specimens. **PROPAGATION:** seeds germinate readily and growth rate under optimal conditions is fast. Offsets (suckers) are not produced in plants of this cycad. **LANDSCAPE USES:** suitable and safe for in-ground or container planting indoor and outdoor. **HARDINESS ZONE:** 9B–13B, perhaps erroneously reported to be relatively cold hardy and even cold-tolerant in zone 8A.

COMMENTS One of few cycads not formally considered threatened or endangered. The two species, *Lepidozamia proffskyana* and *L. hopei*, are ideal subjects for landscape plantings and the most popular in their native country. Absence of spines or prickles is a major advantage. The species can be readily distinguished by the significantly larger size of plants and strobili as well as arched leaflets of *L. hopei* and darker and broader leaves of *L. proffskyana*. Seeds of both species have poisonous properties.

Lepidozamia peroffskyana specimen branched from base (above left), typical unbranched specimen (above right), and large female strobilus below left. ***Lepidozamia hopei*** (middle right) is the tallest known cycad. Species of *Lepizamia* possess the largest strobilus of any known extant cycad. Photographed at Bogor Botanical Garden, Indonesia; female strobilus of *L. hopei* photographed at Tenerif Botanical Garden.

MACROZAMIA

About 40 Species

ETYMOLOGY [From Greek *makros*, large and the genus name *Zamia*]

GROWTH HABIT Dioecious palm-like caudiciform shrubs or more or less arborescent with age. **STEM/BARK:** cylindrical trunk with persistent leaf bases. **LEAVES:** few to many, pinnate, spirally arranged; leaflets simple or dichotomously divided, inserted near the edges of the rachis towards the lower side; lower leaflets often reduced to spines; while leaflets of most species are straight and flat (subg. *Macrozamia*), some of the smaller species have twisted leaflets (subg. *Parazamia*); pubescent when young; veins numerous and parallel, without a distinct midrib; petioles covered with few to many spines. Leaves are often produced one or few at a time as opposed to annual flushes. **STROBILI:** microstobilus spirally aggregated into stalked glabrous male strobilus which terminate with a spine; numerous microsporangia cover the lower side of the mirosporophylls; peltate mgasporophylls with an upward leaning spine are also spirally aggregated on stalked glabrous female strobili; ovules generally two, sessile, inserted on the inner surface and directed inward. **SEEDS:** oblong to ellipsoidal, with red or less commonly yellow, orange, or brown sarcotesta; cotyledons two, joined at the tip and remain within the sclerotesta (cryptocotylar germination). **NATIVE HABITAT:** most endemic to eastern Australia but a few to central and southwest Australia.

CULTURE Nearly all species prefer full sun, although they can tolerate partial shade, but not full shade. Selection of appropriate species for any given climate is important. For example, *Macrozamia communis* is said to be cold-tolerant and can be readily grown in most well-drained soils, with adequate water and fertilizer. Large specimens transplant without difficulty. Scales and mealybugs may become troublesome. **PROPAGATION:** seed requires longer period than most other cycads to germinate (6–24 months!). **LANDSCAPE USES:** can be grown as specimen plants in landscapes or as container plants for which some species are particularly suited. Macrozamias have pointed leaflets and it is advisable that they be located away from close proximity of people. **HARDINESS ZONE:** 9A–12B (species-dependent).

COMMENTS Several species were reportedly food sources for Australian aborigines, after processing to remove toxins, and some species are reportedly poisonous to livestock. Generally larger *Macrozamia* species are more common and available in the trade, such as **M. communis, M. diplomera, M. dyeri, M. fraseri, M. johnsonii, M. longispina, M. lucida, M. macdonnellii, M. miquelii, M. montana, M. moorei, M. riedlei**, and **M. spiralis** (!), Among the smaller species **M. concinna, M. douglasii, M. fawcettii, M. flexuosa, M. parcifolia, M. pauliguilielmi** (pineapple zamia), **M. stenomera**, and perhaps others are also offered for sale.

Macrozamia moorei grown as container plant (above left, photographed at Hortus Botanicus, Leiden), as landscape specimen (above right, photographed at Fairchild Tropical Garden, Coral Gables, Florida), and female strobilus with seeds (below); *Macrozamia* **sp**. female strobilus to illustrate the single apical spine on megasporophylls (below middle), photographed in Royal Botanical Garden, Sydney, Australia, and microsporophyll with sporangia (below right).

Field collected specimen of *Macrozamia moorei* at a nursery. This is a perfect example of how plants become endangered.

MICROCYCAS

1 Species
ETYMOLOGY [From Greek *micro*, small and the genus *Cycas*]

GROWTH HABIT Dioecious palm-like monotypic genus (***Microcycas calocoma***), is essentially an arborescent plant that reaches a height of 28–30 ft. (8½–9 m) and 12–15 in. (30½–38 cm) in trunk diameter in its natural habitat, but usually smaller in cultivation; branched or unbranched from the trunk and occasional offsets (suckers) at base; leaves emerge in flushes. **STEM/BARK:** persistent leaf bases; the cylindrical trunk is marked with conspicuous rings. **LEAVES:** pinnate, erect, with truncate apices (appear as if cut near the terminus of the leaves), leaflets inserted towards the upper part of the rachis, entire, lanceolate, leathery, lower ones not reduced to spines; veins numerous, branched, but lacking a distinct midrib; petiole, rachis, and leaflet margins all without spines. **STROBILI:** microsporophylls spirally aggregated into cylindrical, pubescent, yellowish-brown sessile male strobili, each with numerous microsporangia on their lower surface; megasporophylls also spirally aggregated into broadly cylindrical creamy whitish-brown female strobili; sporophylls appear peltate, flattened on top and with two apical projections; ovules two, sessile, inserted on the inner face of the sporophylls. **SEEDS:** oblong to ellipsoidal, with red sarcotesta; cotyledons two, united at the tips and remain within the sclerotesta after germination (cryptocotyoar germination). **NATIVE HABITAT:** endemic to Cuba but uncommon, restricted to a small area in province of Pinar del Rio in the western part of the island in lowland and ravines and steep banks of valley and mountain regions.

CULTURE Full sun, relatively fast-growing plant that prefers well-drained but moist soil (under natural conditions this plant is sometimes inundated); application of micronutrient rich slow-release fertilizers is necessary to maintain appearance and color of the plant. **PROPAGATION:** seed germinate readily but seedlings are apparently susceptible to fungal rot. **LANDSCAPE USES:** exceptionally attractive landscape and container plant which exists in the trade but unfortunately not yet as widely available as one would hope. Roots are used as rat poison. **HARDINESS ZONE:** 10B–13B. Sensitive to cold, the leaves may be damaged by sudden frosts.

COMMENTS The name Microcycas is in fact an inappropriate designation for an arborescent plant that reaches up to 30 ft (nearly 10 m) tall. It was originally described as *Zamia calocoma* by the Dutch botanist F. A. W. Miquel (1811–1871), who was aware of its similarity with *Cycas*. The species was later designated by the Swiss botanist A. P. de Candole (1806–1893), as the generic name *Microcycas*.

With an estimated 600 remaining plants in the wild and poor natural reproduction, it is considered highly endangered (CITES Appendix 1), but it is becoming increasingly available in nurseries.

Landry, G. P. 1991. Portrait of a species: *Microcycas calocoma*. http://www.cycad.org/documents/TCN-Focus-Nov-1991-Microcycas-calocoma.pdf

Microcycas calocoma unbranched male plant with strobilus (above left), uncommonly branched female specimen (above right), specimen illustrating growth habit and leaf characteristics (below left), and plant with pollinated female strobilus (below middle), that had to be supported because of potential breakage as a result of its heavy weight. photographed at Montgomery botanical Center, Coral Gables, Florida.

Original female (left) and male (middle). The female plant at Fairchild Botanical Garden research area was pollinated by the male specimen located at Montgomery Botanical Center (middle) and the resulting seeds have been successfully grown to mature specimens, shown in the previous page photos, and distributed to major botanical gardens; an excellent example of conservation. Apparently, the original female specimen is no longer alive, presumably struck by lightning.

STANGERIA

1 Species

ETYMOLOGY [After William Stanger (1812–1854), English naturalist, medical man, and Surveyor General of Natal Province who collected the plant in 1851 and sent it to Chelsea Physic Garden in London] [from Greek *erio-*, wooly and *pus*, footed, in reference to wooly petiole bases]

GROWTH HABIT Monotypic (*Stangeria eriopus*), dioecious caudiciform, fern-like shrub, with carrot-shaped and coralloid roots (apogeotropic roots with nodules containing nitrogen fixing bacteria). **STEM/BARK:** subterranean, dichotomously branched. **LEAVES:** pinnate, folded upward towards the rachis as it expands, terminating with a single leaflet; petiole, rachis, and leaflet margins without spines; petioles approximate one-half the length of the leaves; leaflets opposite or subopposite, flat, sometimes slightly serrate-crenate, with a midrib and penniparallel bifurcating laterals; leaves not appearing in flushes and often one or rarely two at a time. **STROBILI:** microsporophylls spirally aggregated into stalked silvery pubescent cylindrical male strobili, each with numerous microsporangia on their upper surface; megasporophylls also spirally aggregated into stalked egg-shaped silvery pubescent female strobili, each with two sessile ovules on the inner surface and directed inwards. **SEEDS:** somewhat spherical to ellipsoidal with dark red sarcotesta, turning brown as the seed matures; the two cotyledons united at the tip and remain within sclerotesta as the seed germinates (cryptocotylar germination). **NATIVE HABITAT:** endemic to Southern Africa: South Africa and southern Mozambique. It occurs from wet forest to dry grasslands, resulting in variation in longer leaves and wider leaflets in wetter regions.

CULTURE It is sensitive to both too much sun or too much shade; as in other cycads, *Stangeria* also prefers well-drained but moist media. It is a fast-growing, comparatively early maturing plant that looks greener and more attractive with some regular fertilization. **PROPAGATION:** seed. **LANDSCAPE USES:** excellent for container and landscape use, as one of the smaller cycads it is also a very good candidate for conservatory and greenhouse use. **HARDINESS ZONE:** 10A–13B.

COMMENTS Endangered (IUCN Appendix 1), primarily as a result of habitat destruction and collection for medical use. Originally identified as a fern (*Lomaria coriacea*), but corrected as a cycad in 1829. The error is understandable since there is a superficial vegetative similarity between *Stangeria eriopus* and *Angiopteris evecta* (see ferns).

Stangeria eriopus specimen plant in the landscape (above left), close-up of leaves to illustrate the midrib and penniprallel laterals (above right), female strobilus with a seed from the previous year (below left), and a male strobilus (below right), of which there is usually more than one. Photographed at a private garden in Sarasota, Florida

ZAMIA

±50 Species

ETYMOLOGY [Name derived from *zamiae*, a false rendering for *azaniae*, referring to pine cones]

GROWTH HABIT Palm-like dioecious, evergreen shrubby plants, with subterranean dichotomously branched or often branched aerial stems, few to densely foliated. **STEM/BARK:** subterranean contractile stems are continuously pulled into the ground; aerial stems also often branched and eventually without leaf bases. **LEAVES:** pinnately compound, from 2–4 ft ($\frac{2}{3}$–$1\frac{1}{3}$ m), petioles and infrequently the rachis variously with prickles; leaflets glabrous to densely pubescent, soft and fern-like to extremely tough and leathery; without a distinct midrib, veins parallel, bifurcating, and upraised and quite noticeable in some species; from almost round to long linear, 2–10 in. (5–15 cm) or more in length and ½–3 in. (1¼–7½ cm) wide; margins entire or partly to completely and sometimes sharply serrate. Leaves may emerge annually in flushes or less often one or few at a time; green or deep bronze, eventually turning green. **STROBILI:** microsporophylls spirally aggregated into narrow cylindrical to ovoid-cylindrical stalked or subsessile male strobili with sporangia along both edges of the microsporophylls; megasporophylls spirally aggregate into cylindrical ovoid stalked to subsessile female strobili that are often much larger in diameter and possess larger more or less peltate sporophylls in vertical rows, each bearing two ovules. Cones are from whitish-yellow to rusty brown to silvery-grey tomentose. **SEEDS:** ovoid to ellipsoidal, with yellow to orange or scarlet sarcotesta which may be thick and fleshy at maturity and can be removed easily or with difficulty; cones ripen in fall. **NATIVE HABITATS:** New World, from Southern Georgia to Central and South America, and the Caribbean.

COMMENTS The number of known and described Zamia species exceeds 50 and several of these are offered in the trade: *Zamia amblyphyllidia*, *Z. fairchildiana*, *Z. inermis*, *Z. integifolia* (offered as *Z. floridana*), *Z. lindenii* (confused name with *Z. poeppigiana*), *Z. loddigesii*, *Z. lucayana*, *Z. muricata*, *Z. muricata* var. *picta*, *Z. paucijuga*, *Z. roezlii*, *Z. skinneri*, *Z. spartea*, *Z. splendens*, *Z. standleyi*, *Z. vazquezii* (offered as *Z. fischeri*), and *Z. verschaffeltii*, among others.

Zamia furfuracea
CARDBOARD PLANT

SYNONYMS *Palma pumila, Plamifolium furfuraceum, Zamia crassifolia, Z. gutierrezii, Z. vestita*

ETYMOLOGY [Name derived from *zamiae*, a false rendering for *azaniae*, referring to pine cones] [Mealy, scurfy, in reference to bronze leaflet pubescence].

GROWTH HABIT Dioecious, evergreen shrubby plant, with subterranean stems, only the upper portion that bears the leaves exposed; usually densely foliated; from 2–4 ft. ($\frac{2}{3}$–$1\frac{1}{2}$ m); achieves mature size in a few years in landscape plantings; clumps spread slowly. **STEM/BARK:** subterranean contractile stems are continuously pulled into the ground; dichotomously branched. **LEAVES:** pinnately compound, entirely glabrous to densely pubescent, extremely tough and leathery leaflets, hence the common name. A variable species, leaflets range from almost round to long linear, 2–6 in. (5–15 cm) in length and ½–3 in. ($1\frac{1}{4}$–$7\frac{2}{3}$ cm) wide; marginal serration may be present in the upper half of the leaflets or concentrated near the apex; may emerge green or deep bronze, eventually turning green; petioles and sometimes the rachis variously with prickles. **STROBILI:** microsporophylls aggregate into narrow cylindrical to ovoid-cylindrical stalked male strobili, with sporophylls sporangia bearing pollen; megasporophylls aggregate into cylindrical ovoid stalked female strobili that are larger in diameter and possess larger sporophylls, each bearing two ovules; strobili are rusty brown to silvery grey tomentose. **SEEDS:** ovoid, with scarlet, thick, fleshy integument at maturity that can be removed easily; strobili mature in fall, releasing the colored seeds. **NATIVE HABITAT:** east coast of Mexico in the state of Veracruz on arid thorn scrub and sand dunes.

CULTURE Full sun is preferable to shadier conditions for best appearance, requires good drainage, otherwise, in a variety of soils. As a houseplant, the sunniest location available is preferred; scales and mealy bug may become problematic. **PROPAGATION:** the fleshy seed coat should be removed, at which time the seeds can be scarified and planted; often germinate where they fall by the mother plant. **LANDSCAPE USES:** excellent specimen or foundation plant for larger buildings and makes excellent indoor foliage plant for bright spots in cooler climates. **HARDINESS ZONE:** 10A–13B.

COMMENTS Hybrids with other Mexican species, notably *Z. loddigesii*, are common in the trade, so individual plants should be handpicked to accomplish the desired effect in the landscape. This species is also visually unique in many respects and few other similar taxa may be found. However, as in other *Zamia* spp. this taxon (really taxa, if one considers variations of the plants in nature and trade) is also in need of some taxonomic attention and clarification.

Zamia furfuracea a typical landscape specimen (above left), female cone with seeds (above middle), and male strobilus (above right); newly developing leaves to illustrate the reddish marginal pubescence (middle left), and a mature leaf (middle right), unusual unnamed cultivated variants (below).

Zamia integrifolia
COONTIE, FLORIDA ARROWROOT, SEMINOLE BREAD

SYNONYMS *Zamia floridana*; *Z. silvicola.*

ETYMOLOGY [Name derived from *zamiae*, a false rendering for *azaniae*, referring to pine cones] [With entire or uncut leaves]

GROWTH HABIT Dioecious, shrubby plant, evergreen, with subterranean dichotomously branched tuberous stems, usually only the leaf-bearing tip exposed; densely foliated and, depending on the individual, may have arching to fully erect, ranges from 1–3 ft. (30–90 cm) in height; growth rate is variable, clumps spread slowly; **STEM/BARK:** subterranean stems are contractile: as the stems grow, they are pulled deeper into the ground. **LEAVES:** pinnately compound, deep glossy green, range 3–5 ft. (1–1½ m) in length; petioles with stipules and without prickles; leaflets variable in shape, size, and number: ¼–½ in. (7–12 mm) wide and about 4–10 in. (10–25 cm) long; possess small serrations near the apex; they may be variously twisted or flat; there is no midrib, veins are parallel and sometimes bifurcate. **STROBILI:** microsporophylls peltate, aggregate into stalked cylindrical male strobili with numerous microsporangia on each side of the micrsporophylls; megasporophylls shield-shaped and peltate, aggregating into stalked ovoid female strobili, with each sporophyll possessing two ovules; strobili of both sexes are heavily tomentose and rusty to dark brown. **SEEDS:** ovoid, with a thick scarlet to orange fleshy coat (outer integument) which is uncommonly difficult to remove from the bonny (inner integument) layer; these ripen in fall-winter. **NATIVE HABITAT:** as far north as St. Johns County and extreme southern Georgia and throughout east and west coast of Florida in pine and oak woodlands and hammocks. The species distribution range has recently been expanded to include many of the Caribbean islands and various ecological niches.

CULTURE Best in partial shade, in well-drained various soils, but with preference for sandy soils; scales and mealy bugs may be problematic. **PROPAGATION:** seeds may be planted after the fleshy layer is removed and the stony layer scarified with sulfuric acid or some mechanical means (Dehgan and Johnson 1983); subterranean stems that are often referred to as "tubers" or "bulbs," may be divided. **LANDSCAPE USES:** foundation or hedge planting or in mass as ground cover. It has become a favorite subject in landscaping of roadway medians because it can be easily propagated, grown, and maintained. **HARDINESS ZONE:** 8B–12B.

COMMENTS Foliar variability of this species lends itself to selection, although not much attention has been paid to this potential; wider or narrower leaflets, arching to erect leaves, and leaflets that are twisted vs. those that lie in a single plane. Similar species include: **Z. debilis, Z. portoricensis, Z. pumila** (restricted to the Dominican Republic), **Z. lucayana** (Bahamas), and the dwarf species **Z. pygmaea** (Cuba). **Zamia integrifolia,** as it is the current suggested name, is widely available and sold as **Z. floridana** and sometimes as **Z. pumila.** The reality is that there is an enormous diversity in this species and a careful thorough study of various populations may be necessary. For example, the known population of the species in Ocala National Forest with tall leaves and significantly wider leaflets and those with much shorter leaves or very narrow leaflets of northern taxa may require subspecific nomenclatural assignments.

Zamia integrifolia typical landscape specimen with developing leaves in spring (above left), one with mature summer leaves (above right), female strobili (middle left), male strobili (middle right), specimen with upright longer leaves and narrow leaflets more typical of northern populations (below left), and peltate megasporophylls with seeds (below right).

Zamia vazquezii
FERNLEAF ZAMIA, HELECHO

ETYMOLOGY [Name derived from *zamiae*, a false rendering for azaniae, referring to pine cones] [After Mario Vasquez Torres, Mexican botanist who first discovered the species].

GROWTH HABIT Dioecious, shrubby, evergreen, seemingly herbaceous plant that resembles ferns such as *Cyrtomium falcatum*; from 1–4 ft. (30–120 cm); mature size is achieved slowly and clumps also spread slowly. **STEM/ BARK:** subterranean stems, only the apex of which remains at ground level; subterranean stems are contractile, continuously pulled into the ground, and infrequently branch dichotomously. **LEAVES:** pinnately compound, glabrous, with fernlike toothed leaflets, hence the common name "helecho" (Spanish for "fern"); leaflets simple, ovate, dark green, and no spines; there is no midrib, veins dichotomously branched. **STROBILI:** microsporophylls aggregate to stalked slender male strobili and have sporophylls bearing pollen in many small sacs; megasporophylls aggregate into stalked female strobili that are larger in diameter and possess larger megasporophylls, each bearing two ovules. Cones are densely tomentose and range from light to dark brown in color. **SEEDS:** with scarlet sarcotesta at maturity; strobili ripen in the fall and release the seeds which are attractive to birds. **NATIVE HABITAT:** apparently known only from two small populations in northern Veracruz, Mexico, in semievergreen forests; also found growing in cornfields, hence the common name "amigo del maiz".

CULTURE Grow best in lighter partial shade and progressively less under shadier conditions, require good drainage but otherwise grow well in a variety of soils; adept to adversity in nutritional, soil, and light conditions, but appear more luxuriant when irrigated and fertilized occasionally; insect infestations, notably scale and mealybug can be problematic. **PROPAGATION:** seed germinate relatively quickly after the fleshy coat is removed; division of large plants is risky but may be successful if the cuts are treated with fungicide and plants watered carefully to avoid rot. **LANDSCAPE USES:** the fernlike foliage is attractive in foundation plantings and individual plants are good specimen plants when in decorative containers. **HARDINESS ZONE:** 9A–13B. Plants are known to survive in zone 9A–9B if under evergreen trees.

COMMENTS The fleshy seed layer contains carcinogenic agents, so they should be handled carefully during attempts at propagation. Avoid planting in areas where livestock could graze because they often eat emerging foliage which can cause nerve disorders. *Zamia vazquezii* is more or less unique in its general appearance, although it is often confused with *Zamia fischeri*, which has fewer leaves and lanceolate leaflets, acute apices, and no prickles as compared to numerous ovate leaflets, and sometimes with few small prickles of *Z. vazquezii*.

Zamia vazquezii a typical landscape specimen (above left), mature female strobili (above right), multi-branched specimen with developing leaves (middle left), multi-branched specimen with solitary female strobilus on each branch (middle right), leaf with narrow leaflets which is said to be more typical of *Z. fischeri*, except fewer (below left), leaf with typical wide leaflets (below middle), a pair of male strobili (below right).

ADDITIONAL SPECIES

Zamia poeppigina is a good example of a species with aerial branched stem (left) with typical foliage (right). Photographed at Golden Gate Park Conservatory, San Francisco, California

Zamia inermis (left), as the name implies, is without spines or prickles on its petiole and rachis, stiff erect leaves and coriaceous leaflets with sharp apices. ***Zamia cremnophila*** (right), by contrast, has prickled petiole and rachis and lower part of the overlapping leaflets, but it represents a species with reddish young leaves.

Zamia pseudoparasitica is the only known epiphytic cycad and, as in majority of the cycads, it is endangered; specimen plant (above left) and close-up of female strobilus (above right), male plant (below left), and female strobilus (below right).

Zamia loddigesii is a variable species for which varieties have been recognized, including var. ***latifolia***
which is presented here (above left and right). It is said to be one of the most common and widespread
Zamia species in Mexico and Guatemala, although such statements probably require determination of a
typical example. The photographs (below left and right) best illustrate the identity problem. Plant on
the left with densely pubescent leaflets and plant on the right with long narrow leaflets (***Z. loddigesii***
var. ***angustifolia***?)

Zamia cunaria (native to Panama) Photographed in Nong Nooch Tropical Botanical Garden, Thailand. Listed by IUCN as appendix II, as only three small populations are known. Named after Cuna Indians who use its seeds to make necklaces.

GINKGOIDAE

GINKGOACEAE

1 Genus and 1 Species

Ginkgo biloba

MAIDENHAIR TREE

ETYMOLOGY [*Ginkgo* is a misinterpretation of the Japanese *gin*, silver and *kyō*, apricot, used in seventeenth-century Japan but now obsolete] [Two-lobed, in reference to the leaves]

GROWTH HABIT A monogeneric, monotypic (***Ginkgo biloba***), dioecious, excurrent, deciduous tree of medium texture; irregular shape and branching pattern; to 120 ft (36½ m) tall. **STEM/BARK:** long and short shoots; lateral branches arranged more or less horizontally often in a ring around the trunk. **LEAVES:** spiral or pseudoverticillate (false whorled) arrangement on short spur shoots; fan-shaped, incised, and with dichotomously branched veins. **STRIBILI:** microsporophylls terminal on elongated axis, male trees bear clusters of long peduncles; megasporophylls axillary, with a branched or single stalk bearing a pair of sessile, naked ovules (seeds) on a long peduncle; both sexes appear in early spring, ovules persisting until mature in autumn. **SEEDS:** spherical, drupe-like, about 1 in. (2½ cm) in diameter, yellowish, similar to cycads with a fleshy sarcotesta and a hard bony sclerotesta, occur in pairs but often only one develops. **NATIVE HABITAT:** perhaps extinct in the wild, originally found in SE China cultivated in a temple garden. It is, however, widely cultivated in China and Japan and probably best known from cultivation, and introduced into Europe and United States in eighteenth century.

CULTURE Although more suitable for northern cooler climates, it does grow in northern subtropical regions and only marginally in the central zones. Initially slow growing but more rapidly in later years. Female trees produce abundant quantities of seed with putrid fleshy seed coat (the principal odoriferous component is butyric acid, the smell of rancid butter). The problem can be avoided by planting known males or grafting/budding known males onto seedling rootstocks. It is very resistant to insect pests and diseases. **PROPAGATION:** grafting/budding, cuttings, layering, stratified seed; usually grafted with known males to avoid foul smell of ripened fleshy coat, when used in landscapes but female plants are grown for their edible seeds which are considered a delicacy. **LANDSCAPE USES:** ornamental, noncommercial timber. Trees used in street and in landscapes primarily because of their tolerance to hard-packed, poor soils, and air pollution. **HARDINESS ZONE:** 5A–10A.

COMMENTS Seeds are edible once the fleshy disagreeable portion is removed. Commonly used for food in the Orient. Several cultivars have been reported for growth habit, including fastigiate and weeping, and for leaf shape, color, and size. Examples include: *Ginkgo biloba* **'Autumn Gold', 'Barbits Nana', 'Beijing Gold', 'Blagon', 'Chi', 'Elmwood', 'Emperor', 'Fairmont', 'Fastigiata', 'Globosa', 'God Spire', 'Golden Colonade', 'Gresham', 'Halka', 'Horizontalis', 'Jade Butterfly', 'Kew', 'Laciniata', 'Lakeview', 'Maniken', 'Mayfield', 'Pendula', 'Presidential Gold', 'Princeton Sentry', 'Saratoga', 'Shangri-La', 'The President', 'Tremonia', 'Troll', 'Tubiforme', 'Umbrella', 'Variegata', 'Windover Gold', 'Woodstock',** and probably others.

Ginkgo biloba typical growth habit (above left), bark characteristics and short shoots (above right), clclose-upf leaves showing two lobes and veins (middle left), and stalked paired seeds (middle right), and mature seeds (below).

GNETIDAE

GNETACEAE

1 Genus and About 40 Species

Gnetum **spp.**

ETYMOLOGY [From new Latin, alteration of *gnemon*, modification of Moluccan Malay *ganemu*. The word is probably in reference to the reproductive structure of the plants that resemble the arm of a sun dial]

GEOGRAPHY Malaysia, SE Asia, Philippines, China, tropical West Africa, Fiji, and northern South America.

The monogeneric **Gnetum** species are a group of tropical lianas, vining shrubs, or small to medium size tree (*Gnetum gnemon*) with simple, opposite, entire leaves that often emerge bronze-colored, but are glossy dark green at maturity. The reproductive structure is spike-like: "**Strobili**" consist of a series of collars, each containing small, simple, male and female "flowers," and seeds are nut-like.

Gnetaceae differ anatomically from gymnosperms by presence of vessel elements in their xylem, an angiosperm characteristic. The plants have been variously considered an evolutionary link between gymnosperms and angiosperms. Stevens (APG IV 2017) calls Gnetales as "perennial aggravation", because of uncertainty of its position. Some species are valued as nutritious green vegetables and/or for medicinal purposes in central Africa where excessive harvesting has resulted in disappearance of the forests. **Gnetum** species are under study and planted for economic reasons (for example **Gnetum africanum**) in the tropics, but they are not common in cultivation and often seen only in botanical conservatory collections. Examples are presented here for information purposes.

Gnetum gnemon, arborescent growth habit (left), and **Gnetum montanum,** a liana, leaves (middle) and close up of "cones" (right). They can be propagated by seeds or cuttings.

WELWITSCHIACEAE

1 Genus and 1 Species

Welwitschia mirabilis

WELWITSCHIA

ETYMOLOGY [After Dr. Fredrich Martin Joseph Welwitsch (1806–1872), Austrian botanist and explorer who introduced the plant to Sir Joseph Dalton Hooker of the Linnaean Society in Britain, in 1859] [From Latin meaning wonderful, remarkable]

GEOGRAPHY Endemic to the arid Namib Desert of southwest Africa, in Namibia and Angola. This unusual and truly remarkable monogeneric monotypic dioecious, long-lived (2000 years) plant, which is sometimes referred to as a living fossil, has a large, erect, unbranched carrot-shaped subterranean woody stem that grows to 3 ft. (1 m) in diameter and deep into the ground. From the above ground level, stem apex arise only two strap-shaped leaves that grow across from one another. Unlike most leaves that grow from their meristematic tip, welwitschia leaves grow from their meristematic base and are limited in their length only by splitting as they become older. Their cone-like reproductive structures (strobili) arise from the woody center near leaf bases.

Welwitschia can be readily propagated from seeds, particularly if the fleshy coat is removed prior to planting, as in cycads. The plants are essentially of no landscape value, but are of interest more as curiosity and of course for botanical interest. The plants are usually grown in long pipes or deep containers to accommodate the elongated subterranean stem and root.

For a detailed discussion and specific list of references see: http://en.wikipedia.org/wiki/Welwitschia

Welwitschia mirabilis growth habit which consists of a central meristematic region that gives rise to cones (strabili?) and only two leaves (above left, at Vienna Palmenhaus Schönbrunn), specimen in pipe (above middle, at the University of California Davis Consrvatory); male cones with fertile stamens and female cones (below left and right, at U.C. Davis).

NOTE: The third member of the order Gnetales, the monogeneric Ephedraceae (***Ephedra* spp**.), is widespread in temperate climate regions. Although of medical interest (Mormon Tea), plants are of little horticultural value or curiosity other than botanical-evolutionary interest. Interested readers should consult APG IV (2016) for technical details, or: http://en.wikipedia.org/wiki/Ephedra_(plant)

CONIFERS

(PINOPHYTA, CONIFEROPHYTA)

Subclass PINIDAE

cone scale
seed

    Pine cones with spirally arranged symmetrical microsporophylls (male, above), and female with spirally arranged flattened bract-scales (female, below); seeds with a long terminal wing and cone-scales (right). shown as example of conifer reproductive structures.

PINIDAE

PINACEAE

PINE FAMILY

11 GENERA AND ABOUT 250 SPECIES

GEOGRAPHY The largest and most diverse family of conifers with its genera all natives of the entire northern hemisphere, predominantly in mountainous temperate zones of southwest China, West Indies and Central America, central Japan, and California. *Pinus radiata* has been introduced for timber production and naturalized in some southern hemisphere regions.

GROWTH HABIT Monoecious coniferous resinous and aromatic trees, rarely shrubs (many dwarf cultivars), from about 6–300 ft. (2–100 m) tall. **STEM/BARK:** branches subopposite or whorled, consisting of short and long shoots; barks smooth, scaly, or fissured. **LEAVES:** mostly evergreen, acicular, linear needle-like, spirally arranged, of 1 (e.g., *Pinus monophylla*) on long shoots or fascicles of 2, 3, 4, or 5 (e.g., *Pinus strobus*), on short shoots. **CONES:** male cones (microsporangiate), often clustered, terminal, and often clustered at branch tips, small, from less than ½ to 3 in. (1½–7½ cm) long, with numerous spirally arranged microsporophylls (pollen-bearing scales) each with two microsporangia on their lower surface; female cones large, woody, solitary or in groups of few, with numerous spirally arranged megasporophylls (ovuliferous) scales 1–24 in. (2½–60 cm) long, each with two ovules. **SEEDS:** winged only on one side or without wings (in some species of *Pinus*), with 3–24 (18?) cotyledons.

ECONOMIC USES Largest and economically most important conifers, used as ornamentals, most important source of softwood timber, and numerous other uses, including resins and chemicals. Pine seeds are edible and essential diet of birds, squirrels, and other rodents as well as humans.

CULTIVATED GENERA All the genera assigned to this group are native to and cultivated in cooler climate areas of Northern Hemisphere: *Abies, Cathaya, Cedrus, Keteleeria, Larix, Nothotsuga, Picea, Pinus* (100 spp.), *Pseudolarix, Pseudotsuga,* and *Tsuga*. It should be noted that *Larix* and *Pseudolarix* are deciduous.

Subclass PINIDAE

Cedrus deodara

DEODAR CEDAR, HIMALAYAN CEDAR
SYNONYMS *Abies deodara, Cedrus indica, Pinus deodara*

ETYMOLOGY [From original Latin *cēdrus* or Greek *kĕdros*][From Sanskrit *deva*, God and *daru*, wood or "wood of the Gods"].

GROWTH HABIT Monoecious, large coniferous, aromatic evergreen tree of excurrent, pyramidal growth habit and horizontally arranged, nodding branches; to 150 ft. (50 m) tall and with a trunk diameter of 10 ft. (3¼ m), but often much smaller in cultivation. **STEM/BARK:** dark gray, smooth on young trees, fissured and scaly on older ones. **LEAVES:** acicular, blue-green, grouped to about 30 in a cluster, 1½–2 in. (3–5 cm) long on short shoots or solitary on long shoots. **CONES:** male cones 1½–2½ in. (4–6 cm) long and about I in. (2½ cm) wide; female cones one or two together on short shoots, more or less ovate, 3–4 in. (7½–10 cm) long and 2–3½ in. (5–7½ cm) wide, rounded at the apex. **SEEDS:** whitish with light brown wings. **NATIVE HABITAT:** western Himalayas, in eastern Afghanistan, northern Pakistan, northern India and Kashmir, and Tibet and Nepal at elevations of 5000–10,000 ft. (1500–3200 m).

CULTURE Full sun, in large spaces, in well-drained soil. Difficult to transplant when mature. Tip dieback due to borer. **PROPAGATION:** primarily by seed but cuttings root with IBA treatment. **LANDSCAPE USES:** accent in large spaces, particularly attractive because of its form and blue-green color. **HARDINESS ZONE:** 7A–10A. Does not grow to an ideal size in warmer climates.

COMMENTS In India, the tree is considered divine and sacred. The wood is often used in construction of houses, temples, bridges, etc. because of its durability and its rot resistance. Many cultivars are known but are uncommon in warmer climates. A few examples representing various growth forms or color include: **'Albospicata'** (white tip deodar cedar), **'Argentea'** (fast-growing with silver-gray to bluish-gray foliage), **'Aurea'** (10–15 ft. tall with golden-yellow new growth), **'Aurea Pendula'** (weeping habit), **'Bill's Blue', 'Blue Ice', 'Blue Velvet', 'Divinity Blue', 'Compacta'** (globose-conical compact habit with dense branching), **'Emerald Falls', 'Fastigiata'** (columnar habit, branches widely spaced), **'Girard Weeping', 'Gold Cone', 'Prostrate Beauty', 'Pygmaea', 'Silver Cedar', 'Verticilata Glauca', 'Viridis'** (needles conspicuously glossy light green), and many more. Two related species, *C. atlantica* (Atlantic cedar) and *C. libani* (Cedar of Lebanon), are well-known and common in cultivation in the western United States.

Cedrus deodara growth habit of a younger plant (above left), an older plant (above middle), branching habit from the older plant (above right), male cones (below left), and younger female cones (below right).

Cedrus deodara specimens (above), *C. deodara* 'Glauca Pendula' (below, weeping deodar cedar). Photographed at Parc Floral de Paris, France

Cedrus libani is even less suitable for warmer areas but grows best in Mediterranean climate regions. Photographed at Kew Gardens

Pinus elliottii
SLASH PINE

ETYMOLOGY [Classical Latin name for pines] [After Stephen Elliott (1771–1830), botanist at Charleston, South Carolina].

GROWTH HABIT Monoecious, coniferous, evergreen tree, heavy horizontal branches clearing from lower portions of trunk with age, rounded open canopy; to 100 ft. (30 m), rapid growth rate. **STEM/BARK:** to 3 ft. (1 m) in diameter, usually much smaller, tapering, bark dark brown, irregularly cracked and platy, with the thin surfaces of the plates peeling off. **LEAVES:** needles, in fascicles of 2 and 3, to about 8–9 in. (20–24 cm) long, clustered on ends of branches, deep green, stiff, and straight. **CONES:** male cones cylindrical catkin-like, red to yellow, clustered, to 3 in. (7½ cm) long; early immature female cones globose; mature cones red-brown, woody, ovoid, to 6 in. (15 cm) long, with a small spine and stalked; mature 2nd year, not persistent. **SEEDS:** winged. **NATIVE HABITAT:** southeastern United States, from southern South Carolina, south and west to Florida and Louisiana, where summers are hot and wet.

CULTURE Full sun, various soils, moderately salt-tolerant. More tolerant of wet sites than most pines. Problems include pine blister rust, borers, and pine beetle. **PROPAGATION:** seed. **LANDSCAPE USES:** light shade, street plantings (needles are messy). **HARDINESS ZONE**: 8A–12B.

COMMENTS Mainly used for lumber, pulp, etc. *Pinus elliottii* **var.** *densa*, a south Florida endemic (including the Everglades and Florida Keys), with nearly all needles in bundles of two, is considered by some to be a distinct species.

Pinus elliottii growth habit (above, left), needle aggregates at branch tips (above, middle), examples of 2 and 3 needle fascicles (above, right), male cones (below, left), closed cone (below, middle), and open cone (below, right)

Pinus glabra
SPRUCE PINE

ETYMOLOGY [Classical Latin name for pines] [Glabrous, smooth].

GROWTH HABIT Monoecious coniferous, evergreen tree, heavily foliated, much-branched, with bushy pyramidal or irregular canopy; to 80 ft. (25 m) tall, but usually 30–50 ft. (10–12 m); moderate growth rate. **STEM/BARK:** stem erect but may curve or twist when shaded, slender, smooth, and gray in younger trees, but grayish-brown, fissured and with scaly plates, bark finer than *P. elliottii*. **LEAVES:** needles, in fascicles of 2–3 in. (8 cm), clustered on ends of branches, light green when young turning glossy deep green, soft, and twisted when older and remain on the plant for 2–3 years. **CONES:** male cones catkin-like, long and cylindric ½–¾ in.

(1–1½ cm) long; purple-brown; female cones brown, globose, 2–3 in. (4–7 cm) long and to 2.5 in. (5 cm) in diameter, woody; remain on tree for 3–4 years. **SEEDS:** winged, smaller than most pine species. **NATIVE HABITAT:** coastal plains of southern United States, from South Carolina south and west to Florida and Louisiana, in moist woodlands.

CULTURE Full sun, moist fertile soils, slightly salt-tolerant. Retains branches near the ground when not shaded, casts a heavy shade. Pine blister rust, borers may be problems. **PROPAGATION:** seed. **LANDSCAPE USES:** specimen, street plantings, accent. **HARDINESS ZONE:** 8A–10B.

COMMENTS Frequently used as Christmas trees in warmer climates because it is heavily foliated with short needles. Not very useful for timber, but more useful for landscaping. It has greater shade tolerance than other pines.

Pinus glabra growth habit (above), needles (below, left), and closed and open cones (below, right)

Pinus palustris
LONGLEAF PINE

SYNONYMS *Pinus australis*

ETYMOLOGY [Classical Latin name for pines] [Growing in marshes, marsh-loving].

GROWTH HABIT Monoecious, evergreen, coniferous tree; upright; sparse foliage concentrated in upper third on a few large irregular branches; coarse-textured; reaches a height of 100 ft. (30 m), though often much smaller; the spread is 30 ft. (10 m); growth rate is moderate. **STEM/ BARK**: trunk is 1–2 ft. (30–60 cm) wide and covered with thick brown plates of bark. **LEAVES**: needles in fascicles of 3s, to 14 in. (36 cm) long, slender and flexible giving a weeping appearance. Winter buds are silvery-white and quite conspicuous. **CONES**: male cones purplish, catkin-like to 3 in. (7½ cm) long; developing ovulate cones, to ½ in. (1¾ cm) long, borne at the base of the current season's growth in early spring; mature female cones brown, to 10 in. (25 cm) long and 2–2¾ in. (5–7 cm) wide; with a downward prickle on each scale; opening 1½ years after pollination. **SEED**: winged seeds are 2 in. (5 cm) long. **NATIVE HABITAT**: dry flatwoods and sand ridges of the southeastern United States, from eastern Texas to Virginia, including north and central Florida; occasionally found in swampy areas.

CULTURE Full sun; will grow in a wide range of soils, but contrary to its specific epithet (palustris), it prefers well-drained sandy soil. It is not salt-tolerant. It is very slow-growing for the first few years (often called grass-stage because it looks like a clump of tall grass); resistant to fusiform rust and pine bark beetle; fallen needles can be unsightly on other plants. **PROPAGATION**: extract seed from partially open cones, sow on soil surface, cover with cut pine needles, and protect from animals. Long leaf pines transplant poorly because of its extensive root system and must have a good root ball of soil when transplanted. **LANDSCAPE USES**: an excellent specimen for giving an illusion of depth to a planting and for providing a light shade for azaleas and other shade-loving plants. **HARDINESS ZONE**: 7A–10B.

COMMENTS This species may be distinguished from all other pines by its very long needles and conspicuous white winter bud. It is a source of timber, resin, and turpentine (needed for ships and boats), and for this reason, overharvested and not often replanted because of its slow growth.

Pinus palustris mature growth habit (above left), branched specimens in habitat (above, right), catkin-like male cones (middle, left), and female cones (middle, right), plated reddish-brown bark (below left), and conspicuous silvery-white winter bud characteristic of the species (below, right).

Pinus taeda
LOBLOLLY PINE

SYNONYMS *Pinus lutea, P. mughoides*

ETYMOLOGY [Classical Latin name for pines] [Resinous wood].

GROWTH HABIT Monoecious, evergreen coniferous tree, upright, with a dense, rounded crown; medium-textured, can reach a height of 120 ft. (40 m), though normally in the 40–70 ft. (12–21 m) range; spread is about 30 ft. (10 m); fast growth rate. **STEM/BARK:** the straight, 1–2½ ft. (30–75 cm) wide trunks are covered with plates of reddish-brown fissured platy bark. **LEAVES:** needles in fascicles of 3s, to 8 in. (20 cm) long, somewhat soft and flexible, especially in comparison to *P. elliottii.* **CONES:** yellow-green catkin-like male cones to 1–1½ in (2½–4.0 cm) long-form clusters at previous year's branch tips; ovulate cones are ovoid and range from ½–¾ in (1–1½ cm) long, pale green with purplish tips, but vary to some extent with stages of development; these are borne at the base of new growth in early spring; at maturity brown, to 5 in. (13 cm) long, often in pairs, mature in fall, 1½ years after pollination; persist on the tree after opening. **SEEDS:** winged, 1 in. (2½ cm) long. **NATIVE HABITAT:** Southeastern United States, from New Jersey south to central Florida, extending west to Louisiana and eastern Texas. It grows in humid, warm temperate climates with hot summers and mild winters, but it has proven to be an adaptable species and grows on a wide variety of soils.

CULTURE Full sun, in almost any soil of reasonable fertility. It has no salt tolerance; the most serious insect problem is the bark beetles which can cause death of the trees, but to a lesser degree borers and fusiform rust can be troublesome as well. **PROPAGATION:** extract seed from partially opened cones, place on soil surface, and cover with cut pine needles. There is some literature on tissue culture. **LANDSCAPE USES:** employed as a screen/windbreak while young, for giving a depth effect to planting and for providing light shade for azaleas, camellias, and other partial shade-loving plants, but can prove unsightly and troublesome, however, as needles fall and cover the plants. **HARDINESS ZONE:** 7A–12B.

COMMENTS According to the United States Forest Service, loblolly pine is noted for being the second most extensively grown tree (after red maple), known as southern yellow pine, in the United States. The vernacular name loblolly is in reference to lowland swampy areas where the plant grows. There are a few cultivars of loblolly pine mostly densely foliated dwarf forms referred to as "witches' brooms," the best known of which are **'J. C. Raulston'** and **'Little Albert'.** There are also several "improved" clones of the species used in forestry plantings.

Pinus taeda growth habit of mature plant (above left), dichotomously branched specimens ocassionally seen (above right), trunk with platy bark (below left), and typical open and closed female cones (below right).

ADDITIONAL GENERA

Abies procera (noble fir), native to California, Oregon, and Washington but widely cultivated as Christmas tree in eastern US in zones 5A–6B; grows to 50 ft. (16 m).

Picea pungens 'Globosa' (Colorado blue spruce)   *Picea pungens* cv.

Tsuga canadensis (Canadian hemlock, eastern hemlock), native to North America, as far south as applalchian Mountains of Goergia and Alabama; evergreen tree as high as 70 ft. (23 m), but often used as hedge; in zones 3A–7A.

Picea abies (Norway spruce, European spruce), native to Northern, Central, and Eastern Europe). Branches typically pendulous and the largest cone of any spruce (Photographed in New York Botanical Garden)

ARAUCARIACEAE

3 GENERA AND ±40 SPECIES

GEOGRAPHY Tropics of Southern Hemisphere, including Australia, New Caledonia, Indonesia, New Zealand, Philippines, and South America.

GROWTH HABIT Evergreen monoecious or dioecious resinous tall to very tall (±200 ft., 60 m) excurrent, and in time, columnar trees, often with large trunks. **STEM/BARK**: typically branching is horizontal and tiered, arising in whorls. **LEAVES**: spirally arranged and may be laminate (blade-like) and broadly ovate, acicular (needle-like), or small and subulate (scale-like) and appressed to the stem; petiole lacking; multiveined (veins branch dichotomously at base). **CONES**: unisexual, pollen cones axillary or terminal; solitary or clustered on branchlets; microsporophylls spirally aggregated into drooping cylindrical catkin-like comparatively large male cones on branch tips or leaf axils; sporangia are attached to the lower surface; seed cones together with bracts are partly or completely fused to one another to form spherical to ovoid female cones on short shoots or branch tips; ovules connate with ovulate scales or with bracts; bracts of mature female cones deciduous, flattened and woody, each bearing one seed; seed cones may be large and heavy. **SEEDS**: only one per megasporophyll, adnate to the scales, and with a wing-like structure that assists with wind dispersal, comparatively larger than other conifers; cotyledons usually 4.

CULTIVATED GENERA *Agathis* and *Araucaria* are commonly cultivated in warmer regions, but *Wollemia* is currently rare in cultivation and usually found only in botanical collections.

ECONOMIC USES Important ornamental plants for landscape and containers, significant high-quality softwood timber, and a few are sources of edible nuts, resins, and some artifacts.

COMMENTS According to Christenhusz et al. (2011), three genera are recognized: *Agathis, Araucaria,* and the more recently discovered and uncommon monotypic Australian endemic genus *Wollemia*, that until 1994 was known only from fossil.

Araucaria columnaris winged seed, more or less typical for the family.

Agathis robusta

QUEENSLAND KAURI; SMOOTH BARK KAURI; DAMMAR KURI PINE

SYNONYMS *Agathis palmerstonii, Damara bidwilii, D. palmerstonii, D. robusta*

ETYMOLOGY [From Greek *agathis*, a ball of thread, alluding to the shape of the female cone] [Robust, strong grower]

GROWTH HABIT large monoecious evergreen excurrent tree, becoming columnar with age; densely foliated. **STEM/BARK:** smooth or somewhat scaly bark, orange-brown to brown, inner wood a mixture of pinkish-red and brown; lateral secondary shoots spirally arranged. **LEAVES:** broadly laminate, petiolate, in opposite pairs, elliptic, glabrous, coriaceous when mature, veins parallel, weakly visible in young leaves, but distinct in mature leaves; margins entire, apex acute. **CONES:** numerous microsporophylls unite to form terminal ovoid to globose subsessile cylindrical male cones, which are on secondary lateral shoots; seed cones globose-ovoid, large, also made up of numerous scale-like megasporophylls each with a single seed. **SEEDS:** cordate, winged, mature between July to September; cotyledons 2. **NATIVE HABITAT**: eastern and northern Queensland, and on Fraser Island, Australia, in seasonal tropical areas and in deep well-drained soil.

CULTURE Grows best in well-drained deep soils with additional irrigation particularly during summer months, when necessary. **PROPAGATION:** seed and rooting of cuttings; seeds have short viability period. **LANDSCAPE USES**: makes a very attractive landscape and accent plant. **HARDINESS ZONE:** 10B–13B.

COMMENTS *Agathis robusta* is the source of high-quality timber useful in cabinet, furniture making, and boat building. Dammar gum or copal used in manufacture of varnish. Although there are some 22 reported species, only three in addition to *Agathis robusta* are in cultivation, including: *A. australis, A. lanceolata*, and *A. macrophylla*.

Agathis robusta, growth habit (above left), part of the trunk with branches (above right), and leaf shape and arrangement (below). Photographed in Safari Park Cisarua, Indonesia.

Araucaria bidwillii

BUNYA-BUNYA, FALSE MONKEY PUZZLE TREE
SYNONYMS *Columbea bidwillii, Marywildea bidwillii*

ETYMOLOGY [After the Araucani Indians of central Chile] [In honor of the botanist John Carne Bidwill (1815–1853), Australian botanist who became the first director of the Royal Botanic Gardens at Sydney and collected and described the species in 1843].

GROWTH HABIT Monoecious coniferous, tall evergreen tree; excurrent at first but with more or less rounded crown with age, tiered branching, medium texture; to 100+ ft. (30+ m) tall; rapid growth rate. **STEM/BARK:** rough, dark brown to nearly black, bark peeling, scars from old branches; shoots spirally arranged, nodding, terminating in reduced scale-like leaves. **LEAVES:** simple, laminate, spirally arranged around the stems; sessile and attached broadly at base, overlapping, rhomboid (triangular ovate), less than 1 in. (2 cm) long and wide; rigid, terminates with a sharp spine; dark green; entire, translucent margins; veins numerous, parallel, but scarcely visible. **CONES:** male cones 3–5 in. (7½–16 cm) long and ½ in. (1¼ cm) in diameter, solitary terminal at short shoot tips; megasporophylls numerous, rounded and each with a sharp spine, aggregating into a heavy (to 22 lb, 10 kg.), dark green, globose female cones 7–9 in. (17–21 cm) long and 6–8 in. (15–20 cm) in diameter. **SEEDS:** fused to bracts, comparatively large wingless

1–1½ in. (3–4 cm) is produced per megasporophyll (cone scale). **NATIVE HABITAT:** coastal northeastern Queensland in Australian wet tropical rainforests.

CULTURE Full sun, various soils; moderate salt tolerance; scale, sooty mold, and leaf spot may cause problems. **PROPAGATION:** seeds or cuttings of erect shoot tips only; seed germination is cryptogeal (shoots emerge from an underground tuber following germination) and may take several months, the medium should be kept moist. **LANDSCAPE USES:** this species is appropriate as a landscape accent specimen or urn subject; not recommended for indoor use because of its sharp leaf apex. The heavy spiny female cones can pose danger to pedestrians beneath trees; hence trees should be away from public areas. **HARDINESS ZONE:** 9B–13B, can be grown in cooler climates, but needs a protected site.

COMMENTS Seeds are edible. The tree has been considered sacred by the native Aborigines. Logging of trees for timber, which began in the 1860s, ultimately resulted in decimation of forest stands. The existing trees, however, are currently protected. Leaves of *Araucaria* may remain on the plant for 25 years or longer. The timber has been used for manufacture of a number of items. *Araucaria* and to a lesser extent *Agathis* have been major sources of amber. As a point of interest, *Araucaria columnaris*, the Caledonian endemic, leans towards the south in the northern hemisphere and to the north in the southern hemisphere; the higher the latitude, the more pronounced the lean.

Araucaria bidwillii, Photographed in a private property in West Palm Beach, Florida

Araucaria heterophylla

NORFOLK ISLAND PINE
SYNONYM *Araucaria excelsa, Eutassa heterophylla*

ETYNOLOGY [After the Araucani Indians of central Chile] [In reference to differences between juvenile and adult leaves].

GROWTH HABIT Monoecious, coniferous, evergreen tree; single upright trunk (excurrent), tiered branching; pyramidal form; medium texture; to 200 ft. (65 m) tall (usually about 80 feet in cultivation); rapid growth rate in ground but slow growth as indoor container plant. **STEM/BARK:** rough, scars from old branches. **LEAVES:** spirally arranged; juvenile leaves needle-shaped, adult leaves subulate, to ½ in. (13 mm) long; sessile, soft, curved toward stem, dark green. **CONES:** male cones cylindrical, 1½–2 in. (3¾–5.0 cm) long, in clusters and pendent; female cones subglobose to globose, 3–5 in. (7.6–12.7 cm) long and to 6 in. (15 cm) in diameter, and predominantly on the upper parts of the plant.

SEEDS: one seed is produced per cone scale, broadly winged and with an apical outgrowth, edible. **NATIVE HABITAT:** endemic to Norfolk Island.

CULTURE Full sun to partial shade; varied soils; moderate salt tolerance but performs best in proximity to coastal areas; scale, root rot, and bacterial diseases spread by rain; trees have little tolerance to high winds. **PROPAGATION:** seed or cuttings of erect shoot tips only. **LANDSCAPE USES:** specimen or in large containers, but it is also commonly grown as an interior plant and as a tropical Christmas tree. **HARDINESS ZONE:** 10A–13B, south and protected sites in cooler climates.

COMMENTS Apparently the wood is of little industrial or other use. *Araucaria heterophylla* and its listed cultivars **'Majestic Beauty'** and **'Monia'** often suggest subtropical and tropical areas. This species is sometimes confused with *A. columnaris*. Other related species include: *A. araucana* (monkey-puzzle tree) and *A. cunninghamii* (Moreton Bay Pine).

Araucaria heterophylla in the landscape (above), adult foliage and male cones (below left), and close up of the branching habit and trunk (below middle), and branchlets and leaves (below right). Photographed in Jakarta, Indonesia.

Araucaria araucana younger plant (left) and older cone-bearing specimen (right). This species is sufficiently cold hardy to have been planted in Canada and Europe.

Araucaria araucana in backlighted photo to demonstrate its startling silhouette landscape effect (left), ***Araucaria* sp.**, Photographed in native habitat, Brazil (right)

***Araucaria auricana* female cones**

Wollemia nobilis

WELLEMI TREE, WELLOMI PINE

ETYMOLOGY [Named after Wollemi National Parks in New South Wales, Australia; *wollemi* is an aboriginal name, meaning look around you] [Named after David Noble, its discoverer, a National Parks and Wildlife Services Officer].

GROWTH HABIT Monoecious evergreen coniferous tree, 80–100 ft. (27–30 m) tall and 40–65 ft. (12–20 m) wide. **STEM/BARK**: adult and juvenile shoots differ in leaf arrangement, leaf shape; trunk diameter 3 ft. (1 m), sometimes with multiple trunks that arise from the same base (appear as clumps); trunk surfaces bumpy; branched without additional side branches. **LEAVES**: leaves on adult shoots are 4-ranked; stiff, flattened, flat-linear in 5–8 spiral rows, yellowish-green; juvenile leaves fern-like, dark green, waxy on underside. **CONES**: terminal on branches; female cones to 3 in. (8 cm), rounded, green then brown when mature, present all year; male cones narrower than female cones, to 4 in. (10 cm) long, reddish-brown, usually lower on the tree; mature in late September. **SEEDS**: small, brown, papery, winged. **NATIVE HABITAT**: endemic to temperate rainforest in Wollemi National Park in New South Wales, Australia.

CULTURE Part shade, in well-drained moist (particularly during growing season), acidic soil, in wind-protected location and afternoon sun; susceptible to pathogens (*Phytophthora cinnamomic*) and dieback; said to be easy to grow. **PROPAGATION**: seed, tip cuttings. **LANDSAPE USES**: specimen tree when grown outdoor landscape (often placed in a cage or inside conservatories to avoid contact); can be maintained in a container indefinitely and may be used as an indoor foliage plant. **HARDINESS ZONE**: 9A–12A. (8, 7–?). Little is known of behavior of this plant in cultivation. According to Wikipedia, "wollemi pine is extremely hardy and versatile in cultivation."

COMMENTS *Wollemia nobilis* is a monotypic conifer, unrelated to pines, unidentified until 1994, and was known only as a fossil. It is most closely related to *Agathis*. DNA samples from 39 known wild plants have not shown any discernable variation, suggesting the only known population is clonal. No specific use has been reported. It is listed as critically endangered by the IUCN. Plants, however, have been propagated and distributed to various botanic gardens and recently offered for sale.

For further information, see Wikipedia and Missouri Botanical Garden' Plant Finder.
http://en.wikipedia.org/wiki/Wollemia

Wollemia noblis rarity and value are demonstrated by its presence in the conservatory at Hortus Botanicus Leiden in Holland (left and middle close-up), and at the University of Vienna Botanical Garden where it is placed inside a wire cage so as to prevent mishandling by visitors.

PODOCARPACEAE

PODOCARPUS FAMILY

17 GENERA AND ±170 SPECIES

GEOGRAPHY Widespread in wet tropical and subtropical regions, mostly in Southern hemisphere, especially Malaysia, New Caledonia, and New Zealand, extending to the Philippines, but also in southern Japan and China, Central and South America, and West Indies.

GROWTH HABIT Monoecious or dioecious coniferous evergreen trees to 180 ft. (60 m) tall and shrubs; sometimes with leaf-like short shoots (phylloclades). **STEM/BARK**: mostly excurrent with lateral branches at right angle. **LEAVES**: spirally arranged, decussate, scale- or needle-shaped or laminate, lanceolate to ovate. **CONES**: male cones numerous, pendulous, catkin-like, on terminal or axillary branches; with two pollen sacs; female cones with spiral, decussate, or whorled scales, each with one ovule, or expanded to an aril-like structure (= epimatium) in which the single inverted ovule is atop or enclosed within a fleshy aril. **SEEDS**: often subtended by a fleshy aril-like epimatium; cotyledons two.

ECONOMIC USES Ornamentals in tropical and subtropical regions, and a few species important sources of softwood timber.

CULTIVATED GENERA *Afrocarpus, Dacrycarpus, Dacrydium, Nageia, and Podocarpus* are the only five genera commonly used in landscapes, while *Phyllocladus* and possibly species of other genera may be in cultivation in tropical or subtropical botanical gardens as well.

COMMENTS There is some disagreement as to the generic limits of this family. Some authorities recognize up to 12 genera, while others recognize 18 or 19 genera. Several *Podocarpus* species have been relegated to other genera (e.g., *Afrocarpus gracilior, Nageia nagi)*. The currently accepted genera, based on Christenhusz et al. (2011), include: *Acmopyle, Afrocarpus, Dacrycarpus, Dacrydium, Falcatifolium, Halocarpus, Lagarostrobos, Lepidothamnus, Manoao, Microcachrys, Nageia, Parasitaxus, Pherosphaera, Phyllocladus, Podocarpus, Prumnopitys, Retrophyllum, Saxegothaea, Sundacarpus.*

Afrocarpus gracilior

AFRICAN FERN TREE, WEEPING OR FERN PODOCARPUS

SYNONYMS *Decussocarpus gracilior, Podocarpus gracilior*

ETYMOLOGY [African and from Greek *carpos,* a fruit, referring to the aril-like stalk of the seed] [More graceful].

GROWTH HABIT Dioecious, coniferous evergreen tree, 65–130 ft. (20–40 m) tall but usually less in cultivation, upright, pyramidal, densely foliated, graceful, weeping branches; fine-textured; medium growth rate. **STEM/BARK**: dark gray trunk and pendulous branches. **LEAVES**: spirally arranged, lanceolate. **CONES**: male cones, catkin-like, solitary but often few on short shoots; female cones consist of a single seed embedded within a yellow fleshy coat (epimatium), on a short peduncle, which when mature turns purplish-blue. **SEEDS**: mature seeds purple, favored by some birds. **NATURAL HABITAT**: montane evergreen rainforests of northeast and east Africa, from Ethiopia south to Kenya, Tanzania, and Uganda, at altitudes of 5000–8000 ft. (1800–2400 m).

CULTURE Not particular as to its light or soil conditions, but grows very slowly under full shade. It has no salt tolerance; requires regular irrigation during dry periods; pest free. **PROPAGATION**: seed, cuttings. **LANDSCAPE USES:** makes a very beautiful and graceful specimen or screen. **LANDSCAPE USES**: may be used as a shade or street tree or for screening. **HARDINESS ZONE**: 9B–13B.

COMMENTS *Afrocarpus gracilior* is a recent nomenclatural change and its consideration as a distinct genus from *Podocarpus gracilior,* which is currently listed in most plant catalogues as such. Appropriate or not, the recognition of the new genus appears to be more based on its geography than morphological characteristics. *Afrocarpus gracilior* is an important timber tree for use in furniture, door, floor, and wall paneling. It is threatened by deforestation.

Afrocarpus gracilior growth habit and its use along a road (left), fine textured foliage close up (right)

Nageia nagi

ASIAN BAYBERRY, NAGI PODOCARPUS
SYNONYM *Agathis veitchii, Dammara veitchii, Decussocarpus nagi, Nageia caesia, N. cuspidata, N. grandifolia, N. ovata, Podocarpus caesius, P. cuspidatus, P. grandifolius, P. nagi, P. ovata, P. ovatus*

ETYMOLOGY [No etymologies found, probably a local Japanese name] [Native Japanese name].

GROWTH HABIT Dioecious, coniferous evergreen tree, upright, densely foliated, symmetrical, narrow columnar canopy; medium texture; to 90 ft. (28 m), commonly seen 30–50 ft, (9–15 m); medium growth rate. STEM/BARK: young branches (together with leaves) light green, becoming reddish-brown and woody with age; bark exfoliating. LEAVES: simple, subopposite, entire, elliptic-ovate to 3 in. (7½ cm) long, coriaceous when mature, dark green, often glaucous, lighter beneath, many parallel veins but lacking a distinct midrib, stiff, leathery; petiole short. CONES: male cones catkin-like; female scale-like, inconspicuous. SEEDS: drupe-like, green, ovoid, 1 in. (2½ cm) long, on a very small green receptacle, covered with a dusty plume; a conspicuous aril similar to that of *Podocarpus* spp. is not present. NATIVE HABITAT: mountainous areas of southern Japan, China, and Taiwan.

CULTURE Full sun to partial shade, various well-drained but preference for moist soils, tolerates some drought; moderately salt-tolerant; scale, sooty mold, root rot, micronutrient deficiencies in alkaline soils noticeable particularly in female plants. PROPAGATION: seed, cuttings. LANDSCAPE USES: hedge, screen, framing tree, specimen, and as container plant. HARDINESS ZONE: 9A–12B.

COMMENTS Recent publications have recognized *Podocarpus nagi* as *Nageia nagi*, based on the absence of a leaf midrib and of an aril. Its lumber is utilized for construction of houses, bridges, utensils, furniture, and other projects. Apparently, the seed is edible. Branches with leaves can be suitable in flower arranging. It is listed by IUCN as threatened by habitat loss.

Nageia nagi growth habit and landscape use (above), leaves and few seeds (below, left), and numerous fallen seeds (below, right)

Podocarpus macrophyllus

PODOCARP, KUSAMAKI

SYNONYMS *Margbesonia forrestii, M. macrocarpa, Nagia macrophylla, Podocarpus forrestii, P. vertisillatus, Taxus macrophylla, T. makoya*

ETYMOLOGY [From Greek *podos,* a foot, and *carpos,* a fruit, referring to the aril-like stalk of the seed] [Large-leaved].

GROWTH HABIT Dioecious coniferous evergreen tree or shrub (depending on cultivar); upright, narrow columnar canopy; densely foliated, horizontal branching, medium-fine texture; medium rapid growth rate; to 45 ft. (15 m), often maintained as a shrub 4–8 ft. (1–3 m), slow growth rate. **STEM/BARK:** new growth green, woody with age, bark peeling. **LEAVES:** simple, spirally arranged, close on twigs, entire, linear to 2 in. (5 cm) long and about ½ in. (1¼ cm) wide, glossy dark green above, lighter beneath, leathery, with a prominent midrib. **CONES:** male cones catkin-like, yellow, to 1½ in. (3½ cm) long, female cones scale-like, inconspicuous, bear a single seed atop a fleshy receptacle when mature. **SEEDS:** drupe-like, green, ovoid to ½ in. (1½ cm) long, on a fleshy purple, edible receptacle (epimatium, aril). **NATIVE HABITAT:** southern Japan and eastern to southeastern China, in forests and open thickets from sea level to 3000 ft. (1000 m) elevation.

CULTURE Full sun to partial shade, various well-drained soils; poor salt and wet soils tolerance; develops chlorosis in alkaline soils. Although not seriously susceptible, it may be infested with scale, sooty mold, mites, and root rot fungus, but essentially it is a trouble-free plant. **PROPAGATION:** seeds germinate relatively quickly and cuttings root readily. **LANDSCAPE USES:** hedge, screen, accent, framing tree, in containers, and can be easily pruned to various topiary forms. It is one of the most versatile woody plants in the landscape. **HARDINESS ZONE:** 8A–13B.

COMMENTS The lumber is resistant to termites and excessive moisture which is the reason for its use in home construction. It is regarded as a feng shui tree. Many cultivars for habit, leaf form, color, etc. **'Angustifolius'** (with narrow leaves), **'Crusty Drybread'**, **'Dwarf Pringles'**, **'Edgefield'**, **'Meta'**, **'Pringleii'**, **'Variegatus'** are commonly sold. Related species such as *P. neriifolius, P. nivalis*, and *P. salignus* are also available for warmer areas. The aril is edible only when fully ripe (nearly black), otherwise extremely bitter.

Podocarpus macrophyllus is one of the most versatile landscape plants that may be used as an accent plant or various topiary shapes: in the previous page is growth habit of the plant in landscape (above left), in container (above right), pruned to conical shape to fit the formal design (middle left and right), as a hedge (below left), and as topiary form of a swan (below right). On this page, the top two photographs show other examples of how the plant can be manipulated into different shapes, male cones (middle left) and unripe female structures (middle right) showing seeds atop fleshy red arils, mature seed atop a ripe aril (below left), and finally a variegated specimen (below right).

CUPRESSACEAE

CYPRESS FAMILY

31 GENERA AND ±130 SPECIES

GEOGRAPHY Principally of temperate northern hemisphere but with a few in tropical mountains and temperate South America, in low or high altitudes.

GROWTH HABIT Aromatic monoecious or dioecious (in *Juniperus*), resinous evergreen or rarely deciduous (in *Taxodium*) erect, decumbent, or procumbent, much-branched coniferous trees and shrubs, ranging in height from a few inches in some cultivars to 300+ ft. (100+ m). **STEM/BARK**: branches cylindrical or flattened, branchlets deciduous together with leaves; bark of mature trees red to reddish-brown, often smooth or variously exfoliating. **LEAVES**: decussate or in opposite pairs or whorls of 3s; often dimorphic, with juvenile leaves acicular but adult leaves scale-like and adpressed to twigs; mature plants of some genera may have completely juvenile or adult foliage. **CONES**: male cones small, solitary or in clusters, and usually at the end of branchlets; female cones woody with one to several scales or fleshy and drupe-like, globose or elongated; cone scales imbricate or fused at margins so as to resemble a berry, without spines but often hooked; ovules develop at the base of the subtending scale (= subaxillary). **SEEDS**: generally small, flattened, with lateral wings or not winged.

ECONOMIC USES Ornamentals, timber, and *Juniperus* "berries" used for flavoring the alcoholic drink gin. Pollens of most species are highly allergenic (cause allergic reactions).

CULTIVATED GENERA Seven subfamilies have been recognized, any discussion of which serves no useful purpose in this book. The list of genera in cultivation includes: *Anthrotaxis, Austrocedrus, Callitris* (**incl.** *Actinostrobus*), *Calocedrus, Chamaecyparis, Callitropsis, Cryptomeria, Cupressus, Cunninghamia, Cupressocyparis × leylandii, Diselma, Fitzroya, Fokienia, Glyptostrobus, Juniperus, Libocedrus, Metasequoia, Microbiota, Neocallitropsis, Papuacedrus, Pilgerodendron, Platycladus, Sequoia, Sequoiadendron, Taxodium, Taiwania, Tetraclinis, Thuja, Thujopsis, Widdringtonia*, and *Xanthocyparis*.

Cupressaceae is divided into two clades (groups, subfamilies, each derived from common ancestors): the **Cupressoideae** (all from Northern Hemisphere), and **Callitroideae** (all from Southern Hemisphere). It is notable that several of the listed genera were previously placed in **Taxodiaceae** (the Taxus family: *Taxodium, Glyptostrobus*, and *Cryptomeria*).

Cupressus × leylandii

LEYLAND CYPRESS

SYNONYMS × *Cupressocyparis leylandii*; *Callitropsis* × *leylandii*; × *Hesperotropsis leylandii*.

ETYMOLOGY [Originally considered an intergeneric hybrid between *Cupressus macrocarpa* and *Chamaecyparis nootkatensis*, but recent taxonomic works have recognized the generic name *Chamaecyparis* as illegitimate and its species the same as that of *Cupressus*, hence the change to interspecific hybrid between two species of the same genus] [After C.J. Leyland who grew some of the first plants at Hagerstown Hall, England].

GROWTH HABIT Monoecious, evergreen, usually excurrent, more or less conical/pyramidal evergreen coniferous tree with a dense crown; reaches a height of 100 ft. (30 m) or more with a fairly rapid growth rate. **STEM/BARK:** branchlets flatter, finer, longer, and slenderer than *Chamaecyparis*. **LEAVES:** adult leaves opposite, entire, scale-like and juvenile leaves awl-shaped or acicular, spreading. **CONES:** female cones to about 1 in. (2½ cm) in diam., most with 8 scales. **SEEDS:** 5 under each scale. **NATIVE HABITAT:** originated from *Chamaecyparis macrocarpa* hybrid seeds grown by J.M. Naylor at Leighton Hall, Welsh Pool, England in 1911.

CULTURE Tolerates a wide range of conditions but grows best in full sun and well-drained soils; should be irrigated occasionally if no rainfalls. No major problems noted but dead branches appear if under water stress. **PROPAGATION:** cuttings that root relatively easily when treated with 5–10,000 ppm IBA. **LANDSCAPE USES:** useful as accent or specimen tree, as hedge, or for wind break and screens, and as container-grown Christmas tree. **HARDINESS ZONE:** 7A–11A.

COMMENTS Used as timber and for resin, but includes several very interesting cultivars, primarily involving shape and/or color variations Some of the currently available cultivars include: **'Castlewellan'**, **'Emerald Isle'**, **'Golconda'**, **'Gold Cup'**, **'Gold Rider'**, **'Green Spire'** (narrowly pyramidal, very dense, central leader often poorly developed; branch angles variable in relation to the stem; foliage bright green), **'Haggerston Grey'** (branches open, opposite, at right angles to one another and decussate; foliage gray-green), **'Harlequin'**, **'Irish Mint'**, **'Leighton Green'** (with a distinct central leader, irregular branching arrangement, dense, and lying flat; foliage bright green to yellow-green; frequent cone producer), **'Moncal'**, **'Naylor's Blue'** (similar to 'Leighton Green' but with gray-blue foliage, and infrequent coning). **'Robinson's Gold'** (exceptionally fast grower and with more compact and conical shape; densely branched;

foliage bronze-yellow in spring, turning gold-yellow in summer), **'Silver Dust'**, **'Star Wars'**, and **'Variegata',** and probably others.

Based on the nomenclatural change mentioned above, two other interspecific hybrids have been recognized and given correct names by rule: ***Cupressus × nobilis*** (*C. arizonica* var. *glabra* × *C. nootkatensis*) and ***Cupressus × ovensii*** (*C. lusitanica* × *C. nootkatensis*). Both hybrids are noted as being attractive, useful, and more drought-tolerant than ***Cupressus × leylandii***.

Cupressus × leylandii young plant (above left), an old specimen (above right), a main branch (below left), and enlarged segment showing subulate (scale-like) leaf arrangement (below right).

Cryptomeria japonica

JAPANESE CEDAR (SUGI)
SYNONYM *Cupressus japonica*

ETYMOLOGY [From Greek *krypto*, to hide and *meris*, a part, apparently in reference to seeds that are hidden by the cone scales] [From Japan]

GROWTH HABIT Monoecious evergreen coniferous pyramidal excurrent tree that reaches a height of 230 ft. (70 m), spreads 20–30 ft. (7–10 m), and with a trunk diameter of 13 ft. (4 m). In cultivation, the tree is much smaller and as it gets older, it becomes more open and irregular in shape. **STEM/BARK**: bark reddish-brown to dark gray, exfoliating in vertical strips; branches green, whorled, horizontal, tiered, and nodding at the tips. **LEAVES**: spirally arranged, fragrant, awl-shaped with a point but soft to the touch; stomatal bands present. **CONES:** male cones ovoid to ovoid-ellipsoid, on second year branchlets, crowded into terminal racemes, red but turn yellow at maturity; female cones spherical, axillary, sessile, globose, solitary or aggregated in groups of 2–6, ½–¾ in. (1–2 cm) in diameter, ripen the first year but remain on the plant for about two years. **SEEDS:** winged, brown, ellipsoid, 2–5 per cone scale. **NATIVE HABITAT**: endemic to Japan where it grows in deep, well-drained but moist soils. It is, however, also known from China, but some authorities believe it has been introduced for forestry.

CULTURE Full sun, in well-drained, fertile, moist soil; it does not respond well to poor soils and cold dry climates, it is not recommended for hot humid climates but performs well in warm Mediterranean climate regions.

PROPAGATION: seeds, cuttings from younger shoots of mature trees are reported to root in a few months when treated with IBA; layering is also possible. **LANDSCAPE USES**: widely used as landscape specimen and for wind screen in warm temperate climates and is particularly favored in its compact dwarf forms for use in rock gardens and for bonsai. **HARDINESS ZONE**: 7A–10A.

COMMENTS This monotypic genus was previously assigned to Taxodiaceae, but based on DNA studies it is currently considered a member of the Cupressaceae. Some authorities recognize two varieties: *C. japonica* var. *japonica* and *C. japonica* **var.** *sinensis*. It is the national tree of Japan and generously used around temples and houses. The timber is valued for its fragrance, softness and low density, and moisture and insect resistance, and is used in furniture, casket building, and various other uses. Several cultivars are listed, including **'Araucarioides'**, **'Bandai-tsugi'**, **'Benjamin Franklin'**, **'Birdo-sugi'**, **'Chapel View'**, **'Compressa'**, **'Cristata'**, **'Dragon Prince'**, **'Dragon Warrior'**, **'Elegans'** (dense and bushy, permanently retains juvenile foliage), **'Elegans Aurea'**, **'Elegans Compacta'**, **'Giokumo'**, **'Globosa'**, **'Globosa Nana'** (grows 3–4 ft. tall and wide with dark green needles), **'Hino-sugi'**, **'Jindai-sugi'**, **'Kimacurragh'**, **'Knaptonensis'**, **'Komodo Dragon'**, **'Koshyi'**, **'Little Diamond'**, **'Lobbii'** (denser and more compact than the species and with bronze winter foliage; it is recommended for southern regions), **'Lobbii Nana'**, **'Manishmi-sugi'**, **'Monstrosa'**, **'Mushroom'**, **'Nana'**, **'Pygmaea'**, **'Radicans'**, **'Rasen'**, **'Ryoku Gyoku'**, **'Sekkan-sugi'**, **'Spiralis'**, **'Spiraliter Falcata'**, **'Taisho Tama'**, **'Tansu'**, **'Tansu Sport'**, **'Tarhill Blue'**, **'Tenzan'**, **'Vilmoriniana'**, and **'Yoshino'** (remains green in winter), among others.

Cryptomeria Japonica mature plant (above left), branching habit (above right), and young male cones (middle), female cones (below) (photograph courtesy of Professor Walter Judd). [In the seed/ovulate cones note the points at the tip of each cone scale, those are the leaves of the short shoot (= cone scale, or ovuliferous scale), with the larger, triangular structure being the bract, out of the axil of which the short shoot emerged. This photo nicely shows the difference between a cone (where the ovules are borne on short-shoots) and a strobilus (where the ovules are borne on modified leaves)].

Cunninghamia lanceolata

CUNNINGHAMIA, CHINA FIR
ETYMOLOGY [After Dr. James Cunningham (d. 1709), the Scotchman physician and botanist who between 1698 and 1702 collected plants in China and sent to England] [Lanceolate, in reference to the shape of leaves]

GROWTH HABIT Monoecious coniferous evergreen excurrent conical tree, but may produce suckers and become multi-trunked if injured or in case of exceptionally cold weather; it may reach 150 ft. (50 m) in height and about 9 ft. (3 m) in trunk diameter in the wild, but usually 45–90 ft. (15–30 m) in cultivation and much less in trunk diameter. **STEM/BARK**: branches whorled and tiered more or less horizontally and nodding at the tip; bark is fissured and exfoliates in strips revealing the reddish-brown bark beneath. **LEAVES:** spirally arranged around the stems but appearing in two ranks, dark blue-green, sharply pointed; with two distinct whitish stomatal bands prominently on the lower side, but less noticeable on the upper side. **CONES:** male cones in terminal clusters of 10–30; female cones also terminal, ovoid to globose, solitary or sometimes in two or more, 1–2 in. (2½–5 cm) long, scales spirally arranged; turn brown when mature. **SEEDS:** each scale contains 3–5 flat, brownish seeds; cotyledons 2. **NATIVE HABITAT**: moist forests of China, Taiwan, Vietnam, and Laos, from lowlands to higher elevations of about 6000 ft. (2800 m), in sandy loam soils.

CULTURE Full sun to part shade, prefers moist but well-drained somewhat acid soil. No serious problems have been reported, but plants are damaged in exceptional harsh winters and cannot tolerate wet soils. **PROPAGATION**: seed but cuttings taken from the lower branches root. **LANDSCAPE USES**: accent plant but preferably planted at a distance from walkways. Turns bronze with cold. **HARDINESS ZONE**: 8A–10B.

COMMENTS The scented wood of *Cunninghamia* is soft but durable and is highly prized for construction of temples and coffins. The selection *C. lanceolata* **'Glauca'** is available in the trade. *Cunninghamia* is a monotypic genus; however, based on DNA studies, *C. konishi* and *C. lanceolata* var. *konishi* are very similar if indeed not the same as *C. lanceolata*. If so, *C. lanceolate* var. *konishi* "Little Leo" is not acceptable; therefore, *C. lanceolata* **'Little Leo'** is the appropriate designation. Nevertheless, some consider *C. konishi* from Taiwan as a distinct species but perhaps more plausibly, one species with two varieties. *Cunninghamia* is included in the subfamily **Cunninghamioideae** of Cupressaceae.

Cunninghamia lanceolata growth habit (above left), branches with leaves and the distinct central characteristic shoot (above right), male cones on shoots (below left), and 1and 3 young female cones on shoot apeces (below right).

Cupressus sempervirens

ITALIAN CYPRESS
ETYMOLOGY [Classical Greek name *kuparissos*, for *Cuprssus sempervirens*] [Evergreen]

GROWTH HABIT Monoecious, evergreen coniferous tree, upright, narrow, strongly columnar (fastigiate) habit. It reaches to ±100 ft (30–35 m), and 3–6 ft. (1–2 m) in diameter, often much shorter; rapid growth; although not often seen in cultivation, in its native habitats the species is actually more or less decurrent. **STEM/BARK:** needle-like branchlets, round. **LEAVES**: small, scale-like, tightly compressed, whorled, obtuse, dark grey-green color. **CONES:** male ovoid to oblong, small, about ½ in. (15 mm); female cones globose, to 1½ in. (40 mm) in diameter, on separate branches, initially green but turn brown at maturity. **SEEDS:** several per cone scale. **NATIVE HABITAT:** eastern Mediterranean region to southern Europe and western Asia.

CULTURE Full sun, various well-drained soils; moderately salt-tolerant. Spider mites, root rot, bacterial blight, and may suffer from cypress canker; cannot be pruned with success. **PROPAGATION:** seed, cuttings of cultivars. **LANDSCAPE USES:** framing, strong accent around large buildings, formal landscape, and sometimes atop public buildings such as restaurants or roof gardens. **HARDINESS ZONE:** 7A–11A.

COMMENTS *Cupressus sempervirens* '**Stricta**' appears to have been the plant originally described by Linnaeus and has been planted in formal Italian villa gardens of Medici in Florence during the Renaissance period. Several cultivars: '**Glauca**' (narrowly fastigiate, tinted blue leaves); '**Stricta**' (narrowly fastigiate, leaves dark green); '**Monshel**'(compact, dense blue-green foliage); '**Swane's Golden**' (narrow columnar with golden new growth); '**Tiny Tower**' (compact, dense, small, and with rigid columnar form); and others. Related species, such as *C. arizonica* (Arizona cypress) and its cultivars and *C. lusitanica* (Portuguese cypress) and its cultivars, are also known from cultivation.

Cupressus sempervirens growth habit, wide form (above left), *C. sempervirens* **'Stricta'** (above right), narrow form (above right), photographed in Italy, accent specimens at Harry P. Leu Botanical Gardens (below left), and Epcot (below right), both in Orlando, Florida.

Juniperus Foliage Variation

Adult foliage   **juvenile foliage**

Juniperus chinensis

CHINESE JUNIPER

ETYMOLOGY [Classical Latin name for juniper] [From China].

GROWTH HABIT Dioecious, evergreen coniferous trees and shrubs, pyramidal, or horizontal and spreading, branch tips outward-pointing, to 60–65 ft. (18–20 m) tall, most cultivars ½–10 ft. (1½–3 m) tall and variously shaped. **STEM/ BARK:** bark is usually reddish-brown and sheds in thin strips. **LEAVES:** two distinct types – adult (new growth) leaves small ($\frac{1}{16} - \frac{1}{8}$ in., 1–3 mm long), scale-like (= subulate), tightly appressed, decussate, overlapping, obtuse, blue, grey, or green colors, depending on cultivar; juvenile (early growth) leaves awl-shaped (= acicular), $\frac{1}{3} - \frac{1}{2}$ in. (8–12 mm) long, in whorls of 3 or opposite pairs, usually low on the plant; cultivars may have only juvenile, adult foliage, or both. **CONES:** male cones yellow, catkin-like; female cones fleshy, berry-like, with coalesced fleshy scales, sub-globose, to ½ in. (15 mm) in diameter, 2–3 seeded, whitish-blue turning brown to purplish-brown, often glaucous. **SEEDS:** 1–12 per cone, not winged. **NATIVE HABITAT:** temperate East Asia, including China, Japan, Korea, Russia, and others.

CULTURE Full sun, fertile well-drained soils; slight to moderate salt tolerance, good pollution tolerance, but will not tolerate wet feet, newly planted materials should be occasionally irrigated. Often the spreading cultivars are not given enough space and outgrow the site. Most cultivars cannot be pruned severely. Mites (especially in hot, dry sites), bagworms, root rot, *Phomopsis* blight, and others may be troublesome. **PROPAGATION:** cuttings (usually in winter) of ground cover cultivars, grafting of shrubby ones. **LANDSCAPE USES:** variable with cultivar; ground cover, foundation, planter box, border, screening, specimen, and often for bonsai. **HARDINESS ZONE:** 4A–12B.

COMMENTS Many varieties and cultivars. Some of the most popular include: **'Armstrong Aurea'** (old gold Juniper); **'Blaauw'** (blue Juniper); **'Blue Vase'** (horizontal spreading to 8 ft., mostly adult foliage; blue-green, more upright and globose than 'Hetzii', with feathery projecting branches densely foliated on all sides with a tufted "fox tail" appearance); **'Excelsa'** (tall Chinese Juniper); **'Hetzii'** (horizontal, spreading to 15 ft., mostly adult foliage; blue-green, feathery branches project upward, resembles **'Mint Julep'** but color distinctly bluish; **'Hetz Columnaris'**; **'Humphrey's Pride'** (Humphry's pride Juniper)); **'Keteleeri'** (Keteleer Juniper); **'Mint Julep'** (horizontally spreading to 4 ft., mostly adult foliage; mint-green, feathery branches project upward with foliage mostly on upper side of the branch, resembles 'Hetzii' but color distinctly different); **'Parsonii'**, Parson's Juniper - prostrate ground cover to 3 ft. (1 m), mostly adult foliage, grey-green, it is actually *J. davurica* **'Parsonii'** but sometimes sold in Florida as *J. chinensis* **'Parsonii'**; **'Pfitzerana'** is actually a hybrid between *J. chinensis* × *J. sabina* and is grey-green, and *J.* × *pfitzerana* **'Old Gold'** are horizontal spreading to 6 ft. (2 m), have mostly adult foliage, and resemble **'Mint Julep'** and **'Hetzii'** but foliage distinctly finer in texture, feathery branches not as upright, producing a flatter profile; *J.* × *Pfitzerana* **'Aurea'** (golden pfitzer juniper); *J.* × *Pfitzerana* **'Glauca'** (blue pfitzer juniper); *J.* × *Pfitzerana* **'Nick's Compact'** are all similar to the original hybrid except for their color variations; **'Pyramidalis'** (pyramidal Juniper is an upright pyramidal form); **'San Jose'** (San Jose Juniper); **'Sargentii Viridis'**; **'Sea Green'**; **'Shepparadii'** (Sheppard Juniper); **'Spartan'** and **'Spearmint'** are upright and pyramidal; **'Spiralis'**; **'Sylvestris'** (Sylvester Juniper); **'Torulosa'** (upright to 20 ft., predominantly adult foliage; light-green; irregular branching produces a twisted, flame-like habit); *J. chinenis* **var. *procumbens***, **'Nana'** and **'Variegata'** are prostrate ground covers to 2 ft., with dense predominantly juvenile foliage; blue-grey-green, and often mounded; may be attacked seriously by mites and *Phomopsis* in hot humid climates.

There are a number of lesser-known cultivars in the trade whose characteristics are fully noted on the web, such as: **'Ames'**, **'Aquarius'**, **'Aquazam'**, **'Blue Point'**, **'Columnaris'**, **'Eternal Gold'**, **'Etgozam'**, **'Expansa Albovariegata'**, **'Fairview'**, **'Femina'**, **'Fremonti'**, **'Hooks'**, **'Iowa'**, **'Keteleen'**, **'Kuriwao Gold'**, **'Mac's Golden'**, **'Maney'**, **'Mountbatten'**, **'Oxford'**, **'Perfecta'**, **'Plumosa Aurea'**, **'Robusta Green'**, **'Rockery Gem'**, **'San Jose'**, **'San Jose Variegata'**, **'Shimpaku'**, **'Spartan'**, **'Spearmint'**, **'Stricta'**, and **'Winter Green'**, among others.

Juniperus chinensis **'Blue Vase'** (blue vase juniper)

Juniperus chinensis **'Mint Julep'** (mint julep Chinese juniper)

Juniperus chinensis × *pfitzerana* (*J. chinensis* × *J. sabina*), Pfitzer juniper

Juniperus chinensis 'Torulosa' (Hollywood juniper)

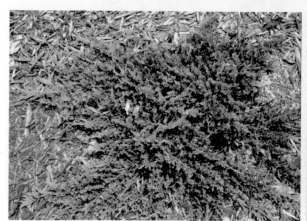

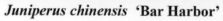

Juniperus chinensis 'Bar Harbor'   *Juniperus chinensis* var. *procumbens*

Juniperus chinensis 'Hetzii'

Juniperus chinensis **'Spartan'** (Chinese Spartan juniper) with upright conical green growth habit. Red-purple specimens are made up of *Alternanthera* cv. in containers. Photographed in Nong Nooch Tropical Garden in Thailand.

Juniperus conferta

SHORE JUNIPER

ETYMOLOGY [Classical Latin name for Juniper] [Crowded or densely pressed together, in reference to the leaves].

GROWTH HABIT Dioecious, evergreen, coniferous groundcover, procumbent, creeping, branch tips verticillate; to 2 ft. (24 cm), with 6–9 ft' (2–3 m) spread; rapid growth rate. **LEAVES:** juvenile foliage only; linear, to ½ in. (1¼ cm) long, in whorls of 3, prickly, white bands on upper surface, light gray-green, often glaucous. **STEM/BARK:** young stems yellowish-green, older branches gray-brown. **CONES:** male cones catkin-like; female cones blue-black, berry-like with coalesced fleshy scales. **SEEDS:** scales with 3 seeds in each. **NATIVE HABITAT:** sea coasts of Japan.

CULTURE Full sun, well-drained soils. Salt and drought tolerant, but will not tolerate waterlogged or wet humid conditions. Mites (in hot, dry locations), bagworm, *Phomopsis* blight, root rot, and others. Tends to be short-lived due to insect and disease problems. **PROPAGATION:** seeds, cuttings of cultivars. **LANDSCAPE USES:** ground cover, embankments, erosion control, planter boxes, draped over walls, seaside plantings, plant in drier areas. **HARDINESS ZONE:** 5A–11B.

COMMENTS Several cultivars offered: **'Blue Lagoon'** (US National Arboretum introduction, forms a tight dense mat, bluish-green); **'Blue Pacific'** (ocean blue-green color, usually less than 1 ft. in height); **'Compacta'** (gray-green, dense with little upright branching, slower growing, 8–10 in. (20–25 cm) in height); **'Emerald Sea'** (US National Arboretum introduction, compact habit, blue-gray foliage, to 1 ft. (30 cm) in height; **'Iron Age'** (=**'Irozam'**, tolerant of harsh dry salty conditions, somewhat upright); **'Silver Mist'** (shade and salt tolerant, turns silver-bronze-green in colder weather), and perhaps others.

Juniperus conferta growth habit used for cascading effect on a short wall (above left), close up of needle-shaped all juvenile leaves (above right), plants used as ground cover (below left), and leaves and seed cones (below right)

Juniperus horizontalis

CREEPING JUNIPER
TRANSLATION [Classical Latin name for juniper] [horizontal].

FORM Dioecious, coniferous evergreen shrub, horizontal-spreading; to 2 ft. (60 cm), most cultivars less than 1 ft. (30 cm), slow-growing. **STEM/BARK:** grey-brown, somewhat peeling bark. **LEAVES:** adult leaves small, about ¼ in. (8 mm) long, scale-like, tightly appressed to the stem, opposite and decussate, bluish, grey or green depending on cultivar; juvenile leaves awl-like; foliage develops with purplish cast in winter. **CONES:** male cones yellow, catkin-like; female cones subglobose, berry-like, to ½ in. (1¼ cm) in diameter, blue-black, glaucous, scales 3–8, fleshy, coalescing. **SEEDS:** mostly 2–4, not winged, rarely seen. **NATIVE HABITAT:** Northeastern United States.

CULTURE Full sun, fertile, well-drained soils; slight to moderate salt tolerance (less so than *J. conferta*). Will not stand waterlogged conditions. Branches easily broken by foot traffic; mites, bagworm, root rot, *Phomopsis* blight may be problems. **PROPAGATION:** seeds, primarily cuttings of cultivars. **LANDSCAPE USES:** planter box, border, ground cover - varies with cultivar. **HARDINESS ZONE:** 2A–11A.

COMMENTS Many cultivars. Some of the most popular include: **'Bar Harbor'** (prostrate ground cover to 8 in. (20 cm)), predominantly adult foliage in plate-like clusters; blue-green. fall color similar to **'Wiltonii'**; **'Wiltonii Blue Rug'** (prostrate ground cover to 6 in. (15 cm)), predominantly adult foliage; blue-green; extremely compact; fall color bluish-purple, difficult to tell from "Bar Harbor"; **'Plumosa'** (Andorra Juniper, more upright ground cover to 18 in. (45 cm)); mostly juvenile foliage; green; more erect than either **'Bar Harbor'** or **'Wiltonii'**, with plume-like branches; fall color reddish-purple. **'Plumosa Compacta'** and **'Youngstown'** are compact selections of **'Plumosa'**.

Juniperus horizontalis (left) and *J. horizontalis* 'Plumosa Compacta' (right)

Juniperus virginiana

RED CEDAR

Juniperus virginiana var. silicicola

SOUTHERN RED CEDAR
SYNONYM *Juniperus silicicola*

TRANSLATION [Classical Latin name for juniper] [From Virginia] [Growing in sand].

GROWTH HABIT Dioecious, evergreen coniferous tree, symmetrical but pyramidal and oval when young and sometimes irregular with age, dense, fine-textured; to 40 ft. (12 m) tall with a 20 ft. (6 m) spread; grows at a fairly rapid pace. **STEM/BARK:** thin, reddish-brown, peeling bark, drooping branchlets. **LEAVES:** opposite or in whorls of 3; juvenile leaves are needle-like, linear, to ½ in. (1¼ cm) long, green, sharp-pointed, prickly, with 2 silvery lines beneath; adult leaves are dark green, scale-like and overlapping. **CONES:** male cones are catkin-like, yellowish, ¼ in. (8 mm) long, and appear at branch tips; female cones are fleshy, drupe-like, one-seeded, purple with a white bloom, aromatic, occurs in late winter and ripen in late fall. **NATIVE HABITAT:** Coastal plain of the southeastern U.S.

CULTURE Plant in full or partial sun on a wide range of soils; tolerates both calcareous soil and seashore salt. It prefers moist soil but can tolerate dry sandy soils as well. Difficult to transplant, except when small. Transplanting shock may be avoided with initial frequent irrigation. Juniper blight, spider mites, bagworms, and cedar-apple rust may be problems. **PROPAGATION:** seed, which germinate faster if planted as soon as the cones mature, or if given a stratification period; IBA-treated tip cuttings will root. **LANDSCAPE USES:** specimen tree, also employed in windbreaks and as hedges; used in the Southeast as Christmas tree, either grown in containers, cut from the wild, or planted in commercial tree farms. **HARDINESS ZONE:** 7A–11A.

COMMENTS *Juniperus virginiana* and *J. virginiana* var. *silicicola* are very similar taxa and in fact often difficult to distinguish. The only difference between them appears to be the greater height and straight branches of the latter, as opposed to drooping branches of the former. In some treatments, however, *J. silicicola* is treated as a variety of *J. virginiana*, primarily based on sympatry of the two taxa but wider geographical spread (said to exist from Ontario Canada to Texas and northern Florida) and its greater cold tolerance. However, there is disagreement between Atlas of Florida Vascular Plants where *J. virginiana* is considered a synonym of *J. silicicola* and Flora of North America in which *J. silicicola* is treated as a variety. This seems to be the accepted view and has resulted in listing the many cultivars under *J. virginiana* var. *silicicola*, including: **'Blue Arrow'**, **'Blue Mountain'**, **'Burkii'**, **'Canaertii'**, **'Glauca'**, **'High Shoals'**, **'Hillii'**, **'Hillspire'**, **'Idyllwild'**, **'Manhattan Blue'**, **'Monbell'**, **'O'Connor'**, **'Pendula'**, **'Prairie Pillar'**, **'Sapphire Sentinel'**, **'Silver Spreader'**, **'Stover'**, and **'Taylor'**, among others.

The wood repels insects and the tree was lumbered extensively for wood for chests and pencils. Apparently, at one time Cedar Key, Florida, had extensive red cedar forests, hence name of locality.

For excellent photographs see: http://lee.ifas.ufl.edu/Hort/GardenPubsAZ/Juniperus_virginiana2.pdf

Juniperus virginiana **var.** *silicicola* growth habit (above left), trunk illustrating reddish color and exfoliating bark (above right), close up of juvenile and adult foliage (below left), and a profusion of young female cones (below right).

Metasequoia glyptostroboides

DAWN REDWOOD

ETYMOLOGY [From Greek *mĕtra*, with, after, sharing, changed in nature; *Sequoia,* to which it is related, in reference to fossil specimen] [Resembling the genus *Glyptostrobus*]

GROWTH HABIT Monoecious coniferous deciduous trees with persistence or deciduous, opposite or irregularly whorled main branches that develop from superposed axillary buds (surrounded by petiole base); reaches a height of 70–100 ft. (21–31 m) and 15–25 ft. (5–8 m) spread. **STEM/BARK**: branchlets subtended by about 4 whorls of salmon-colored, early deciduous scales; lateral branches deciduous together with leaves; trunk somewhat similar to redwood and giant redwood (*Sequoia sempervirens* and *Sequoiadendron giganteum*), but with noticeably complex and thickened vertical grooves; bark deeply fissured. **LEAVES**: decussate, 2-ranked, spirally arranged; linear, flattened, soft, with noticeable midvein and 4–8 stomatal bands on lower side, sessile and twisted at base. **CONES**: male cones ½ in. (1¼ cm) long, in spikes or panicles and with a short peduncle, develop in autumn but release pollen in following spring; microsporophylls 15–20; female cones solitary, subglobose, ¾ in. (2 cm) long, with 16–24 shield-like woody scales; terminal on previous year's growth, becoming pendulous after pollination and ripening the first year; seeds 5–9, obovoid, winged; cotyledons 2. **NATIVE HABITAT**: in valley floors and moist ravines on acidic soil, in central and western China, introduced to US in 1947.

CULTURE Full sun in medium to nearly wet but well-drained soils, usually tolerant of wet soil and air pollution, and essentially trouble-free. **PROPAGATION**: best by seed but leafy cuttings root taken in mid-summer when the plant is with green leaves. **LANDSCAPE USES**: makes for interesting large landscape specimens or street tree. **HARDINESS ZONE**: 5A–9A. Not recommended for warmer climates.

COMMENTS This monotypic genus is listed under Taxodiaceae with two names noted as lower taxa (varieties or subspecies?) in the Flora of China: *Metasequoia glyptostroboides*, and *M. honshuenensis* (big cone down edwood). It was known as a fossil but discovered in 1947 and seeds were distributed to various botanical gardens. Its closest living relatives are the redwood (*Sequoia sempervirens*) and giant redwood (*Sequoiadendron giganteum*) of California. The plant is listed as highly endangered in the wild, but has become increasingly popular in European trade and gardens. No cultivars have been reported. Leaves and bark of *Metasequoia* are used as antimicrobic, analgesic, and anti-inflammatory diseases in Chinese folk medicine. *Metasequoia glyptostroboides* together with *Sequoia sempervirens* (redwood) and *Sequoiadendron giganteum* (giant sequoia) are members of subfamily **Sequoioideae** of Cupressaceae.

Metasequioa glyptostroboides at Keukenhof Gardens, Amsterdam in early spring (above and middle), trunk close up (middle, right), specimens in summer at Missouri Botanical Garden (below)

Platycladus orientalis

ORIENTAL ARBORVITAE
SYNONYMS *Biota* **spp.,** *Chamaecyparis glauca, Cupressus filiformis, C. pendula, C. thuja, Juniperus ericoides, Platycladus chengii, Retinospora* **spp.,** *Thuja orientalis,* and *Thuja* **spp.**

ETYMOLOGY [From Greek *platys*, broad, *cladis*, stem, with flat stem] [Oriental] [Arborvitae from Latin, tree of life, in reference to the Buddhist concept of long life and vitality].

GROWTH HABIT Monoecious, evergreen coniferous tree, densely branched, becoming more open with age, symmetrical, broad conical form; to 40 ft. (12 m) tall, medium growth rate. **STEM/BARK:** branchlets flat, in vertical plane; trunk usually not visible as it is covered with dense branching and flat foliage. **LEAVES:** small, scale-like, to ¼ in. (7 mm) long, glandular, grooved, opposite, closely appressed to branchlets, bright green, and without scent. **CONES:** male cones catkin-like; female fleshy cones to 1 in. (2½ cm) in diameter, erect, blue-green, becoming woody, with 6–12 thick scales, each hooked at apex. **SEEDS:** 1–2 wingless seeds per cone scale. **NATIVE HABITAT:** northwestern China, Korea, and Russian Far East, naturalized elsewhere in Asia.

CULTURE Full sun, various well-drained soils. Somewhat tolerant of heat and dry conditions, but will not tolerate salt or waterlogged soils. It is easily transplantable. Although survives in most locations, it prefers cooler Mediterranean climates. Mites, *Phomopsis* blight, bagworms; none serious but may be potential problems. **PROPAGATION:** seeds, cuttings of cultivars. **LANDSCAPE USES:** roadside plantings, barrier, and large specimen in cemeteries; often overplanted in areas too small to accommodate the full-grown plants that are older and often open and ragged in appearance. **HARDINESS ZONE:** 6A–11A.

COMMENTS Wood is used for construction and for incense burning. Many cultivars exist in the trade: **'Aureus'** (gold tip arborvitae, low, compact, branchlet tips golden yellow); **'Aurea Nana'** (Beckman arborvitae, same as **'Aureus'** except dwarf) and "**Elegantissima**" (yellow column arborvitae), have received the Royal Horticultural Society's "Award of Garden Merit"; **'Baken', 'Bakeri'** (Baker arborvitae, pale green; best adapted to hot, dry sites); **'Blue Cone'** (blue cone arborvitae)**; 'Bonita'** (cone-shaped; leaves tipped golden yellow); **'Excelsus'** (tall arborvitae)**; 'Fruitlandii'** (dwarf, globose form; dark green foliage)**; 'Globosa', 'Golden Rocket', 'Golden Surprise', 'Green Cone', 'Locogreen', 'Minima Glauca', 'Morgan', 'Raffles' 'Sanderi', 'Strictus', 'Sunkist',** and **'Sunlight',** among others.

Platycladus orientalis younger but still several years old specimens (left), close up of a branch to show flat branchlets, adult foliage, and cones of the species. Blockage of the specimens at the doorway will ultimately result in removal or unsightly pruning of the plants.

Taxodium distichum

BALD CYPRESS; SWAMP CYPRESS

SYNONYMS *Cupresspinnata disticha, Cupressus americana, C. disticha, C. montezumae, Glyptostrobus columnaris, Schubertia disticha, Taxodium denudatum, T. knightii, T. pyramidatum*

ETYMOLOGY [From Latin name *Taxus* and Greek *eidos*, resemblance, in reference to similarity of leaves] [2- ranked, in reference to leaf arrangement].

GROWTH HABIT Monoecious, large deciduous coniferous tree, spreading; foliage sparse, pyramidal when young, irregular when much older and often with contorted canopy; fine texture; to ±150 ft. (45 m), moderate growth rate. **STEM/ BARK:** base of trunk often flared (buttressed), bark grey-brown, peeling; aerating projections (pneumatophores or knees), often growing up from roots in wet sites (they do not develop significantly on dry sites); branchlets spirally arranged; wood is durable and decay-resistant. **LEAVES:** simple needle-like, linear, appressed, to ½ in. (1¼ cm) long. Plants also possess deciduous branchlets that are flat, linear, 2-ranked, spreading, to ¾ in. (2 cm) long. Both are light to dark green in spring, coppery brown in fall. **CONES:** male catkin-like cones, clustered, to 4 in. (10 cm) long on branchlet tips; female cones globose, many woody scales to 1 in. (2½ cm) in diameter, usually on branches, green when young turning brown when mature. **SEEDS:** irregularly 3-angled, 2 to each cone scale, and narrowly winged. **NATIVE HABITAT:** on alluvial soils of seasonally wet eastern United States and from Florida west to Texas.

CULTURE Full sun to partial shade, various soils including wet, muck-type; not salt tolerant. Will tolerate drier sites with sufficient irrigation until establishment. No serious pests reported. **PROPAGATION:** seeds or cuttings. **LANDSCAPE USES:** specimen, light shade, street plantings (dense foliage may block traffic visibility in summer), parks, primarily for large wetter areas. **HARDINESS ZONE:** 4A–12B.

COMMENTS Many regional variations *Taxodium distichum* collected in the wild have been given cultivar names, including: **'Autumn Gold', 'Cascade Falls', 'Debonair', 'Falling Waters', 'Fastigiata', 'Lindsy's Skyward', 'Mickelson', 'Monark of Illinois', 'Pendenws', 'Peve Minaret', 'Secrest', 'Shawnee Brave', 'Skyward',** and **'Sofine',** and perhaps others. *T. distichum* var. *imbricarium* **'Nutans'** (Pond Cypress) is now believed to be a distinct species. *T. ascendens* has smaller, awl-shaped leaves that are closely appressed to the pendulous stems. Also, other cultivars such as **'Prairie Sentinel'** and **'Sooner'**. The *T. mucronatum* (Mexican swamp cypress, Montezuma cypress, etc.) is a massive evergreen species native to Mexico, Guatemala, and southern Texas; seen occasionally in the trade. Some consider the latter taxon to be a variety of *T. distichum*.

Taxodium distichum planted near a small lake, two of three plants going dormant the third on the right still green (above left), same plant near the lake the following summer, note the aerial roots at its base (top right), inundated plants (middle left), close up of the aerial roots that are known as pneumatophores or knees (middle right), male cones at branch tips (below left), female cones on inner branches (below middle), and the large buttress of an old specimen (below right).

TAXACEAE

TAXUS (YEW) FAMILY

6 GENERA AND ±30 SPECIES

GEOGRAPHY Northern hemisphere, especially eastern Asia, western China, Europe, United States, and Central America.

GROWTH HABIT Evergreen monoecious or dioecious coniferous much-branched shrubs or small trees; without resin ducts. **STEM/BARK**: shoots cylindrical but ridged with leaf bases; bark scaly or fissured. **LEAVES**: persistence, spirally arranged, occasionally decussate; linear-lanceolate; with green or white stomatal bands on the lower side; apex is acute. **CONES**: male cones globose to ovoid, solitary, or in small clusters in leaf axils; female cones small, solitary or (in *Taxus* and *Torreya*) 2 or rarely more at the end of the short shoots. **SEEDS**: not winged, partly or completely enclosed within a fleshy brightly colored aril.

ECONOMIC USES Ornamentals, used in high-grade furniture, also used for medicinal purposes

CULTIVATED GENERA *Amentotaxus, Austrotaxus, Cephalotaxus, Pseudotaxus, Taxus,* and *Torreya* have been to a greater or lesser extent reported in cultivation, with *Cephalotaxus* and *Taxus* more common.

COMMENTS Some recent publications have included *Torreya* in Cephalotaxaceae. The distinction between Taxaceae and Cephalotaxaceae is based on seed characteristics: smaller and only partially enclosed within the aril in Taxaceae and larger and fully enclosed within the aril in Cephalotaxaceae. With the current classification having all genera in Taxaceae, seed morphology is relevant only at generic levels. Several taxonomists of the past considered the group distinct from other conifers

Cephalotaxus harringtonii

JAPANESE PLUM YEW

ETYMOLOGY [From Greek *kephale,* a head, referring to the resemblance of the plant to *Taxus*] [Earl of Harrington, the first to grow the plant in Europe in 1829].

GROWTH HABIT Dioecious, evergreen, fine-textured shrub; commonly less than 8 ft. (2½ m) in height with a variable spread; slow growth rate. **STEM/BARK:** gray fissured bark, shoots with one row of leaves on each side. **LEAVES:** broadly linear, 1–1½ in. (2½–4 cm) long, with abruptly pointed apices soft to the touch, somewhat coriaceous; with two glaucous bands on the underside, one on each side of the midrib. **CONES:** male plants covered with pairs of short pedunculated cream-colored microstrobili beneath each pair of leaves on shoots and turn brown with age; female plants have two pairs of pale green globose megastrobili on peduncles at the bases of shoots, which also turn brown with age. **SEEDS:** drupelike, green, almond-shaped, to 1 in. (2½ cm) long; often not produced in plants cultivated outside their natural habitat probably for lack of a specific pollinator that can transfer pollen from male to female individuals. **NATIVE HABITAT:** Mount Kiyosumi, Japan.

CULTURE Partial to full shade, on soils of reasonable fertility and drainage. It has no salt tolerance; nematodes and mushroom root rot may become problems. **PROPAGATION:** cuttings. **LANDSCAPE USES:** foundation plantings or in planters. **HARDINESS ZONE:** 6A–10B.

COMMENTS *Cephalotaxus harringtonii* **var**. *drupacea* is a low spreading shrub; two-ranked leaves form a V-shaped trough on horizontal arching branches. *Cephalotaxus harringtonii* **var**. *nana* is a dwarf form not often seen in cultivation. There are a number of selections such as: **'Duke Gardens', 'Fastigiata'** (upright columnar shrub selected in Japan in 1861); leaves dark green and spirally arranged on ascending vertical unbranched shoots that grow to 6 ft. (2 m), and superficially resemble ***Podocarpus macrophyl-*** ***lus***; **'Korean Gold', 'Prostrata'** are also reported from cultivation, as well as ***C. fortunii*** (Chinese plum yew, which has been in cultivation since 1848), and ***C. sinensis*** (Chinese yew) which is probably a synonym of *C. harringtonii*.

Cephalotaxus harringtonia **'Drupacea'** in raised planter (left), branch and leaf characteristics (center), and male cones (right)

Cephalotaxus harringtonia growth habits: **'Fastigiata'** specimen (left); combined **'Drupacea'** in front and **'Fastigiata'** in back (right)

Taxus floridana

FLORIDA YEW

ETYMOLOGY [Classical Latin name of the plant] [From Florida].

GROWTH HABIT Dioecious, evergreen shrub or small tree, spreading, drooping, densely foliated and fine-textured; can reach a height of 25 ft. (8 m), but normally much smaller; spread is variable and growth rate is slow. **STEM/BARK:** short narrow trunks have purplish brown, platy bark; twigs green early but become purplish-brown later, spreading. **LEAVES:** simple, narrowly linear, to 1 in. (2½ cm) long, held in a horizontal plane; dark green above, light green below, soft to the touch and occasionally curve downward and have an acute apex. The new growth is bright green, and sometimes pinkish, making an attractive contrast to the mature leaves. Leaves are poisonous. **CONES:** inconspicuous; male cones globose, in microsporophyll heads beneath branches; female cones berry-like, with bracted ovules within a small red fleshy aril, in March. **SEEDS:** bony, ¼ in. (7 mm), dark brown, ripening in fall. **NATIVE HABITAT:** shaded, moist, wooded ravines and slopes in a limited area in the Apalachicola River Valley in northwestern Florida.

CULTURE Partially shaded site, in fertile, slightly acid, well-drained soil; no salt tolerance; scales and mushroom root rot may become problematic. **PROPAGATION:** seed, if available, may require first warm, then cold stratification; mature wood cuttings taken in winter root well under mist. **LANDSCAPE USES:** useful as a topiary or specimen shrub, hedge or foundation planting. Should be planted more often to keep this rare native plant from extinction. **HARDINESS ZONE:** 8A–10B.

COMMENTS Leaves and seeds are both toxic, but are experimentally used for cancer treatment. The species is listed by IUCN as critically endangered and also listed as endangered species by the US Government. It is of notable interest that other *Taxus* species are commonly cultivated and used extensively in temperate climate regions such as *T. baccata*, *T. cuspidata*, and *T. media*, and their many cultivars among the most popular. A few other species such as *T. breviloba*, *T. chinensis*, *T. meyeri*, and *T. wallichiana* are also reported from cultivation.

Taxus floridana used as hedge (left) and close up of foliage to show the linear leaves arrange on horizontal plane (sometimes appearing spiral) on short shoots.

Taxus baccata (common yew, European yew), native to N.W. Africa, Europe, to northern Iran; a long-lived common tree in European landscapes; with seeds enclosed within bright red arils (baccate means berry, commonly known as yew berries, but it is not a fruit). It is often used for bonsai. Leaves and seeds are deadly poison if eaten. Photographed at Dumbarton Oaks, Washington D.C. Illustration from Otto Wilhelm Thomé (1885).

ADDITIONAL GENUS

Torreya taxifolia (torreya), native to Florida Panhandle, listed by IUCN and US Endangered Species Act as a highly endangered Florida native taxon that suffers primarily from a fungal disease, resulting in decimation of much of its population. Photographed in natural habitat.

SCIADOPITYACEAE

1 Genus and 1 Species

Sciadopitys verticillata

UMBRELLA PINE
SYNONYMS *Pinus verticillata, Taxus verticillata*

ETYMOLOGY [From Greek *skiados*, umbrella and *pitys*, fir tree, in reference to the leaves that appear in whorls like the ribs of an umbrella] [In whorls]

GROWTH HABIT Monoecious evergreen coniferous trees of about 25–30 ft. (7½–10 m) high and 15–20 ft. (5–6 m) wide in cultivation, but significantly taller and wider in its native habitat; excurrent pyramidal for the first several years, but does not maintain this form with age; may be single or multiple trunked. **STEM/BARK**: reddish-brown, with both long and short shoots, exfoliating bark; primary shots terminate in whorls of cladodes at branch tips but they are covered with small scale-like true leaves. **LEAVES**: the radiating umbrella-like needles are actually cladodes that are essentially flattened leaf-like green stems that function as leaves; about 5 in. (12 cm) long. **CONES**: male cones small, about ½ in. (1¼ cm), in dense terminal clusters; female cones subsessile, ovoid, about 4 in. (10 cm) long and initially appear green the first year but turn brown in the second year. **SEEDS**: flattened, winged on both sides,

orange-brown, 5–9 per scale, about ½ in. (1¼ cm) long. **NATIVE HABITAT**: endemic to Japan, at middle altitude wet cloud forests at about 1500–3000 ft. (500–1000 M) elevation.

CULTURE Full sun to partial shade, slow-growing plant performs best in well-drained but moist slightly acid soil. Although it grows best in areas with cooler summers, it can tolerate warmer southern Mediterranean climates. No serious problem has been reported, but despite its requirement for high rainfall and humidity, it tolerates excessive summer heat with humidity but may discolor and damage the leaves. **PROPAGATION**: seed; hardwood cuttings rooted best with 24-h soak in water, to minimize accumulation of the latex, prior to hormone dip. **LANDSCAPE USES**: graceful landscape accent plant in cooler climates but in containers in moderately warmer regions. However, because of its slow growth rate, it may be used in rock gardens and potentially a good subject for bonsai. **HARDINESS ZONE**: 6A–10A.

COMMENTS A few cultivars are listed in the trade, including: *Sciadopitys verticillata* **'Gruene Kugel'** (compact, well branched), **'Joe Kozey'** (narrow, tight branching), **'Mitsch Select'** (dense, slow growing), **'Ossorio Gold'** (with golden needles), **'Picola'** (dense and compact), **'Sternschnuppe'** (straight, excurrent), **'Variegata'**, (yellow and green needles), and **'Wintergreen'** (dark green foliage). The water-resistant wood is used in boat making. The tree is considered sacred in Japan.

Sciadopitys verticillata containerized specimen, photographed at the University of Vienna Botanical Garden.

INTRODUCTION

The astonishing diversity of plant life is best illustrated by the Angiospermae (or Anthophyta = flowering plants). It was the fruit or seeds of these plants and the method of their propagation that made organized human societies possible. Flowering plants are the dominant feature of the natural and manmade landscapes and the main source of food, shelter, clothing, and drugs. Unlike gymnosperms that predominantly occupy cool-temperate and some tropical habitats, angiosperms have adapted to a diversity of environments through modification and/or development of specialized structures for survival in aquatic habitats or the dry deserts and from the tropical rainforest to the arctic tundra. Enclosure of the seed within the **ovary** (the ultimate fruit) is a unique feature of the angiosperms. The ovary is a part of a complex but delicate, short-lived flower which may be born solitary or in groups on an **inflorescence**. Within the flower the ovary is situated at the base of solitary or fused **carpels** (the gynoecium), while the pollen grains are produced in **anthers** (the androecium). The annual and biennial growth habits common in angiosperms are absent in ferns and gymnosperms. Even the perennial habit is modified to include **monocarpic** (flower and fruit once and they perish) and **polycarpic** (flower and fruit many times), do not apply to ferns and gymnosperms because they do not possess fruit (carpus = fruit). Herbaceous plants that constitute the majority of angiosperms are unknown in gymnosperms. Among other unique characteristics of flowering plants are the modification of leaves and stems (succulence, pitchers, trigger traps and sticky leaves with digestive enzymes; tubers, bulbs, corms, etc.) and the parasitic and saprophytic habits.

In reality, our very existence on planet earth depends on flowering plants. We would not survive on pine seeds and fern frond tips alone, we would need grains, green vegetables, and fresh fruits as a source of our nutrition. While ferns and gymnosperms are wonderful additions to our natural surroundings and landscapes, there is no substitute for flowers. There are more than 300,000+ flowering plant species, not to mention thousands of selected cultigens, with an astonishing diversity of forms and structures and colors that feed our physical being as well as our soles.

ANGIOSPERM from Greek *Angeion*, container and *sperm*, seed, hence plants whose seeds are enclosed within a container (the fruit), in contrast to gymnosperms whose seeds are naked (not enclosed in any container) and ferns that are produced by spores. The most distinguishing feature of angiosperms is the flower which consists of **carpel**(s), **stamens**, and the **petals** and **sepals**, the accessory organs, that essentially function as attractant and provide landing platform for pollinators. Pollination biology is an interesting subject but beyond the scope of this book. Leaf and flower characteristics are fully illustrated in the pages that follow and described further in the glossary.

Angiosperms constitute two well-defined groups[a]: **Monocotyledons** (**Monocots**) and **Dicotyledons** (**Eudicots**). The following table provides the distinction between the two groups.

Monocotyledons	Dicotyledons
Embryos with one cotyledon	Embryos with two cotyledons
Flower parts in 3s	Flower parts usually in 4s, 5s
Adventitious fibrous roots	Primary tap roots
Vascular bundles scattered	Vascular bundles in a ring
Predominantly herbaceous	Woody and herbaceous
Leaves often linear and sheathing	Leaves usually broad and not sheathing
Veins usually parallel	Veins pinnate or palmate

[a]The recently established **Eudicots** introduced by James A. Doyle and Carol L. Hutton (1991) is essentially an evolutionary term that refers to most dicotyledons, with particular reference to pollen morphology (tricolpate or derived types in Eudicots). Except for the two families referred to as **Basal Lineages** and seven **Magnoliids**, all other dicotyledonous families are included in Eudicots, that in turn include the **Asterids** (with commonly fused petals forming a corolla tube), the largest group of flowering plants, and the **Rosids** (with commonly free petals). Although botanically significant, these divisions are of little consequence in horticulture and not specifically discussed. Monocotyledons as a whole are considered to be a monophyletic (have a common ancestor) class. Pollen of Magnoliids and Monocots are typically monosulcate

Despite their enormous diversity, monophyly (common ancestry) of both monocots and eudicots has been discussed by several authors. The recent publications by Soltis, et al. (2018), on "Phylogeny and Evolution of Angiosperms," and other recent publications by Judd et al. (2015), and Christenhusz et al. (2017), and of course, APG IV (2016–2021), are excellent notable sources for examination of phylogenetic classification of the two classes. Although a superb contribution toward understanding of flowering plants, Soltis et al. (2018) may be too technical to be useful in horticulture. There is a regularly updated and printable "Angiosperm Phylogeny Poster" (in 20 languages) on the web.

It is natural for humans to want that which is exotic in preference to that which is readily available. It follows that in human communities introduced exotic species in landscapes commonly exceed that of native plants. Accordingly, much of what is covered in this book does not specifically concerns native plants, but that of diverse climates. The intent is to show the diversity of the plants in our world landscapes with direct emphasis on recent DNA-based phylogenetic classification.

The organization of families and genera is based on the most recent APG system (Angiosperm Phylogeny Group IV, 2016–2021) of flowering plant classification based on DNA analyses which recognized 59 orders and 413 families. In part because of some uncertainty (or disagreement) of order delimitations and in part because orders are not used in horticulture and do not serve any useful purpose, they have not been discussed. As with ferns and gymnosperms, for practical reasons in this book only families of selected cultivated genera are discussed.

ANGIOSPERM CHARACTERISTICS

PLANT BODY

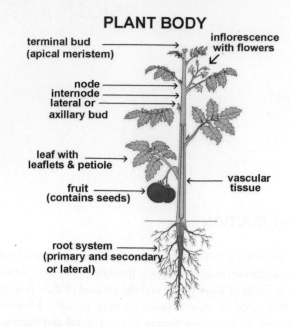

PLANT BODY

Analogous to animals, **vascular plants** have specific parts that are similar in all plant types, despite differences in shape or technical features, such as the flower of angiosperms versus cones of gymnosperms. In all cases, the plant consists of the complex nonreproductive (**vegetative**) as opposed to **reproductive** (flowers, cones, or spores) parts. The illustrated tomato plant is a typical example of a plant and its parts.

VEGETATIVE CHARACTERISTICS

LEAF STRUCTURE: SIMPLE

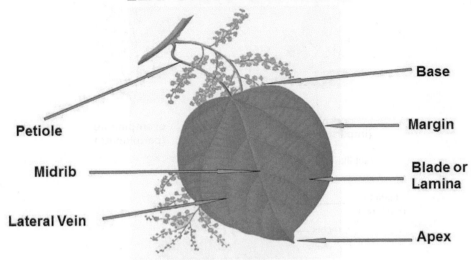

Base

Margin

Petiole

Blade or
Lamina

Midrib

Lateral Vein

Apex

SIMPLE ENTIRE LEAF

LEAF CHARACTRISTICS: SIMPLE LOBED

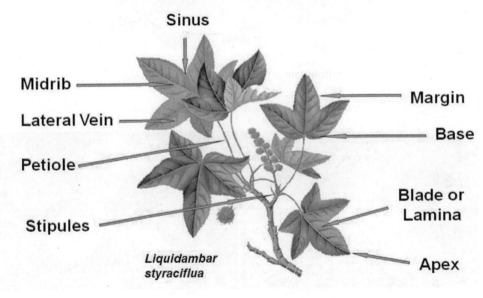

Sinus

Midrib

Margin

Lateral Vein

Base

Petiole

Stipules

Blade or
Lamina

*Liquidambar
styraciflua*

Apex

COMPOUND LEAVES

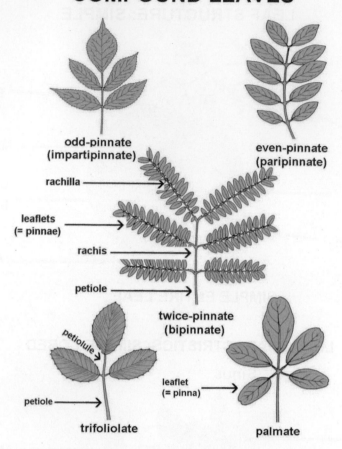

odd-pinnate
(impartipinnate)

even-pinnate
(paripinnate)

rachilla

leaflets
(= pinnae)

rachis

petiole

twice-pinnate
(bipinnate)

petiolule

petiole

leaflet
(= pinna)

trifoliolate

palmate

Trillium sp. with three leaves
(a trifoliate plant)

Arisima ringens
with leaves of three leaflets
(a trifoliolate plant)

LEAF SHAPES

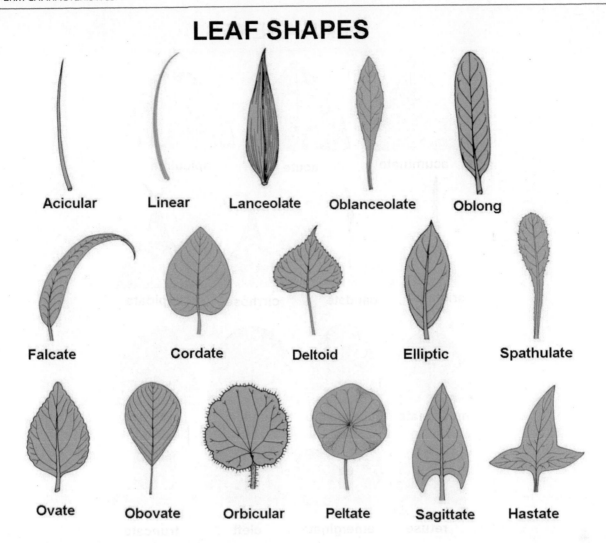

Acicular   Linear   Lanceolate   Oblanceolate   Oblong

Falcate   Cordate   Deltoid   Elliptic   Spathulate

Ovate   Obovate   Orbicular   Peltate   Sagittate   Hastate

LEAF APEXES

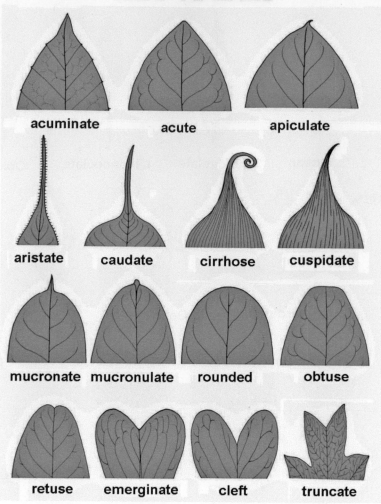

LEAF BASES

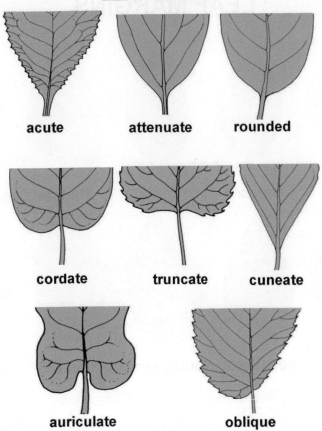

LEAF MARGINS

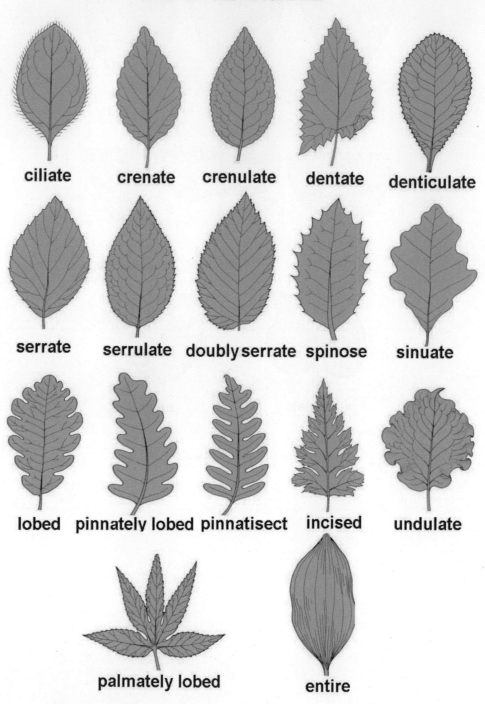

ciliate   crenate   crenulate   dentate   denticulate

serrate   serrulate   doubly serrate   spinose   sinuate

lobed   pinnately lobed   pinnatisect   incised   undulate

palmately lobed   entire

LEAF ATTACHMENT

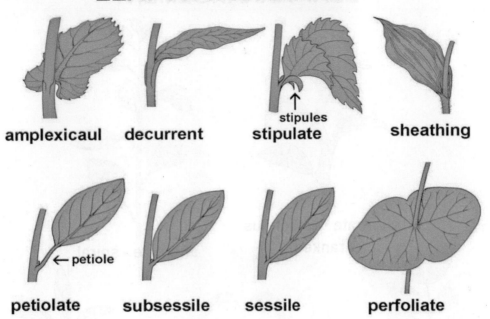

amplexicaul   decurrent   stipules stipulate   sheathing

← petiole

petiolate   subsessile   sessile   perfoliate

LEAF VENATION

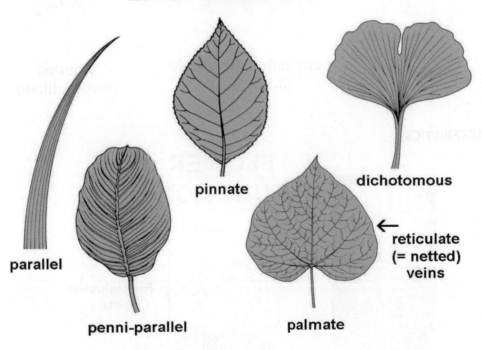

parallel

penni-parallel

pinnate

palmate

dichotomous

← reticulate (= netted) veins

LEAF ARRANGEMENT

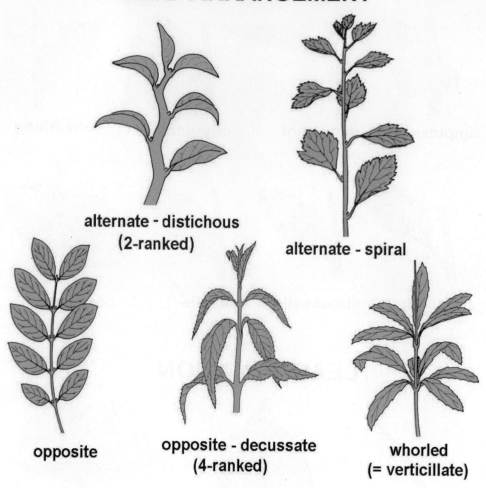

alternate - distichous
(2-ranked)

alternate - spiral

opposite

opposite - decussate
(4-ranked)

whorled
(= verticillate)

FLORAL CHARACTERISTICS

FLOWER FUNCTION

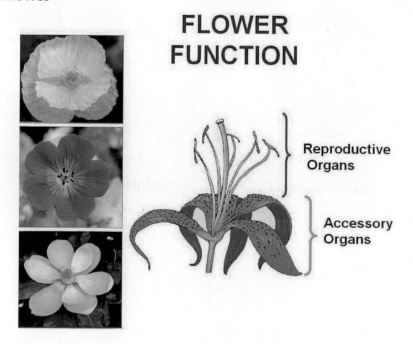

Reproductive Organs

Accessory Organs

FLOWER STRUCTURE

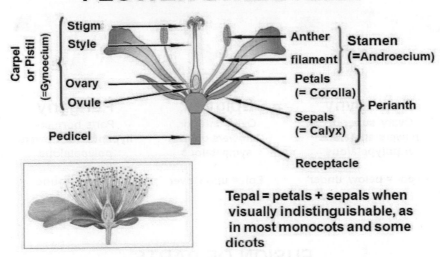

Tepal = petals + sepals when visually indistinguishable, as in most monocots and some dicots

Flower on the left is intended to illustrate a typical complete flower used as model for the above, not to illustrate tepals.

FLOWER STRUCTURE

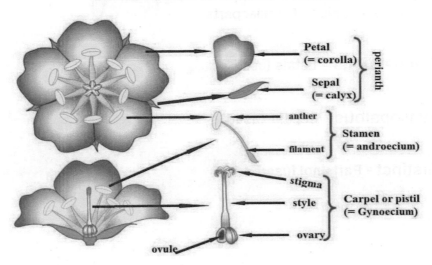

ARANGEMENT OF PARTS

Hypogyny
**Ovary superior,
flowers apetalous
or polypetalous**

Hypo- = below, under

Epigyny
**Ovary inferior,
flowers often
sympetalous**

Epi- = upon, over

Perigyny
**Parts born on
hypanthium, flowers
polypetalous**

Peri- = around

FUSION OF PARTS

Adnate - **Fusion of dissimilar parts**

Connate - **Fusion of similar parts**

Sympetalous - **Petals fused**

Synsepalous - **Sepals fused**

Distinct - **Parts not fused**

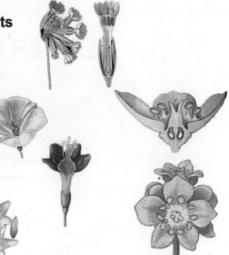

PRESENCE / ABSENCE OF PARTS

Complete - Sepals, petals, stamens, and carpel are present

Incomplete - One or more of the four organs missing

Perfect - Both reproductive organs present

Imperfect - One or both reproductive organs absent

Apetalous - Petals lacking

Asepalous - Sepals lacking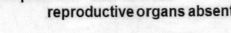

FLOWER SHAPES

Parts free   **Rotate**   **Tubular**

 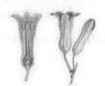

Campanulate   **Papilionaceous**   **Urceolate**

Funnelform   **Salverform**   **Bilabiate**

FLOWER CHARACTERISTICS: GENDER

Paeonia "Sweet May"

Magnolia grandiflora

Bisexual
(Hermaphroditic plants)

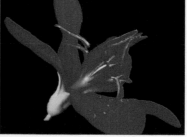

Jatropha weddeliana

pistillate   **staminate**
(female)   **(male)**

Unisexual
[monoecious (unisexual) or dioecious plants]

INFLORESCENCE STRUCTURE

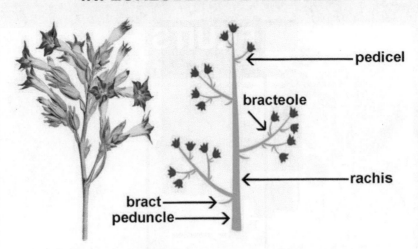

Definition: the arrangement of flower(s) and associated parts on an axis.

INFLORESCENCE TYPES

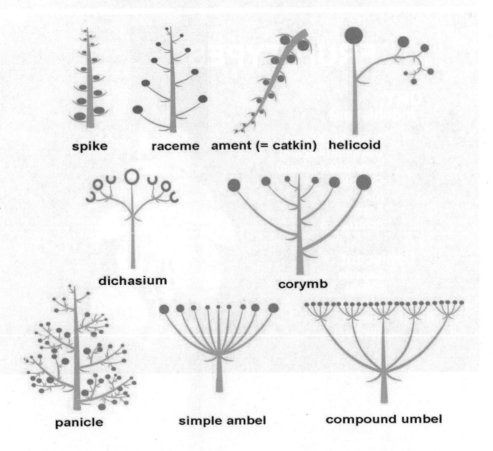

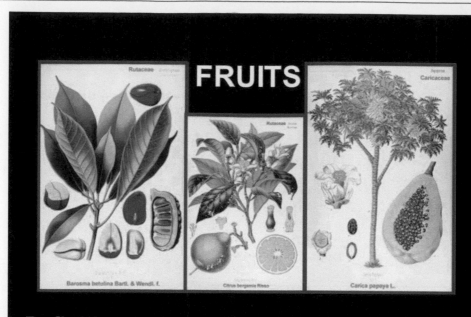

FRUITS

Definition: In angiosperms, seed-bearing organ or mature ovary of angiosperms, with or without adjoining parts

FRUIT TYPES

DRY:
Indehiscent

Achene (= Akene)
Calybium (= Acorn)
Caryopsis (= Grain)
Cypsella
Nut
Samara
Schizocarp

Dehiscent
Capsule
Follicle
Legume
Loment
Silique

FRUIT TYPES

DRY:

Indehiscent
Achene (= Akene)
Calybium (= Acorn)
Caryopsis (= Grain)
Nut
Samara
Schizocarp

Dehiscent
Capsule
Follicle
Legume
Loment
Silique

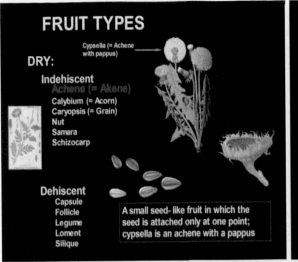

Cypsella (= Achene with pappus)

A small seed- like fruit in which the seed is attached only at one point; cypsella is an achene with a pappus

FRUIT TYPES

DRY:

Indehiscent
Achene (= Akene)
Calybium (= Acorn)
Caryopsis (= Grain)
Cypsella
Nut
Samara
Schizocarp

Dehiscent
Capsule
Follicle
Legume
Loment
Silique

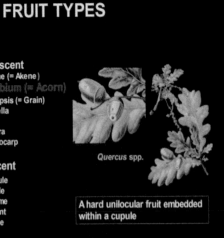

Quercus spp.

A hard unilocular fruit embedded within a cupule

FRUIT TYPES

DRY:

Indehiscent
Achene (= Akene)
Calybium (= Acorn)
Caryopsis (= Grain)
Cypsella
Nut
Samara
Schizocarp

Dehiscent
Capsule
Follicle
Legume
Loment
Silique

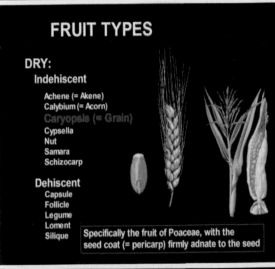

Specifically the fruit of Poaceae, with the seed coat (= pericarp) firmly adnate to the seed

FRUIT TYPES

DRY:

Indehiscent
Achene (= Akene)
Calybium (= Acorn)
Caryopsis (= Grain)
Cypsella
Nut
Samara
Schizocarp

Dehiscent
Capsule
Follicle
Legume
Loment
Silique

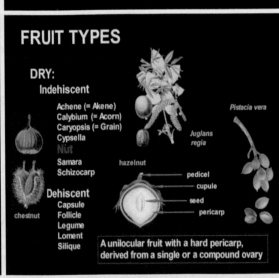

Pistacia vera
Juglans regia
hazelnut
chestnut
pedicel
cupule
seed
pericarp

A unilocular fruit with a hard pericarp, derived from a single or a compound ovary

FRUIT TYPES

DRY:

Indehiscent
Achene (= Akene)
Calybium (= Acorn)
Caryopsis (= Grain)
Cypsella
Nut
Samara
Schizocarp

Dehiscent
Capsule
Follicle
Legume
Loment
Silique

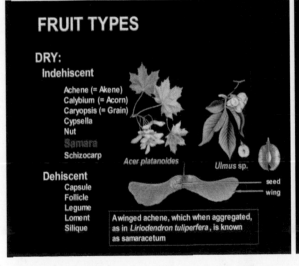

Acer platanoides
Ulmus sp.
seed
wing

A winged achene, which when aggregated, as in *Liriodendron tuliperfera*, is known as samaracetum

FRUIT TYPES

DRY:

Indehiscent
Achene (= Akene)
Calybium (= Acorn)
Caryopsis (= Grain)
Cypsella
Nut
Samara
Schizocarp

Dehiscent
Capsule
Follicle
Legume
Loment
Silique

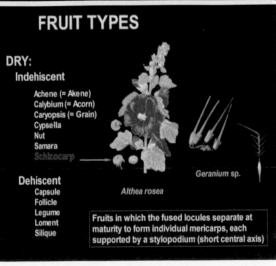

Althea rosea
Geranium sp.

Fruits in which the fused locules separate at maturity to form individual mericarps, each supported by a stylopodium (short central axis)

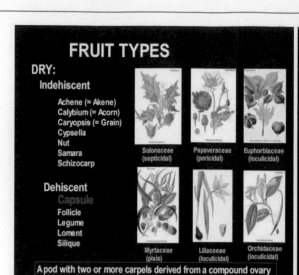

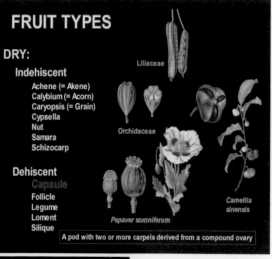

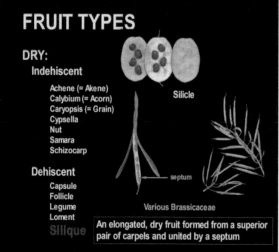

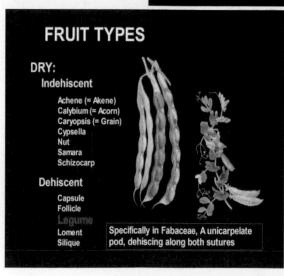

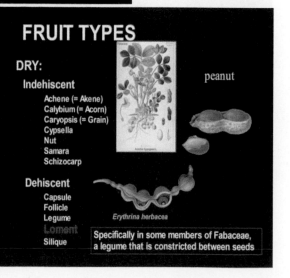

FRUIT TYPES

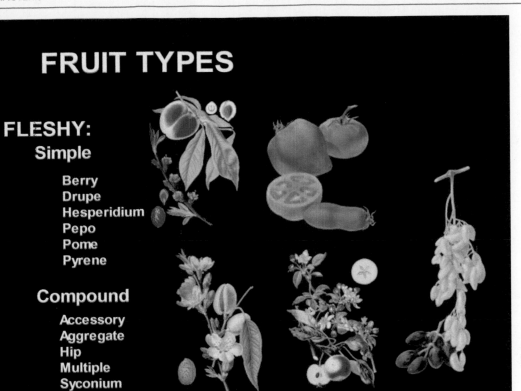

FLESHY:

Simple

Berry
Drupe
Hesperidium
Pepo
Pome
Pyrene

Compound

Accessory
Aggregate
Hip
Multiple
Syconium

FRUIT TYPES

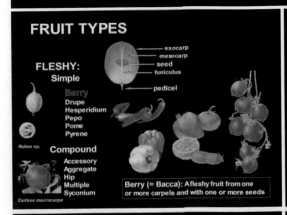

FLESHY:
Simple

Berry
Drupe
Hesperidium
Pepo
Pome
Pyrene

Rubus sp.

Compound

Accessory
Aggregate
Hip
Multiple
Syconium

Carissa macrocarpa

exocarp
mesocarp
seed
funiculus
pedicel

Berry (= Bacca): A fleshy fruit from one or more carpels and with one or more seeds

FRUIT TYPES

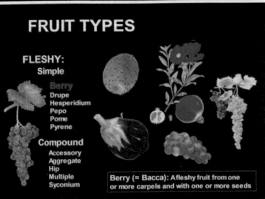

FLESHY:
Simple

Berry
Drupe
Hesperidium
Pepo
Pome
Pyrene

Compound

Accessory
Aggregate
Hip
Multiple
Syconium

Berry (= Bacca): A fleshy fruit from one or more carpels and with one or more seeds

FRUIT TYPES

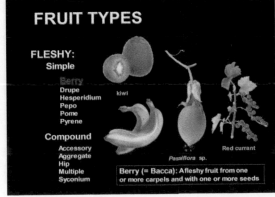

FLESHY:
Simple

Berry
Drupe
Hesperidium
Pepo
Pome
Pyrene

kiwi

Compound

Accessory
Aggregate
Hip
Multiple
Syconium

Passiflora sp.

Red currant

Berry (= Bacca): A fleshy fruit from one or more carpels and with one or more seeds

FRUIT TYPES

FLESHY:
Simple

Berry
Drupe
Hesperidium
Pepo
Pome

Compound

Accessory
Aggregate
Hip
Multiple
Syconium

A stone fruit; fleshy fruit with a single seed is enclosed within a stony endocarp

FRUIT TYPES

FLESHY:
Simple

   Berry
   Drupe
   Hesperidium
   Pepo
   Pome

Compound

   Accessory
   Aggregate
   Hip
   Multiple
   Syconium

Coffee

Dates

A stone fruit; fleshy fruit with a single seed enclosed within a stony endocarp

---

FRUIT TYPES

FLESHY:
Simple

   Berry
   Drupe
   Hesperidium
   Pepo
   Pome

Compound

   Accessory
   Aggregate
   Hip
   Multiple
   Syconium

Hedera helix

Olea europea

Nyssa ogiche

A stone fruit; fleshy fruit with a single seed enclosed within a stony endocarp

---

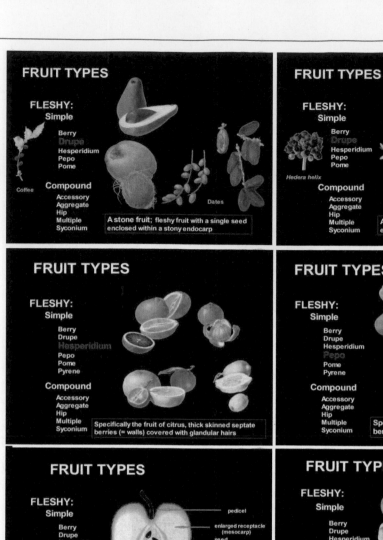

FRUIT TYPES

FLESHY:
Simple

   Berry
   Drupe
   Hesperidium
   Pepo
   Pome
   Pyrene

Compound

   Accessory
   Aggregate
   Hip
   Multiple
   Syconium

Specifically the fruit of citrus, thick skinned septate berries (= walls) covered with glandular hairs

---

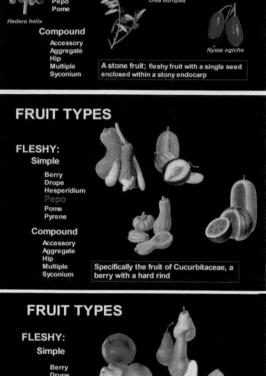

FRUIT TYPES

FLESHY:
Simple

   Berry
   Drupe
   Hesperidium
   Pepo
   Pome
   Pyrene

Compound

   Accessory
   Aggregate
   Hip
   Multiple
   Syconium

Specifically the fruit of Cucurbitaceae, a berry with a hard rind

---

FRUIT TYPES

FLESHY:
Simple

   Berry
   Drupe
   Hesperidium
   Pepo
   Pome
   Pyrene

Compound

   Accessory
   Aggregate
   Hip
   Multiple
   Syconium

pedicel
enlarged receptacle (mesocarp)
seed
remains of ovary vascular bundles
endocarp
remnants of the perianth and stamens

Fruits in which a major portion consists of tissue derived from other flower parts, such as the receptacle

---

FRUIT TYPES

FLESHY:
Simple

   Berry
   Drupe
   Hesperidium
   Pepo
   Pome
   Pyrene

Compound

   Accessory
   Aggregate
   Hip
   Multiple
   Syconium

Malus sp.   *Pyrus* sp.   *Cydonia oblonga*

Fruit in which the fleshy part is the enlarged axis (receptacle or disk) of the flower, rather than the ovary

---

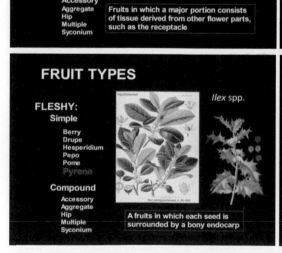

FRUIT TYPES

FLESHY:
Simple

   Berry
   Drupe
   Hesperidium
   Pepo
   Pome
   Pyrene

Compound

   Accessory
   Aggregate
   Hip
   Multiple
   Syconium

Ilex spp.

A fruits in which each seed is surrounded by a bony endocarp

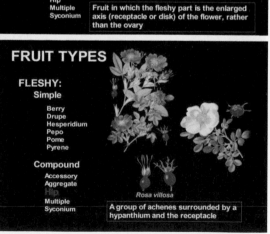

FRUIT TYPES

FLESHY:
Simple

   Berry
   Drupe
   Hesperidium
   Pepo
   Pome
   Pyrene

Compound

   Accessory
   Aggregate
   Hip
   Multiple
   Syconium

Rosa villosa

A group of achenes surrounded by a hypanthium and the receptacle

FRUIT TYPES

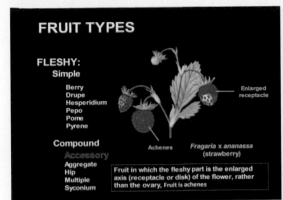

FLESHY:
Simple

Berry
Drupe
Hesperidium
Pepo
Pome
Pyrene

Compound
Accessory
Aggregate
Hip
Multiple
Syconium

Achenes

Enlarged receptacle

Fragaria x ananassa (strawberry)

Fruit in which the fleshy part is the enlarged axis (receptacle or disk) of the flower, rather than the ovary, Fruit is achenes

FRUIT TYPES

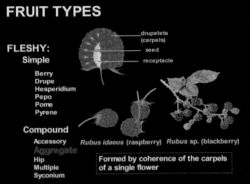

FLESHY:
Simple

Berry
Drupe
Hesperidium
Pepo
Pome
Pyrene

Compound
Accessory
Aggregate
Hip
Multiple
Syconium

drupelets (carpels)
seed
receptacle

Rubus idaeus (raspberry) *Rubus* sp. (blackberry)

Formed by coherence of the carpels of a single flower

FRUIT TYPES

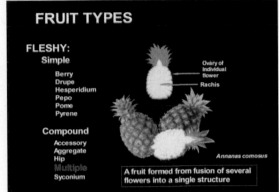

FLESHY:
Simple

Berry
Drupe
Hesperidium
Pepo
Pome
Pyrene

Compound
Accessory
Aggregate
Hip
Multiple
Syconium

Ovary of Individual flower
Rachis

Annanas comosus

A fruit formed from fusion of several flowers into a single structure

FRUIT TYPES

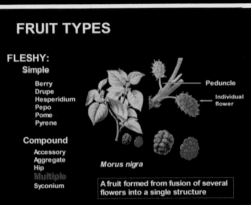

FLESHY:
Simple

Berry
Drupe
Hesperidium
Pepo
Pome
Pyrene

Compound
Accessory
Aggregate
Hip
Multiple
Syconium

Peduncle
Individual flower

Morus nigra

A fruit formed from fusion of several flowers into a single structure

FRUIT TYPES

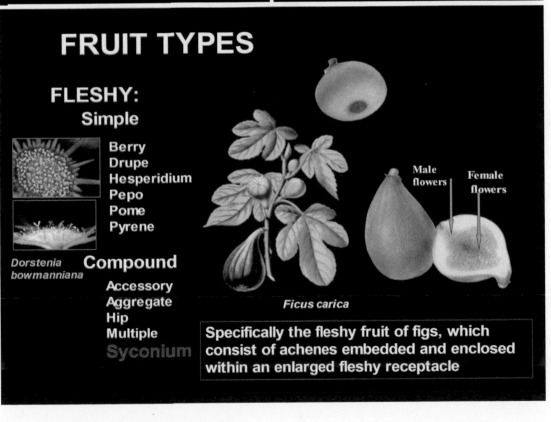

FLESHY:
Simple

Berry
Drupe
Hesperidium
Pepo
Pome
Pyrene

Dorstenia bowmanniana

Compound

Accessory
Aggregate
Hip
Multiple
Syconium

Male flowers
Female flowers

Ficus carica

Specifically the fleshy fruit of figs, which consist of achenes embedded and enclosed within an enlarged fleshy receptacle

MODIFIED STEMS: BULBS

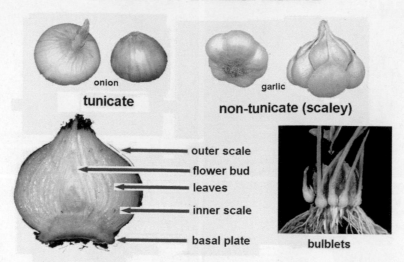

onion

tunicate

garlic

non-tunicate (scaley)

outer scale
flower bud
leaves
inner scale
basal plate

bulblets

MODIFIED STEMS: CORMS

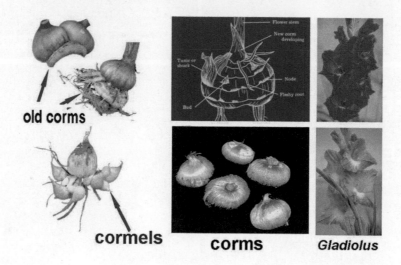

old corms

cormels

corms

Gladiolus

OTHER MODIFIED STEMS

pseudobulb   caudex   stolon

cladophyll   tuber   rhizome

MONOCOTYLEDONS (MONOCOTS)

ACORACEAE

SWEET-FLAG FAMILY

GEOGRAPHY East and Southeast Asia, naturalized in Europe and North America.

GROWTH HABIT Aromatic herbaceous perennial, with creeping rhizomes, much branched; roots on lower side of rhizome and leaves at apex. **LEAVES**: distichous, unifacial (oriented toward a single face), bases overlapping, ensiform (having sharp edges and a narrow point), not differentiated into petiole and blade; venation reticulate, with or without a central vein (species-dependent). **FLOWERS**: spadix cylindrical, lateral, densely covered with flowers, not enclosed in the spade; flowers bisexual; tepals 6, orbiculate, curved; stamen filaments linear, flat; ovary superior, conical, ovules numerous, pendulous, trichomes as well as ethereal oil cells on micropyle of the ovules. **FRUIT**: Berry with a lathery skin; seeds oblong, with perisperm (nutritive tissue derived from the nucellus and deposited outside of the embryo sac) and endosperm (perisperm is absent in Araceae).

ECONOMIC USES Ornamentals, medicinal.

CULTIVATED GENERA Monogeneric *Acorus* with 2 species: *A. gramineus* and *A. calamus*.

COMMENTS The genus *Acorus* was originally placed in Araceae, but recent phylogenetic treatments recognized it as a monogeneric family. Acoraceae was already established in 1820. Pieces of rhizome as well as seeds are dispersed by water along rivers and creeks. The rhizomes are used for a number of ailments, including chronic bronchitis, diarrhea, abdominal distention, cold, and a number of other health issues. Both species are grown in bog gardens; they differ in:

Acorus calamus has distinct midribs, stout rhizomes, seeds without long bristles, and foveolate testa (with small pits).
Acorus gramineus does not have distinct midribs, rhizomes are slender, seeds have long bristles, and testa is smooth.

ETYMOLOGY [From Latin *acorus*, derived from Greek *achorous*. Pupil of an eye, because the juice from rhizome is used as a remedy for eye disease] [From Greek *kalamos*, stalk or reed, used as pen] [similar to grasses]

Acorus gramineus 'Oborozuki' (grassy-leaved dwarf sweet flag)

Acorus calamus (sweet flag). Illustrations from Otto Wilhelm Thomé (1885).

ARACEAE

ARUM FAMILY

113 (144?) GENERA AND ±3300 (3645?) SPECIES

GEOGRAPHY Predominantly tropical with a few temperate.

GROWTH HABIT Mostly herbaceous, cauline or with subterranean tubers or rhizomes, many epiphytic vines and aquatic species; roots adventitious, usually unbranched, growing downward or in epiphytic species clasping trees for support. **LEAVES**: simple or compound, pinnately or palmately divided, basal or on stems, with penni-parallel or palmate venation, and often petiolate. **FLOWERS**: inflorescence scapose, bracts large (= spathe), flowers densely spicate (= spadix), small and numerous, stamens usually 6 but may be reduced to as few as 1 and perianth often lacking; unisexual or bisexual, with the male on the upper and the female on the lower part of the spadix which is situated in or enclosed within a petaloid, often brightly colored, conspicuous spathe. Flowers often have a disagreeable odor. **FRUIT**: berry, with 1-many seeds.

ECONOMIC USES The tuberous species are economically important as a source of starch in the tropics and subtropics: *Alocasia*, *Colocasia esculenta* (taro or dasheen), *Amorphophallus*, etc.

CULTIVATED GENERA *Aglaonema*, *Alocasia*, *Amorphophallus*, *Anthurium*, *Caladium*, *Colocasia*, *Dieffenbachia*, *Epipremnum*, *Monstera*, *Philodendron*, *Scindapsus*, *Spathiphyllum*, *Syngonium*, *Thaumatophyllum* (= *Meconostigma*), *Zamioculcas* and *Zantedeschia*, among others.

COMMENTS A relatively large family consisting of 106–113 genera and more than 3500–4205 cosmopolitan species. Most authorities, including APG IV (2016), recognize 8 subfamilies that were in fact recognized by some prior to DNA-based phylogenetic treatments. Subfamilies are not noted by Judd et al. (2015) and Soltis et al. (2018), and Christenhusz et al. (2017) recognized only four with nearly all of the cultivated genera included in **Aroideae**. Any discussion of subfamilies and tribes serves no useful purpose in the scope of this book. Suffice it to note that some of the subfamilies contain one to very few genera. Examples of representative genera and species in cultivation include: *Amydrium medium*, and its variegated cultivar are rare in cultivation; *Anubias barteri*, and *A. coffeefolia*, and its cultivars are attractive aquarium plants; *Arisarum proboscideum*, *A. vulgari* (friar's cowl), with "miniature cobra lily" arum type flowers, rhizomatous perennial herbs native to Mediterranean Europe; *Arum creticum*, *A. hygrophyllum*, *A. italicum* (Italian arum), *A. maculatum* (Adam and Eve), *A. orientale*, *A. palaestinum* (Salmon's lily, black calla lily), and *A. pictum* (black calla), and others are acaulescent herbs with oblong to obovate, long-petiolate leaves and broad, showy spathe, native to the Old World, used as garden plants; *Callopsis volkensii* is a monotypic genus with white spathe and yellow spadix, native to eastern Africa; *Cercestis mirabilis* (= *Rhektophyllum mirabilis*) is a herbaceous perennial climbing epiphyte with adhesive roots and attractive green leaves with silver markings, used for medicinal purposes, native to tropical Africa; *Cryptocoryne affinis* (native to Malayan Peninsula) , *C. balansae*, *C. ciliata*, *C. cordata*, *C. longicauda*, *C. lutea*, *C. nevillii*, *C. purpurea*, *C. spirallis*, and *C. willisii*, among others, are small rhizomatous and often stoloniferous, aquatic herbs with simple, long-petiolate leaves, native to Southeast Asia, commonly grown in aquaria; *Dracunculus medinensis* and *D. vulgaris* are tuberous perennials with large purple spathe and spadix that are often fetid, native to the Mediterranean Region and Canary Islands; *Helicodiceros muscivorus* (dead horse arum lily) is a monotypic thermogenic (flower with higher temperature) genus, with rotten meat smell that attracts pollinating flies, native to parts of the Mediterranean region; *Homalomena humilis*, *H. lindenii*, *H. polyandra*, *H. pygmaea*, *H. roelzii*, *H. rubescens*, and *H. wallisii* are herbs, often with short, erect stems, entire leaves with sheathing petioles, and green, white or red spathe, native to tropical Asia and America, grown as landscape plants in the tropics; *Holochlamys beccarii* is a monotypic genus, native to New Guinea, closely related to *Spathiphyllum*, but rare in cultivation; *Lagenandra dewitii*, *L. meeboldii*, *L. tanekasi*, and others are aquarium plants endemic to the Indian Subcontinent; *Lysichiton americanum* (skunk cabbage) and *L. camtschatcensis* are rhizomatous acaulescent herbs of swamps with leaves arising from a thick rhizome, and attractive inflorescences before the leaves, used as landscape plants in wet locations; *Nephthytis afzelii*, *N. hallaei*, and *N. swainei*, and a few others are evergreen rhizomatous perennials, native to western tropical Africa; *Orontium aquaticum* (golden club, floating arum) grows in ponds and streams, endemic to eastern United States; *Peltandra sagittifolia*, and *P. virginica* (green arrow arum) are herbaceous aquatic or marshes plants, native to Canada, Cuba, and eastern United States; *Rhaphidophora cryptantha*, *R. korthalsii*, *R. obesa*, *R. tenuis*, and *R. tetrasperma* (split-leafed tropical wonder) are evergreen usually terrestrial hemiepiphytic climbing plants native from tropical Africa, Malaysia, and Australia; *Rhodospatha* includes 20 Central and South American species but *R. dammeri* seems to be the only taxon in cultivation, listed by IUCN as threatened; *Sauromatum guttatum* and *S. venosum*, and a few others are commonly known as voodoo lily, another example where flowers heat up and emit strong foul odor, native to tropical Africa, Asia, and Arabian Peninsula; *Schismatoglottis* is a large genus similar to *Dieffenbachia* but with few species in cultivation: *S. asperata*, *S. gangsta*, *S. prietoi*, and *S. roseopatha*, sometimes used in the landscape, native to tropical Southeast Asia; *Spathicarpa gardneri*, *S. hastifolia*, and *S. lanceolata* are unusual taxa in having their spadix fused to the spathe, resembling a double stripe of small flowers attached to a leaf; these are endemic to South America; *Stenospermation arborescens*, *S. brachypodium*, *S. gracile*, *S. hilligii*, and *S. interruptum*, *S. peripense*, and *S. subellipticum*, among others are epiphytic, hemiepiphytic, or terrestrial herbs native to Central and South America; *Steudnera colocaiaefolia*, *S. kerrii*, and a few others are evergreen erect or creeping stems and uncommonly attractive inflorescences that are yellow, yellow and red, or dark purple within, native to rainforests of southern China, the Himalayas, and Indochina, infrequently seen in cultivation; *Symplocarpus foetidus* (skunk cabbage), *S. renifolius*, and few others are rhizomatous perennials with deep contractile roots and fetid leaves when crushed, native to United States, Canada, and eastern Asia; *Taccarum caudatum*, *T. peregrinum*, *T. warmingii*, and *T. weddellianum* are seasonally dormant tuberous perennials that develop a single leaf and solitary inflorescence, native to open tropical woodlands of South America; *Xanthosoma atrovirens*, *X. daguense*, *X. lindenii* (Indian kale), *X. sagitifolium*, *X. violaceum* (blue taro), and others are usually large caulescent or acaulescent, rhizomatous or tuberous herbs with entire, sagittate or hastate, or pedately dissected, succulent, long-petiolate leaves, native to tropical America, cultivated primarily in the tropics for their edible tubers, but also occasionally as landscape plants.

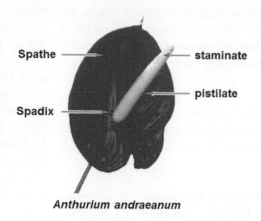

Anthurium andraeanum

Aglaonema commutatum

AGLAONEMA; POISON DART PLANT
SYNONYM *Aglonema marantifoium* var. *commutatum*

ETYMOLOGY [From Greek *aglaos*, bright, clear; *nēma*, a thread (bright thread), in reference to the stamens] [Change or changing, often referring to a closely related species].

GROWTH HABIT Monoecious, evergreen herbaceous perennial shrub; round, with leaves arranged at stem ends; medium texture; 3–5 ft. (1–1 ½ m) tall with variable spread; coarse texture; medium growth rate. **STEM/BARK**: erect, round, green, or variegated stems. **LEAVES**: simple, oblong-elliptic to lanceolate; usually less than 12 in. (30 cm) long and 4 in. (10 cm) wide; obtuse to subcordate; dark glossy green, variously marked gray-green but in newer cultivars red or yellow along primary veins; petioles are equal to or shorter than blades, sheathing. **FLOWERS**: pale green spathe 1 ½–3 ½ in. (4–10 cm) long with whitish shorter spadix. Unisexual, with inconspicuous male flowers at top of spadix; females, below. Blooms appear on peduncles, to 6 in. (15 cm) long, during warm months. **FRUIT**: brown, glossy, oblong, to 2 in. (5 cm) wide; 1–3; turning yellow, then bright red in late summer; often hidden by leaves; seeds large covered by a thin layered exocarp. **NATIVE HABITAT**: Philippines, northeastern Celebes, in tropical and subtropical humid regions as understory shade plants.

CULTURE Low light preferred (leaves burn in direct sun), in fertile well-drained but moist nematode-free growing medium; no salt tolerance; nematodes, Pythium root rot under some conditions, and mites could present problems. **PROPAGATION**: cuttings, division, and air layers. **LANDSCAPE USES**: outstanding for areas of reduced light, may be used in north side foundation or as container specimens indoors. **HARDINESS ZONE**: 11A–13B. No tolerance to cold.

COMMENTS Numerous cultivars of diverse colors with confusing and often multiple names have been introduced. There are extreme and often minor color variations also present. Enumeration of cultivar names in this case would add even more confusion, particularly since almost without exception parental species names are not noted. Cultivar names listed under figures may be potentially misrepresentations of the names. A total of 24 *Aglaonema* spp. are known of which in addition to *Aglaonema commutatum*, other species in the trade include: *A. costatum* (spotted evergreen), *A. crispum*, *A. marantifolium*, *A. modestum*, which are rela-

tively common in the trade and may be involved in hybridization and selection of cultivars, *A. nitidum*, *A. pictum*, *A. simplex*, and others are herbs with oblong to ovate, petiolate, mostly coriaceous, often spotted leaves, native to tropical Asia. These are among the most popular house plants because of their tolerance of low light conditions. Leaves and particularly stems of *Aglaonema* contain calcium oxalate that poses danger to children and animals.

Aglaonema commutatum (above, left), *A. commutatum* **'Pride of Sumatra'** (above right), *A. commutatum* **'Siam Aurora'** (middle left; may have originated from *A. modestum*), *A. commutatum* **'Red Sumatra'** (middle right), *A. commutatum* **'Red Cochin'** (below left), *A. costatum* (below right) – sometimes listed as *A. commutatum*.

Alocasia spp.

ELEPHANT'S EAR

ETYMOLOGY [Without *Colocasia*, an allied genus from which it was separated].

GROWTH HABIT Tuberous or rhizomatous perennial shrubs; irregular form; coarse texture; varies with species; to 15 ft. (5 m) tall with a variable spread; some are vigorous growers. **STEM/BARK**: in addition to rhizomes or corms in a few species, a short stem-like segment is formed by leaf bases. **LEAVES**: simple; entire to pinnatifid, often cordate or with basal lobes (sagittate); frequently marked or colored; blades as long as 12–35 in. (30–90 cm); petioles cylindrical to 3 $\frac{1}{2}$ ft. (1 $\frac{1}{3}$ m) long; size varies greatly among species. **FLOWERS**: greenish spathe and spadix; monoecious; both male and female flowers occupy the same spadix. Blooms during warm months. **FRUIT**: fleshy berries, infrequent. **NATIVE HABITAT**: from tropical and subtropical Asia to eastern Australia.

CULTURE Full sun or broken shifting shade, depending on the species or cultivar, in moisture-retentive, rich organic soil; no salt tolerance; spider mites and soil-borne fungus diseases may be problematic. **PROPAGATION**: Division of underground storage organs, from stem pieces, and seeds. **LANDSCAPE USES**: outdoors as specimen or accent plants to enhance the feeling of the tropics, indoors, smaller fancy-leaved types serve as container specimens. **HARDINESS ZONE**: 8A–13B; Cold tolerance varies with species but generally not extended cold-tolerant.

COMMENTS About 80 species are listed of which many are in cultivation as are numerous cultivars: *Alocasia cucullata*, *A. macrorrhiza* and its cultivars '**Lutea**', '**New Guinea Gold**', (Taro, Poi), and the closely related *Colocasia esculenta* (Dasheen) are cultivated for their edible rhizome. Several species including *Alocasia × mortfontanensis* (*Alocasia × A. amazonica*, Amazonian elephant ear), and its cultivars '**Compacta**', '**Polly**', and '**Purpley**', as well as *A. cucullata* (Chinese Taro), *A. advincula*, *A. cadieri*, *A.*

cuprea, *A. imperialis*, *A. longiloba*, *A. odora* and its cultivar '**California**', *A. plumbea*, and its cultivar '**Nigra**', *A. ringula* and its cultivar '**Black Velvet**', *A. sanderiana*, and *A. veitchii* and others are listed as ornamentals. A few notable cultivars include '**Ellaine**', '**Frydek**', '**Hilo Beauty**', '**Porfora**', '**Portodora**', and '**Igrina Superba**' and are among others listed. These are erect rhizomatous herbs with entire to pinnatifid, often strikingly marked or colored, long-petiolate leaves and without perianth but green spathe, native to tropical Asia. Species of *Alocasia* that are commonly grown as greenhouse subjects are usually of hybrid origin. Similar to other members of Araceae, species of *Alocasia* contain calcium oxalate which may cause serious skin irritation and breathing difficulty if chewed.

Note There is an inherent difficulty in distinguishing the genera *Alocasia*, *Colocasia*, and *Xanthosoma*. A simplified explanation is that of leaf shapes and petiolar attachments: Leaves of *Alocasia* and *Colocasia* are both primarily cordate (heart-shaped), but they differ in the mode of attachment of the petiole to the blade; that of *Alocasia* is attached at the base (the V junction) of the blade and seemingly continue through it, resulting in an upright or horizontal position while that of *Colocasia* is attached beneath and often closer to the center of the blade, hence more or less peltate, resulting in blade appearing in a drooping or nearly vertical position. Leaves of *Xanthosoma* are sagittate (arrow-shaped) but also nearly, though not quite peltate. There is also a noticeable difference in characteristics of their corms: *Alocasia* has relatively thin and long corm and without banding while that of *Colocasia* is large, swollen, and banded in contrast to the much smaller corm of *Xanthosoma*. There are of course other technical differences in inflorescences and flowers which are actually the primary reasons for their recognition as separate genera. Having noted the differences, however, their visual distinction remains somewhat problematic.

Colocasia esculenta (taro, elephant's-ear) and its many cultivars, *C. fallax*, *C. gigantea*, and few others, are large tuberous herbs with ovate, peltate, long-petiolate leaves, native to tropical Asia. These are widely grown for their edible tubers and as landscape plants in warm, usually wet regions.

Alocasia macrorrhiza

Alocasia micholitziana

Alocasia macrorrhiza **'Variegata'** (upright elephant ear)

Colocasia esculenta

Alocasia sanderiana (kris plant), native to the Philippines

Xanthosoma sagittifolium 'Lime Ginger'

Xanthosoma nigrum

Amorphophallus paeoniifolius

STINK LILY; ELEPHANT FOOT YAM
SYNONYM *Amorphophallus campanulatus*

ETYMOLOGY [From Greek *amorphous*, shapeless or deformed; *phallus*, penis, in reference to the reproductive structure of the flower] [Peony-like foliage]

GROWTH HABIT Monoecious deciduous herb. STEM/ BARK: a large corm that weighs as much as 18 lb (8 kg). LEAVES: significantly large single attractive leaf of up to 6 ft. (2 m) tall and 3 ft. (1 m) wide, with numerous leaflets, arises after flowering; the fleshy petiole may be about 5 in. (13 cm) in diameter and noticeably marked with darker green spots. FLOWERS: large inflorescence with red-purple spathe (bract) and a cylindrical spadix that is covered by a purple hood, to which are attached many small male and female flowers. Flowers of *Amorphophallus* species are pollinated principally by carrion beetles which are attracted to them because of their foul odor. Fruit: red ovoid-subglobose berries. NATIVE HABITAT: tropical seasonally dry areas from India, to Southeast Asia, and Australia.

CULTURE Sunny but sheltered humid location in well-drained but rich nematode-free soil. The soil should be kept moist during active growth period, but allowed to dry out once plant begins dormancy. PROPAGATION: seeds and corm offsets (cormels). LANDSCAPE USES: may be planted in ground or in containers where in either case it represents an impressive specimen, though regretfully only for a relatively short time period. HARDINESS ZONE: 8B–13B. Corm does not tolerate cold temperatures.

COMMENTS The corm of this plant is cultivated in several tropical counties as food crop and sometimes as medicine for a number of ailments. Other species in cultivation include *Amorphophallus bulbifer* (voodoo lily, devil's-tongue), *A. campanulatus* (Telingo potato), *A. konjac* (also known as voodoo lily, used for medicinal purposes, particularly for weight loss), *A. mulleri*, *A. rivieri* (leopard palm), and *A. titanum* (Titan arum, well-known species because of its very large fetid flower), and probably others in botanical and private collections. These are aculescent herbs, often with large corms, large bipinnatifid, 3-parted, spotted leaves with a stout petiole, and a large, foul-smelling inflorescence which appear before the leaves, native to the Old-World tropics. Grown mostly for curiosity in greenhouses or outdoors in warmer climates.

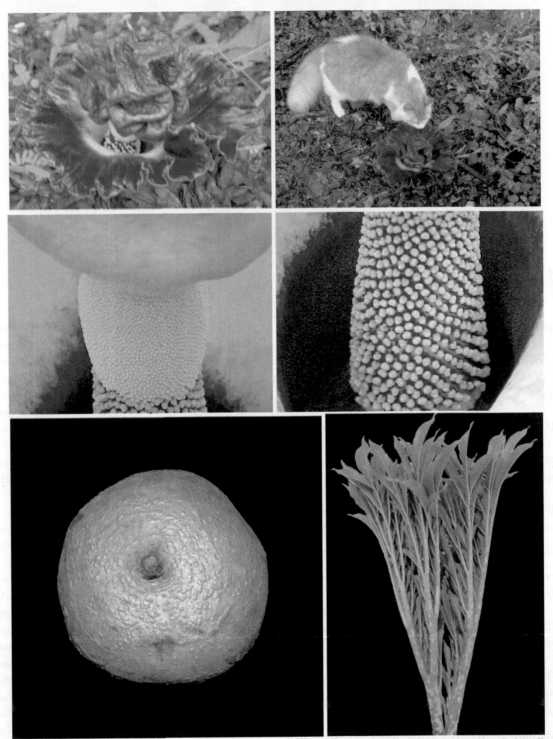

Amorphophallus paeoniifolius flower (above, left), cat seeking the fetid smell of the flower (above, right), male (middle left) and female (middle right) flowers, corm (below, left), and developing single large leaf (below, right).

Amorphophallus muelleri
Photographed at US National
Conservatory, Washington D. C.

Amorphophallus titanium
Photographed at Fairchild Tropical
Garden, Florida

Anthurium andraeanum

TAIL FLOWER, FLAMINGO FLOWER
SYNONYMS *Anthurium andraearum* var. *divergens*, *A. venustum*

ETYMOLOGY [From Greek *anthos*, flower and *oura*, a tail, in reference to the tail-like spadix of the flower] [After Édouard François André (1840–1911), well-known French landscape architect and park designer].

GROWTH HABIT Monoecious, epiphytic evergreen herbaceous perennial; compact, round, upright; to 3 ft. (1 m) tall with a 4 ft. (1 $\frac{1}{2}$ m) spread; medium-coarse texture; slow growth rate. **STEM/BARK**: essentially stemless but sometimes leaf bases of older plants create a thicket. **LEAVES**: simple, medium green, reflexed, ovate, and sagittate; to 10 in. (25 cm) long and half as wide; petioles somewhat longer; primary lateral veins are mostly connected by a well-defined vein running along inside and parallel to the margin. **FLOWERS**: bisexual, spathe, and spadix on peduncle 15–20 in. (35–50 cm) long; spathe is spreading, orbicular-ovate, 3–5 in. (7 $\frac{1}{2}$ –12 $\frac{1}{2}$ cm) long, cordate, puckered, polished, and salmon-red; basal lobes are semicircular, sometimes shortly united; spadix is sessile, recurved, to 2 $\frac{1}{2}$ in. (7 cm) long, golden with an ivory zone, and densely covered with inconspicuous flowers. **FRUIT**: fleshy red berries. **NATIVE HABITAT**: Colombia, Ecuador.

CULTURE low light conditions but will tolerate bright diffused light; high humidity is necessary for flowering; in very well-drained but moist, rich, and high in organic matter; no salt tolerance; may require repotting and division; mites, nematodes, scales, mealy-bugs, and grasshoppers are potential problems. **PROPAGATION**: seeds, cuttings, division, and tissue culture. **LANDSCAPE USES**: important cut flower greenhouse crop often imported from Central America; specimen in frost free shady outdoor spots; accent plant in pots and interiorscapes. **HARDINESS ZONE**: 10B–13B. No tolerance to low temperatures.

COMMENTS There are nearly 900 *Anthurium* species, including terrestrial and epiphytic ones, by far the largest genus in the family and important floriculture, foliage, and garden plants. This species is one of the parents of a group of hybrids with large, showy spathes, from dark-red to red, salmon, pink, white, yellow, and yellowish-green, that are often incorrectly referred to as varieties of the species. A few listed cultivars include: '**Julia**', '**Kohara Double**', '**Lady Ann**', '**Lady Beth**' '**Lady Jane**', '**Mary Jane**', '**Orange Hot**', '**Pink Frost**', '**Royal Flush**', '**Southern Blush**', '**White Frost**', and others. *Anthurium scherzerianum* (flamingo flower), and its cultivars of hybrid origin are preferable to *Anthurium* × *hortulanum*. Examples of other species in the trade include: *A. barclayanum*, *A. clarinervium* (heart-shaped, velvety dark leaves with distinct white venation pattern), *A. clavigerum*, *A. cordifolium*, *A. crenatum*, *A. crystallinum* (crystal anthurium), *A. digitatum*, *A. fraternum* (?), *A. hookeri* and its cultivars '**Alicia**' and '**Ruffles**',

A. palmatum, *A. sagittatum*, *A. sellowianum* (= *A. longilaminatum*, actually a *Philodendron*), *A. veitchii*, and *A. warocqueanum* (queen anthurium), among many others. The Bird's Nest Anthuriums are sold for outdoor landscape use in warmer climates. There are also a number of interspecific hybrids, such as **Anthurium × carneum** (*A. andraeanum × A. nymphaeifolium*), **A. × chelsiense** (*A. andraeanum × A. veitchii*), **A × cultorum** (a complex hybrid group), **A. × roseum** (*A. andraeanum × A. lindenianum*). These are epiphytic and terrestrial plants with short or long stems, mostly large, glabrous, petiolate, often sagittate leaves with the primary parallel veins to the margin and reticulate secondary veins, and flowers with stiff, persistent, often showy white to deep red spathe, native to tropical America.

Anthurium andraeanum and cultivars

Anthurium offersianum (above left), ***A. scherzerianum*** fruit (above right)**, *A. trinerve*** (middle left), ***A.* 'Florino'** fruit, (middle right), ***A.* 'Small Talk'** (below left), and ***A. clarinervium*** (below right)

Anthurium sp. in natural habitat, Bolivia   **Anthurium 'Kozohara'**

Anthurium superbum (left), photographed at Longwood Botanical Gardens, Pennsylvania;
Anthurium cupulispathum (right), photographed at Huntington Botanical Gardens, California.

Arisaema ringens

JAPANESE COBRA LILY
SYNONYMS *Alocasia ringens, Arisaema arisanensis, A. glaucescens, A. praecax, A. serotinum, A. sieboldii, A. sierotium, A. taihokense, Arum ringens, Ringentiarum glaucescens, R. ringens*

ETYMOLOGY From Greek *aris* or aron, arum and *aima*, blood, in reference to being akin to arums or in allusion to the red blotches on leaves of some species] [Gaping or open, in reference to the open mouth of the tubular spathe]

GROWTH HABIT Herbaceous tuberous deciduous perennial, 12–15 in. (25–30 cm) tall and 6–10 in. (15–20 cm) wide. **STEM/BARK**: single stalk that produces a pair of leaves (bifoliate). **LEAVES**: trifoliolate (each leaf with 3 leaflets), with glossy green ovate to elliptic leaflets, 6–8 in. (15–20 cm) long and less wide. **FLOWERS**: unisexual; green with purple interior and striped hood of a tubular spathe on the outside and yellow and white spadix on the inside; staminate and pistillate on same or different spadix; petals absent. **FRUIT**: cluster of bright red berries. **NATIVE HABITAT**: southern Japan, Southern Korea, and Eastern China.

CULTURE Part shade but tolerates full sun, in well-drained but moist organic soil; plant goes dormant during summer months at which time soil should be kept on the dry side. **PROPAGATION**: seed (slow to reach maturity and bloom) or tuber offsets. **LANDSCAPE USES**: in groups or against walled borders in moist locations. **HARDINESS ZONE**: 7A–10A.

COMMENTS A genus of about 180 species including the American natives *Arisaema triphyllum* (jack in the pulpit) which also has a pair of trifoliolate leaves, while *Arisaema dracontium* (jack in the pulpit) has a single leaf with 7–13 leaflets. Species in cultivation include *A. formosanum, A. serratum, A. sikokianum,* and *A. taiwanense,* and probably others in botanical and private collections. Tuber and roots of *Arisaema* contain calcium oxalate which can cause serious health problems if eaten and best to handle with gloves to avoid skin irritation.

Arisaema ringens (Japanese cobra lily), with a pair of trifoliolate leaves (photographed at US National Arboretum)

Arisaema dracontium (Jack-in-the-pulpit), with a solitary palmately compound (ternate) leaf

Caladium bicolor

FANCY LEAFED CALADIUM; ANGEL WINGS

SYNONYM *Caladium* × *hortulanum*, *Alocasia rex*, *A. roezlii*, *Arum pulcheri*, and many more, consisting primarily of varieties and forms.

ETYMOLOGY [Latinized for of the Malay plant name *kaladi*] [Two colors].

GROWTH HABIT Monoecious tuberous perennial herb; ground cover; upright, spreading; from 8–36 in. (20–115 cm) tall, more typically 12 in. (30 cm), with similar spread but variable with cultivar; coarse texture; rapid growth rate. **STEM/BARK**: tuber Irregular, dark brown, otherwise stemless. **LEAVES**: simple, basal, sometimes peltate; ovate to cordate in outline or mostly sagittate; to 14 in. (30 cm) long; basally truncate; flat, undulate, or ruffled; variously variegated red, rose, white, green, pink, and often with colored veins or spots; entire margins; petioles as long as or several times longer than the blades. **FLOWERS**: unisexual, spathe white, pink, or green; spadix with male flowers on top and females below. **FRUIT**: berries, densely packed along spadix. **NATIVE HABITAT**: tropical America; hybrid origin.

CULTURE Partial shade but can tolerate full sun for short periods, in organic well-drained but moist soils; no salt tolerance; essentially trouble-free but mealy bugs, caterpillars, and fungal diseases may sometimes cause problems. **PROPAGATION**: seeds, division of tubers. **LANDSCAPE USES**: bedding plant for foliage, planter box, edging, potted plants. **HARDINESS ZONE**: 8B–12B. Tubers may be left in ground throughout the year, although digging is recommended in colder areas each year before frost.

COMMENTS Until recently, hybrids and cultivars of caladium were recognized under the name *Caladium* × *hortulanum*. The two main groups included "fancy leaf" (= *C. bicolor* subsp. *pictoratum*) and "strap leaf" (= *C. bicolor* subsp. *marmoratum*). Within each of the two groups, numerous cultivars are included. In addition to the many cultivars, other species such as *C. humboldtii*, *C. lindenii* and its cultivar 'Magnificum', *C. mormoratum*, *C. picturatum*, *C. schomburgkii*, *C. sagittifolium*, and *C. zamifolium* are in the trade. These and several new hybrid cultivars are tuberous acaulescent herbs with basal, long-petiolate, variously colored and marked leaves, native to tropical America. They are popular bedding and potted subjects.

Caladium bicolor cultivars in the landscape

Representative cultivars of **_Caladium bicolor_** and a typical tuber (below, right)

Caladium lindenii (= *Xanthosoma lindenii*)

Dieffenbachia spp.

DUMB CANES

ETYMOLOGY [In honor of Joseph F. Dieffenbach (1796–1863), who was in charge of the gardens of the Royal Palace of Schönbrunn, in Vienna] [species].

GROWTH HABIT Monoecious, herbaceous perennials; ascending branches; to 8 ft. (2 $\frac{1}{2}$ m) in height with a 5 ft. (1 $\frac{1}{2}$ m) spread; coarse texture; moderately rapid growth rate. **STEM/BARK**: typically, straight and stout, unbranched, bright green, and ringed with leaf scars. **LEAVES**: simple, alternate but mostly clustered at stem apex; elliptic to oblong; up to 20 in. (50 cm) in length; dark green with irregular zones of creamy white along, between, or over the primary lateral veins; petiole light green, sheathing; about 12 in. (30 cm) long extending into the midrib. **FLOWERS**: spathe and spadix, typical of the family in having male flowers at the apex of the spadix and female flowers basally; the greenish spathe may be up to 5 in. (12 cm) and is constricted in the middle; flowers are not significant. **Fruit**: reddish berries covered by a persistent bract. **NATIVE HABITAT**: tropical America, from Mexico and West Indies to northern Argentina, usually as understory plants.

CULTURE Partial to bright shade, foliage burns if exposed to prolonged full sunlight; tolerate many soil types; prefer moderate moisture and freedom from nematodes; slight salt tolerance; mites, mealy bugs, and bacterial stem rot may be problems. **PROPAGATION**: cuttings, cane sections, air layering, and tissue culture. Stems should be handled with gloves to avoid irritation. **LANDSCAPE USES**: accents in outdoor plantings or as interior specimens; lends a tropical effect. **HARDINESS ZONE**: 10B–13B. Cold-intolerant.

COMMENTS The name "dumb cane" is given because the leaves and particularly the stem contain insoluble calcium oxalate crystals which, when combined with enzymes, cause irritation and swelling of tissues, especially vocal cords, if ingested. Dogs are known to chew on stems resulting in serious problems and death. Selected cultivars include: '**Alex**' (or 'Alix'), '**Carina**', '**Compacta**', '**Hilo**' (has large leaves with prominent white veins), '**Panther**', '**Parachute**', '**Paradise**' (yellowish leaves covered with green speckles), '**Rudolph Roehrs**' (creamy yellow leaves with white spots and green marginal veins), '**Sarah**', '**Tiki**' (a large cultivar with silvery leaves mottled with white and green), and '**Triumph**', among others. About 50 species of *Dieffenbachia* are recognized of which a few are in the trade, such as *D. amoena*, *D.* × *bausei* (probably involved in several of the cultivars noted above), *D. maculata* and its cultivars '**Camille**' (creamy yellow leaves with green margins), '**Exotica Compacta**', '**Exotica Perfection**' (cream with white veins), and '**Snow Queen**' (pale yellow with white veins and dark green edges), '**Tropic Snow**' (dark green leaves blotched with creamy white veins), *D. memoriacorsii*, and *D. picta*. These are usually large, erect herbs, with thick cane-like, sometimes branched (at least in cultivation when the apex is removed) stems, and entire, often spotted or variegated leaves and sheathing petioles, native to tropical America. Extensively grown and among the most popular foliage house plants.

Dieffenbachia maculata (above left), ***D. cv.*** inflorescence (above right), ***D. picta*** (below left), ***D. amoena*** (below right),

Epipremnum pinnatum

GOLDEN POTHOS

SYNONYMS *Epipremnum aureum, Pothos aureus; Scindapsus aureus; Rapidophora aurea, Philodendron nechdomii.*

ETYMOLOGY [From Greek *epi*, upon and *premnum*, tree-trunk, in reference to the epiphytic climbing habit of the plant (up on trees)] [Featherlike, having leaflets arranged on each side of a common stalk].

GROWTH HABIT Evergreen herbaceous vine; vigorous, climbs into trees by means of aerial roots, to 45 ft. (15 m) or more; Juvenile plants with smaller leaves of medium texture, mature plants have larger and coarse texture; rapid growth rate. **STEM/BARK**: green or striped, somewhat flattened and may be as wide as 2 in. (5 cm). **LEAVES**: simple, alter-nate, ovate-cordate. Juvenile (most frequently seen): entire to 12 in. (30 cm); mature (tree-climbing) pinnatifid or perfo-rate to 30 in. (75 cm); blades glossy bright green or irregu-larly variegated with yellow or white; petiole channeled at base, clasping. **FLOWERS**: bisexual (?), densely packed on spadix; surrounded by white spathe. **FRUIT**: berries with 2 to 4 seeds. **NATIVE HABITAT**: Southeastern Asia, Solomon Islands, but naturalized in most tropical and sub-tropical forests including Hawaii.

CULTURE Full sun to dense shade, varied soils; slight to moderate salt tolerance; scale, mites, mealy bugs. **PROPAGATION**: cuttings. **LANDSCAPE USES**: ground cover, potted foliage, hanging baskets, planters, totems, grown on trees. **HARDINESS ZONE**: l0A–13B.

COMMENTS Several variegated cultivars are very popu-lar as house plants and sometimes sold as *Scindapsus aureus*.

However, cultivars '**Aureum**', '**Goldilocks**', '**Jade**' '**Marble Queen**', '**Neon**', and '**Wilcoxii**', are selections of *E. pinnatum*. As in other members of the Araceae, this species also contains calcium oxalate and is toxic, particularly harmful to cats and dogs. It is considered an invasive species in many tropical and subtropical regions where it can overtake trees and grow as dense ground cover. Scandent vines with entire or segmented, petiolate leaves which are often spotted or marbled or variegated yellow or white, native to Southeast Asia. Commonly grown as foliage plants.

Epipremnum pinnatum

Epipremnum pinnatum '**Marble Queen**'

Epipremnum pinnatum young plant with entire leaves (left) and well-established plant with pinnatifid adult leaves (right)

Monstera deliciosa

SPLIT-LEAF PHILODENDRON, SWISS CHEESE PLANT
ETYMOLOGY [Derivation obscure, but perhaps from Latin *monstrum*, a marvel, or *monsterous*, probably in reference to the unusual leaves] [Delicious, in reference to the edible fruit].

GROWTH HABIT Monoecious, epiphytic evergreen herbaceous vine; vigorous, climbing; to 30 ft. (10 m) or more; coarse texture; rapid growth rate. **STEM/BARK**: ringed leaf scars; cylindrical, $2\frac{1}{2}$ –3 in. (6–7 cm) thick, many aerial roots that twist around tree brunches for support. **LEAVES**: simple, alternate, ovate, leathery, dark green; juvenile leaves cordate; base cordate; margins entire, adult leaves pinnatifid halfway to midrib, perforate with oblong holes (= fenestrated), about 3 ft. (1 m) long including the petiole and nearly as wide; petiole channeled at base, clasping. **FLOWERS**: unisexual, creamy white; densely packed on spadix; spathe to 1 ft. (30 cm) long. **FRUIT**: densely packed cluster of yellowish berries; edible with pineapple/banana flavor; dark area on fruit segments contains calcium oxalate which must be removed before consumption. **NATIVE HABITAT**: tropical rainforests of Mexico and Central America.

CULTURE Partial to deep shade, in moist fertile soils; no salt tolerance; scale, mites, and mealy bugs may become problems. **PROPAGATION**: cuttings, seeds (uncommon), tissue culture. **LANDSCAPE USES**: specimen, totems, grown on trees in tropics, and commonly used as indoor foliage plant. **HARDINESS ZONE**: 10B–13B.

COMMENTS The genus *Monstera* includes about 50 species of which the most familiar is *M. deliciosa*. Only two variegated cultivars have been introduced: ‘**Albo-variegata**’ (white variegations) and ‘**Variegata**’ (yellow variegations). Two varieties have also been recognized, *M. deliciosa* **var**. *borsigiana* and *M. deliciosa* **var**. *sierrana* both apparently with smaller leaves. Related species sometimes seen include: *Monstera acuminata* (shingle plant), *M. epipremnoides*, *M. karwinskyi*, *M. punctulata*, and *M. standleyana*, among others, are large epiphytic vines with entire or pinnatifid, often perforated, coriaceous, petiolate, strongly dimorphic leaves and large spadix surrounded by a boat-shaped spathe, native to tropical America. Primarily used as foliage plants but outdoors in warmer regions at the base of and on trees.

Monstera deliciosa growth habit (above and below left) and close-up of leaves and fruit (right). The brown attachments are remnants of the spathes.

Philodendron Species and Cultivars

ETYMOLOGY [From Greek *philo*, to love and *Dendron*, a tree, in reference to the tree climbing habit of some species]

GROWTH HABIT Herbaceous epiphytic and some terrestrial or hemiepiphytic (start as epiphytes and become terrestrial once they touch the ground or vice versa) evergreen perennials; rosettes to 3 ft. (1 m); vines variable depending on support. **STEM/BARK**: varies with species and cultivars, some species are stemless while others have a distinct stem with aerial roots and leaf scars. **LEAVES**: most species are either dimorphic, with juvenile leaves small and often cordate in shape while adult leaves very large and entire, lobed, or pinnatifid, whereas the heterophyllous species have both small entire and large pinnatifid leaves simultaneously; leaves are generally petiolate; developing leaves are usually covered by a boat-shaped cataphyll that falls off after leaf expansion; colors vary depending on species or manmade hybrids, ranging from red, burgundy, yellow, to variegated forms. **FLOWERS**: typical Araceae, with inflorescence of various colors and sizes, spathe and spadix with numerous small unisexual flowers, males above, females on the lower portion, and often sterile flowers in the middle, said to prevent self-pollination. Pollination is accomplished by beetles. **FRUIT**: berries of various colors. **NATIVE HABITAT**: Humid tropical Central and South America and the West Indies. Probably many are also of artificial or natural hybrid origin and some have naturalized in tropical zones in other countries.

CULTURE Tolerate low light very well; slight salt tolerance; generally, with few problems, mainly mealy bugs, scale, mites indoors. **PROPAGATION**: cuttings, layering, and tissue culture. **LANDSCAPE USES**: vining forms for trellises; rosettes as accents or small specimens; mass plantings. **HARDINESS ZONE**: 10B–13B. With the exception of *P. bipinnatifidum*, which can be grown in zone 9A.

COMMENTS A large genus of ±450 species of diverse growth habits. There is disagreement as to the number of species by various authorities. Younger leaves usually have most intense color and are generally more commonly known in the trade. Many of the best cultivars are patented. A number of hybrids are often sold as cultivars: '**Anderson Red**', '**Autumn**', '**Black Cardinal**', '**Burle Marx**', '**Emerald Duke**', '**Emerald Prince**', '**Emerald Queen**', '**Kaleidoscope**', '**Moonlight**', '**Prince Albert**', '**Red Empress**', '**Royal Queen**', and '**Xanadu**', among others. Among the several species in cultivation are *P. erubescens* and its cultivars '**Imperial Green**', and '**Imperial Red**'. **Philodendron** is the second largest genus in Araceae, but taxonomically the most difficult genus in the family (cf. *Anthurium*), and contains several commonly cultivated taxa, both as foliage as well as landscape plants. Some of the more common species include: *P. angustisectum*, *P. auriculatum*, *P. bipennifolium* (fiddle-leaf philodendron), *P. bipinnatifidum* (= *P. selloum* (relatively cold hardy and with edible fruit), *P. cordatum* (heart-leaf philodendron), *P. cruentum* (redleaf), *P. × dometicum* (spade-leaf philodendron), *P. erubescens* (redleaf philodendron), *P. × evansii*, *P. giganteum* (giant philodendron), *P. guttiferum* (leather-leaf philodendron), *P. hastatum*, *P. imbe*, *P. hederaceum* (= *P. scandens*, heart-leaf philodendron), *P. insigne*, *P. lingulatum*, *P. melanochrysum* (balck-gold philodendron), *P. ornatum*, *P. pedatum*, *P. pinnatifidum*, *P. radiatum*, *P. sagittifolium*, *P. speciosum*, *P. squamiferum*, *P. tripartitum*, *P. verrucosum*, *P. wendlandii*, and *P. williamsii*. These are epiphytic or less often terrestrial herbs with aerial roots and often climbing stems, usually large, entire or variously lobed or pinnatifid leaves with parallel venation, and sometimes attractive inflorescences with white to red spathes, native to tropical America.

Philodendron erubescens   *Philodendron erubescens* 'Royal Queen'

Philodendron hederaceum (= *Philodendron scandens*)

Philodendron 'Red Princess'

Philodendron 'Moonlight'

Philodendron 'Prince of Orange'

Philodendron 'Xanadu'

Philodendron gloriosum

Philodendron bipennifolium

FIDDLELEAF OR HORSEHEAD PHILODENDRON
SYNONYMS *Philodendron wayombense*

ETYMOLOGY [From Greek *philo*, to love and *Dendron*, a tree, in reference to their tree climbing habit] [with bipinnate leaves].

GROWTH HABIT Epiphytic heterophyllous evergreen herbaceous vine; of variable height depending on support; coarse texture; rapid growth rate. **STEM/BARK**: stems are green; aerial roots present and used for climbing. **LEAVES**: alternate, five-lobed, to 18 in. (45 cm) long; basal lobes extended and central lobes narrow towards the middle; glaucous olive green with a leathery texture. Juvenile leaves may be cordate and entire, but adult foliage are lobed and shaped like a violin, hence the common name fiddleleaf. **FLOWERS**: unisexual; a greenish spathe 4 $\frac{1}{2}$ in. (12 cm) long encloses a spadix-bearing inconspicuous flowers. **FRUIT**: Berries. **NATIVE HABITAT**: rainforests of southeastern Brazil.

CULTURE Bright light or partial shade but not direct sun, in well-drained soil; slight salt tolerance. A variety of pests including mites, scale, thrips, and mealy bugs may cause problems. **PROPAGATION**: cutting or layering. **LANDSCAPE USES**: a totem subject on a bark slab, a focal point in dish gardens, or hanging on a trellis or trees. **HARDINESS ZONE**: 10B–13B.

COMMENTS Sometimes incorrectly sold as *Philodendron panduriforme*.

Philodendron bipennifolium

Philodendron bipinnatifidum

SELLOUM
SYNONYM *Philodendron pygmaeum*, *P. selloum*, *P. sphincterostigma*.

ETYMOLOGY [From Greek *philo*, to love and *dendron*, a tree, in reference to their tree-climbing habit] [twice pinnately divided].

GROWTH HABIT Evergreen shrub in cultivation but has the potential to become epiphytic vine; densely foliated, stout, round, erect; arborescent (self- heading); to l5 ft. (5 m) tall and as wide spread; coarse texture; rapid growth rate. **STEM/BARK**: stout more or less woody trunks to 10 ft. (3 m), with conspicuous leaf scars and many strong adventitious roots. **LEAVES**: simple, alternate, spirally arranged; sagittate in outline, 3–5 ft. (1 = 1 $\frac{1}{2}$ m) long; undulate; deeply pinnatifid, many segments overlapping; deep green, more or less leathery. **FLOWERS**: unisexual; spadix with thick, usually green, boat-shaped spathe to 12 in. (30 cm), persistent. There is a short period of slightly higher constant temperature within the spade accompanied by a fetid odor which is intended to attract the pollinating beetle. This phenomenon is similar to that of female cycad cones. **FRUIT**: berries; edible. **NATIVE HABITAT**: Tropical South America: from southern Brazil, to northern Argentina, and eastern Bolivia, and Paraguay.

CULTURE Full sun to partial shade, prefers rich moist, but well-drained soils but tolerates less than ideal conditions; only slightly salt tolerant; generally, trouble free but scale, mites, and leaf spot may pose problems. **PROPAGATION**: seed and tissue culture. **LANDSCAPE USES**: foundation planting for large buildings; screen, specimen, urn plant; probably inappropriate for indoor use because of its size but may be used in large interiorscapes. **HARDINESS ZONE**: 8B–12B.

COMMENTS *Philodendron bipinnatifidum* is a significant landscape species in warmer climates. Only few cultivars are known, including 'German Selloum' (graceful, finely cut leaves), 'Miniature Selloum' (dwarf with small leaves), 'Uruguay' (leaves large and thick), and 'Variegatum' (leaves marbled light green to yellow). It is widely used as parent for arborescent hybrids. As in other members of the Araceae, *Philodendron bipinnatifidum* contains harmful calcium oxalate in all its parts and should be handled with gloves.

Philodendron bipinnatifidum (above and middle), specimens below are similar related taxa.

Philodendron hederaceum

HEART-LEAF PHILODENDRON
SYNONYMS *Arum hederaceum*, *Philodendron acrocardium*, *P. cuspidatum*, *P. harlowii*, *P. microphyllum*, *P. oxyprorum*, *P. scandens*, *P. subsessile*, *Pothos hederaceus*, etc.

ETYMOLOGY [From Greek *philo*, to love and *Dendron*, a tree, in reference to their tree climbing habit] [Resembling *Hedera*, ivy].

GROWTH HABIT Evergreen herbaceous vine; vigorous, climbing; medium texture; climbs to 30 ft. (10 m) or more; rapid growth rate. **STEM/BARK**: green, round. **LEAVES**: simple, alternate; cordate, to 12 in. (25 cm) long but much smaller on young plants; glossy-green; base cordate; margins entire; leaf sheaths dry and persist on stem; petiole round, not clasping. **FLOWERS**: unisexual, densely packed spadix; spathe greenish-white. **FRUIT**: berries. **NATIVE HABITAT**: tropical Central and South America, and the Caribbean.

CULTURE Partial to deep shade, in well-drained fertile moist soils; slight salt tolerance; an easy plant to maintain, except for occasional leaf spot, but without any serious problem if kept clean. **PROPAGATION**: stem cuttings. **LANDSCAPE USES**: ground cover, potted foliage, hanging baskets, totems. **HARDINESS ZONE**: 10B–13B.

COMMENTS *Philodendron hederaceum*, commonly known as *P. scandens* in the trade, is probably the most popular indoor foliage and is often used in combination with more upright plants. *Philodendron hederaceum* var. *oxycardium* (= *P. scandens* subsp. *oxycardium*) is often sold as *P. oxycardium* or *P. cordatum*. *P. hederaceum* f. *micans* (the underside of leaves is purple); 'Variegatum' has marbled, off-white and green-gray on dark green leaves. As in other members of the family, calcium oxalate is present throughout the plant body; cats have been mentioned as being most susceptible.

Philodendron hederaceum

Philodendron martianum

FLASK PHILODENDRON
SYNONYMS *Caladium crassipes*, *C. macropus*

ETYMOLOGY [From Greek *philo*, to love and *Dendron*, tree, in reference to their tree climbing habit] [After Carl Friedrich von Martius (1794–1868), German botanist and explorer and chief author of Flora Braziliensis].

GROWTH HABIT Self-heading, evergreen epiphytic herbaceous perennial vine but usually grown as a shrub (growth vine-like but growth rate is so slow that plant functions as a shrub); to 6 ft. (2 m) or taller. **STEM/BARK**: green with short internodes. **LEAVES**: simple, lanceolate to ovate; to 18 in. (45 cm) long, 6–8 in. (15–20 cm) wide; acute-acuminate, basally cuneate to truncate; petiole unusually full and spongy, with a shallow depression on the upper front part, dark green. **FLOWERS**: solitary, unisexual, with male, female, and sterile flowers on the spadix; spade 5–6 in. (12–15 cm) long; whitish-green inside with cherry-red base; cream outside. **FRUIT**: berries. **NATIVE HABITAT**: endemic to subtropical rainforests of southeastern Brazil.

CULTURE In bright light or shady locations; plants may be grown as either epiphytes on tree branches or as terrestrials in well-drained moist fertile organic soils; mites, scale, mealy bugs, and fungal leaf spot diseases may be problems; some salt tolerance. **PROPAGATION**: seeds, offsets, and tissue culture. **LANDSCAPE USES**: accent or specimen in containers or on trees. **HARDINESS ZONE**: 11B–13B.

COMMENTS Mainly used in breeding, but could be used in landscapes more often. Although *Philodendron cannifolium* is sometimes listed as a synonym of *P. martianum*, it is considered an illegitimate name and not a synonym of any *Philodendron* species.

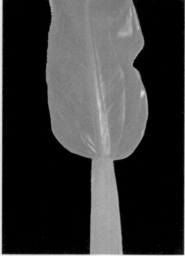

Philodendron martianum

Scindapsus pictus

POTHOS; SILVER VINE
SYNONYMS *Pothos argetius*, *P. argyraea*, *Scindapsus argyraeus*

ETYMOLOGY [Greek name for an ivy-like plant] [Painted].

GROWTH HABIT Hermaphroditic evergreen herbaceous vine; to 40 ft. (15 m), depending on support; medium texture; slow growth rate. **STEM/BARK**: green, with aerial roots. **LEAVES**: simple; obliquely ovate to oblong, juvenile leaves entire but adult foliage often pinnately lobed, cordate base, to 6 in (15 cm) long or longer in adult leaves; dark green splotched silvery-gray above and very pale beneath; petioles short and sheathing. **FLOWERS**: bisexual, inconspicuous, on spadix enclosed within a 3 in, (7 $\frac{1}{2}$ cm), white spathe. **FRUIT**: berries with a single seed. **NATIVE HABITAT**: widely distributed in Southeast Asia.

CULTURE Prefers bright indirect light to partial shade; well-drained soils; moderate salt tolerance. Root rot if kept too wet. **PROPAGATION**: Cuttings. **LANDSCAPE USES**: As ground cover or more commonly as interior foliage plant in hanging basket. **HARDINESS ZONE**: Zone 10–13.

COMMENTS The cultivar '**Argyraeus**' is known as Silver or silver pothos, has larger dark green leaves with silver spots.

Scindapsus pictus

Spathiphyllum wallisii

PEACE LILY, SPATHE FLOWER

ETYMOLOGY [From Greek *spathe*, spathe and *phylon*, a leaf, in reference to the leaf like spathes] [After Gustave Wallis, the German plant collector who introduced it from Colombia in 1824]

GROWTH HABIT Monoecious, evergreen, rhizomatous perennial herb; acaulescent and clumping; 3 ft. (1 m) tall and as wide; medium-coarse texture; moderately slow growth rate. **STEM/BARK**: rhizome. **LEAVES**: simple, narrow blades to 12 in. (30 cm) long and 4–6 in. (10–15 cm) wide; base attenuate; margins entire; apex acuminate; glossy dark green with depressed venation; petioles slender and longer than the blades and not channeled. **FLOWERS**: typical Araceae with inconspicuous, densely covered spadix; spathe is cream colored when young but turns greenish with age (in "Clevelandii" and others it turns white), erect, and ovate, to 6 in. (15 cm); flowers are usually well above foliage on slender conspicuous peduncles, during warm months. **FRUIT**: berry-like, dry. **NATIVE HABITAT**: Central America.

CULTURE Light intensities between deep and partial shade; in well-drained moist fertile organic soils; no salt tolerance; mites, mealy bugs, scale, foliar diseases, and *Cylindrocladium* root rot are potential problems. **PROPAGATION**: division, tissue culture. **LANDSCAPE USES**: course herb in shaded areas of warmer climates, potted foliage, and a good specimen for interiorscapes. It also tolerates aquariums. **HARDINESS ZONE**: 10B–13B. Frost-free areas.

COMMENTS There are about 52 species in the genus as well as many cultivars, mostly of unknown origin but probably at least some are of hybrid origin, with varying plant and flower sizes: 'Clevlandii' (perhaps the best known and most common), 'Daniel', 'Deneve', 'Domino', 'Emerald Swirl', 'Illusion', 'koutKnoc', 'Leprechaun', 'Linda', 'Petite', 'Pride', 'Sensation' (with significantly larger leaves and flowers), 'Sonya', 'Starlight', 'Supreme', 'Sweet Dario', 'Sweet Pablo', 'Symphony', 'Tasson', 'Viscount', and 'Viscount Prima', among others. Also available are *S. floribundum* (snowflower) and its cultivars 'Mauna Loa' and 'Mauna Loa Supreme'. *Spathiphyllum commutatum*, *S. floribundum* (snowflower), *S. × hybridum* (*S. cannifolium* × *S. patinii*), *S. patinii*, and others are rhizomatous herbs with clustered, oblong, elliptic, long-petiolate leaves, white or green spathe and white fragrant spadix, native mostly to the American tropics. Commonly grown as foliage plants but also for their attractive, long-lasting flowers.

***Spathiphyllum wallisii* 'Clevelandii'** flower (above, left), developing fruit (above, right), typical *S. wallisii* (below, left), and ***Spathophyllum* 'Sensation Jr.'** (below, right)

Syngonium podophyllum

NEPHTHYTIS, SYNGONIUM
SYNONYM *Nephthytis triphylla* (?)

ETYMOLOGY [From Greek *syn*, with or together and *gone*, reproductive organs, womb, in allusion to the united ovaries] [Foot-like leaf, referring to sagittate shape].

GROWTH HABIT Monoecious, evergreen herbaceous vine; varies in height depending upon support, medium-coarse texture; rapid growth rate; 3–6 ft. (1–2 m) tall. **LEAVES**: in juvenile plant ovate, to about 5 in (13 cm) long and nearly as wide; base cordate; margins entire, adult leaves arrow-shaped (sagittate), becoming palmately divided (pedate) at maturity when the plant will have an odd number of leaflets or divisions; blade size ranges from 2–12 in. (5–30 cm); leaf colors may be green or variegated with silver, cream, or white, or pinkish is newer cultivars. **FLOWERS**: whitish spathe to 4 in. (10 cm), with a shorter spadix appearing during the warm months. Unlike most other members of Araceae where the inflorescence is usually acaulescent and arise directly from rhizomes, in this case they appear in small groups in leaf axils. **FRUIT**: brownish-black berries. **NATIVE HABITAT**: tropical America, from Mexico and Costa Rica to Bolivia and naturalized in a number of other regions including Florida.

CULTURE Bright indirect light (not sun) to partial shade; fertile well-drained fibrous acid compost soil; prefers high humidity; has no salt tolerance; mites, scale, and a host of different disease organisms may become problems without proper air circulation. **PROPAGATION**: cuttings; primarily tissue culture. **LANDSCAPE USES**: ground cover or vine in frost-free areas. Juvenile forms are popular for interior use as hanging baskets and potted plants. **HARDINESS ZONE**: 10B–13B.

COMMENTS *Syngonium podophyllum* 'White Butterfly' is the most commonly available cultivar. However, several other cultivars are often listed, such as: 'Banana Cream Allusion', 'Bing Cherry Allusion', 'Butterfly Allusion', 'Confetti' (pinkish leaves), 'Capricorn' (orange leaves), 'Emerald Gem', 'Emerald Gem Variegated', 'Freckles Allusion', 'Gold Allusion', 'Lemon & Lime', 'Mango Allusion', 'Maria Allusion', 'Maya Red', 'Pink Allusion', 'Pixie', and others, particularly in the **Allusion Series**. Virtually all cultivars are produced by tissue culture. Other *Syngonium* species in the trade include: *S. angustatum*, *S. auritum* (five-fingers), *S. erythrophyllum*, *S. hoffmannii*, *S. podophyllum* var. *peliocladum*, *S. standeyanum*, *S. wendlandii*, and probably others. These are vines with simple juvenile leaves and 3-parted or pedate, long-petiolate, often variegated mature leaves, native to tropical America. Commonly grown as foliage plants but also as ground cover in warmer regions.

Syngonium podophyllum 'White Butterfly'

Zantedeschia aethiopica

CALLA LILY; ARUM LILY
SYNONYMS *Arodes aethiopicum*, *Calla aethiopica*, *C. ambigua*, *Colocasia aethiopica*, *Otosma aethiopica*, *Pseudohomalomena pastoensis*, *Richardia aethiopica*, *R. africcna*

ETYMOLOGY [After Giovanni Zantedeschi (1773–1848), Italian physician and botanist in Verona. Some sources, however, mention Francesco Zantedeschi (1797–1873), the Italian priest and professor of physics and philosophy at University of Padua] [From Ethiopia]

GROWTH HABIT Monoecious rhizomatous evergreen clumping perennial herb; 2–3 ft. ($\frac{2}{3}$ –1 m) tall, in large clusters. **STEM/BARK**: fleshy rhizomes. **LEAVES**: simple dark green sagittate (arrow-shaped), coriaceous leaves, with entire margins, to 18 in. (45 cm) long. **FLOWERS**: typical acaulescent inflorescence arising from the rhizome, with white spade and yellow spadix in the species but various colors in cultivars, which are initially green but change color as they develop. **FRUIT**: cluster of green and orange berries. **NATIVE HABITAT**: Lesotho, Swaziland, and South Africa, in stream banks and ponds. It has naturalized in a number of other regions where is some cases considered invasive.

CULTURE Full sun to part shades, in well-drained but moist fertile soil or may be planted in shallow water; rhizome fungus rot may be a problem but generally it is a trouble-free plant. **PROPAGATION**: division, seed, and tissue culture. **LANDSCAPE USES**: landscape plants in borders and mixed plantings, in containers outdoors or as house plants in winter, as well as cut flowers, also in wet locations such as pond edge and water gardens. **HARDINESS ZONES**: 8B–12B.

COMMENTS Eight species are recognized of which *Zantedeschia aethiopica* is the most common in cultivation. A number of cultivars of various colors are primarily the result of hybridization and grown as floriculture crops, including: '**Blush**', (reddish spathe), '**Captain Safari**' (dark orange spathe), '**Crowborogh**' (yellow spathe with red margins), '**Glow**' (pinkish in spathe throat), '**Green Goddess**' (with green stripes on the spathe), '**Kiwi Blush**' (lighter pink in spathe throat), '**Little Gem**' (lighter pink), '**Neon Blush**' (dark pink spathe), '**Pink Mist**' (with pinkish spathe and pink spadix), '**Red Desire**' (with yellow and pinkish-red spadix), '**Super Gem**' (pink spathe and white margins), '**White Sail**' (tall and with broad spathe), among other diverse solid colors and color combinations, ranging from pure white to nearly black, but not blue. Several species are also in cultivation, including: *Z. albomaculata* (leaves with white spots), *Z. elliottiana*, *Z. odorata*, *Z. rehmannii*, and its cultivars '**Alba**' and '**Superba**' (dwarf plants with lanceolate leaves). *Zantedeschia* spp. contains calcium oxalate and may cause severe problems if ingested. These are acaulescent, cormous herbs with entire, often spotted leaves, long petioles, and attractive, white, yellow or reddish-pink spathes, native to South Africa. Commonly grown for cut flowers or as garden plants.

Zantedeschia aethiopica

Zamioculcas zamiifolia

ZANZIBAR GEM; ZZ PLANT
SYNONYM *Caladium zamiifolium*, *Z. lanceolate*, *Z. lodegesii*

ETYMOLOGY [From resemblance of its leaves to that of *Zamia* spp. leaflets, and kinship to the genus *Colocasia*; Plant Finder refers to culcas as translation from Arabic referring to elephant's ear] [Zamia-like leaf]

GROWTH HABIT Monoecious evergreen succulent perennial herb, 20–25 in. (50–60 cm) tall; medium texture; rapid growth rate. **STEM/BARK**: stemless but with stout succulent rhizome. **LEAVES**: pinnately compound, with 6–8 pairs of elliptic glossy dark green, alternately arranged coriaceous-succulent leaflets of 2–5 in. (5–13 cm) long and nearly as wide; margins entire; petiole thick and succulent. **FLOWERS**: typical for the family but with green spathe and yellowish-white spadix that become brownish with age, arising from the rhizome. **FRUIT**: white berries, containing 1–2 brown seeds. **NATIVE HABITAT**: dry grasslands and lowland forests of eastern Africa, from Kenya to northeastern South Africa.

CULTURE Bright indirect light or partial shade, in well-drained organic soil that should not be allowed to remain wet for long (too much water causes the rhizome to rot), otherwise no serious problems. **PROPAGATION**: single leaves or separation of the rhizome. **LANDSCAPE USES**: container plant for outdoors in warmer climates or indoors as foliage for relatively short time. **HARDINESS ZONES**: 10B–13B. It does not tolerate temperatures below 45 °F (15 °C).

COMMENTS *Zamioculcas* is a monotypic genus with *Z. zamiifolia* as its only known species. All parts of the plant are said to contain calcium oxalate and are poisonous.

Zamioculcas zamiifolia

ADDITIONAL GENUS

Synandrospadix vermitoxicus, A monotypic genus native to southern South America.
Photographed in Cordillera, Bolivia

ALISMATACEAE

WATER-PLANTAINS

17 GENERA AND 85 to 120 SPECIES

GEOGRAPHY Cosmopolitan but predominantly tropical or subtropical but also temperate northern hemisphere.

GROWTH HABIT Hermaphroditic or monoecious herbaceous perennial aquatic or marsh plants; sap milky. **STEM/BARK**: rhizomatous, corm-like, or stoloniferous. **LEAVES**: simple, basal, entire, often dimorphic with submerged juvenile leaves usually linear but mature leaves may be broader linear, ovate, sagittate, or hastate; mostly petiolate, sheathing at base; veins reticulate. **FLOWERS**: bisexual or unisexual, 3-merous, stamens and carpel 3 or multiples, mostly white, on acaulescent or paniculate inflorescences, or solitary; petals are essentially ephemeral, lasting only for a day; stamens distinct, 0–30; ovary superior; pistils 0-numerous. **FRUIT**: achenes or follicles; seeds without endosperm.

CULTIVATED GENERA *Alisma*, *Sagittaria*, and *Echinodorus* are the only genera of ornamental importance.

COMMENTS Interesting freshwater or marine aquatic, wetland, and marsh plants, predominantly in northern hemisphere but essentially cosmopolitan distribution. Several species are grown in pools or aquaria and at least one species for its edible tubers: *Alisma plantago-aquatica* (European water plantain) and its varieties are the most commonly cultivated taxa, have lanceolate to broad, cordate leaves with basally sheathing petioles and paniculate inflorescences with white or pale rose flowers, native primarily to the northern hemisphere. *Echinodorus*, with ±30 species, is the largest genus in the family and has the highest number of species in cultivation, including: *E. berteroi* (upright burhead), *E. bleheri* (often used in fish tanks), *E. cordifolius* ('**Marble Queen**', Texas mudbaby), *E. grandifolius*, *E. grisebachii* (= *E. bleherae*), *E. Osiris* (red melon Amazon sword), *E. ozelot* (ozelot sword plant, '**Ozelot Red**', used in fish tanks), *E. paniculatus* (Amazon sword plant), *E. peruensis*, *E. tenellus* (pygmy chain sword plant), and *Echinodorus × beleheri* '**Red Melon**', and others, are annual or perennial herbs with markings on the leaves and flowers in globose heads, native to North and South America and Africa. Species of *Echinodorus* are mostly submersed and commonly used in aquaria. Additional cultivated taxa in Alismataceae include: ×*Alismatodorus* (*Alisma × Echinodorus*), *Hydrocleys*, *Limnocharis*, and *Luronium* (floating water plantain).

Sagittaria montevidensis

RUBY-EYE ARROWHEAD; GIANT ARROWHEAD
SYNONYMS *Sagittaria alismifolia*, *S. andina*, *S. chilensis*, *S. incrassata*, *S. multinervia*, *S. oblonga*, *S. ovifolia*, *S. teniifolia*

ETYMOLOGY [From Latin *sagitta*, an arrow, in reference to the shape of leaves] [From Montevideo, capital of Uruguay]

GROWTH HABIT Monoecious aquatic annual or rhizomatous perennial, with submerged and floating or petiolate emersed leaves. **STEM/BARK**: stemless but with submerged horizontal rhizome (not stolon). **LEAVES**: submerged, absent, or sessile, narrow, ribbon-like, aerial (emersed) leaves, hastate to sagittate, glabrous; petiolate spongy, to $2\frac{1}{2}$ ft. ($\frac{3}{4}$ m) long and about 3 in. ($7\frac{1}{2}$ cm) thick. **FLOWERS**: unisexual, staminate long pedicellate, pistillate also pedicellate or sometimes sessile or subsessile; petals white with purple spots at their base; sepals green; in whorls or opposite pairs on acaulescent inflorescence. **FRUIT**: small, dry, one-seeded indehiscent achene. **NATIVE HABITAT**: widespread in wetlands of North and Central America, but noted as being originally native to South America.

CULTIVATION Prefers Pond edges where the water is shallow. **PROPAGATION**: seeds and division of rhizomes. **LANDSCAPE USES**: in ponds or pools; smaller cultivars are marketed for use in aquaria. **HARDINESS ZONE**; 2B–12A. It is adaptable to nearly all climatic conditions.

COMMENTS The genus *Sagittaria* includes about 30 species of which *S. montevidensis* seems to be best known and most common. Three subspecies have been recognized in Flora of North America (Mexico, US, and Canada), of which the typical subspecies *S. montevidensis* **subsp.** *montevidensis* is perennial, *S. montevidensis* **subsp.** *spongiosa* is an annual, and *S. montevidensis* **subsp.** *calycina* is annual or perennial. The differences are essentially one of geographical distribution and tidal ecology. Rhizome is edible and used for consumption by both human and animals. Other *Sagittaria* species reported from cultivation include: *S. engelmanniana*, *S. graminea*, *S. lancifolia*, *S. latifolia* (duck potato), *S. sagittifolia* (arrowhead, swamp potato), *S. subulata* (awl-leaf arrowhead), and others and are stoloniferous, tuber-bearing herbs with conspicuous emergent leaves, and racemose-paniculate scapose inflorescences, primarily native to the New World. These are cultivated in ponds and aquaria, but especially *S. sagittifolia*, an herbaceous perennial native of Europe, is cultivated in most oriental countries for its edible tubers.

Sagittaria montevidensis, growth habit with immersed leaves and acaulescent flower stalk (above left), male flowers (above right), and female flowers (below)

ADDITIONAL GENUS

Hydrocleys nymphoides (water poppy), native to Central and South America, naturalized in Australia, New Zealand, Africa, and elsewhere. Photographed at Fairchild tropical Botanical Gardens, Coral Gables, Florida.

DIOSCOREACEAE

YAM FAMILY

4 GENERA AND 870 SPECIES

GEOGRAPHY Tropical and warm-temperate; primarily in Africa, Madagascar, and Southeast Asia.

GROWTH HABIT Plants monoecious or dioecious; mostly herbaceous vines, with thick rhizomes (in *Stenomerida*) or subterranean and cauline tubers (in *Dioscorea*). **LEAVES**: alternate or rarely opposite, simple or palmately compound, usually entire, broad, petiolate, venation palmate and reticulate, stipule-like appendages sometimes present. **FLOWERS**: unisexual or bisexual, actinomorphic, usually small and inconspicuous, epigynous; in axillary racemose inflorescences which are subtended by a bract; perianth of 6 tepals, biseriate, sometimes fused at the base, tubular or campanulate; stamens 6 or rarely 3 and 3 staminodia, filaments sometimes fused into a tube; carpels 3, fused, trilocular or rarely unilocular, ovary inferior, ovules 2 to many per locule; style simple, stigma trilobate. **FRUIT**: 3-winged capsule or rarely a samara or berry; seeds flat and winged or globose, embryo small, straight, endosperm present.

ECONOMIC USES Tuberous edible roots of *Dioscorea* (yams) are cultivated and used for food, and also medicinally, and certain species with caudices are collected by succulent fanciers.

CULTIVATED GENERA *Dioscorea*

COMMENTS Two subfamilies were recognized in earlier classification of Dioscoreaceae (**Stenomeridae** and **Dioscoridea**). In the most recent APG IV (2016), three genera are listed as *Stenomeris* (2 spp.), *Tricopus* (2 spp., *T. zylanicus* is used medicinally), and *Dioscorea* (634 spp., may be in cultivation or invasive). The monogeneric Taccaceae (*Tacca*) has been included in APG II (2003) through IV (2016), but is conserved as a distinct family because the authors of APG IV admit the phylogeny of this group is undecided.

Dioscorea spp.

Vines or trailing plants with woody (caudiciform) or fleshy (subterranean and/or cauline) tubers; leaves simple to digitately compound; flowers unisexual; staminate flowers with epipetalous, filaments attached to the base of the perianth, pistillate flowers with rudimentary stamens; fruit a 3-valved capsule. With more than 600 species, it is by far the largest and the most widely distributed genus in the family and includes several species with edible tuberous roots (not to be confused with *Ipomea batatas*, sweet potato), such as *D. alata* (Chinese yam), *D. bulbifera* (air potato, some forms are not edible), *D. batatas* (Japanese yam), *D. ×cayenensis* (yellow or attoto yam), and others; some species such as *D. discolor*, *D. trifida* (cush-cush yampee), and others are grown as foliage plants; several species known in the trade are caudiciform plants grown by succulent growers, such as *D. elephantipes* (elephnt's foot), *D. glauca*, *D. macrostachya*, *D. Mexicana*, *D. sylvatica*, etc.; while some species are used medicinally (e.g., in contraceptive pills) and others are highly poisonous.

Dioscorea bulbifera, the air potato, a noxious invasive species, forms tubers along its stems (not shown).

Testudinaria elephantipes* f. *montana (elephant's foot, turtle's back), deciduous, dioecious caudiciform plant with deeply fissured caudex surface, that grows to a significant size, and leafy climbing shoots with greenish flowers in winter; native to arid regions of South Africa

TACCACEAE

BAT FLOWERS

1 GENUS AND 10–12 SPECIES

GEOGRAPHY Pantropical, widespread in Africa, China, Australia, Southeast Asia, and several islands, and with a single species in South America. Five of the known species, including *T. chantrieri*, are indigenous to Thailand.

GROWTH HABIT Hermaphroditic acaulescent rhizomatous perennial herbs. **LEAVES**: simple or compound, alternate, spirally arranged into a basal group, long petiolate, entire or variously dissected; pinnately or palmately veined. **FLOWERS**: bisexual, actinomorphic, with two whorls of 6 petaloid tepals that are fused into a cup-shaped structure; stamens 6, in 2 whorls and all fertile, adnate to the tepals; ovary inferior, with 3 carpels; inflorescence cymose umbellate scapes, long pedunculated, with two or more (up to 12) conspicuously colored bracts. **FRUIT**: berry or rarely a loculicidal capsule containing many non-arillate seeds.

ECONOMIC USES Cultivated as ornamentals primarily for their unusual flowers and attractive foliage. Starch-rich rhizomes of some species are sold and used as arrowroots.

CULTIVATED GENERA *Tacca*

COMMENTS The 2003 APG II System has combined members of Taccaceae together with those of Dioscoreaceae and is listed as such in The Plant List. The 2016 APG IV, however, recognizes Taccaceae as a distinct monogeneric family.

Tacca chantrieri

BLACK BAT FLOWER; TIGER'S WHISKERS; DEVIL FLOWER
SYNONYMS *Clerodendrum esquirolii, Schizocapsa breviscapa, S. itagakii, Tacca esquirolii, T. garrettii, T. macrantha, T. minor, T. paxiana, T. roxburghii, T. vespertillo, T. wilsonii*

ETYMOLOGY [The Latinized form of *taka*, the Indonesian name] [Named after Chantrier Freres, the French brothers Adolphe and Ernest, who created the plant nursery in 1871]

GROWTH HABIT Perennial rhizomatous herb to about 2 ft. ($\frac{2}{3}$ m) tall as wide; course texture; medium growth rate. **STEM/BARK**: without cauline stem; rhizomes cylindrical vertical, 3–6 in. (7 $\frac{1}{2}$ –15 cm) long. **LEAVES**: basal, elliptic-ovate, cordate base, entire, penniparallel venation. **Flowers**: bisexual, with two dark brown to black bracts in pairs at right angle to each other with the larger horizontal pair resembling wings of a bat in flight, hence the common name; inflorescence scapose, long and thread-like filamentous bracteoles that resemble cat's whiskers; flower buds and one-day old flowers are held erect during the day, but bend downward at night and at anthesis flowers become pendent. **FRUIT**: kidney-shaped, glossy purplish-black with deep furrows; many seeds. **NATIVE HABITAT**: shaded understory of seasonally dry lowland tropical forests of Bangladesh, China, and Southeast Asia.

CULTURE As an understory plant in its **NATIVE HABITAT**, it grows mostly in full bright shade, well-drained organic moist soil, and prefers warm humid environment with good air circulation; generally, pest free. **PROPAGATION**: seeds (slow to germinate), or division of rhizomes. **LANDSCAPE USES**: best in greenhouse or conservatory but may be used as container plant indoors when growth conditions are proper. **HARDINESS ZONES**:10B–13A.

COMMENTS Despite their suggestive appearance, flowers of *Tacca chantrieri* do not produce nectar or any detectable odor and contrary to suggestions that they are fly pollinated, they have been shown to be selfers and the unique structure of the flowers supports self-pollination. Other species of *Tacca* are also in cultivation including *T. integrifolia* (white batflower – white bracts) and *T. nivea* (also with a large pair of white bracts that resemble butterfly wings).

Tacca chantrieri

Tacca integrifolia

VELLOZIACEAE

7 GENERA AND ±300 SPECIES

GEOGRAPHY Primarily South America with one species extending to Panama, but also in Africa- Madagascar, Arabia, and one species in China.

GROWTH HABIT Hermaphroditic, evergreen xerophytic perennial herbs or shrubs, from few inches to 18 ft. (6 m), with fibrous dichotomously branched stems, usually covered with persistent leaf sheaths on upper and adventitious roots on lower portions at nodes. **LEAVES**: simple, alternate in basal or rosettes at branch tips, linear, flat, entire, or spinulose along the margins, often pungent, to 3 ft. (1 m) long, sessile, sheathing. **FLOWERS**: bisexual; solitary or on axillary inflorescences; solitary or few in terminal panicles; epigenous, actinomorphic, 3-merous, tepals 6, in two petaloid whorls; stamens 6 or more, free, adnate to the perianth; various colors from white to violet; carpels 3, fused, trilocular, ovary inferior, ovules numerous; style long, slender, stigmas capitate or 3-lobed, peltate (in *Vellozia)* or capitate (in *Barbacenia*); ovules many per carpel. **FRUIT**: woody verrucate, septicidally or irregularly dehiscent capsule; seeds numerous and small, embryo ovoid in shape, endosperm copious.

ECONOMIC USES Tropical ornamentals.

CULTIVATED GENERA All 5 (7?) genera in cultivation but uncommon: *Acanthochlamys*, *Aylethonia*, *Barbacenia*, *Barbaceniopsis*, *Burlemarxia*, *Vellozia*, *Xerophyta*

COMMENTS A distinct family with unusual growth habits of fibrous woody stems and attractive flowers daylily-like bright colors. Often transplanted from the wild. Species of this family are known to thrive in nutritionally impoverished (particularly phosphorus) soils, seasonal low water availability, and exposed rocky locations. Despite their attractive flowers and some hybridization work, members of this family are uncommon in the trade probably because of their tropical nature. With 100 species each, the two American genera *Barbacenia* and *Vellosia* are the largest. Velloziaceae have not been studied in great detail and species are difficult to identify. These are uncommon but several listed in the trade. http://www.kew.org/science/tropamerica/neotropikey/families/Velloziaceae.htm

Vellozia sp.

***Vellozia* spp.**: Flowers bisexual, tepals predominantly violet, corona absent, stamens usually 6 or sometimes more (to 76); style longer than stigma lobes, stigmas horizontal and fused at center. Native to Bolivia, central and eastern Brazil, Colombia, Guyana, Venezuela, and Panama. The gentleman in the photograph is the well-known botanist professor Grady L. Webster. (Photographed in Bahia, Brazil)

Barbacenia graminifolia

Barbacenia spp.: flowers bisexual; corona present; tepals yellow, orange, red, or greenish; stamens 6, stigmas vertical, fused at apex or free. Native to central and eastern Brazil and Venezuela. (Photographed in vicinity of Recife, Brazil)

Xerrophyta elegans (= *Barbacenia elegans, Talbotia elegans, Vellozia elegans*); evergreen with leathery leaves and white flowers with yellow stamens, has high level of tolerance to desiccation; endemic to South Africa.

PANDANACEAE

SCREW-PINE FAMILY

5 GENERA AND ±770 SPECIES

GEOGRAPHY Tropics and subtropics of the Old World, from western Africa to the Pacific Ocean, usually in coastal or marshy areas, solitary or in thickets.

GROWTH HABIT Dioecious branched woody trees, shrubs, and vines, with distinct leaf base scars and aerial adventitious roots. **LEAVES:** appearing spirally arranged but 3-ranked, long and strap-shaped, often coriaceous and spiny, margins spinose-serrate; sometimes twisted and may be aggregated at stem ends. **FLOWERS:** unisexual; spadix is subtended by a spade; flowers of both sexes apetalous and asepalous; male flowers with numerous free or fused stamens; female flowers with superior ovary consisting of usually numerous but rarely only one carpel; inflorescences subtended by bracts. **FRUIT:** berry or cone-like multiple drupes which resemble a pineapple.

ECONOMIC USES Ornamentals, but several species of *Pandanus* are used as food, flowers of some used in perfumery, and leaves are used for thatching and weaving.

CULTIVATED GENERA *Freycinetia*, *Pandanus*, *Sararanga*

COMMENTS *Sararanga* species are not reported from cultivation but are similar to *Pandanus* in growth habit.

Freycinetia cumingiana

CLIMBING PANDANUS; FLOWERING PANDANUS
SYNONYM *Freycinetia luteocarpa, F. luzonensis; F. membranifolia.*

ETYMOLOGY [Admiral Louis Claude de Saulces de Freycinet (1779–1841), of French voyage to Australia and other regions] [Hugh Cuming (1791–1865), conchologist and botanist who collected in South Africa and Philippines]

GROWTH HABIT Evergreen climbing or scrambling shrub, to 6 ft. (2 m) tall; medium texture; relatively rapid growth rate. **STEM/BARK:** stems grow $1\frac{1}{2}$ –6 ft. ($\frac{1}{2}$ –2 m) long, with prominent annular leaf scars; roots growing along the stem and used for attachment to tree or object near the plant. **LEAVES:** alternate, amplexicaul (bases surround the stem), linear-oblong, coriaceous, glabrous, to about 10 in. (25 cm) long and 1 in. ($2\frac{1}{2}$ cm) broad; margins serrulate-spinose; apex acuminate. **FLOWERS:** unisexual; arranged in dense terminal cylindrical spike inflorescences; pinkish-orange bracts subtend a cluster of flowers. **FRUIT:** fleshy aggregate (syncarp), $\frac{1}{4}$ – $\frac{2}{3}$ in. (1–2 cm) long and containing many small seeds. **NATIVE HABITAT:** Asian, Malaysian, and Philippine tropics.

CULTURE Bright filtered sunlight, in well-drained sandy-peat soil; best when supported by an adjoining tree. **PROPAGATION:** seed, rooting of young shoot cuttings or removal of basal suckers. **LANDSCAPE USES:** landscape and container plant. **HARDINESS ZONE:** 10B–13A.

COMMENTS The genus *Freycinetia* includes 180–200 species distributed throughout the Old-World tropics. They are pollinated by bats, flying foxes, and rats. Other species reported from cultivation include *F. arborea, F. banksii, F. baueriana, F. excelsa, F. formosana, F. javanica, F. percostata, F. reineckei, F. scandens*, and probably others. The plant in the trade under the name *F. multiflora* is noted as misidentified.

Freycinetia cumingiana general plant appearance (above), and close-up of bracteate inflorescences (below). Photographed at Longwood Botanical Gardens, Pennsylvania.

Pandanus utilis

SCREW PINE

SYNONYMS *Hasskarlia globosa*, *Marquartia globosa*, *Pandanus distichus*, *P. elegantissimus*, *P. flabelliformis*, *P. maritimus*, *P. nudus*, *P. vacqua*, *Vinsonia stephanocarpa*, *V. utilis*

ETYMOLOGY [Latinized form of Malayan *pandan*] [Useful].

GROWTH HABIT Palm-like evergreen tree; multiple-branched; pyramidal, to 25 ft. (8 $\frac{1}{3}$ m) tall and 15 ft. (5 m) spread; coarse texture; slow growth rate. **STEM/BARK**: trunk and multiple branches ringed by old leaf scars; trunk stout, 8–10 in. (2 $\frac{2}{3}$ –25 cm), braced at the bottom by adventitious prop roots. **LEAVES**: simple; sword-shaped, arranged in spiral tufts at the end of each stout branch; stiff, sessile, keeled, to 3 ft. (1 m) long and 3 in. (8 cm) wide; margins lined with small reddish spines; lacking a petiole and broadly attached to the stem at base. **FLOWERS**: unisexual, male inflorescence 4 in. (10 cm) long branched spadix; female inflorescence, a dense head 1–2 in. (2 $\frac{1}{2}$ –5 cm) in diameter.

FRUIT: spherical syncarp (aggregation of drupes); to 8 in. (20 cm) long; composed of 100–200 prismatic drupes, 1 $\frac{1}{2}$ in. (4 cm) long; drupes green with a red-banded basal end; yellowish when ripe; contains a small amount of edible pulp. **NATIVE HABITAT**: Madagascar, Mauritius, and Seychelles islands.

CULTURE Full sun for fruiting specimens, young plants endure shade; wide range of soils; high tolerance of soilborne salt, no tolerance of salt spray; scale insects and fungal pathogens may become troublesome. **PROPAGATION**: seed, which should be soaked for 24 h before planting; basal sucker division; large cuttings. **LANDSCAPE USES**: free-standing specimens give a tropical effect; used as potted patio plants; could be used along intercostal shorelines for erosion control. **HARDINESS ZONE**: 10B–13A.

COMMENTS Leaves used to make mats and baskets in the tropics. *Pandanus* species number about 450. Species in cultivation include *Pandanus amaryllifolius*, *P. dubius*, *P. odoratissimus*, *P. sanderi*, *P. tectorius* and its cultivar "**Aureus**," *P. veitchii*, and probably others.

Pandanus utilis, free-standing in the landscape (above left), stout prop roots that anchor the plant (above right), plants used in such manner as to prevent soil erosion (middle left, probably a different species), palm-like leaf close-ups (middle right), and typical aggregate fruit (below)

Pandanus baptistii **'Aureus'**

Pandanus odorifer (= *Pandanus fascicularis,* fragrant screw pine, umbrella tree), native to Polynesia, Australia, the Philippines, etc., with bright orange clusters of edible fruit. Photographed in Bogor Botanical Garden, Indonesia

MELANTHIACEAE

TRILLIUM FAMILY

17 GENERA AND 173 (186?) SPECIES
GEOGRAPHY Northern Hemisphere: Eurasia and North America.

GROWTH HABIT Hermaphroditic or rarely dioecious perennial or rarely annual herbs with short or long rhizomes or rarely corms or bulb-like; caulescent or acaulescent and scapose. **LEAVES**: spiral or distichous, more or less flat, sometimes enciform (long, narrow, and pointed), sheathing at the base, linear-lanceolate or rarely ovate, sometimes scale-like. **FLOWERS**: bisexual or rarely unisexual, hypogynous or half-epigynous, actinomorphic or exceptionally zygomorphic (in *Chmaelirium* and *Chionographis*), in simple or compound racemose inflorescences or rarely in panicles or spikes; perianth of 6 similar tepals, biseriate, free or sometimes connate at base, campanulate or tubular, inconspicuous, variously colored; stamens 6, biseriate or rarely 9–12 (in *Pleea*), filaments free; carpels 3, fused, trilocular or rarely unilocular, ovary superior or half-inferior, ovules 2 to many per locule; styles 3, distinct or simple, stigma capitate or rarely trilobate. **FRUIT**: loculicidal, septicidal, or apically dehiscent capsule; seeds rounded and isodiametric, with wings or terminal appendages, embryo small, straight, endosperm present.

ECONOMIC USES Primarily as ornamentals and for medicinal use.

CULTIVATED GENERA *Chamaelirium*, *Paris*, *Toxicoscordion*, *Trillium*, *Veratrum*, *Zigadenus*, and probably others.

COMMENTS Genera currently included in Melianthaceae were placed in Liliaceae by many authors. In fact, such genera as *Trillium*, *Zigadenus*, and others are included in Liliaceae in Flora of North America and Melianthaceae is not mentioned. Five tribes (**Heloniaceae**, **Chionographideae**, **Melianthieae**, **Xerophylieae**, and **Parrideae**) have been recognized to accommodate the 17 genera, any detail discussion does not serve any useful purpose in this book. Examples of genera reported from cultivation include: *Amianthium muscitoxicum*, is infrequently cultivated bulbous herb native to eastern U. S.; *Chamaelirium luteum* (Devil's-bit), is a herb with tuberous roots, native to eastern North America; *Chionographis japonica* and *C. koidzumiana*, are rosette plants native to Japan; *Helonias bullata* (swamp pink), is a herb with tuberous rhizomes native to eastern North America; *Heloniopsis orientalis*, and few other species are rhizomatous herbs native to Korea, Japan, and Taiwan; *Melanthium hybridum* and *M. virginicum*, are rhizomatous herbs native

to North America; *Schoenocaulon* (= *Sabadilla*), with 10 species of herbs with bulb-like corms, native from Florida to Peru, not reported in cultivation but are of interest; *Stenanthium gramineum* (feather-bells) and *S. occidentale*, have a bulbous base, native to America and Asia; *Veratrum album* (European white hellebore), *V. californicum*, *V. insolitum*, (corn lily), *V. nigrum*, *V. stamineum*, and others, are rhizomatous perennial herbs native to North America and Eurasia; *Xerophyllum asphodeloides* (turkeybeard) and *X. tenax* (elkgrass, Indian basket grass), are grass-like plants with dense terminal racemes.

Trillium grandiflorum

SHOWY TRILLIUM, WHITE TRILLIUM, WOOD LILY
SYNONYMS *Trillium chandieri*, *T. erythrocarpum*, *T. linoides*, *T. obcordatum*, *T. scouleri*

ETYMOLOGY [From Latin *tres*, three, in reference to the leaves and floral parts which are all in 3s] [Large-flowered]

GROWTH HABIT Rhizomatous herbaceous scapose unbranched perennial, to $1\frac{1}{2}$ ft. (45 cm) tall and about 1 ft. (30 cm) wide; delicately coarse-textured; slow growth rate. **STEM/BARK**: rhizome short, thick and fleshy, with distal end bearing a large terminal bud from which arises a thick, glabrous scape; bracts sessile or subsessile, in whorls of 3. **LEAVES**: simple, dark green, ovate-rhombic, prominently veined, to 8 in. (20 cm) long and 6 cm (15 cm) wide; base rounded, margins entire; apex acuminate. **FLOWERS**: bisexual; 3-merous, parts distinct; sepals green, lanceolate, spreading flat; petals ovate, recurving above the middle to form a more or less funnelform corolla, white or pink, overlapping at base; margins undulate, stamens 6, straight, same height as the pistil and with a pair between the stigma lobes; blooming in late spring. **FRUIT**: globose 6-angled fleshy capsule, pale green; seeds many, elliptic, arillate. **NATIVE HABITAT**: Northeast United States to Quebec, Canada, south to Georgia and Alabama.

CULTURE Part to full shade, in well-drained but consistently moist organic fertile soils; spreads slowly; difficult to transplant; no serious insect of diseases reported. **PROPAGATION**: seed (difficult). **LANDSCAPE USE**: natural woodland landscapes, under trees. **HARDINESS ZONE**: 4A–8B.

COMMENTS *Trillium* is placed in Melanthiaceae by APG IV (2016); it is considered as Liliaceae by the Flora of North America and by other authors but accepted by most and even in earlier treatments as Melanthiaceae. As usual there is no disagreement with phylogenetic approach, but it is more convenient to adhere with what is commonly understood. There are about 50 species of *Trillium*. Some natural and several cultivars, '**Flore Pleno**' (double flowers), '**Eco Double Gardenia**' (large full double flowers), '**Elgin Form**' (full foliose double), '**Jenny Rhodes**' (white 6-petals and some smaller central petals), '**Charles O. Rhodes**' (full double, but 3 rows are sepal-like green), '**Julia**' (foliose double), '**Lundel Form**' (double green petalled flowers), '**Snow bunting**' (full double, gardenia-flowered), '**Bressingham**' (Flore plenum, from natural origin), '**Otis Bigelow**', '**Quicksilver**', '**McDaniels**', '**Pink Chiffon**', and '**Smith's Double**' have also been reported.

Trillium grandiflorum (great white trillium). Photographed at Longwood Botanical Garden, Pennsylvania

Trillium grandiflorum **f.** *roseum* (rose-colored trillium). Photographed at US National Arboretum, Washington D.C.

Trillium decumbens (prostrate toadshade), Photographed at US National Arboretum (left);
Trillium cuneatum (little sweet Betsy, purple toadshade). Photographed at Kanapaha Botanical Garden, Gainesville, Florida (right)

Trillium declinatum (= *Trillium catesbaei*; Catesby's trillium, bashful trillium), native to southeastern US Photographed at US National Arboretum, Washington D. C.

Trillium ovatum **var.** ***ovatum*** (western trillium, Pacific trillium). Photographed at US National Arboretum, Washington D. C.

Trillium luteum (yellow trillium), native to Tennessee, N. Carolina, Kentucky, and Georgia, has twisted petals and lemony fragrance. (Photographed at US National Arboretum, Washington D. C.)

Paris polyphylla

LOVE APPLE

SYNONYMS *Daiswa polyphylla*, *Paris biodii*, *P. daiswus*, *P. debeauxii*, *P. kwantungensis*, *P. taitungensis*

ETYMOLOGY [From Latin *par*, equal, alluding to regularity or symmetry of the plant parts] [Many-leaved]

GROWTH HABIT Rhizomatous perennial herb, to 3 ft. (1 m) tall and 1 ft. (30 cm) wide; coarse texture; rapid growth rate. **STEM/BARK**: rhizome $\frac{1}{3}$ –1 in. (1–2 $\frac{1}{2}$ cm) thick. **LEAVES**: simple; long petiolate, blades variable, usually oblong to lanceolate but may be ovate, to 6 in. (15 cm) long and 2 in. (5 cm) broad (larger in cultivation); base rounded to cuneate; in terminal whorl; with 3 primary and anastomosing secondary veinlets. **FLOWERS**: bisexual; solitary; terminal; long pedunculate; tepals in 2 whorls, free; outer ones green to yellow-green, narrowly ovate to ovate-lanceolate, to 3 in. (±7 cm) long and to 1 $\frac{3}{4}$ in. (±4 cm) wide; inner tepals linear, also yellow-green; both series appear to be 6 or more each (12 in the specimen photographed); stamens 2–12 or more (as many as the tepals), filaments to about $\frac{1}{2}$ in. (1 cm) long; ovary subglobose, ribbed, 1-loculed, sometimes tuberculate; style short, base enlarged, purple to white; stigma lobed 4 or 5; blooms in early spring through summer and early fall. **FRUIT**: globose berry-like indehiscent capsule, sometimes tuberculate; seeds many, covered with red fleshy aril. **NATIVE HABITAT**: bamboo forests and grassy and rocky slopes, in China, India, Laos, Myanmar, Nepal, Sikkim, Thailand, and Vietnam.

CULTURE Partial to full shade, in organic fertile moist soils; do not like to be disturbed; no serious pests or diseases reported. **PROPAGATION**: seed, rhizome sections. **LANDSCAPE USE**: woodland natural gardens and in tree shades. **HARDINESS ZONE**: 5A–10A. Prefers cooler climates.

COMMENTS Although five varieties of *Paris polyphylla* have been accepted by The Plant List: (*P. polyphylla* var. *alba*, *P. polyphylla* var. *chinensis*, *P. polyphyla* var. *latifolia*, *P. polyphylla* var. *stenophylla*, and *P.* var. *yunnanensis*), five additional ones are recognized by the Flora of China (*P. polyphylla* var. *minor*, *P. polyphylla* var. *pseudothibetica*, *P. polyphylla* var. *kwantungensis*, *P. polyphylla* var. *nana*, and *P. polyphylla* var. *polyphylla*). Recognition of 5 or 10 varieties is indicative of the species morphological diversity. Rhizomes of this (and other) species are important high value medicinal plants in the Indian Himalayan Region, resulting in excessive exploitation and its placement on the endangered list. The genus consists of 27 species of which 22 are listed in Flora of China, with 12 being endemic. The genus is generally found in Asia and Europe. In addition to *P. polyphylla*, *P. quadrifolia* and *P. japonica* are reported from cultivation.

Paris polyphylla (photographed at St. Petersburg Botanical Garden Conservatory in Russia)

Anticlea (Zigadenus) elegans

MOUNTAIN DEATHCAMAS, WHITE DEATHCOMAS

SYNONYMS *Anticlea alpina*, *A. chlorantha*, *A. coloradensis*, *A. elegans*, *A. glauca*, *A. gracielenta*, *A. longa*, *A. mohinorensis*, *Gomphostylis bracteata*, *Helonias bracteata*, *Melanthium glaucum*, *M. hultgreenii*, *Zigadenus alpinus*, *Z. bracteatus*, *Z. canadensis*, *Z. chloranthus*, *Z coloradensis*, *Z. commutatus*, *Z. dilatatus*, *Z. glaucus*, *Z. gracilentus*, *Z. longus*, *Z. mohinorensis*, *Z. speciosus*, *Z. washakie*

ETYMOLOGY [From Greek *zygos*, a yoke and *aden*, a gland, in reference to the 2-lobed glands at the base of the perianth] [Graceful, elegant]

GROWTH HABIT Herbaceous deciduous bulbous perennial with contractile roots. **STEM/BARK**: tunicate bulb. **LEAVES**: Usually few, basal, simple, linear, 4–12 in. (10–30 cm) long $\frac{1}{8}$ – $\frac{1}{2}$ in. (3–15 mm) wide, glabrous; margins entire; base often sheathing; often nearly folded from base to half their length, much reduced on the inflorescence stalk. **FLOWERS**: bisexual (sometimes functionally staminate); on terminal racemose bracteate 3 ft. (1 m) high glabrous inflorescence; flowers few to several, rotate, actinomorphic, long pedicellate, subtended by bracteoles; to $\frac{3}{4}$ in. (2 cm) across; tepals 6, connate at base, each with a bilobed nectary gland on the outer base (abaxial); stamens 6; pistil 1, stigmata 3, creamy white; blooming January to August. **FRUIT**: septicidally dehiscent capsule, 3-locular; seeds elongate, many. **NATIVE HABITAT**: widespread in various habitats in much of United States, extending into Mexico.

CULTURE Full to part sun, in well-drained medium to dry sandy soils; no major insect or disease problem reported; cold- and drought-tolerant. **PROPAGATION**: seed (cold treatment for 4 weeks is recommended). **LANDSCAPE USE**: natural wild garden, rock garden, alpine gardens, borders, potted, meadow. **HARDINESS ZONE**: 3A–11B. Widely adaptable to cold and warm climates

COMMENTS Flora of North America has described *Zigadenus elegans*, but noted *Anticlea elegans* as a synonym, while The Plant List notes ***Anticlea elegans*** as the accepted name (as does Wikipedia) and *Z. elegans* as synonym. As the aim of this book is not study of historical nomenclatural issues, no judgment is made regarding accuracy of either name. *Anticlea elegans* is used here despite its uncommonliness in horticultural trade. Eleven species of *Anticlea* are listed by the Plant List. To state that taxonomy of *Zigadenus-Anticlea* is confusing would be an understatement. In addition to mix-up between the two genera, some of the previously known taxa have been transferred to other genera. For example, ***Toxicoscordion brevibracteatum*** (= *Zigadenus brevibracteatus*, desert deathcomas, native to Baja California), ***T. paniculatus*** (= *Z. paniculatus*, sand corn, native to western U.S.), ***T. venenosus*** (= *Z. venenosus*, death camas), and others. These are bulbous or rhizomatous herbs native from Siberia to North America and Mexico. Genus *Zigadenus* was previously included in Liliaceae. Flora of North America does not make a distinction between listed varieties of *Zigadenus*. As the vernacular names suggest, *Zigadenus* spp. are known to be dangerously poisonous if eaten.

Anticlea elegans (photographed in Santa Cruz Island, California)

ALSTROEMERIACEAE

ALSTROMERIA FAMILY

4 GENERA AND 253 SPECIES

GEOGRAPHY Native to Central and South America, from Mexico to Chile and from Panama to Brazil.

GROWTH HABIT Herbaceous erect or twining rhizomatous and tuberous perennials. **LEAVES**: simple, linear to lanceolate, alternate, spirally arranged, petiolate (not sheathing), margins entire, veins parallel, twisted at base resulting in inverted blade; infrequently in basal rosette. **FLOWERS**: bisexual, actinomorphic or zygomorphic, aggregated in umbellate bracteate cymose inflorescence; 3-merous, tepals in two distinct (3 + 3) whorls; stamens 6, in two whorls; carpels unilocular or trilocular, ovary inferior. **FRUIT**: loculicidally dehiscent trilocular capsule that is sometimes explosive or fleshy indehiscent berry; seeds globose, without endosperm, and sometimes with a reddish aril.

CULTIVATED GENERA *Alstroemeria*, *Bomarea*, and *Luzuriaga*

COMMENTS Genera *Drymophila* and *Luzuriaga* from the recently published Luzuriagaceae (1998 and 2003 APG I and II Systems) are included in Alstroemeriaceae (Chacone, et al. 2012; APG III and IV, 2009 and 2016).

Alstroemeria spp.

PERUVIAN LILY, LILY OF THE INCAS

ETYMOLOGY [Swedish Baron Clas Alströmer (1736–1794), friend of Carl Linnaeus]

GROWTH HABIT Herbaceous perennials with tuberous roots (except the Atacama Desert native *A. graminea* which is an annual), to 5 ft. (1 $\frac{1}{2}$ m) high. **STEM/BARK**: the flowering shoots. **LEAVES**: simple, alternate, spirally arranged, linear to lanceolate, petiole twisted so that the lower side of the blade is facing up (= resupinate), margins entire. **FLOWERS**: solitary or several on umbellate inflorescence; tepals 6, in distinct 3 + 3 arrangement, mostly zygomorphic and variable in size; usually speckled or striped primarily in the upper petal, flowers are otherwise red, orange, purple, green, and white; androecium of 6 stamens and gynoecium of 3 lobed stigma; ovary inferior. **FRUIT**: loculicidally dehiscent trilocular capsule. **NATIVE HABITAT**: Central and South America, particularly in Chile and Brazil, but some have become naturalized elsewhere.

CULTURE Sunny location in moist but well-drained soil. They do not tolerate excessive heat. Higher temperatures result in production of more tuberous roots as opposed to flowers. **PROPAGATION**: to maintain integrity of the cultivars, tissue culture appears to be the primary method of reproduction. **LANDSCAPE USES**: largely as floriculture crops but also as garden flowers where they bloom in late spring to early summer. **HARDINESS ZONES**: 8B–11A.

COMMENTS Over 120 species of *Alstroemeria* are listed. The Chilean species grow in winter months, while those of Brazil do so in summer. Nearly 200 cultivars of various colors have been selected. Apparently, these are the result of hybrids between Chilean and Brazilian species that can withstand cooler temperatures and are grown as floriculture cut flower crops throughout the year. The most commonly listed species include: *A. aurea* (lily of the Incas, cultivars '**Lutea**', '**Orange King**', '**Splendens**'), *A. caryophyllacea* (Brazilian lily, **Premier Series**), *A. haemantha* (purplespot parrot lily), *A. hookeri*, *A. psittacina* (white alstroemeria), *A. pulchella* (red parrot lily), among others.

Alstroemeria 'Little Miss Vanessa'

Alstroemeria 'Summer Break'

Alstroemeria 'Princess Fabiana'

Alstroemeria 'Fougere'

Alsroemeria cv.

Alstroemeria 'Summer Saint'

Alstroemeria **'Pink Star'**

Alstroemeria seedings illustrating early spiral rosette leaf arrangement (below left).

COLCHICACEAE

COLCHICUM FAMILY

15 GENERA AND ±285 SPECIES

GEOGRAPHY Except for the genus *Uvularia*, members of this family are all of the Old-World Mediterranean climate origin, and Australia, Tasmania, and New Zealand.

GROWTH HABIT Monoecious erect or twining herbs with corms or sometimes stolons that bear apical tuberculate corms; sometimes with tuberous roots (in *Burchardia)*. **LEAVES**: few and basal, often appearing after the flowers (as in *Colchicum*) or many and alternately arranged on herbaceous or stiff stems, linear or lanceolate, sessile or rarely with a pseudopetiole, sheathing at the base, rarely cirrhose (as in *Gloriosa*); venation parallel. **FLOWERS**: bisexual, actinomorphic, hypogynous; perianth of 6 similar tepals, biseriate, free or sometimes basally connate, forming a long, narrow basal tube (as in *Colchicum*), colors various; stamens 6, biseriate, free, filaments short, narrow or broad at the base, carpels 3, fused, trilocular, ovary superior, ovules several to many per locule; style 3 and distinct or simple, stigma 3-branched. **FRUIT**: septicidally or loculicidally dehiscent capsule; seeds globose or rarely ovoid, sometimes with apical appendages, embryo short and straight, endosperm present.

ECONOMIC USES Species of several genera are cultivated as ornamentals. The alkaloid Colchicine, which is obtained from *Colchicum autumnale* and others, is used in plant breeding to double chromosome numbers, often resulting in larger flowers and changes in other features.

CULTIVATED GENERA *Androcymbium*, *Baeometra*, *Burchardia*, *Camptorrhiza*, *Colchicum*, *Gloriosa*, *Hexacyrtis*, *Iphigenia*, *Kuntheria*, *Littonia*, *Merendera* (sometimes included in *Colchicum*), *Ornithoglossum*, *Sandersonia*, *Schelhammera*, *Tripladenia*, *Uvularia*, and *Wurmbea*.

COMMENTS Recent phylogenetic classification confirms recognition of the family, with 15 or 16 genera, as distinct from Liliaceae by previous authors. Species are native to summer rainfall regions of South Africa, the Mediterranean, western Asia, and Australia. The latest (2007) classification of the family recognizes six tribes: **Anguillarieae, Burchardieae, Colchiceae, Iphigenieae, Tripladenieae**, and **Uvularieae**, any discussion of which does not serve a useful purpose in this book. Most members of this family are desirable attractive ornamentals and to a greater or lesser extent in cultivation. All are known to be highly toxic to livestock. Genera in cultivation include: *Baeometra uniflora* is a deciduous cormous plant, native to South Africa; *Buchardia congesta*, *B. monantha*, *B. multiflora*, *B. rosea*, and *B. umbellata* are endemic to western Australia; *Camptorrhiza indica* and *C. strumosa* with a single unbranched inflorescence stalk and edible tuber, native to Africa and India; *Colchicum autumnale* (autumn crocus), *C. giganteum*, *C. speciosum* and its cultivars, *C. variegatum*, and many others are herbs with basal leaves and variously colored showy flowers, important as the source of "Colchicum" and "Colchicine" drugs from *Colchicum autumanle* and also as ornamentals; *Hexacyrtis dickiana* (Namib lily), native to Namibia and South Africa; *Iphigenia magnifica*, *I. mysorensis*, and *I. socotrana* are native to Africa, Arabia, India, and Australia; *Kuntheria merenderapeduculata*, monotypic taxon endemic to Queensland, Australia; *Littonia rigidifolia* is a prostrate or erect herb with leafy stem and axillary, solitary campanulate flowers, native to Africa and Arabia; *Baeometra uniflora* is a deciduous cormous plant, native to South Africa; *Ornithoglossum dinteri*, *O. parviflorum*, *O. undulatum*, *O. viride*, and few others are widely cultivated as ornamentals, native to southern Africa; *Sandersonia aurantiaca* (Chinese-lantern lily) has erect leafy stems and axillary globose-campanulate flowers, native to South Africa, often grown in greenhouses; *Schelhammera multiflora* and *S. undulata* are attractive ornamentals with thin rhizomes bearing small tubers, native to New South Wales, Australia, and New Guinea; *Tripladenia cunninghamii*, monotypic taxon, native to New South Wales and Queensland, Australia; *Uvularia floridana*, *U. grandiflora*, *U. perfoliata U. puberula*, and *U. sessiflora*, commonly known as bellworts or bellflowers because of their hanging flowers, are with stolons or stoloniferous rhizomes, endemic to North America from north Florida to Nova Scotia; *Wurmbea capensis*, *W. elongata*, and *W. recutvata*, are native to South Africa, primarily in Cape Province.

Gloriosa superba

GLORIOSA LILY, GLORY LILY
SYNONYMS *Clinostylis speciosa*, *Eugone superba*, *Gloriosa angulata*, *G. rothschildiana*, *G. simplex*, *Methonica abyssinica*, *M. gloriosa*, *M. superba*, etc.

ETYMOLOGY [From Latin precious, glorious] [superb, in reference to the flower].

GROWTH HABIT Perennial herb, climbing to 8 ft. ($2\frac{2}{3}$ m), usually seen 4–5 ft. ($1\frac{1}{4}$–$1\frac{1}{2}$ m); medium texture; moderate growth rate. **STEM/BARK**: tuberous underground stem (recently referred to as a "hypopodial tuber" rather than corm; an elongated tuber); above-ground stem green, succulent, easily broken. **LEAVES**: simple, alternate, opposite or whorled, ovate-lanceolate, 5–7 in. (12–18 cm) long, with a coiled tendril-like (cirrhose) tip which is used for climbing; margins entire. **FLOWERS**: bisexual; solitary in upper leaf axils, lily-like, crimson, yellow, and whitish on margins and at base, perianth segments 6, oblong-lanceolate to 3 in. ($7\frac{1}{2}$ cm) long, strongly recurved, margins wavy and crisped; stamens exerted and to 2 in. (5 cm) long; style 1, usually about the same length as the stamens but reflexed; uncommonly attractive and showy; from late spring throughout summer. **FRUIT**: loculicidal 3-valved fleshy capsule; seeds red. **NATIVE HABITAT**: Tropical Africa and Asia in various vegetation types.

CULTURE Full sun to partial shade, various moisture-retentive organic fertile soils; not salt-tolerant (despite mentioned that it grows on sand dunes); chewing insects, aphids. The underground structures said to be brittle and difficult to remove, hence recommended to permanent planting in containers sunk into the ground. **PROPAGATION**: seeds, offsets, division of tuber, as well as tissue culture. **LANDSCAPE USES**: fence, trellis, specimen, cut flowers, and known to climb on trees. **HARDINESS ZONE**: 8B–12B.

COMMENTS Plant of unusual growth habit; spectacular when in bloom. It has long been used as a treatment in a range of ailments, from gout to arthritis, snake bite, and infertility. However, the plant, which contains the alkaloid gloriocine as well as colchicine, is extremely toxic to humans and animals if ingested. Related species in cultivation include *Gloriosa aurea*, *G. baudii*, *G. flavovirens*, *G. lindenii*, *G. littonioides*, *G. modesta*, *G. revoilii*, *G. rigidifolia*, and *G. sessiflora*. These are clambering or twining herbs with cirrhose leaves and showy flowers which have spreading and reflexed tepals, popular greenhouse subjects, or outdoors in warmer areas.

Gloriosa superba (above) and *G. superba* ‘Rothchildiana’ (below). Note the cirrhose leaf tip
on the right side of the photograph below.

ADDITIONAL GENERA

Uvularia grandiflora (large-flowered bellwort, merrybells), native to eastern and central North America, clump-forming rhizomatous herb with pendent perfoliate leaves and attractive bell-shaped pendent yellow flowers and typical trilocular fruit.

Uvularia grandiflora (large-flowered bellwort), clump-forming, erect herbaceous perennial, with lanceolate perfoliate leaves, and pendulous partially twisted yellow tepals; native to rich woodlands of central North America.

PHILESIACEAE

CHILEAN-BELLFLOWER FAMILY

2 GENERA AND 2 SPECIES
GEOGRAPHY Endemic to central and southern Chile.

GROWTH HABIT Evergreen unarmed shrubby and climbing plants with woody stems arising from rhizomes with swollen roots

LEAVES Alternate, more or less sessile and somewhat sheathing at base, veins parallel but reticulate secondary veins; stipules lacking. **FLOWERS**: inflorescences few-flowered, terminal or axillary cymes; flowers bisexual, actinomorphic, pendulous, tepals in two whorls of 3, forming a tube, outer whorl shorter than the inner thicker and more pronounced inner overlapping whorl; stamens in 2 whorls of 3; ovary superior, 3-locular (fused carpels); style filiform, stigma capitate. **FRUIT**: berry, red; seeds many

ECONOMIC USES Fruit of *Lapageria*

CULTIVATED GENERA *Lapageria*, *Philesia*

COMMENTS **Philesiaceae** genera had traditionally been included in Liliaceae

Lapageria rosea

CHILEAN BELFLOWER
ETYMOLOGY [Named after Marie Joséphine Rose Tascher de la Pagerie (1763–1814), also known as Napoleon's Empress Josephine, who was an avid plant collector for her garden at Chateau de Malmaison] [pinkish-red]

CHARACTERISTICS Essentially the same as that of the family. They are climbing plants that reach up to 30 ft. (10 m) high, leaves are lanceolate, flowers with pink white-spotted tepals. Fruit is an elongated red berry with tough skin containing numerous seeds that are covered with edible fleshy arils.

CULTURE Full sun to partial shade, in moist but well-drained, somewhat acidic soil. **PROPAGATION**: fresh seed, cuttings, and layering; flowers may require hand pollination for fruiting and seed (pollinated by hummingbirds in the wild). **LANDSCAPE USES**: best grown among trees or in greenhouses and conservatories, where cool moist conditions may be created. **HARDINESS ZONE**: 7A–9A. (?). In cultivation, *Lapageria* requires equable climate zones, where there is little fluctuation in night/day and seasonal temperatures, hence the reason for successful cultivation of the species and its cultivars at University of California at Berkeley Botanical Garden.

COMMENTS *Lapageria* is the national flower of Chile. The intergeneric hybrid ×*Philageria veitchii* (*Lapagreia rosea* × *Philesia magellanica*) is apparently more similar to *Lapageria*. University of California Botanical Garden has the largest collection of *Lapagreia* cultivars. Cultivars vary greatly in size, color, number per cluster, and semidouble. Some examples include: 'Nash Court' (older variety with larger flowers), 'Penheale' (red, foliage more lanceolate), 'Picotee' (white with pink around the edges), 'Beatrix Anderson' (red with much white spotting), 'Flesh Pink' (pale pink), and many more of Chilean and RHS origin. For an extensive list, see lapageria pages (roselandhouse.co.uk)

Lapageria rosea (photographed at the Univreisty of California Davis Conservatory)

LILIACEAE

LILY FAMILY

15 (18?) GENERA AND ±700 SPECIES

GEOGRAPHY Widely distributed in temperate regions of Northern Hemisphere but extend to Africa, India, China, Taiwan, and Luzon (Philippines).

GROWTH HABIT Hermaphroditic herbaceous geophytic bulbous or rarely rhizomatous perennials. **LEAVES**: simple, linear-lanceolate, alternate or sometimes whorled on flowering stems, and in some cases in rosettes atop the bulbs; veins usually parallel. **FLOWERS**: bisexual, actinomorphic; 3-merous with six similar tepals but occasionally readily distinguishable into sepals and petals; solitary or sometimes in cymes or umbellate heads; stamens usually 6; ovary superior and of 3 fused carpels. **FRUIT**: dry trilocular loculicidal capsule and sometimes a berry.

ECONOMIC USES Many ornamentals as garden plants and cut flowers; and lily flower buds used in Chinese cooking.

CULTIVATED GENERA *Calochortus*, *Erythronium*, *Fritillaria*, *Gagea*, *Lilium*, *Lloydia*, *Nomocharis*, *Tulipa*, among others.

COMMENTS Although in its current circumscription the family includes about 15 genera, in previous classifications it was treated as a pigeon-hole for a very large and diverse group of plants that are now dispersed among several other families. Most genera are not suitable for cultivation in warmer climates and only two, *Lilium* and *Tulipa*, are commercially produced in cooler climates and sold as potted or cut flower floriculture crops elsewhere. *Fritillaria* is mentioned because of their interesting features and popularity in temperate landscapes. Genera *Trillium* and *Zigadenus* have been transferred to Melanthiaceae.

Fritillaria imperialis

CROWN IMPERIAL
SYNONYMS *Fritillaria aintabensis*, *F. corona-imperialis*, *Imperialis comosa*, *I. coronata*, *I. superba*, *Lilium persicum*, *Petilium imperial*

ETYMOLOGY [From Latin *Fritillus*, dice box, in reference to the checkerboard pattern of petals of *Fritillaria meleagris*] [Imperial, majestic, regal]

GROWTH HABIT Herbaceous bulbiferous perennial, with fleshy mealy tunicate bulbs; plants 3–4 ft. (90–120 cm) tall and about $1\frac{1}{2}$ ft. (45 cm) wide: fine texture, rapid growth from bulb. **STEM/BARK**: stem erect, stout, leafy; **LEAVES**: simple; opposite or whorled; basal leaves petiolate, upper leaves sessile, oblong-lanceolate, to 6 in. (15 cm) long; fetid. **FLOWERS**: bisexual, campanulate, nodding (pendent), several flowers in umbellate inflorescence, bracts foliar cluster; tepals 6, free, solid orange, red, or yellow colors; stamens 6, inserted at base of tepals; styles 3-lobed, caduceus; stigmas short; flowering early spring. **FRUIT**: loculicidally dehiscent capsule, 3-locular, 6-angled; seeds flat, arranged in two rows in each locule. **NATIVE HABITAT**: southwestern Asia (Turkey, Iran, Afghanistan) to Himalayas.

CULTURE Full sun to part shade, in well-drained organic moist fertile soil; bulbs should be planted 6 in. (15 cm) deep and slightly sideways in autumn; should not be disturbed once dormant; lily beetle, bulb rot, leaf spot, rust, and mosaic virus are mentioned as potential problems. **PROPAGATION**: separation of bulblets or seed, and tissue culture. **LANDSCAPE USE**: usually planted in groups or among tulips and other winter/spring flowers. **HARDINESS ZONE**: 5A–9A. Generally best in cooler Mediterranean climates, intolerant of hot humid environments.

COMMENTS Bulbs are used medicinally. Few cultivars such as 'Auromarginata' (variegated leaves, red flowers), 'Primere', 'Lutea' (yellow), and 'William Rex' (orange). *Fritillaria* is a relatively large genus of 140 species, of which only a few are occasionally reported from cultivation, such as *F. affinis* (= *F. lanceolata*, dark purple flowers, native to western North America), *F. kurdica* (= *F. karadaghensis*, yellow-green flowers, native to Iran), *F. meleagris* (petals checkerboard, native to Europe), *F. persica* (flowers dark purple, native to Iran), *pluriflora* (pinkish-purple flowers, native to California), *F. pudica* (orange-yellow flowers, native to western North America), *F. recurva* (scarlet with yellow markings, western North America), and probably others.

Fritillaria imperialis 'Rubra' (above left), *F. imperialis* 'Aureomarginata' (above right), *F. imperialis* 'Lutea Maxima' (below)

Fritillaria persica

Lilium Species and Cultivars

LILIES

ETYMOLOGY [Latin form of the Greek name *lerion*, used for the white Madonna Lily (*Lilium candidum*)]

GROWTH HABIT Bulbous herbs, to 7 ft. (2 $\frac{1}{3}$ m) tall. **STEM/BARK**: bulbs non-tunicate (scaly), similar to those of garlic; aerial stems green with whorled leaves and sometimes bulbils (in leaf bases). **LEAVES**: whorled linear to elliptic-lanceolate, to about 6 in. (15 cm) long. **FLOWERS**: bisexual, large cup- or funnel-shaped, in a range of solid or spotted colors, on racemes or umbels, horizontally or downward facing, with 6 tepals, 6 free stamens, a single style, and superior ovary. **FRUIT**: trilocular capsule. **NATIVE HABITAT**: occurs in moist woods and meadows of temperate Northern Hemisphere to subtropical northern Philippines

CULTURE "Heads in the sun feet in the shade" is the expression used in recommendations for growing lilies; in well-drained organic media, with bulbs planted twice as deep as the height of the bulb; potential problems may include lily beetle, weevil, and botrytis, the latter apparently caused by wet foliage. **PROPAGATION**: seed, bulbils, and rooting bulb segments. **LANDSCAPE USES**: borders, mixed plantings, and most attractive when planted in groups; various cultivars, particularly *Lilium longiflorum* (Easter lily), are a major cut flower or potted plant crops. **HARDINESS ZONE**: 5A–11A. Species-dependent, most will not tolerate hot humid climates and a few may not be suitable for colder regions.

COMMENT It is important to note that true lilies have been noteworthy plants in history and literature as well as in holy books of major religions. But, the name lily as a vernacular has been applied to a host of unrelated and in some cases related plants of other Liliaceae genera or hybrids (e.g., daylilies, water lilies, etc.). The word lily actually means and signifies purity (for an interesting religious, historic, and mythological account of lilies, see Pizzetti and Cocker, 1968). There are thousands of cultivars that have resulted primarily from hybrid selections and too many to enumerate here. Various species and cultivar lists may be found on the web and in references. Suffice it to note that *Lilium* hybrids have been grouped into divisions based on flowering period, color, and growth habits into 9 Divisions: **Asiatic Hybrids**, **Mortagone Hybrids** (Turkscap Lilies), **Candidum Hybrids**, **American Hybrids**, **Longiflorum Hybrids**, **Trumpet Lilies** (Aurelian lilies), **Oriental Lilies**, **Interdivisional Lilies**, and **species Lilies**. A few of the reported species from cultivation include: *Lilium amabile*, *L. auratum*, *L. bolanderi*, *L. brownii*, *L. bulbiferum*, *L. candidum*, *L. davidii*, *Lilium × elegans* (= *L. duaricum*), *L. formosanum*, *L. henryi*, *L. humboltii*, *L. japonicum*, *L. leucanthum*, *L. longiflorum*, *L. martagon*, *L. parvum*, *L. regale*, *L. rubescens*, *L. speciousm*, *L. sulphureum*, *L. superbum*, *L. tigrinum*, *L. washingtonianum*, and others. For an interesting and detailed treatment of lilies, see Wikipedia.

Representative examples of lily cultivars. The white flowering specimen is that of ***Lilium candidum***.

Representative examples of lily cultivars (above and middle), *L. longiflorum* and its bulb (below)

Tulipa Species, Hybrids, and Cultivars

TULIPS

TRANSLATION [From Turkish *tulband*, a turban, in reference to shape of the flowers]

GROWTH HABIT Herbaceous perennials, tunicate bulbs, low growing, erect stem in some; variable with cultivar, 12 in. (30 cm) tall, rapid growth. **STEM/BARK**: true stem lacking; typical tunicate bulbs generally pointed and brown tunic in color. **LEAVES**: simple, basal, or a few on the flower stalk in some tall cultivars, smooth, undulate, or folded, to 18 in. (45 cm) long, strap-shaped, generally bluish-green, thick, margins entire. **FLOWERS**: all colors except true blue, many combinations; campanulate to rotate, mostly acaulescent erect stalks, petals, and sepals (tepals) indistinguishable, as such totaling 6 except in double-flowered forms, 6 short stamens; variously pubescent to glabrous inside; usually solitary on smooth stalk often without but sometimes with few leaves, to 3 ft. (1 m); flowers in early spring. **FRUIT**: loculicidal, 3-valved capsule, seeds flat, many. **NATIVE HABITAT**: widespread from Central Asia, particularly Turkey and Iran to north Africa and also to northwest China, common in Mediterranean climate regions.

CULTURE Full sun to partial shade, various soils; slightly salt tolerant; grey bulb rot, *Botrytis*, maggots; Mosaic Virus may become troublesome. **PROPAGATION**: seeds, separation of bulblets of cultivars. **LANDSCAPE USES**: bedding plant, planter box, pot plant, cut flowers. **HARDINESS ZONE**: 3B to 7B. Can be grown in warmer areas if treated as an annual, with the understanding that tulip bulbs flower only once unless their cold requirement is met and flowers last for only a few days. Usually imported to warmer areas as potted plants.

COMMENTS Over100 species and 4000 named varieties and cultivars. Hybrids and cultivars have been divided into 15 groups based on their flowering time and morphological features. Inclusion of specific examples is of little value in the context of this book since tulips are not particularly suited to warmer climates and there are too many to mention. The dazzling display of spring flowers in cooler climate regions such as northeastern US and much of Europe, particularly the Netherlands, is essentially unmatched by any other flower.

Tulipa acuminata   *Tulipa saxatilis*   *Tulipa clusiana*
(All potographed at Hortus Botanicus, Leiden, Netherlands)

Representative examples of single flower tulip cultivars
(photographed at Keukenhof Botanical Garden, Netherlands)

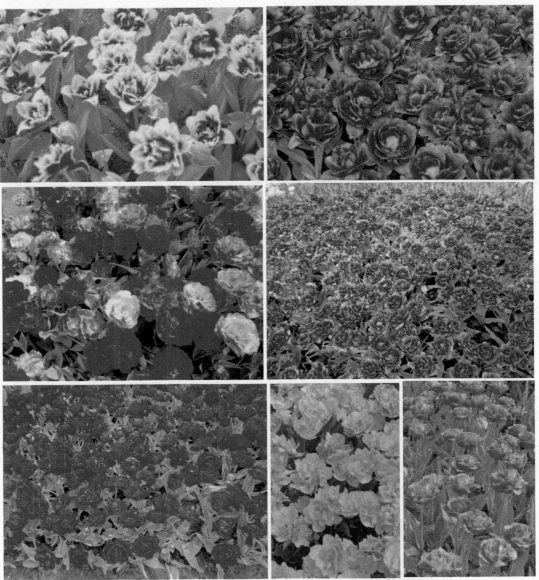

Representative examples of double tulip cultivars
(photographed at Keukenhof Botanical Garden, Netherlands)

Close-up of two cultivars to illustrate structure of the tulip flower. (Photographed at Montreal Botanic
Garden, Canada by F. Almira)

Tulip bulb production in Holland

Example of tulips in design in Keukenhof Botanical Garden, Netherlands

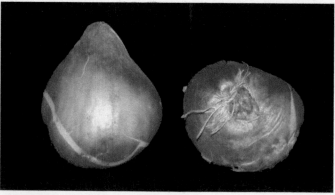

Tulip bulbs

'**Monsella**' tulip (flowers open in response to brightness of the sun)

ADDITIONAL GENERA

Erythronium revolutum (mahogany fawn lily), tuberous perennial, usually with 4 nodding green to red flowers with yellow anthers; native to from California to southern British Columbia.

Tricyrtis hirta (=*Tricyrtis japoica, Ttricyrtis clinata*; toad lily), herbacious perennial, with large and wide leaves clasping around the stem, flowers whitish to pale purple; native to Japan

ORCHIDACEAE

ORCHID FAMILY

750 (763) GENERA AND ±28,000 SPECIES ±125,000 HYBRIDS

GEOGRAPHY Cosmopolitan, except in the Arctic and extreme dry deserts, found in a diversity of habitats.

GROWTH HABIT Hermaphroditic terrestrial, epiphytic, lithophilic, or sometimes saprophytic herbs; monopodial (usually epiphytes, in which the main axis grows continuously, flowering on lateral branches), or sympodial (usually terrestrials, in which the primary axis produces successive superposed branches, beginning with scale leaves and terminating in flowers), and rarely clambering vines (in *Vanilla*); with rhizomes or pseudobulbs; from minuscule to nearly 30 ft. (10 m); fleshy velamen-covered adventitious aerial roots universally present. **LEAVES**: simple, fleshy, basal, alternate or rarely opposite, often distichous, sometimes reduced to scales (in saprophytic species) or rarely absent; and sheathing at their base around the stem in cauline species. **FLOWERS**: bisexual or rarely unisexual, predominantly very showy but some inconspicuous; zygomorphic but rarely actinomorphic, usually resupinate (twisted 180°), solitary, or in spikes or racemes or panicles; 3-merous, with the sepals similar but petals dissimilar: two lateral petals (wings) elongated but the dorsal petal (labellum or lip) of various shapes, sizes, and colors in a multitude of forms; reproductive organs represented in a column, with the stigmatic surface below and 1 or 2 pollinia-bearing stamens around it, although it may vary considerably among various taxa; ovary inferior. **FRUIT**: capsule containing numerous minute seeds without endosperm, hence difficulty in germination.

ECONOMIC USES Except for fruit of vanilla (*Vanila planifolia*), orchids are used only as ornamentals.

CULTIVATED GENERA Nearly all known orchid genera are in cultivation: *Acacallis, Acampe, Acineta, Ada, Aerangis, Aeranthes, Aerides, Alamania, Amblostoma, Amesiella, Ancistrochilus, Angraecopis, Angraecum, Anguloa, Ansellia, Arachnis, Arpophyllum, Arundina, Ascocentrum, Aspasia, Barbosella, Barkeria, Bifrenaria, Bletia, Bollea, Bolusiella, Brassavola, Brassia, Broughtonia, Bulbophyllum* (**1867 spp.**)*, Calanthe, Calyptrochilum, Capanemia, Catasetum, Cattleya, Caularthron, Chiloschista, Chysis, Cirrhopetalum, Cleisostoma, Cochleanthes, Cochlioda, Coelogyne, Comparettia, Cryptophoranthus, Cyclopogon, Cychnoches, Cymbidium, Cynorkis, Cypripedium, Cyrtopedium, Cyrtorchis, Dendrobium* (**1509 spp.**)*, Dendrochilum,*

Diaphananthe, Diploprora, Disa, Doritis, Dresslerella, Eleanthus, Encyclia, Epidendrum (**1143 spp.**)*, Epigeneium, Eria, Erycina, Esmeralda, Euanthe, Eulophia, Galeandra, Gastrochilus, Gomesa, Gongora, Goodyera, Grammangis, Graphorkis, Haemeria, Habenaria* (**835 spp.**)*, Helcia, Hexisea, Huntleya, Ionopsis, Isochilus, Jummellea, Kefersteinia, Laelia, Lanium, Leochilus, Lepanthes* (**1085 spp.**)*, Leptotes, Leucohyle, Liparis* (**426 spp.**)*, Listrostachys, Lockhartia, Ludisia, Luisia, Lycaste, Macodes, Macradenia, Masdevallia* (**589 spp.**)*, Maxillaria* (**658 spp.**)*, Meiracyllium, Microcoelia, Miltonia, Miltoniopsis, Mormodes, Mormolyca, Nageliella, Neobathiea, Neomoorea, Notylia, Octomeria, Odontoglossum, Oeceoclades, Oeonia, Oeoniella, Oncidium, Ophrys, Orchis, Ornithocephalus, Otoglossum, Pabstia, Palumbina, Paphiopedilum, Pecteilis, Pelatantheria, Pescatorea, Phaius, Phalonopsis, Pholidota, Phragmipedium, Physosiphon, Pleione, Pleurothallis* (**552 spp.**)*, Polycycnis, Polystachya, Ponthieva, Porroglossum, Promenaea, Rangaeris, Renanthera, Restrepia, Rhynchostylis, Robiquetia, Rodriguezia, Rodrigueziella, Rossioglossum, Rudolfiella, Sarchochilus, Sarcoglottis, Scaphosepalum, Schoenorchis, Schomburgkia, Scuticaria, Sigmatostalix, Sobralia, Solenidium, Sophronitella, Sophronitis, Spathoglottis, Sphyrarhynchus, Stanhopea, Stelis* (**879 spp.**)*, Stenia, Stenoglottis, Stenorrhynchus, Symphyglossum, Tainia, Telipogon, Thunia, Trichocentrum, Trichoglottis, Trichopilia, Tridactyle, Vanda, Vandopsis, Vanilla, Xylobium,* and *Zygopetalum,* among others. It should be noted that not all the genera listed above are of general appeal. Several are primarily collector's items. Also, many intergeneric hybrids and grexes that are common in cultivation have not been included here but noted in photographs. Detailed classification of orchids below subfamilies (tribes, subtribes, etc.) is not discussed. Listed number of larger genera and subgeneric assignments (below) is from Christenhusz et al. (2017). Several specific orchid references are listed in the reference list.

COMMENTS This book is not entirely about orchids and inclusion of descriptions for a large number of taxa would be beyond its current scope. Moreover, considering the number and enormous diversity of the multitude of species and hybrids, no other group of plants has been as extensively documented as that of orchids. Hence, publication of the largest number of books and published literature on orchids is perfectly understandable. Although there are several comprehensively treated and extensively illustrated volumes currently available, the more recently published small book by Sheehan and Black (2007) is highly recommended for beginners. Also see: International Orchid Commission. 1994. Handbook of Orchid

Nomenclature and Registration (4th ed.). Identification of orchid genera is a relatively simple task, but assignment of correct specific epithets and particularly grexes and cultivars is troublesome even for expert orchidologists. Insect pollination of orchids, including pseudocopulation (erroneous assumption by male insets that attack and pollinate flowers), is of particular interest and there is extensive literature on the subject.

ORCHID FEATURES

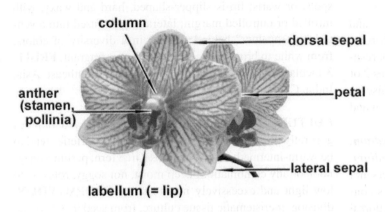

Phalaenopsis "Kalaidoscope"

Typical orchid fruit

Oncidium pseudobulb – Sympodial

Ascocentrum garayi – Monopodial

ORCHIDACEAE SUBFAMILIES

Orchids have been divided into five monohyletic (common ancestry) subfamilies (see Dressler, 1993, Chase et al. 2003, and 2015, Alrich, et al. 2008, Judd et al. (2015), Christenhursz et al. (2017), APG IV (2016–2021), and for a brief historical review and list of tribes, subtribes, alliances, and included genera, etc., see Wikipedia):

SUBFAMILY Apostasioideae (*Apostasia* and *Neuwiedia*). The most basal of orchids. Roots tuberous, tubers with irregular papillae; leaves spiral; flowers not resupinate (not upside down); tepals pointed; fertile stamens 2 or 3; stigma papillate, lobs spreading; fruits baccate, dehiscing irregularly; seeds dark color. The genera are uncommon and only rarely seen in cultivation.

SUBFAMILY Cypripedioideae (*Cypripedium*, *Maxipedium*, *Paphiopedilum*, *Phragmipedium*, *Selenipedium*). Plants epiphytic or terrestrial; flowers not resupinate when single, lateral tepals of outer whorles connate, labellum saccate (pouch-like), median lobe (dorsal sepal) largest; fertile stamens 2, conspicuous single staminode shield-shaped; stigma lobes spreading, papillate (with warts).

Paphiopedilum

LADY'S SLIPPER ORCHID

ETYMOLOGY [From Greek *paphia*, an epithet of Aphrodite and *pedilon*, slipper), in reference to the slipper-shaped lip of the flower].

GROWTH HABIT Evergreen, sympodial (fan-type), mostly terrestrial orchids but some epiphytic or lithiphytic (on rocks); moderate foliage density; 6–18 in. (15–45 cm) tall with an equal or wider spread; growth rate slow (some medium). **STEM/BARK**: stemless, without pseudobulb. **LEAVES**: fans of 6–10 folded leathery leaves, alternate, elliptic to lanceolate, to 15 in. (37 cm) long and 2 in. (5 cm) wide, may be green or green with darker green mottling. **FLOWERS**: terminal, single, or few-flowered scapes in any season, depending on species; hard waxy, long-lasting flowers; dorsal sepal usually large and showy with spots, stripes, etc. Petals spread laterally, may have undulations, hairs, spots, or warts; lip is slipper-shaped, hard and waxy, with inrolled or outrolled margin; lateral sepals fused into a ventral "synsepalum" behind slipper; in a diversity of colors, from white to browns, greens, or pink; not fragrant. **FRUIT**: 3-locular capsule. **NATIVE HABITAT**: Southeast Asia, India, China, Philippines, and New Guinea.

CULTURE Low light; temperature: green-leaved varieties generally intermediate-cool; mottled-leaved varieties tend to be warm-intermediate, in peat moss, tree fern, perlite, sphagnum, or any combination; keep moist, not soggy, rots due to low light and excessively moist media. **PROPAGATION**: division, meristematic tissue culture, from seed in 3–5 years. **LANDSCAPE USES**: potted plants for both the showy flowers and the attractive mottled foliage. **HARDINESS ZONE**: 9B–12B. Usually grown in greenhouses.

COMMENTS About 107 species and hundreds of hybrids. Common examples include: *Paphiopedilum callosum* (popular mottled-leaved species with brown/green/purple/white patterned flowers); *P. glaucophyllum* (bears flowers one at a time over several months, flowers are white and green with a pink slipper). It should be noted that "Lady Slipper Orchids" constitute five groups of species with familiar generic names, all with a pouch-like labellum. Of these, *Mexipedium xerophyticum* and *Selenipedium* spp. are uncommon but *Paphiopedilum* (tropical Asia-epiphytic or terrestrial), *Phragmipedium* (Central and South American–epiphytic or terrestrial), and *Cypripedium* (north temperate-terrestrial) are well-known and common in the trade.

Paphiopedilum druryi

Paphiopedilum concolor

Paphiopedilum druryi × *gratixianum*

Paphiopedilum Grand Vaux

Paphiopedilum gratrixianum

Paphiopedilum Irish Lullaby
'Lehua Erin Green'

Paphiopedilum Lippewunder

Paphiopedilum maudiae

Paphiopedilum 'Magic Voodoo'

Paphiopedilum transvaal

Paphiopedilum niveium

Paphiopedilum violaine 'Narine'

Paphiopedilum 'Armeni White'

Paphiopedilum delenatii × malipoense

Paphiopedilum delenatii

Paphiopedilum liemianum × delenatii

Paphiopedilum lowii 'Light Sun'

Paphiopedilum micranthum

P. philippenense

P. glanduliferum

P. sanderianum

P. Angelina Kruger

P. rothschildianum
(*P. philippenense* × *P. rothschildianum*)

P. Saint Swithin (*P. philippenense* × *P. rothschildianum*)

Paphiopedilum primulinum

Paphiopedilum wilhelminae

Paphiopedilum **Leonard Smith** (Leonard Smith × rothchildiana)

Phragmipedium

SLIPPER ORCHID

ETYMOLOGY [From the Greek *phragma*, partition and *pedilon*, slipper, in reference to the slipper-shaped lip and division in the ovary]

GROWTH HABIT Large terrestrial, epiphytic, or lithophytic sympodial herbs. **STEM/BARK**: pseudobulbs lacking, stems short. **LEAVES**: fans of 6–8 tightly arranged distichous flat leathery leaves with inconspicuous veins, to about 30 in. (75 cm) or more in height. **FLOWERS**: terminal inflorescences arise from leaf axils and produce from 1–15 flowers; lateral sepals united at their apeces to form the pouch-like lip which appears inflated and folded inward at the margins; petals elongate up to 15 in. (45 cm) long, mostly linear but some wider; column short with two fertile anthers each with a pair of pollinia, and a sterile staminode; flower colors predominantly green and brownish but brilliant reds and lavender species are also known. **FRUIT**: 3-locular capsule. **NATIVE HABITAT**: tropical Central and South America, from Guatemala and Panama south to Brazil and Bolivia.

CULTURE Medium-light intensity; in well-drained epiphytic or terrestrial medium depending on species, keep moist; abscission of turgid flowers is not necessarily indicative of disease problem. **PROPAGATION**: seed, division, meristem culture. **LANDSCAPE USES**: flowering potted plants for temporary use indoors or in greenhouse and conservatories. **HARDINESS ZONE**: 10B–13A.

COMMENTS There are about 25 known species and many interspecific and intergeneric hybrids. Among the more commonly listed are *Phragmipedium besseae*, *P. boissierianum*, *P. calorum* (hybrid), *P. cardinale* (hybrid), *P. caudatum*, *P. hirtzii*, *P. kovachii*, *P. longifolium*, *P. schlimii*, and others.

Phragmipedium besseae
(named in honor of Libby Bessy)

Phragmipedium **Andean Fire**

Phragmipedium **Memoria Dick Clements**

Phragmipedium schroderae

Phragmipedium longifolium

Phragmipedium **Eric Young**

Phragmipedium calorum

Phragmipedium **Elizabeth March**

Phragmipedium grande

SUBFAMILY EPIDENDROIDEAE (520 genera and 21,100 species, including *Bulbophyllum, Brassia. Catasetum, Cattleya, Ceratostylis, Coelogyne, Dendrobium, Dracula, Encyclia, Epidendrum, Gomesa, Liparis, Masdevallia, Maxillaria, Oncidium, Pleurothalis, Polystachya,* and *Sobralia,* among many others. Includes 80% of orchid species). Mostly epiphytic, monopodial; roots with pneumatophores (aerial roots); stems and leaves fleshy; leaves unifacial; anther ascending (bent forward), with beak; pollinia hard, 2–12, with waxy surface.

Acineta superba (Native to Venezuela, Colombia, and Peru)

Aerangis hyaloides (glossy aerangis, endemic to Madagascar)

Ancellia africana (leopard orchid)   *Angraecum praestans* (native to Madagascar)

Brassia

SPIDER ORCHIDS

ETYMOLOGY [Named to honor William Brass, eighteenth-century British plant collector and illustrator. He collected in West Africa for Sir Joseph Banks in 1782–1783, and died at sea in 1783]

GROWTH HABIT Epiphytic with a stout rhizome; vary from small to comparatively very large sympodial plants. **STEM/BARK**: pseudobulbs subglobose-ovoid to oblong-cylindrical. **LEAVES**: linear, elliptic to oblong-lanceolate; usually with one or two and sometimes three leathery leaves per pseudobulb, subtended by two or more pairs of imbricating sheaths; there is one main vein below. **FLOWERS**: bisexual; from small to large, on lateral basal few to many-flowered racemose or paniculate inflorescences that are usually as tall as the leaves; sepals linear-lanceolate, free and similar; petals shorter than the sepals but similar to them; lip sessile, not attached to the column, spreading and shorter than the sepals and petals; column short, stout, and without a foot; pollinia 2. May flower more than once each year. **FRUIT**: fleshy trilocular capsule with numerous small seeds. **NATIVE HABITAT**: wet tropical America, from Central America (Mexico) to northern South America (Venezuela), and the Caribbean. *Brassia caudata* is native to Florida and elsewhere.

CULTURE In bright shade in compost or bark mixes, keep on dry side during rest period; should not be disturbed for several years. **PROPAGATION**: division of pseudobulbs, seed in sterile media, or meristem tissue culture. **LANDSCAPE USES**: in containers or on tree trunks preferably in greenhouses or conservatories. **HARDINESS ZONE**: 12–13A.

COMMENTS *Brassia* consist of 34 species. Their long tepals make brassias somewhat unique among orchids, hence the vernacular spider orchids. In addition to the cultivated species such as *B. caudata*, *B. lanceana*, *B. maculata*, *B. pumila*, *B. warscewiczii*, and others, there are several inter-specific as well as intergeneric hybrids with such familiar genera as *Cattleya*, *Miltonia*, *Odontoglossum*, and *Oncidium*, among others.

Brassia aurantiaca (native to Colombia, Ecuador, and Venezuela)

Brassia gireoudiana

Brassia cochleata

Brassia Orange Delight 'Starbeck'

Brassia Memoria Bert Fields

***Brassavola* 'Little Stars'**
(B. nodosa × B. subulifolia)

***Burrageara* Nelly Isler 'Swiss Beauty'**

Bulbophyllum

Bulbophyllum eberhardtii (= *B. longiflorum* ?)
The largest genus in Orchidaceae (1,867) and among the largest in flowering plants. Epiphytic and lithophilic orchids in diverse habitats in warmer areas of Africa, Asia, Latin America, etc.

Bulbophyllum blumei (= *B. masdevaliacum*; native to peninsular Asia and Australia; rhizomatous, with small pseudobulbs, each with a single leaf.

Calanthe discolor (left and middle) and *Calanthe striata* (right), both native to Korea, China, and Japan, adapted to cooler climates. (Christmas orchids). Photographed at US National Arboretum, Washington D. C.

Cattleya

CATTLEYA

ETYMOLOGY [After William Cattley (1788–1835), botanist and one of the first to successfully grow *Cattleya labiata* and epiphytic orchids].

GROWTH HABIT Evergreen sympodial epiphytes, 3–18 in. (8–50 cm), average: 12 in. (25 cm); growth rate slow. **STEM/BARK**: pseudobulbs average 8 in. (20 cm) tall, 1in. (2 $\frac{1}{2}$ cm) wide; some short and others tall and reed-like. **LEAVES**: unifoliate or bifoliate, 1 or 2 leathery or fleshy leaves at apex of pseudobulb; 4–12 in. (10–25 cm) long, 1–4 (2 $\frac{1}{2}$ –10 cm) wide. **FLOWERS**: terminal, one inflorescence stalk from each pseudobulb; one to few-flowered, 3–7 in. (7 $\frac{1}{2}$ –21 cm) across; sepals are relatively narrow, free, and more or less equal; petals much broader than sepals, lip sessile, conspicuous, and usually free, folded at base but expanded at the apex; column long and semi-waxy; occur in all colors except blue, but purple is the dominant color, and white, yellow, red and bicolor and tricolor are also available; some species are fragrant; most species flower in fall, winter, and spring but some may flower in summer. **FRUIT**: trilocular capsule. **NATIVE HABITAT**: Central and South America, from Costa Rica to Argentina.

CULTURE Filtered light; warm but not hot temperature, in bark; tree fern, or any freely draining epiphytic mix; allow surface of the media to dry out briefly between watering; Botrytis spotting of flowers; virus causing flower break or distorted flower shape and color, as well as brown rot and scale are potential problems. **PROPAGATION**: division into clumps of 4 or more growths plus lead; meristem tissue culture; seed in 5–7 years to flower, usually in sterile media; backbulbs (pseudobulbs remaining after flowering) may also be used. **LANDSCAPE USES**: cut flowers in the past but currently more popular as decorative pot plant when in bloom, popular corsage orchid. **HARDINESS ZONE**: 12A–13A. Usually greenhouse grown.

COMMENTS Various authors disagree as to the number of species, although 151 species and hundreds of intergeneric and interspecific hybrids, grexes, and cultivars commonly listed in the trade. *Cattleya labiata* is the origin of common purple cattleyas; *C. dowiana* is the origin of yellow/red lip hybrids; while *C. skinneri* is the origin of purple flower which is 3 in. (8 cm) across and produces as many as 5–10 flowers per stalk. In addition to the interspecific hybrids, *Cattleya* has been hybridized with species of several genera, particularly *Brassavola*, *Sophronites*, and *Laelia* have been combined with *Cattleya* by some recent authors, increasing the accepted number of species to 113.

Cattleya amethystoglossa

Brassolaeliocattleya **Castle Treasure**

Cattleya Earl **'Imperialis'**

Cattleya gaskelliana

Cattleya lueddemanniana
'Standley' × Orquimeil

Cattleya **Kagaribi Dawn 'Red Star'**
(Sophrolaeliocattleya Kagaribi Dawn 'Red
Star')

Cattleya latifolia subsp. *rubra*

Cattleya trianae var. *semi-alba*

Cattleya trianaei 'The President'

Cattleya Hwa Yuan Gold 'Yhuan Kuan # 2'

Cattleya massiae

Cattleya intermedia

Brassolaeliocattleya Mem Anna Balmores
'Convex'

Brassolaeliocattleya Blc Hua Yuan Gold (?)

Brassolaeliocattleya Kawale Gem 'Xin Xin'

Brassolaeliocattleya Sanyung Ruby 'Shin Mei'

Brassolaeliocattleya Dal's Rapture
'Water Colors'

Brassolaeliocattleya Pamela Finney
'Pink Beauty'

Brassocattleya thalie (Bc deesse × Cat. odalisque)

Brassocattleya Irene Holguin

Brassocattleya 'Roman Holiday'

Cattleya (*Laelia*) *purpurata* var. *carnea*

Laeliocattleya g. Longwood Gardens

Laeliocattleya Casitas Springs 'Linden'

Brassidium 'Shooting Star'   *Cattlianthe* Trick or Treat 'Orange Beauty'

Cochleanthes amazonica

Coelogyne cristata 'Wossen'   *Coelogyne pandurata*
                                                 Bessey Creek

Dendrobium

DENDROBIUM

ETYMOLOGY [From Greek *Dendron*, a tree and *bios*, life, in reference to their epiphytic habit].

GROWTH HABIT Evergreen or deciduous; sympodial; epiphytic or lithophytic (on rocks); variable foliage density, 5 in. (10 cm) to 5 ft. (1 $\frac{1}{2}$ m) or more; growth rapid during warm seasons but slow to none in cooler seasons. **STEM/BARK**: uncommonly variable, rhizomatous or pseudobulbous: from a single node small and rounded to multi-nodes and 5 ft. (1 $\frac{1}{2}$ m) tall canes. **LEAVES**: highly variable depending on type; 1- many adult leaves; linear-lanceolate to oblong-ovate 1–15 in. (2 $\frac{1}{2}$ –37 cm) long and 1–2 in (2 $\frac{1}{2}$ –5 cm) wide; leathery (evergreen), terete, or papery, bunched at the apex or distichously arranged along stem. **FLOWERS**: terminal or axillary racemose inflorescences bearing 1–100+ flowers, from less than 1 (2 $\frac{1}{2}$ cm) to 3 (7 $\frac{1}{2}$ cm); sepals and petals uniform but petals may be broader than sepals, undulated, or helically twisted; lip entire or 3-lobed, with lateral lobes encircling the column, apical lobe constricted and attached to spur at base of column; pollinia 4, in 2 pairs; colors of white, purple, and yellow; not commonly fragrant. **FRUIT**: trilocular fleshy capsule containing numerous minute seeds without endosperm. **NATIVE HABITAT**: in much of Asia, the Philippines, Australia, and Pacific islands, from wet tropical forests to dry desert of central Australia.

CULTURE In medium light intensities; in epiphytic mix of bark and/or tree fern; temperature requirements for flowering vary from cool-growing *D. nobile* types and deciduous, cool temperature in fall to initiate flowers to *Dendrobium phalaenopsis* types that are warm-growing. Many others are intermediate; allow surface of medium to dry briefly between watering; most taxa are relatively problem-free. **PROPAGATION**: division, meristem tissue culture, from seed in 3–5 years to bloom, offsets from canes. **LANDSCAPE USES**: potted plants when in bloom but inflorescences and flowers (particularly *D. phalanopsis*) are used in flower arranging and various ceremonial events. **HARDINESS ZONE**: 10A–13B, but primarily greenhouse plants.

COMMENTS With 1217 species plus hybrids, *Dendrobium* may be the largest genus in the Orchidaceae. Number and morphological diversity of the species has been and continues to be a source of confusion and disagreement among orchid taxonomists. Hundreds of species and hybrids are in the trade. *Dendrobium aggregatum* (clustered 2 in., 5 cm pseudobulbs, round yellow flowers in spring); *D. nobile* (deciduous; white and purple blooms in 2's and 3's at nodes in winter-spring); *D. phalaenopsis* (sprays of "butterfly" flowers in Fall in colors of white through purple). A few examples of other species in the trade include *Dendrobium antennatum*, *D. atroviolaceum*, *D. bigibbum*, *D. caniculatum*, *D. bracteosum*, *D. finisterrae*, *D. kingianum*, *D. lawesii*, *D. secundum*, *D. speciosum*, *D. spectabile*, *D. superbum*, *D. tetragonum*, *D. topaziacum*, *D. victoriareginae*, and *D. woodsii*.

Dendrobium victoria-reginae

Dendrobium lindleyi   *Dendrobium palpebrae*

Dendrobium BKK Money Orange

Dendrobium bigibbum

Dendrobium **Den Mini Pink**

Dendrobium densiflorum

Dendrobium **Emerald Gold**

Dendrobium loddigesii

Dendrobium Woon-Leng hybrid

Dendrobium snowman × *delicatum*

Dendrobium Salaya Mimist

Dendrobium Stardust Gold Medal

Dendrobium 'Miss Singapore'

Dendrobium phalaenopsis

Dendrobium chrysotoxum

Doritaenopsis 'Chain Xen Diamond'   ×*Doritaenopsis* Sogo Berry

Epidendrum

EPIDENDRUM, REED-STEM EPIDENDRUMS

ETYMOLOGY [From Greek *epi*, upon and *Dendron*, a tree, in reference to epiphytic habit of some species].

GROWTH HABIT Evergreen, sympodial, epiphytic, or lithophytic; variable foliage density and size, from 1 in (2 $\frac{1}{2}$ cm) to 5 in (1 $\frac{2}{3}$ m) tall, usually around 1 ft. (30 cm) or less. **STEM/BARK**: caulescent, erect or creeping, sometimes branching, to 5 ft. (1 $\frac{2}{3}$ m) tall and 1 in (2 $\frac{1}{2}$ cm) in diameter, bearing few to many leaves. **LEAVES**: linear-lanceolate to ovate; alternate, thick and fleshy or leathery, to 4 in. (10 cm) long and 1 (2 $\frac{1}{2}$ cm) wide. **FLOWERS**: terminal inflorescence bearing few to many flowers, each ranging from less than 1 (2 $\frac{1}{2}$ cm) to 3 in. (7 $\frac{1}{2}$ cm) wide; uniform sepals and uniform petals which are slightly narrower than sepals; lip 3-lobed with lateral lobes attached to the column; center lobe of lip may be broad, lobed, or fringed; pollinia 4; colors include green, brown, yellow, red, purple, and white; many are fragrant. **NATIVE HABITAT**: North Carolina to south Florida; Central and South America, from tropical forests to dry sandy back dunes.

CULTURE Bright light, intermediate temperature range of about 50 °F (10 °C) night and 70°–75 °F (21–24 °C) day time, in epiphytic mix of bark and/or tree fern fiber or in some cases in organic materials with addition of some coarse sand; allow surface of media to dry between watering; red spider mites, leaf spots and virus may prove troublesome. **PROPAGATION**: division, offsets, meristematic tissue culture; from seed in 4–5 years to bloom. **LANDSCAPE USES**: naturalized on trees in areas with 45 °F (7 °C) or warmer nights. Potted plant when in bloom; reed stem types can be grown in outdoor beds in frost-free climates. **HARDINESS ZONE**: 9B–13A.

COMMENTS Listed number of species 1335, exceed that of *Dendrobium*, in addition to hybrids. *Epidendrum ibaguense* (= *E. radicans*) (reed stem type; terminal inflorescence produces red/orange and yellow 1″ flowers for months); *E. nocturnum* (stems to 3 ft. (1 m) tall, bear few 3–4 in (7 $\frac{1}{2}$ –10 cm) greenish-white flowers that are fragrant at night, Florida native); examples of other species in the trade include *E. anceps*, *E. cinnabarinum*, *E. eburneum*, *E. ilense*, *E. paniculatum*, *E. secundum*, *E. violescens*, among others.

Epidendrum radicans

Epidendrum Hokulea 'Lava'

Epidendrum cinnabarinum

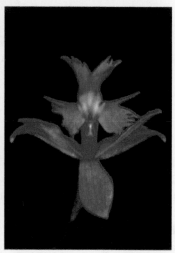

Epidendrum secundum

Epidendrum viviparum

Epidendrum anceps
(= *Coilostylis vivipara*)

Epidendrum **'Pink Cascade'** (photographed at Busch Gardens, Tampa Florida)

Epidendrum magnoliae (green fly orchid), the only epiphytic orchid in the United States that grows north of Florida (photographed in Archer Florida)

Epidendrum ilense (Ila epidendrum; native to Ecuador)

Masdevallia

MASDEVALLIA

ETYMOLOGY [Named in honor of José Masdeval Terrades Llobet Berenguer (1730 or 1740–1801), physician and botanist in the court of Carlos III and IV of Spain.]

GROWTH HABIT Compact sympodial epiphytic or terrestrial herbs, 1–12 in. (2 $\frac{1}{2}$ –30 cm) tall, and species with vegetatively very similar morphology; medium growth rate. **STEM/BARK**: without pseudobulbs, hence lacking cauline stems, but they possess elongated creeping rhizomes from which lateral flowers arise. **LEAVES**: leathery ovate to lanceolate, usually pendent, in tight clusters. **FLOWERS**: corolla reduced, sepals the most conspicuous part of the flowers; these are fused to form an open cup or sometimes a tube, the three elongated tail-like apices are characteristic of the genus; lips are tongue-shaped and attached to the free column; there is an astonishing diversity of flower shapes and colors in this relatively large genus. **FRUIT**: trilocular capsule containing numerous minute seeds. **NATIVE HABITAT**: from southern Mexico to southern Brazil with the greatest concentration in the Andean cloud forests of Colombia, Ecuador, and Peru, in elevations from sea levels to more than 13,000 ft.

CULTURE Best in lower light intensities; cool to intermediate temperatures with a 10°°F–15 °F (8 °C–10 °C) differences between day and night; in typical well-drained orchid bark, tree fern, or similar media; respond negatively to high light intensities, over watering, and lack of air movement. **PROPAGATION**: seed, division, tissue culture. **LANDSCAPE USES**: potted plants or on mossy tree branches in greenhouses and conservatories; not useful as house plants for long periods. **HARDINESS ZONE**: 11A–13A.

COMMENTS With 586 species *Masdevallia* is a large and diverse genus in its floral characteristics. However, despite their unique features, species of the genus are not among the more popular orchids and are grown primarily by collectors and enthusiasts. The genus is included here to represent their unique flowers and the range of orchid diversity. Examples of the more commonly listed species include *M. attenuata*, *M. caudata*, *M. coccinea*, *M. militaris*, *M. uniflora*, and *M. veitchiana*, among others. There are, however, a significant number of intrageneric hybrids with its two closely related genera, *Dracula* and *Porroglossum*.

Masdevallia coccinea var. *harriana*

Masdevallia ignea

Masdevallia veitchiana

Masdevallia Mary
Staal 'Sun Gold'

Masdevallia
'Peach Allure'

Masdevallia Ibanez Behar

Masdevallia uniflora

Masdevalia Prince
Charming 'Speckles'

Masdevallia caudata (native to South America)

Maxillaria irrorata

Maxillaria tenuifolia

Miltonia and *Miltoniopsis*

PANSY ORCHIDS

ETYMOLOGY [After Charles Fitzwilliam Viscount Milton (1786–1857), a devoted evangelical Christian, a politician, and a horticultural patron] [Based on the genus name, and the Greek ending *–opsis*, indicating resemblance to *Miltonia*]

GROWTH HABIT Sympodial epiphytic or sometimes lithophytic herbs, to 20 in. (50 cm) tall. **STEM/BARK**: scaly rhizomes with light green flattened pseudobulbs, about 4 in. (10 cm) tall and separated by rhizomes in *Miltonia*, but tightly packed in *Miltoniopsis*. **LEAVES**: elliptic-lanceolate to linear, coriaceous, *Miltonia* with a pair of light green leaves atop each pseudobulb, but only one in *Miltoniopsis*; apexes round or sometimes only slightly pointed. **FLOWERS**: erect unbranched racemose inflorescences are axillary, arising from the base of the newest maturing pseudobulb and may have one to many medium size to large flowers; pedicellate, showy, flat (all parts in one plain); in a range of colors, from white to dark purple; petals and sepals are usually similar to each other; labellum simple or only slightly lobed and fused to the column; pollinia two, yellow; mostly from spring to late summer. **FRUIT**: 3-locular capsule. **NATIVE HABITAT**: *Miltonia* predominantly on coastal forest regions of central to southern Brazil, with one species extending to Argentina; *Miltoniopsis* from cooler regions of the Andean slopes.

CULTURE These do not respond well to direct sun; should be planted in well-drained epiphytic mixes and not allowed to dry out between watering. **PROPAGATION**: seed, division, and meristem culture. **LANDSCAPE USES**: container plants and particularly suitable for conservatories, on tree branch setups. **HARDINESS ZONE**: 11–13A.

COMMENTS Despite recognition of **Miltonia** and **Miltoniopsis** as two distinct genera since 1899, dissimilarity between them remains obscure, perhaps to all but the expert orchidologists. Nomenclature of the taxa in the accompanying plates is based on those provided by the identification labels, literature, or websites. *Miltoniopsis*, however, consists of 5 epiphytic species with 1-leaved pseudobulbs and racemes of showy flowers that resemble pansies, hence the common name. These were previously included in *Miltonia* with 18 species and with two leaves on each pseudobulb. Description of the two genera is combined because there was no access to the actual plants at the time of writing. There are a significant number of interspecific and intergeneric hybrids in both genera. Nomenclature of similar taxa is troubling. Understandably, the vernacular name "pansy orchid" is used interchangeably for all.

Miltonia **Pluto "Katie"**

Miltonia **Red Knight "The King"**

Miltonia **Lady Snow Paper Doll**

Miltonia **Bel Royal (Hannover × Red Knight)**

Miltonia **Alger Flamingo Queen**

Miltonia **Red Knight "The King"** (?)

Miltoniopsis Michoacan "Mariposa"

Miltoniopsis Maui Mist "Golden Gate"

Miltoniopsis "Herr Alexander"

Miltoniopsis Storm "Red Lightning"

Miltoniopsis Eros Kensington

Miltoniopsis White Summer "Angel Heart"

Miltoniopsis **Emotion "Red Breast"**

Miltoniopsis **Robert Strauss**

Miltoniopsis spectabilis

Miltoniopsis **"Jersey"**

Miltoniopsis **Waterfall**

Miltoniopsis **Echo Bay "Midnight Tears"**

Bratonia barkeria (= Miltassia barkeria)

×Odontocidium catatante "Pacific Sun Spots"

Guarianthe aurantiaca

Guarianthe guatemalensis

(Photographed at Selby Gardens, Sarasota, Florida)

×Laeliocatonia = (Broughtonia × Cattleya × Laelia Lctna Why Not

Lycaste skinneri **Adulte**   *Lycaste skinneri* **var.** *alba*

Oncidium

DANCING LADY ORCHID, ONCIDIUM
ETYMOLOGY [A diminutive of Greek *onkos*, a tumor, in reference to the swelling on the lip of these orchids].

GROWTH HABIT Evergreen, sympodial, epiphytic; variable foliage density 5–24 in. (13–60 cm) tall; growth rate slow (some medium-fast). **STEM/BARK**: pseudobulbs from very small to very large, subtended by sheaths, 8 in. (20 cm) tall and 5 in. (13 cm) across. **LEAVES**: small, soft and thin, papery, terete, or large, thick and leathery ("mule-ears"); light or dark green, or green with brown or red stippling. **FLOWERS**: terminal or axillary inflorescences bearing few to many 1–4 in. (2 $\frac{1}{2}$ –10 cm) flowers, "Dancing Lady" shape: sepals usually similar; petals similar to dorsal sepal; lip usually large and spreading, attached to base of column; primary colors are yellow and brown but also pink, white, and green tones; flowering fall, winter or spring. **FRUIT**: 3-locular capsule. **NATIVE HABITAT**: Florida (*Oncidium ensatum*), West Indies, Central and South American tropics.

CULTURE Light requirements vary from shade to full sun, depending on individual species, in well-drained bark and/or tree fern mix; temperature of mostly intermediate but some species may require warm and humid conditions; allow 1–5 days (depending on growth type) between watering; red spider, virus, and leaf spot may become problematic.

PROPAGATION: seed, division, and meristem culture. **LANDSCAPE USES**: some species may be naturalized or mounted on trees, containers, and hanging baskets. **HARDINESS ZONE**: 10B–13A. The species grown outdoors may have to be protected below 45 °F. (7 °C)

COMMENTS *Oncidium* is a complex genus of about 324 species and hundreds of interspecific and intergeneric hybrids. *Oncidium ampliatum* ("turtle shell"-shaped pseudobulbs, with many yellow "dancing ladies" in spring); *Oncidium papilio* (butterfly orchid, at end of 2 ft. (60 cm) inflorescence and with attractive red stippled foliage). A few other commonly cultivated species include *Oncidium baueri*, *O. concolor*, *O. crispum*, *O. haephamaticum*, *O. hastatum*, *O. isthmi*, *O. microchilum*, *O. onustum*, *O. ornithorhynchum*, *O. sphacelatum*, *O. trulliferum*, **and** *O. velutinum* (= *Tolumnia velutinum*), among others. Equitant oncidiums, the popular small, fast-growing, 4 in. (10 cm) plants that bloom profusely, are now segregated as the distinct genus *Tolumnia*, differing from *Oncidium* in having a fan of 6–8 succulent v-shaped leaves to 6 in. (15 cm) long, with toothed margins. In fact, similarity of *Oncidium*, *Tolumnia*, *Odontoglossum*, *Miltonia*, *Miltoniopsis*, and other related genera has been undergoing changes in recent years and interested students should examine the recently published literature for specific information. Species currently placed in the genus *Psychopsis* were previously included in *Oncidium*. Photographs of two species are included to illustrate their unusual structure.

Oncidium sphacelatum

Oncidium ampliatum

Oncidium maculatum

Oncidium Sharry Baby "Yellow"

Oncidium "Dancing Lady"

Oncidium colmanara "Tricolor Butterfly"

Oncidium **Wildcat**
"Golden Red Star"

Oncidium **Mendenhall "Hildos"**

Oncidium **Sharry Baby**
"Sweet Fragrance"

***Psychopsis × papilio* "Kalihi"**
(= *Oncidium papilio* "Kalihi")

Psychopsis versteegiana
(= *Oncidium versteegiana*)

Phaius tankervilleae

NUN'S ORCHID, PHAIUS, SWAMP ORCHID
SYNONYM *Phaius wallichii.*

ETYMOLOGY [From Greek *phaios*, dusky or gray, in reference to the dark flowers of the type species or dark color of old or injured flowers] [After Lady Emma Colebrooke, Countess of Tankerville (1752–1836), who amassed a large collection of exotic plants in Tancarville, in Normandy France].

GROWTH HABIT Evergreen, sympodial, semiterrestrial, densely foliated, $1\frac{1}{2}$ –4 in. (4–10 cm) tall and half as wide; growth rate slow. **STEM/BARK**: pseudobulbs are 2–3 in. (5–8 cm) across, thick and stocky, hidden by sheathing leaf bases. **LEAVES**: 2–10, alternate or appearing spiral, thin and broadly elliptic, ribbed, and folded. **FLOWERS**: the 3–4 ft. (1–$1\frac{1}{3}$ m) tall inflorescence arises from a leaf axil near the rhizome, bearing many 2–3 in. (5–8 cm) flowers in the spring; inflorescence bears large leafy bracts at the nodes and subtend the flowers; sepals and petals are similar: creamy-white on the outside, yellow/rusty-brown on the inside; the 3-lobed lip has two lateral lobes encircling the column, the middle lobe spreading, giving a tubular appearance; lip is yellow-brown and has a small spur at the base; flowers darken with age. **FRUIT**: trilocular capsule with numerous minute seeds. **NATIVE HABITAT**: widespread in tropical Asia, from India to the Philippines and Australia, naturalized in Florida and Hawaii.

CULTURE Prefers somewhat bright light, intermediate-cool temperature, in peat, perlite, and soil in equal amounts; keep moist; relatively problem-free but red spider and rust may present problems. **Propagation**: division usually; also, old inflorescence placed on moist sphagnum moss will produce offsets from nodes, seed, tissue culture. **LANDSCAPE USES**: decorative pot plant for flowers or in border in warmer areas. **HARDINESS ZONE**: 10A–13A, may be grown in protected areas in cooler climates.

COMMENTS *Phaius* consists of 46 species, of which *Phaius tankervilleae* is commonly grown and widely sold member of the genus. There is also the white variety, *Phaius tankervillaea* **var**. *alba*, *P. tankervillaea* **var**. *australis*, and *P. robertsii*, which are in the trade.

Phaius tankervilleae

Phalaenopsis

MOTH ORCHID, PHALAENOPSIS

ETYMOLOGY [From the Greek *phalaina*, moth and *opsis*, appearance, in reference to the delicate moth-like flowers of some species].

GROWTH HABIT Evergreen, monopodial, mostly epiphytes, less than 1 ft. (30 cm) tall and wider than tall; slow to medium growth rate. **STEM/Bark**: very short, apparently stemless and lacking pseudobulbs; aerial adventitious roots common. **LEAVES**: Three to several, leathery, succulent, usually 4–12 in. (10–30 cm) long and 2–5 in. (5–13 cm) wide, solid green or silver-green with dark green mottling, purplish underside. **FLOWERS**: axial (from leaf axils) racemose or paniculate inflorescences produce few to many 1–6 in. (2 $\frac{1}{2}$ –15 cm) flowers mostly in winter-spring but in some cases all year; sepals are uniform; petals uniform, equal to or broader than sepals; lip is 3-lobed, lateral lobes upright, middle lobe may have two antennae at end; callus between lateral lobes; basic colors are white, pink, yellow, and variously lined or mottled; a few are fragrant. **FRUIT**: trilocular capsule with numerous minute seeds. **NATIVE HABITAT**: Tropical Asia, Australia, and India.

CULTURE Low light intensity and warm temperatures (do well as houseplants); in bark and/or tree fern; keep moist but avoid crown rot which is caused by water in crown of plant; bud blasting due to change in environment or drafts and mesophyll collapse with virus. **PROPAGATION**: offsets from lower "stem" (keikis) or from some nodes on inflorescences; meristem tissue culture; from seed in 2–4 years to blooming. **LANDSCAPE USES**: potted plants with long-lasting display of flowers primarily in winter-spring; with proper care, after first set of flowers, inflorescence may branch and produce more flowers (only on *P. amabilis* types). **HARDINESS ZONE**: 11B–13A, generally grown in greenhouses.

COMMENTS About 69 species and numerous hybrids, grexes, and cultivars of small and large, solid or mottled leaves and flowers, rose, cream, yellow, red, and other colors. Flower color variation in the numerous cultivars of this genus is awe-inspiring and sometimes even difficult to describe. *Phalaenopsis* cultivars are without doubt the most popular of all orchids, the most widely grown and sold.

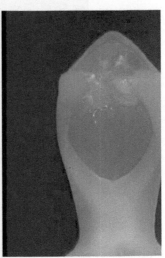

Phalaenopsis amabilis, sepals uniform; petals uniform, broader than sepals (left); lip is 3-lobed, lateral lobes upright, middle lobe with two antennae (middle), and close-up of column with two pollinia (right)

Phalaenopsis Pink Lady "Kayo"

Doritaenopsis (*Dorites* × *Phalaenopsis*) **Little Gem**

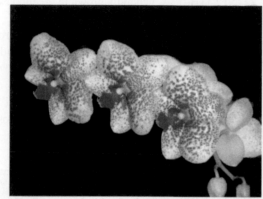
Phalaenopsis Alysha's Dots "Rosemarie"

Phalaenopsis Art Quest "Best Yet"

Phalaenopsis "Brave Heart"

Phalaenopsis Brother Kaiser

Phalaenopsis "Frank Sarris"

Phalaenopsis "Spring Stripe"

Phalaenopsis Spotted Harlequin   *Phalaenopsis* **Dtps Minho Princess**   *Phalaenopsis* Quevedo

Phalaenopsis **magenta striped**   *Phalaenopsis* "Leopard Prince"   *Phalaenopsis aphrodite* subsp. *formosana*

Phalonopsis Alysha'' Rose "Shelden"   *Phalaenopsis* Hawaian Legend   *Phalonopsis* Dtsps KV Charmer
                                         ''Maria Theresa''

Promenaea xanthina (= *Zygopetalum xanthium*), native to Brazil

Galeottia fimbriata

Prosthechea radiata

Rhyncholaelia glauca (= *Brassavola glauca*), native to Central America

Rhynchostylis gigantea (foxtail orchid), native to Southeast Asia

Sobralia decora

× ***Sophrolaeliocattleya*** × ***Poinara***
(*Sophronitis* × *Laelia* × *Cattleya* × *brassavola*)

Sophronitis coccinea

Stanhopea costaricensis

Spathoglottis plicata (Philippine ground orchid)

Vanda

VANDA
ETYMOLOGY [From the Sanskrit or Hindi name for one of the species (*Vanda tessellata*)].

GROWTH HABIT Upright, evergreen, monopodial, mostly epiphytic, 1–6 ft. ($\frac{1}{3}$–2 m) tall or taller, 6–24 in. (15–60 cm) wide; coarse texture; fast growth rate in sunny tropical environments. **STEM/BARK**: sheathed by leaf bases; thick aerial roots present. **LEAVES**: two types: distichous and strap-leaved, 2–12 in. (5–30 cm), folded, leathery, clasping leaf base, "torn" apex; terete-leaved size of a pencil, 4–6 in. (10–15 cm) long, sharp or blunt tip. **FLOWERS**: axial inflorescence produces two to many 1–4 in. (2 $\frac{1}{3}$–10 cm) flowers, in any season; appear rounded with similar sepals and petals; petals may have a 180° twist; lip is commonly tongue-shaped with a spur at base of column; column is short and cylindrical; colors of lavender, blue, pink, yellow, brown, white, and green; some are fragrant. **FRUIT**: 3-locular capsule. **NATIVE HABITAT**: Tropical Asia.

CULTURE Lower medium light in greenhouse, brighter outdoors. Terete-leaved varieties require brighter light than strap-leaved varieties; warm conditions; very airy media, in bark, charcoal, tree fern; and water when growing fast; leaf spot, brown rot, virus. **PROPAGATION**: tip cuttings, after removing top portion bottom may produce offsets, meristem tissue culture, from seed in 4–7 years to blooming. **LANDSCAPE USES**: potted plant, specimen plant for patio and pool areas. **HARDINESS ZONE**: 10A–13B. May be grown in outdoor beds in full sun in Hawaii and South Florida.

COMMENTS There are 56 species and hundreds of hybrids. ***Vanda coerulea*** (smaller-growing strap-leaved, produces many 2 $\frac{1}{2}$ in. (7 cm) lavender-blue flowers in the fall). *Vanda teres* (terete-leaves with delicate looking pink flowers, 3–4 (8–10 cm) across. *Vanda tricolor* (strap-leaved plant with many 2 in. (5 cm) white, reddish-brown spotted flowers with purple lip). As one of the more commonly sold orchids, other *Vanda* species and hybrids include ***V. cristata***, ***V. denisoniana***, ***V. hookeriana***, ***V. insignis***, ***V. lamellata***, ***V. luzonica***, ***V. tessellata***, and others as well as a whole host of interspecific and intergeneric hybrids, particularly with species of ***Ascocenda*** with which it shares considerable similarities and is represented here in a few photographs to illustrate the point.

***Vanda* Fuchs Delight Dr. Anek** × ***Ascosenda* Kaltana Gems**

Vanda Motes Goldpiece "24K"   Vanda Fuchs Violetta "Crownfox"

Vanda coerulea

Vanda sanderiana

Vanda luzonica N Chavanand ×
V. Misty Shinsato "Mary Motes"

Vanda Robert's Delight
"Ink Blue"

Ascocenda **John De Biase "Angela"**

Ascocenda **Miami "Mandarin"**

Ascocenda **Somsri Gold**

Ascocenda **Kulwadee "Fragrance"**

Vanda **Kunwadee "Fragrance"**   *Ascocenda* **Princess Mikasa**

Zygopetalum

ZYGOPETALUM
ETYMOLOGY [From Greek *zygon*, yoke and *petalon*, petal, in reference to holding together flower segments by the callus]

GROWTH HABIT Sympodial epiphytic herbs, to about 2 ft. (60 cm) all; medium growth rate. **STEM/BARK**: tightly clustered pseudobulbs, egg-shaped, 2–3 in. (5–8 cm) tall, ensheated by leaf bases that usually deteriorate with age. **LEAVES**: to 6 stiff strap-shaped–lanceolate leaves arise from each pseudobulb, folded at base. **FLOWERS**: to 10 flowers on tall erect inflorescences, from axils of a sheath; flowers are 2–3 in. (5–8 cm) wide; sepal and lateral petal colors are predominantly brown and labellum pink-lavender and sometimes white or bluish-purple; exceptionally strongly fragrant; blooming all year depending on species. **FRUIT**: 3-locular capsule. **NATIVE HABITAT**: tropical South America, from Brazil to Peru, Paraguay, Bolivia, and Argentina.

CULTURE Partial shade (no direct sun, except early or late); in well-drained medium; best in intermediate temperatures and humid air; will withstand summer heat if shaded and well-watered; intolerant of wet media and cold; leaf spotting as a result of moisture on leaves at night when cold; should be repotted in fresh medium every 2 or 3 years. **PROPAGATION**: seed, division of pseudobulbs, and tissue culture. **LANDSCAPE USE**: best in shallow pots and baskets. **HARDINESS ZONE**: 9B–13B.

COMMENTS A total of 15 *Zygopetalum* species is listed by The Plant List. In addition, there are many intergeneric hybrids. The waxy flowers are long-lived and favored also because of their pleasant fragrance. As in most orchids, *Zygopetalum* species hybridize with species of other genera such as *Cochleanthes*, *Lycaste*, *Zygosepalum*, and others.

Zygopetalum louisendorf   *Zygopetalum* **Artur Elle Wossen**

zygopetalum blackii "**Kathy**"   *Zygopetalum maculatum*   *Zygopetalum* "**Big Country**"
(= *Z. mackayi*)

SUBFAMILY ORCHIDOIDEAE (217 genera and 4965 species, including *Caladenia*, *Disa*, *Habernaria*, *Microchilus*, *Platanthera*, and *Pterostylis*, among others). Roots tuberous or not; leaves soft, spiral, herbacious; anther erect, apex acute; staminodes of inner whorl reduced; pollinia soft; fertile stamen 1.

Ophrys sphegodes (early spider orchid). Photographed in France

Orchis simia (photographed in France)

Ludisia discolor

JEWEL ORCHID
SYNONYM *Haemaria discolor.*

ETYMOLOGY [From Latin *ludisia*, dancer, actor, probably in reference to the usually slightly leaning inflorescences] [Variegated, in reference to color of leaves].

GROWTH HABIT Evergreen terrestrial, sympodial plants with moderate foliage density, to 10 in. (25 cm) tall (usually procumbent); slow growth rate. **STEM/BARK**: succulent, brittle, dark red-brown, procumbent. **LEAVES**: ovate, 4–6 tightly clustered at apex of stem, each to 3 in. (8 cm) long and half as wide; dark velvety-green (sometimes with red or yellow veins) above and blood red below. **Flowers**: terminal hairy inflorescence bears up to 12 flowers in winter/spring; each $\frac{3}{4}$ in. (2 cm) white flower has a yellow anther cap and is subtended by a large leafy bract; sepals and petals are similar, and the lip has a sac at the base; column twisted. **FRUIT**: fleshy trilocular capsule containing numerous minute seeds without endosperm. **NATIVE HABITAT**: Indonesia and Southeast Asia, including Burma (Myanmar), Southern China, Thailand, Philippines, and Malaysia.

CULTURE Best colors in bright shade (no direct sun), but tolerate lower light intensities as well; intermediate cool temperature; in peat moss, perlite, and soil in equal proportions; keep relatively moist; fairly problem free but may lose foliage occasionally. **PROPAGATION**: division and tip cuttings. **LANDSCAPE USES**: attractive pot or hanging basket plant year-round because of its eye-catching foliage; may be planted in-ground at greenhouse and conservatories. **HARDINESS ZONE**: 10B–13B.

COMMENTS Thought to have one species and one variety, *Ludisia discolor (= Haemaria discolor)* – velvety green leaves, sometimes with dark red veins. *Ludisia discolor* **var.** *dawsoniana* (= *Haemaria discolor* var. *dawsoniana*) has red or yellow veins. Recent discovery of a second species *Ludisia ravanii* from the Philippines brings the number of species to two.

Ludicia discolor in hanging basket (left) and *L. discolor* "**Nigrescens**" in a conservatory ground bed (right and below)

Grammatophyllum speciosum
(Giant orchid, 25 ft., 7⅔ m)
World's tallest orchid native to
Indonesia

Schoenorchis pachyacris
(miniature orchid, 1½ in., 3½ cm)
World's smallest orchid native to
Java and Sumatra

(Photographed in Bogor Botanic Gardens, Indonesia)

SUBFAMILY VANILLOIDEAE (14 genera, 245 species, including *Vanilla*): plants monopodial, glabrous; leaf venation reticulate; tepals often carinate (end in a point), margins of labellum fused with column; staminodes 2; anther bent forward by massive expansion of apical column; tepals persistent in fruit; fruit baccate; seeds relatively large, winged. Pantropical.

Vanilla planifolia

FLAT-LEAVED VANILLA
ETYMOLOGY [From the Spanish *vaina*, a small pod, in reference to the shape of the fruit] [With flat leaves].

GROWTH HABIT Evergreen monopodial epiphytic vine with variable heights. **STEM/BARK**: green, succulent, and sheathed by leaf bases. **LEAVES**: alternate, ovate, fleshy, shiny, up to 9 in. (22 cm) long and 3 in. (7 ½ cm) wide. **FLOWERS**: axillary clusters of few to many 3 in. (7 ½ cm) yellow-green flowers appear in spring; sepals and petals are similar; a ruffled tubular lip encircles the column; flowers

last but a day. **FRUIT**: the 6 in. (15 cm) or more long, ½ in. (1.3 cm) wide; seed pod is the original source of vanillin, or vanilla flavoring. **NATIVE HABITAT**: West Indies and Central America, but widely cultivated throughout the tropics.

CULTURE Bright light, but not direct sun; cool-intermediate conditions; in organic compost or epiphytic mix; support is more important. Essentially problem-free. **PROPAGATION**: stem cuttings; tissue culture. **LANDSCAPE USES**: as a curiosity and for vanilla "beans." **HARDINESS ZONE**: 11–13B.

COMMENTS Although there are about 103 species, only *Vanilla planifolia* is commonly grown and is also available as a variegated cultivar. It is cultivated for its fruit which is produced only by hand pollination outside its natural range. *Vanilla fragrance* is the only other species occasionally listed.

Vanilla planifolia

From Herman Adolf Köhler
Medizinal-Pflanzen (1887-1898)

Vanilla fragrance

IRIDACEAE

IRIS FAMILY

66 GENERA AND ±2244 SPECIES

GEOGRAPHY Cosmopolitan, but especially S. Africa and E. Mediterranean in a diversity of climates and habitats.

GROWTH HABIT Hermaphroditic perennial herbs with rhizomes, corms, or less often bulbs. **Leaves:** predominantly deciduous; equitant (overlapping at base) distichous, narrow, linear, mostly 2-ranked and often appearing in one plane, resembling a fan; sheathing at the base, venation parallel. **Flowers:** bisexual, actinomorphic or zygomorphic, epigynous or rarely hypogynous, in terminal cymose, paniculate or spike inflorescences; perianth of 6, biseriate, all petaloid but sometimes conspicuously different in shape and color (as in in *Iris*), united at the base to form a long or short tube, nectaries usually present at the base of tepals; stamens usually 3, filaments narrow and free or sometimes partially or entirely connate, often epitepalous (as in *Tigridia*, *Sisyrinchium*, etc.); carpels 3, united, trilocular or rarely unilocular (in *Hermodactylus*), ovary inferior or only exceptionally superior (in *Isophysis*), ovules many or rarely few or solitary; style usually trifid and often more or less petaloid (as in *Iris*, *Dietes*, and *Moraea*). **FRUIT:** loculicidally dehiscent capsule; seeds usually semi-globose or angular, embryo small and straight, endosperm present.

ECONOMIC USES Ornamentals, saffron (from stigmas of *Crocus sativus*), and perfumes from some *Iris* species.

CULTIVATED GENERA The more common include: *Babiana*, *Crocus*, *Crocosmia*, *Dierama*, *Dietes*, *Freesia*, *Gladiolus*, *Iris*, *Ixia*, *Moraea*, *Neomarica*, *Sisyrinchium*, *Sparaxis*, *Tigridia*, *Trimezia*, and *Watsonia*, among others.

COMMENTS **Iridaceae** is a large and diverse family with many important and common taxa in cultivation. Past classifications had recognized as many as 80 genera. The APG IV (2016) System recognizes eight subfamilies of which two are of greater horticultural interest. Species of the genera in Iridaceae are among the most popular garden flowers, and it may not be farfetched to assume that many species of most, if not all, 66 (80?) genera are in cultivation. A few examples include:

SUBFAMILY IRIDOIDEAE *Alophia amoena*, *A. drummondii*, *A. pulchella*, and probably others are herbs with tunicate corms, basal distichous leaves, and blue to violet terminal flowers with spreading tepals, native from Texas to South America; *Cypella herbertii*, *C. plumbea*, and others are cormous plants with pleated leaves, showy flowers with spreading (the inner recurved) tepals in a cymose inflorescence on a scape, native to South America; *Diplarrena moraea*, is a herb with short rhizomes, narrow, stiff, mostly basal leaves, and large zygomorphic flowers, native to Australia and Tasmania; *Moraea collina* (= *Homeria breyniana*), *M. polyanthos* (= *Homeria lilacina*), *M. ochroleuca* (= *Homeria ochroleuca*), and others are perennial herbs with tunicate corms, narrowly lanceolate leaves, and fugacious, variously colored flowers that are produced successively in terminal clusters, endemic to South Africa; *Orthrosanthus* *chimboracensis* (morning flag), is similar to *Sisyrinchium*, except for the free filaments, native to Central and South America, *O. multiflorus* (morning iris, morning flag), *O. polystachys*, although related species occur in Australia; *Sisyrinchium alatum*, *S. angustifolium*, *S. bellum* (California blue-eyed grass), *S. bermudiana*, *S. californicum* (golden-eyed grass), *S. cuspidatum*, *S. laxum*, *S. macrocephalum*, *S. mucronatum*, *S. striatum*, and *S. vaginatum*, among others, are clump-forming, grass-like herbs with actinomorphic, solitary or paniculate, blue, yellow or white flowers, native to North and South America; *Tigridia pavonia* (tiger flower) and its many cultivars, as well as other species, are bulbous herbs with simple or forked stems, linear to lanceolate leaves, and erect or nodding, campanulate flowers of dissimilar series and various colors, native to Central and South America; *Schizostylis coccinea* (Crimson flag, kaffir lily), a clump-forming taxon that grows from rhizome as opposed to corm, native to South Africa, Swaziland, and Zimbabwe.

SUBFAMILY CROCOIDEAE *Babiana ambigua*, *B. hypogea*, *B. nana*, *B. plicata*, *B. purpurea*, *B. sambucina*, and *B. stricta* are herbs with tunicate corms, ribbed or strongly veined, sometimes petiolate leaves, and actinomorphic or zygomorphic, often fragrant flowers on distichous, spicate inflorescences, native primarily to South Africa; *Crocus sativus* (saffron) is extensively cultivated in the Mediterranean region and some Middle Eastern countries for its stigmas, *C. angustifolius* (cloth-gold crocus), *C. asturicus*, *C. balancae*, *C. biflorus* (Scotch crocus), *C. candidus*, *C. chrysanthus*, *C. korolkowii* (celandine crocus), *C. laevigatus*, *C. nudiflorus*, *C. speciosus*, *C. tomasinianus*, *C. vernus* (Dutch crocus), *C. versicolor*, and many other species, hybrids, and cultivars are cultivated as ornamentals, these are spring or fall flowering herbs with tunicate corms, grass-like leaves, and slender tubular, white, yellow, or lilac to purple flowers, native mostly to the Mediterranean region; *Dierama gracilis* (wandflower), *D. pendulum* (grassy-bells), *D. pulcherrimum*, and others are cormous herbs with tall stems and funnelform mostly nodding, spreading flowers, native to South Africa; *Freesia alba*, *F. × hybrida* (a hybrid complex of unknown origin), *F. refracta*, and probably others, are herbs with tunicate corms, linear distichous leaves, and fragrant funnelform flowers of various colors on the upper side of a spike, native to South Africa, common in cultivation; *Geissorhiza aspera*, *G. erosa*, *G. melanthera*, *G. monanthos*, *G. radiance*, and others are endemic to Cape Province of western South Africa; *Hesperantha baurii*, *H. inflexa*, and others are cormous herbs with grass-like leaves and nocturnal flowers on bracteate spikes, native to South Africa; *Ixia campanulata* (corn lily), *I. incarnata*, *I. leucantha*, *I. maculata*, *I. monodelpha*, *I. patens*, *I. speciosa*, and *I. viridiflora*, are slender herbs with fibrous corms, mostly basal grass-like leaves, and actinomorphic flowers in spikes or panicles, native to South Africa; *Lapeirousia lacta*, *L. plicta*, *L. neglecta*, *L. corymbosa*, *L. oreogena*, endemic to Namaqualand and northwestern Cape Province. *Melaspherula racemosa* (= *M. graminea*, fairybells); *Romulea bulbocodium*, *R. rosea*, and *R. sabulosa*, are crocus-like cormous herbs with pedunculate tubular flowers, native to the Old World; *Sparaxis bulbifera*, *S. grandiflora*, and *S. tricolor* (wandflower); *Tritonia crocata*, *T. hyalina*, *T. rubrolucens*, and *T. squalida*, are cormous herbs with linear leaves and funnelform, dilated flowers of various colors in simple or branched inflorescences, native to South Africa; *Watsonia borbonica*, *W. densiflora*, *W. fulgens*, *W. marginata*, *W. meriana*, *W. pillansii*, *W. tabularis*, *W. wilmaniae*, and others, are large herbs with fibrous corms, ensiform (like sword blade) leaves, and large, tubular, brightly colored flowers in spikes, native to South Africa; *Aristea capitata*, *A. ecklonii*, and *A. thyrsiflora*, are fibrous-rooted herbs with basal distichous leaves, and blue flower is the sole member of **ARISTEOIDEA**; and *Patersonia glauca*, *P. occidentalis*, *P. sericea* (purple flag), *P. umbrosa*, and others, are herbaceous plants, sometimes with linear woody stem, distichous basal, stiff leaves, and blue or yellow flowers on a leafless stem, native to Australia, Borneo, New Guinea, and the Philippines, the only genus in **PATERSNIOIDEAE**.

Crocosmia aurea

FALLING STARS; VALENTINE FLOWERS
SYNONYMS *Babiana aurea*, *Crocanthus mossambicensis*, *Crocosmia cinnabarina*, *C. maculata*, *Tritonia aurea*, *T. cinnabarina*

ETYMOLOGY [Greek *krŏkŏs* "saffron"; *ŏsmē*, "smell," referring to the saffron smell of dried flowers placed in warm water] [Latin "golden"]

GROWTH HABIT Clump-forming cormous perennial herbs, 2–4 ft. (0.75–1.25 m) tall and as wide or wider when the plant spreads. **STEM/BARK**: corms form a vertical chain with the youngest at the top and oldest deeply buried in soil; roots are contractile (they pull the corms deeper into the soil). **LEAVES**: alternate, distichous, lanceolate; margins entire. **FLOWERS**: bisexual, sessile, orange, on horizontally branched terminal cymose inflorescences; flowers are born unilaterally on one side of the stems. **FRUIT**: dehiscent trilocular capsules. **NATIVE HABITAT**: grasslands of southern and eastern Africa.

CULTURE An easy plant to grow in full or partial sun and well-drained, preferably organic soils; requires occasional irrigation during drought periods; known to be slow to sprout; spider mites can become troublesome. **PROPAGATION**: seed or separation of cormels.

LANDSCAPE USES: ground cover, bedding plant, mixed border, containers, other imaginative uses, and may be used as cut flower. **HARDINESS ZONE**: 5B–10B. Corms should be dug in colder climates.

COMMENTS *Crocosmia* is a genus in Crocoideae: *C. aurea* is known to become invasive in some areas by breakage and separation of the cormels, *Crocosmia* 'Lucifer' seems to be the most popular, although there are other cultivars such as '**Miss Myrtle**', '**Orangerot**', and '**Plaisir**', which are listed as well. The hybrid *Crocosmia* × *Crocosmiiflora* has provided several choice selections, including '**Babylon**', '**Bressingham Beacon**', '**Buttercup**', '**Carmine Brilliant**', '**Citronella**', '**Emberglow**', '**George Davidson**', '**Jenny Bloom**', '**Jupiter**', '**Queen Alexandra**', '**Spitfire**', '**Twilight Fairy Crimson**', and '**Venus**', which are but a few of the available cultivars. Also reported from cultivation are such species as *C. masoniorum*, *C. paniculata*, *C. pottsii*, and *Crocosmia aurea* subsp. *pauciflora*.

Crocosmia aurea

Dietes iridioides

AFRICAN IRIS; FORTNIGHT LILY
SYNONYMS *Dietes vegeta*, *Moraea iridioides*, *M. vegeta*.

ETYMOLOGY [Greek *di*, two, and *etes*, affinities, in reference to its similarity to *Iris* and also to *Moraea*] [Iris-like].

GROWTH HABIT Evergreen clumping perennial herb; reaches a height of 2 ft. (60 cm), with an equal spread; of medium texture; grows at a medium rate. **STEM/BARK:** spreads by creeping stout rhizomes. **LEAVES:** simple, sword-shaped, linear, rigid, to 2 ft. long (60 cm) long, 2-ranked, held in a vertical, fan-like plane. **FLOWERS:** bisexual, white with yellow or brown spots toward tips of the outer petals, to 3 in. (8 cm), with 3–4 flowers per head. The style is marked with blue; short-lived flowers in spring. **FRUIT:** capsules, small, 3-valved, ellipsoidal, 1–1 $\frac{1}{2}$ in. (2 $\frac{1}{2}$–4 cm) long. **NATIVE HABITAT:** eastern Africa, from Kenya to South Africa.

CULTURE Full sun but tolerate partial shade, in fertile well-drained soil; not salt-tolerant; nematodes can be troublesome. **PROPAGATION:** seed or division of rhizome. **LANDSCAPE USES:** ground cover, mixed plantings, used in foundation plantings, mass plantings, and planters. **HARDINESS ZONE:** 8B–11A.

COMMENTS A member of Iridoideae, *Dietes* consists of six species, of which five are African and one is from Lord Howe Island. *Dietes iridioides* 'Johnsonii' has erect leaves and tall flowering stems. Oakhurst Hybrids such as 'Lemon Drops' and 'Orange Drops' have spreading habit, cream-white flowers blotched brown-yellow with a purple center. *Dietes grandiflora* and its cultivar 'Variegata' have larger flowers. Other species reported from cultivation include: *Dietes bicolor*, *D. robinsoniana*, and *D. vegeta*. These are Iris-like herbs with creeping rhizomes, linear-lanceolate, distichous leaves, and flowers with fugacious tepals, native to topical and South Africa.

Dietes iridioides

Dietes bicolor

Gladiolus spp. and Cultivars

GLADIOLUS

ETYMOLOGY [Latin "small sword," alluding to the shape of the leaves]

GROWTH HABIT Perennial caulescent herb with corm, stems erect, unbranched, leafy, sturdy; variable with species and cultivar, to 5 ft. (1 $\frac{2}{3}$ m) tall, rapid growth rate; can be left in ground year-round in warmer areas, but plants flower more profusely if the corms are dug and reset annually. **STEM/BARK**: green, leafy, to 6 ft. (2 m) tall, stiff, unbranched. **LEAVES**: simple, basal and cauline (on flower stalks), linear, sword-shaped or cylindrical to 2 ft. ($\frac{2}{3}$ m) long, often prominently veined or ribbed, light green, often rigid to somewhat papery texture. **FLOWERS**: with 28 recognized basic colors through yellow, whites, oranges, reds, lavenders, purples, browns, variously shaded and particolored; on 1-sided spikes; flowers are irregular (zygomorphic), tepals 6, united into a curved funnelform tube spread wide open to 8 in. (20 cm) across, upper 3 segments larger than lower 3; stamens 3, flowers in summer, fall, winter dependent on species, planting time, and climatic region. **FRUIT**: capsule, usually flattened or winged, 3-valved. **NATIVE HABITAT**: Africa, Madagascar, and the Mediterranean Region, but many of horticultural origin; species primarily native to tropical and South Africa.

CULTURE Full sun, in various fertile soils; not salt-tolerant; thrips, scab, heart rot, *Fusarium*, *Penicillium* fungus diseases may be problems. **PROPAGATION**: seeds, more often separation of cormels. **LANDSCAPE USES**: primarily for cut flowers, also bedding plants, and planter box. **HARDINESS ZONE**: Varies with cultivar and species; generally, 8A–12B. May be used as summer annuals.

COMMENTS About 10,000 cultivars and in excess of 300 related wild species. Important cut flower crop and garden plants. For an excellent treatment of *Gladiolus*, see "The American Horticultural Society A–Z Encyclopedia of Garden Plants" (2004). Three groups have been recognized: **Grandiflorus Group** (late spring to early fall), **Nanus Group** (early summer) and **Primulinus Group** (early to late summer). Cultivated taxa are too numerous to mention, but some of the more common species and hybrids include: *G. alatus*, *G. brevifolius*, *G. carneus*, *G. carinatus*, *G. × colvillei* (*G. cardinalis* × *G. tristis*), *G. dalenii*, *G. × ganadavensis* (*G. cardinalis* × *G. natalensis* or *G. natalensis* × *G. oppositiflorus*), *G. × hortulanus* (garden gladiolus), *G. illyricus*, *G. tristis*, and *G. undulatus*, among others. These are small to tall herbs with tunicate corms, tubular or funnelform, actinomorphic or zygomorphic flowers of various colors on long bracteate inflorescences, native to Africa, Asia, and the Mediterranean region. *Gladiolus* is said to symbolize strength, moral integrity, and passionate infatuation.

New *Gladiolus* corm of current year with previous year's spent corm attached at its base (left), a corm having given rise to several cormels (right)

Gladiolus communis subsp. ***byzantinus*** (above left), ***G. communis*** (above middle and right), and various cultivars (middle and below)

Iris spp. and Hybrids

IRIS, FLEUR-DE-LIS

ETYMOLOGY [After the mythological Greek goddess *Iris*, who came to earth via a rainbow, in reference to flower colors].

GROWTH HABIT Upright rhizomatous or bulbous herbaceous perennials; to about 2 ft. ($\frac{2}{3}$ m). **STEM/BARK**: caulescent, essentially the erect flower stalks, simple or branched. **LEAVES**: basal, but with a few on the flowering stalk, flat, linear, often sword-shaped and distichous, sometimes grass-like. **FLOWERS**: showy and of various sizes and colors, including nearly black; terminal on racemose or paniculate inflorescences, one to several on erect stems, usually of short duration but opening sequentially over a period of time; flower segments united into a long or short tube; the three outer segments (= "falls") usually reflexed and with a row of hairs (= "bearded") on the upper surface in many species or sometimes a serrated crest ("crested" or "Evansia" irises), or it may be smooth; the three inner segments (= "standards") erect and often narrowed to a "claw"; style branches are petal-like. **FRUIT**: oblong, 3 or 6-sided, 2–3 in. (5–7 $\frac{1}{2}$ cm) long capsule with many seeds. **NATIVE HABITAT**: widespread in the north temperate regions mostly in dry or cold rocky slopes or grasslands.

CULTURE Full sun or partial shade, species-dependent: in well-drained, loose soils or in moist to wet soils, no major problems, but it is imperative to know the requirements of the specific taxa being grown. **PROPAGATION**: seeds germinate readily when fresh; cultivars usually by separation of bulbs or division of rhizomes. **LANDSCAPE USES**: excellent in rock gardens, as ground covers, for spring or early summer flowering, or as accent plants for cluster of foliage, and a few species in water. **HARDINESS ZONE**: 5A–10B.

COMMENTS With more than 280 species and numerous hybrids and cultivars, irises are among the most popular garden plants around the world. The most commonly cultivated types are the lilac and purple forms which are "bearded," commonly known as the "blue flag" irises and sometimes referred to as the Japanese or German irises. The bulbous irises include "Dutch," "English," and "Spanish" irises and are members of the subgenera *Xiphium* and *Scorpiris* or their hybrids. The "Bearded Iris" are of the Mediterranean origin (*I. × germanica*) and are more adapted to drier regions. The "Louisiana" irises (including *I. fulva*, *I. brevicaulis*, *I. nelsonii*, and *I. giganticaerulea* and their hybrids) occur in the southern states (35 species are native to North America), although they are known to perform well in other regions. These require moist to wet soils. In general, *Iris* (bulbous and rhizomatous) species and hundreds of hybrids and cultivars are by far the largest genus in the family, and among the most widely planted ornamentals. For information on horticultural or botanical classification of the genus, the reader should consult several books specifically written about *Iris*. A few of the more commonly cultivated taxa include: *I. brevicaulis* (lamance iris), *I. chrysographis*, *I. chrysophylla*, *I. clarkei*, *I. cristata*, *I. delavayi*, *Iris domestica* (= *Blamcanda chinensis*, blckberry lily), *B. flabellata* (= *Belamcanda flabellata*), *I. foetidissima* (scarlet-seeded iris), *I. × germanica* (flur-de-lis), *I. histrioides*, *I. hoogiana*, *I. humilis*, *I. kaempferi* (Japanese iris), *I. magnifica*, *I. missouriensis* (western blue iris), *I. orchidoides* (orchid iris), *I. orientalis*, *I. pallida* (orris iris), *I. pseudacorus* (yellow iris), *I. pumila*, *I. reticulata*, *I. siberica* (Siberian iris), *I. spuria* (butterfly iris), *I. susiana* (morning iris), *I. unguicularis*, *I. versicolor* (wild iris), *I. × vinicolor* (*I. fulva × I. giganticaerulea*), *I. xiphioides* (English iris), and *I. xiphium* (Spanish iris). These are rhizomatous or bulbous herbs with equitant, ensiform, or linear-lanceolate leaves, and sometimes crested or bearded outer tepals of various colors including nearly black (but excluding true red), flowers. Native primarily to north temperate regions.

An unusual presentation of *Iris* in Park Floral de Paris; Iris at the famous abstract painter, Claude Monet's Garden in Giverny, France (below left). Typical Iris rhizomes (below right)

Representative sample of *Iris* flowers showing diversity of shapes and colors.

Representative sample of *Iris* flowers showing diversity of shapes and colors.

Iris cultivars in landscapes (above), photographed at U. S. Botanic Garden, Washington D. C., and ***Iris pseudacorus*** (flag iris, below), photographed at Claude Monet's Garden in Giverny, France (below).

A bed of *Iris* in Parc Floral de Paris

Neomarica caerulea

BRAZILIAN IRIS; WALKING IRIS; APOSTLE PLANT
ETYMOLOGY [Greek *neos*, new, and the genus name *Marica* or in reference to the mythological Roman nymph] [dark blue, in reference to color of the flower]

GROWTH HABIT Evergreen clump-forming herbaceous perennial; upright, 3–5 ft. (1–1 $\frac{2}{3}$ m) tall. **STEM/BARK**: relatively thick rhizome. **LEAVES**: dark green sword-like, stiff, 12–60 in. (30–160 cm) long and about $\frac{1}{8}$ to 2 in. (1–5 cm) wide that arise from the rhizome or on the flower stalk. **FLOWERS**: purple blue to violet, 3–4 in. (7 $\frac{1}{2}$ –10 cm) wide; flowers last for only one day but are replaced by the ready supply of developing flower buds; tepals marked with white and yellow spots; plantlets on the flower stalk eventually become large enough to cause bending and touching the ground resulting in development of new plants, hence the common name "walking iris." **FRUIT**: brown, square-shaped trilocular capsule. **NATIVE HABITAT**: tropical America, Brazil.

CULTURE Full sun but leaves are darker and more attractive in part shade; does well in well-drained organic rich moist soil, but said to be drought- and salt-tolerant; snails may become problem. **PROPAGATION**: seed or separating plantlets that develop on the flowering stalks, or division of the rhizome. **LANDSCAPE USES**: similar to *Iris* but with the understanding that the plant may occupy larger space in shorter time period; can be grown in borders, as specimen, as well as in containers. **HARDINESS ZONE**: 9B–13B. Rhizomes must be protected or dug during winter months in cooler climates.

COMMENTS There are 21–28 reported species of *Neomarica*. The vernacular name apostle is based on the presumption that there have to be 12 leaves before flowers develop. The cultivar '**Regina**' has wider leaves and is the one available in the trade. Other cultivated species include *N. candida*, *N. gracilis*, *N. guttata*, *N. longifolia*, and *N. northiana*, among others. These are rhizomatous perennial herbs with lanceolate leaves and flowers of various colors in clusters on flat, leaf-like scapes, native to topical America and West Africa.

Neomarica caerulea (photographed at Kampong, National Tropical Botanical Garden, Florida)

ADDITIONAL GENERA

Freesia laxa (= *Anomatheca laxa*) photographed at Filoli Gardens, California

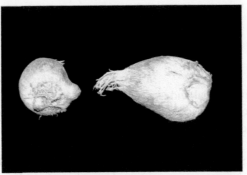

Freesia corms

Sisyrinchium californicum (yellow-eyed grass, golden-eyed grass), native to moist-wet proximity of ponds; native to the pacific coast, from California to British Columbia.

Trimezia sincorina, with elongated corm or rhizome from which leaves grow from base, with sepals (the outer series) larger than the petals (inner series), both with purple markings; native to Brazil

ASPHODELACEAE

ASPHODEL FAMILY

40 GENERA AND 900 SPECIES

GEOGRAPHY Widely distributed in tropical and temperate regions, predominantly South African, but Mediterranean to Central Asia, Australia, and New Zealand.

GROWTH HABIT Hermaphroditic; open rosette-forming leaf succulents, with rhizomatous or tuberous roots or sometimes arborescent pachycauls (plants with thick trunks). **LEAVES**: basal or terminal rosettes on branches, spirally arranged or sometimes distichous (as in *Aloe*), often succulent, sheathing or not at the base, margins usually serrate or dentate, sometimes sharp-pointed, veins indistinct in succulent leaves. **FLOWERS**: bisexual, actinomorphic or zygomorphic, hypogynous, in terminal, usually bracteate simple or compound racemes or spikes; perianth of 6 tepals, the inner may be different from the outer, free or nearly so (in subfamily Asphodeloideae) or more or less fused into a tube and slightly bilabiate (in *Kniphofia*) or zygomorphic, variously colored (except blue or violet); stamens 6, biseriate, inserted at the base of the ovary, filaments linear, free, rarely with hairs; carpels 3, fused, trilocular, ovary superior, ovules 1-many per locule, style simple, long, stigma small. **FRUIT**: loculicidal capsule or rarely fleshy (in *Eremurus*); seeds usually arillate, rarely winged (in *Eremurus*), embryo straight, endosperm present.

ECONOMIC USES Species of several genera are used as ornamentals and at least two *Aloe* species are grown for medicinal and cosmetic purposes.

CULTIVATED GENERA *Aloe*, *Asphodelus*, *Astroloba*, *Bulbine*, *Dianella*, *Eremurus*, *Gasteria*, *Haworthia*, *Hemerocallis*, *Kniphofia*, *Phormium*, *Xanthorrhoea*, and other cultivated genera.

COMMENTS Asphodelaceae was known as **Xanthorrhoeaceae** and recognized in APG III (2009). The more recent APG IV System (2016–2021), however, recommended conservation of the original name Asphodelaceae. In its new circumscription, it includes 3 subfamilies that can be recognized on the bases of their gross morphological features:

SUBFAMILY ASPHODEOIDEAE (mostly rosette-forming leaf succulent rhizomatous herbs to pachycaulos trees): *Aloe*; *Asphodelus acaulis*, *A. albus*, *A. cerasiferus*, *A. delphiniensis*, and *A. tenuifolius*, are herbs with thickened roots, basal leaves, and white to pink flowers on scapes, native from Canary Islands to India, used as bedding plants; *Astroloba aspera*, *A. deltoidea*, *A. foliolosa*, and *A. pentagona*, are succulent plants with spirally arranged leaves, similar to *Aloe* but smaller, native to South Africa, grown in succulent gardens and by collectors; *Bulbine alooides*, *B. caulescens*, and *B. semibarbata*, are annual or perennial herbs with woody or tuberous base, narrow lanceolate, fleshy leaves, and yellow or white flowers in racemes, native to Africa and Australia, sometimes grown as garden flowers; *Bulbinella angustifolia*, *B. floribunda*, *B. latifolia*, and *B. nutans*, and another 14 species are attractive deciduous perennials with various flower colors on dense unbranched racemes, native to Cape Province of South Africa; *Eremurus; Gasteria; Haworthia*; and *Kniphofia*.

SUBFAMILY XANTHORRHOEIDEAE (thick woody stem, resiniferous), *Xanthorrhoea johnsonii*.

SUBFAMILY HEMEROCALLIDOIDEAE (various growth habits, often with swollen roots, spirally arranged but appearing as 2-ranked leaves): *Dianella*; *Hemerocallis*; *Phormium*; *Simethis mattiazzii* (Kerry lily) is the only species in the genus, although other taxa have been described, is native of dry sites from Europe to Africa; *Trachyandra divericata*, *T. flexifolia*, and *T. tortiles*, are with strange snake-like leaves, native to eastern and southern Africa and Madagascar.

Aloë spp.

ALOE

ETYMOLOGY [Probably from the Arabic *allock*, or the Hebrew *allal*, meaning bitter]

GROWTH HABIT Perennial herbs, shrubs, or small trees, solitary or in clumps, with thickened roots; medium texture; medium growth rate. **STEM/BARK**: in some arborescent species with clear smooth barks, but in smaller species usually short to very short stems that are often covered with sheathing leaf bases. **LEAVES**: crowded, spirally arranged into rosettes, fleshy and often noticeably succulent; sap colorless, yellow, brown, or purple, sometimes pungent and sometimes poisonous; linear, lanceolate, or triangular, with sheathing base; apex and margins usually spinose; glabrous or sometimes rough or occasionally spinose; uniformly green or spotted or striated. **FLOWERS**: pedicellate, tubular, or sometimes narrowed about the middle (salverform) on simple or branched racemose inflorescences with stout peduncles; colors variously yellow to red. **FRUIT**: 3-angled subglobose to oblong capsules; seeds many in two rows. **NATIVE HABITAT**: in drier regions of Arabian Peninsula throughout Africa, from Sudan to West Africa and south to Cape Province and also Madagascar. Some species have naturalized several subtropical regions, including the Americas.

CULTURE Easy plants to grow in warmer areas as long as the soil is well-drained and the plant is a mostly sunny location. **PROPAGATION**: seeds, cuttings of the more woody taxa, and division of spreading species. **LANDSCAPE USES**: in landscapes and several species and hybrids as house plants. **HARDINESS ZONE**: 9A–13B. Hardiness depends on species' native origin.

COMMENTS *Aloe*, with more than 500 species, is by far the largest genus in the family and includes many cultivated taxa, some becoming economically increasingly important: the yellow juice of *A. vera* is extensively used in soaps, shampoos, suntan lotions, and other cosmetic products, as well as soft drinks; a very large number of species are commonly grown in gardens or containers, including the tree-like species such as *A. arborescent* (tree aloe), *A. bainesii*, *A. dichotoma*, *A. ferox*, and others are also cultivated and used for pharmaceutical purposes; while many *Aloe* are rosette plants such as *A. africana*, *A. aristata*, *A. brevifolia*, *A. excelsa*, *A. marlothii*, *A. nobilis*, *A. strata* (coral aloe), *A. variegata* (partridge-breast), etc.; and several are used against intestinal worms, wounds, and for other medical problems, such as *A. vera*, *A. saponaria*, and *A. ferox*. In addition to their attractive succulent leaves and growth forms, aloes possess yellow to red flowers on long sometimes branched spikes. They are native to the Old World, but are primarily African. There are also several hybrids and selected cultivars common in the trade including some with blueish leaves such as '**Blue Boy**' and '**Blue Elf**' and others with profuse flowers and intense colors such as '**Always Red**' and '**Super Red**', mostly sold as house plants. The genus *Aloe* has recently been divided into several genera: tree *Aloe* are now *Aloidendron*, and others into *Kumara*, *Aloidampelos*, *Aloe*, *Aristaloe*, and *Gonialoe*. In addition to several interspecific hybrids, there are also a few intergeneric hybrids such as ×*Gasteraloe* (*Aloe* × *Gasteria*), and ×*Aloloba* (*Aloe* × *Astroloba*). Several species are endangered and have been placed on CITES Appendix I.

Aloe **sp.** (*A. vera* **?**) herbaceous spreading growth habit (above left), *A.* **'Pink Perfection'** (above right), *A. dichotoma* (middle left), and close-up to show leaf arrangement (middle right), *A. elegans* (below left) and close-up of inflorescence (below right).

Aloe* sp.**, a seemingly arborescent taxon (left), ***A. plicatilis, woody species with dichotomous branching (right).

Aloe vera in a landscape (left) and ***A. vera*** (= *A. barbadensis*) in Baja California

Dianella tasmanica

TASMANIAN FLAX LILY
SYNONYMS *Dianella archeri*, *D. denssa*, *D. hookeri*

ETYMOLOGY [Diminutive of the mythical Roman goddess of the chase (= hunt), *Diana* (divine, heavenly), who had the power to talk and control animals] [Tasmania, Australia]

GROWTH HABIT Clumping herbaceous perennial, to 6 ft. (2 m) tall; fine texture; rapid growth rate. **STEM/BARK**: thick spreading rhizome. **LEAVES**: linear green with white or yellow finely serrulate margins, to 30–40 in. (80–100 cm) tall. **FLOWERS**: small bluish-white, zygomorphic, on branched inflorescences to about 5 ft. (1 $\frac{2}{3}$ m) tall. **FRUIT**: blue-violet berries of about $\frac{1}{2}$ in. (1 $\frac{1}{4}$ cm) in diameter; remain on plant from fall through winter. **NATIVE HABITAT**: southern Australia and Tasmania, in shady areas of wet forests.

CULTURE Partial sun or shady locations, in well-drained but moist fertile soil; generally pest free but must be kept moist at all times when in sunny locations, otherwise leaf tip burn is the result. **PROPAGATION**: seed, division of rhizomes. **LANDSCAPE USES**: mass plantings or as accent plants near pools, mixed beds, or borders; may also be used as indoor or outdoor potted plants. All parts of the plant are suitable for flower arrangements. **HARDINESS ZONE**: 8A–12B.

COMMENTS A few interesting cultivars have been introduced, including '**Gold Strip**' (yellowish leaves and golden yellow leaf margins), '**Little Devil**' (a dwarf compact form with reddish-pink lower parts of leaves), '**Variegata**' (white margins and few white streaks). There are related species in cultivation as well, such as *D. caerulea*, *D. encifolia*, *D. intermedia*, *D. prunina*, and *D. revoluta*, all with selected cultivars.

Dianella tasmanica

Eremurus robustus

FOXTAIL LILY; DESERT CANDLE
SYNONYMS *Eremurus afghanicus*, *E. alaicus*, *E. albertii*, *E. azerbajdzhanicus*, *E. candidus*, *E. chinensis*, *E. himalaicus*, *E. luteus*, *E. spectabilis*, and many more.

ETYMOLOGY [From Greek *ēremos*, desert and *oura*, tail, in reference to the tall inflorescent in desert habitat] [Robust]

GROWTH HABIT Clumping perennial 6–10 ft. (2–3 m) tall, 2–3 ft. ($\frac{3}{4}$ –1 m) wide; fine texture; slow growth rate. **STEM/BARK**: acaulescent, starfish-shaped bulbs. **LEAVES**: basal, blue-green, strap-shaped, to 4 in. (10 cm) wide and about 4 ft. (1 $\frac{1}{2}$ m) tall. **FLOWERS**: dense numerous showy fragrant white and various shades of pink and yellow tepals and bright yellow stamens in 3–4 ft. (1–1 $\frac{1}{2}$ m) tall leafless terminal raceme (flowers open gradually from base to top), that well exceed the height of leaves. **FRUIT**: trilocular loculicidal capsule. **NATIVE HABITAT**: Central Asia, Afghanistan, in dry high-altitude grasslands.

CULTURE Full sun, in well-drained sandy organic soils that should be kept on the dry side, particularly after flowering since the bulb goes dormant and leaves die in mid-summer. It has no salt tolerance and cannot withstand wet soils. The tall inflorescences may be damaged by strong winds. **PROPAGATION**: seeds and division of offsets. **LANDSCAPE USES**: background plant in mixed borders or in front of tall hedges or walls; in exceptional situations it is probably inappropriate as cut flower crop due to its height. **HARDINESS ZONES**: 6A–10A. Not recommended for wet climate regions, but does well in warm Mediterranean areas.

COMMENTS The genus encompasses about 35 species of which *E. robustus* is the most common. Other species, however, are also seen but less commonly: *Eremurus atlanticus* (including '**Shelford Pink**', '**Cleopatra**', and probably others), *E. elwesii* (king's-spear), *E. himalaicus*, *E. lactiflorus*, *E. olgae*, *E. spectabilis* (desert-candle), *E. stenophyllus* (flowers yellow), and other species as well as *Eremurus* × *isabellinus* (*E. stenophyllus* × *E. olgae*), attractive cultivars '**Cleopatra**' and '**Romance**' with yellow to orange flowers. These are rosette plants with narrow leaves, and flowers of various colors on racemose scapes, native to west and central Asia, used as border plants.

Eremurus robustus (desert candle), Photographed in Parc André Citroën, in Paris, France

Gasteria spp.

OX TONGUE; COW TONGUE

ETYMOLOGY [From Greek *gaster*, stomach, in reference to the pouch-shaped perianth] [pl. species]

GROWTH HABIT Evergreen solitary or clump-forming succulent perennial herbs 6–30 in. (15–75 cm), usually on the smaller size, solitary or may branch at base; medium texture; slow growth rate. **STEM/BARK**: acaulescent or with a short mostly subterranean stem. **LEAVES**: distichous (two ranked) mottled green, tongue-shaped succulent early but become rosettes with age. **FLOWERS**: tubular, belly-shaped, on pendulous recurved pedicels of spreading racemes. **FRUIT**: trilocular loculicidal capsule. **NATIVE HABITAT**: endemic to the dry savanna region of Eastern Cape Province of South Africa.

CULTURE Full sun or partial bright shade, in well-drained sandy soil; generally, a trouble-free plant that should be kept on the dry side. **PROPAGATION**: seed, offsets, or leaf cuttings. **LANDSCAPE USES**: greenhouse grown for house plants, in landscape only in frost-free dry areas, but mostly not suitable as ground covers. **HARDINESS ZONES**: 10B–12B.

COMMENTS *Gasteria* is said to consist of 23 species, most of which are in cultivation by succulent collectors, but a few species are more readily available in the trade, including: *Gasteria acinacifolia*, *G. armstrongii*, *G. batesiana*, *G. bicolor*, *G. carinata*, *G. excelsa*, *G. glomerata*, *G. liliputana* (= *G. bicolor* var. *liliputana*), *G. maculata*, *G. pilansii*, and *G. verrucosa* (= *G. carinata* var. *verrucosa*). There are also a few listed cultivars, such as '**Big Brother**', '**Brisbane**', '**Fugi Yuki**', '**Green Ice**', '**Little Warty**', '**Silver Swirls**', '**Variegata**', '**White Ghost**', and probably others. The intergeneric hybrid ×*Gasteraloe* '**Midnight**' and ×*Gasteraloe* '**Goliath**' (*Gasteria* × *Aloe*) are also known.

Gasteria bicolor

Haworthia angustifolia

ZEBRA HAWORTHIA

SYNONYMS *Aloe stenophylla*, *Catevala angustifolia*, *Howarthia albanensis*

ETYMOLOGY [In honor of Adrian Hardy Haworth (1768–1833), English authority on succulent plants] [From Latin, narrow leaves].

GROWTH HABIT Evergreen clump-forming succulent perennial herb, to 8 in (20 cm) tall, but usually much smaller, medium texture; slow growth rate. **STEM/BARK**: short-stemmed. **LEAVES**: simple, alternate, appearing whorled in rosettes, narrowly triangular-lanceolate, to about 5 in. (12 ½ cm) long, incurved toward apex, succulent, margins serrulate, slightly glossy, without white tubercles on both sides. **FLOWERS**: white, with green stripes, perianth tubular, 2-lipped, stamens 6, racemes on a scape to 12 in. (30 cm); **FRUIT**: trilocular loculicidal, capsule. **NATIVE HABITAT**:

eastern Cape Province of South Africa, where it grows in grassy areas or rock crevices.

CULTURE Full sun or very light shade; in sandy, very well-drained soils; not salt-tolerant; essentially trouble-free plant but scales and root rot due to overwatering may become problems. **PROPAGATION**: seeds, offsets. **LANDSCAPE USES**: greenhouse grown for house plant, in landscape only in frost-free dry areas where it forms large clumps and can be used as ground cover. **HARDINESS ZONES**: 10B–12B.

COMMENTS *Haworthia angustifolia* is apparently somewhat variable, with at least two varieties mentioned: *H. angustifolia* **var**. *liliputana* (brownish leaves that may be caused by poor growing conditions) and **var**. *altissima* (with upright narrower leaves). Many related species of primary interest to succulent lovers and collectors: *Howarthia fasciata*, *H. cymbiformis*, *H. margaritifera*, *H. pumila*, and *H. reinwardtii* are the other species reported in cultivation.

Haworthia angustifolia

Haworthia angustifolia 'Zebra'

Hemerocallis fulva

DAYLILY, TAWNY LILY, LEMON LILY
SYNONYMS *Gloriosa luxuriance*, *Hemerocallis crocea*, *H. kwanso*, *H. maculata*

TRANSLATION [From Greek *hemera*, *day* and *kallos*, beauty, beautiful for a day, in reference to flowers that last only for a day]

GROWTH HABIT Evergreen or deciduous, grass-like, and usually clump-forming perennial acaulescent herbs. STEM/BARK: rhizomatous to slightly fleshy bulbous (not corms). LEAVES: simple, distichous (2-ranked), linear to strap-shaped, from less than 12–36 in. (25–75 cm) long, wider at the base tapering gradually to the apex. FLOWERS: in clusters on long scapes to 5 ft. (1 $\frac{1}{3}$ m) tall, tepals 6, subtended by a bract; flowers of various sizes and colors, funnelform to campanulate, each subtended by a bracteole; some cultivars are fragrant. FRUIT: trilocular loculicidal capsule with few seeds. NATIVE HABITAT: China, Japan, and Korea, but the typical variety (*H. fulva* var. *fulva*) is naturalized in Europe and United States and considered invasive.

CULTURE Full sun to part shade, in well-drained fertile deep soil; generally, a pest free plant with little salt tolerance; only certain cultivars can be grown in warmer climates. PROPAGATION: seeds or by division of cultivars that should be done every few years, as overcrowding usually results in reduced number of flowers. Some cultivars of particular interest are reproduced by tissue culture; offsets (bulbils?) on flower stocks when present may also be used for PROPAGATION. LANDSCAPE USES: ground cover, borders, mixed plantings, in planter boxes, and massed for effective visual effect. HARDINESS ZONE: 3B–11A, depending on cultivars. Some of the primary commercial sources of daylily production are located in north-central Florida, zones 8B–9A.

COMMENTS The wide range of flower colors and sizes and plant heights provide opportunities for selection of appropriate cultivars for warmer climates. However, in addition to their ornamental use, daylilies have been cultivated and utilized for food and medicine for more than 2000 years. Flower buds of *H. esculenta* were called gum tsoy (golden vegetable). The Chinese also referred to daylilies as hsuan t'sao (plant of forgetfulness) because it is said that when eaten the leaves are slightly intoxicating and result in forgetfulness. Daylilies were introduced into North America during Colonial times and have since escaped and become naturalized in several states. Although relatively few (20) species are known, several hundred cultivars ranging in heights from dwarf to medium and tall and nearly all flower colors (except blue), singles and doubles, and different sizes. Linneus recognized two groups of daylilies: *H. fulva*, the widespread naturalized "orange or tawny daylily" and *H. flava*, the "yellow or lemon lily." Much of the several thousand cultivars are tetraploids that resulted from colchicine treatment. According to some sources, there are about 35,000–60,000 registered cultivars; interested individuals should refer to published books and various web sites, including that of the Hemerocallis Society. It is noteworthy that daylilies were until recent phylogenetically based classification (APG III) included in the Liliaceae, together with the familiar typical lilies, as it seemingly shares none-tunicate or scaly bulbs similar to the common lilies.

Representative examples of *Hemerocallis fulva* cultivars

Kniphofia uvaria

RED-HOT POKER

SYNONYMS *Aletris uvaria*, *Aloe longifolia*, *A. uvaria*, *Kniphofia alooides*, *K. bachmanii*, *K. burchellii*, *Tritoma canary*, *T. glauca*, *T. nobilis*, *T. uvaria*, *Veltheimia speciose*, *V. uvaria*

ETYMOLOGY [After Johann Hieronymus Kniphof, (1704–1763), German botanist and physician] [From Latin *uva*, a bunch of grapes, in allusion to the grape-like fruits]

GROWTH HABIT Semievergreen clump-forming herbaceous perennial, 3–4 ft. (1–1.35 m) high with a 2–3 ft. (0.75–1 m) spread. **STEM/BARK**: rhizomatous. **LEAVES**: bluish-green linear tufted grass like. **FLOWERS**: upright racemose leafless (scape) inflorescence with dense tubular terminal flowers, 6–10 in. (15–25 cm) long, red on the upper two-thirds and yellow on the lower portion, and sometimes orange in the middle. **FRUIT**: loculicidal capsule. **NATIVE HABITAT**: Cape Province of South Africa, naturalized in a number of regions, in some cases considered invasive by forming thick clumps.

CULTURE Full sun, in well-drained rich organic humus soil; does not tolerate wet soil and disturbance, otherwise for the most part it is a trouble-free plant. **PROPAGATION**: seed or division of rhizomes, but tissue culture has become the primary source of the cultivar production. **LANDSCAPE USES**: in mixed and border plantings, for cut flowers and container plants. **HARDINESS ZONES**: 6A–11A, in drier Mediterranean climates, not suitable for wet humid zones.

COMMENTS Some 72 species of *Kniphfia* have been reported with the *K. uvaria* and its cultivars the most common in landscapes, including '**Early Hybrids**', **Eco Series** ('**Eco Duo**', '**Eco Mango**', '**Eco Rojo**', etc.), '**First Surprise**', '**Flamingo**', and '**Primrose Beauty**'. In addition, there are a number of cultivars of unknown origin that most likely have had *K. uvaria* as one of the parents: '**Alcazar**', '**Amsterdam**', '**Bressingham Comet**', '**Cobra**', '**Fire Flame**', '**Gold Mine**', '**Little Maid**', '**Nancy's Red**', '**Nobilis**', '**Pensil White**', '**Pfitzeri**', '**Red Hot Popsicle**', '**Rosea Superba**', '**Royal Castle**', '**Royal Standard**', '**Samule's Sensation**', '**Springtime**', '**Vanilla**', and '**White Fairy**', among others. The reported species in cultivation are: *K. galpinii*, *K. gracilis* (torch lily), *K. hirsuta*, and *K. schimperi*. These are rhizomatous perennial acaulescent herbs with numerous red or yellow flowers in spikes or dense racemes terminating the scape, native to Madagascar and tropical and South Africa, often used as specimen plants in the landscape. Flowers produce copious nectar to which hummingbirds, butterflies, and sometimes bees are attracted.

Kniphofia uvaria (photographed at Wisely (RHS) Botanical Garden, England

Phormium tenax

NEW ZEALAND FLAX

ETYMOLOGY [from Greek *phŏrmiŏn*, mat, alluding to the use of the fiber from the plant] [From Latin, strong, tough, matted]

GROWTH HABIT Evergreen perennial, clumping, potentially to 12 ft. (4 m) tall and 7 ft. (2 $\frac{1}{3}$ m) wide. **STEM/BARK**: none, grows from rootstocks. **LEAVES**: long linear strap-like to 6 ft. (2 m) long; margins entire, midrib distinct and sometimes reddish or yellowish, original taxon green but many color variations are available. **FLOWERS**: long paniculate inflorescence with yellow or red flowers, but infrequent flowering. **FRUIT**: Trilocular capsule. **NATIVE HABITAT**: New Zealand and Norfolk Island. It has naturalized in some of the Pacific Islands and Australia.

CULTURE Sunny location, in well-drained but moist soils and benefits from occasional fertilization; otherwise except for uncommon mealybug infestation, it is a trouble-free plant. **PROPAGATION**: division or seed, when available. **LANDSCAPE USES**: accent plants, containers, and in mixed or mass plantings, particularly useful in coastal areas. **HARDINESS ZONE**: 8B–12B. Does not tolerate exposure to freezing cold, but with some protection it may survive in less humid zone 8A.

COMMENTS The plant has been used for production of fiber for rope and sail. Leaves are currently used in basket making. A highly bitter and toxic chemical cucurbitacin is contained in leaves. A number of selections primarily for color are available in the trade, including: '**Apricot Queen**', '**Black Adder**', '**Black Rage**', '**Bronze Baby**' (arching bronze leaves), '**Carousel**', '**Chocolate Baby**', '**Dazzler**' (arching bronze leaves with red and pink stripes), '**Duet**', '**Evening Glow**', '**Flamingo**', '**Gold Ray**', '**Hunter Green**', '**Lime Light**', '**Maori Maiden**', '**Pink Flamingo**', '**Pink Panther**', '**Rainbow Sunrise**', '**Red Heart**', '**Shiraz**', '**Sundowner**' [6 ft, (2 m) tall with striped rose and pink leaves), '**Surfer Bronze**', '**Variegatum**', '**Wild Wood**', and '**Yellow Wave**', among others.

***Phormium tenax* 'Jester'** (above), **'Bronze Beauty'** (left below), and ***Phormium tenax*
'Purpureum'** (below right)

Xanthorrhoea johnsonii

JOHNSON'S GRASS TREE

ETYMOLOGY [From Greek *xanthos*, yellow and *rheo*, to flow, in reference to the yellow resinous gum that is extracted from the stem] [Named after L.A.S. Johnson (1925–1997), the Australian taxonomic botanist at Sydney Royal Botanic Garden]

GROWTH HABIT Long-lived evergreen arborescnet perennial with tufted spreading crown; fine texture; slow growth rate. **STEM/BARK**: stem solitary, arborescent ($\frac{1}{3}$ –5 m) high; covered with leaf bases. **LEAVES**: simple, grass-like, linear, $\frac{1}{10} - \frac{1}{8}$ in. ($1\frac{1}{4}$ –3 mm) wide, quadrate rhombic (widest access at midpoint and with straight margins), in a terminal crown, green, not glacous; young leaves spreading in upright tuft, old leaves often reflexed. **FLOWERS**: bisexual; in spirally arranged clusters subtended by a cluster of bracts and filled with bracteoles between the flowers; on woody cylindrical spike-like inflorescence; flowers actinomorphic, 3-merous; tepals 6, free, in 2 whorles; stamens 6, longer than the tepals; ovary superior, 3-locular; ovules several per locule; styles simple, stigma entire. **FRUIT**: obtuse pointed capsule; seeds 1–2 per locule, usually black. **NATIVE HABITAT**: all known species of *Xanthorrhoea* are endemic to scrophyllous forest in well-drained sites, widespread in NSW subdividions and Queensland, Australia.

CULTURE Full sun, in very well-drained soils; no serious pests or diseases but suffers from prolonged exposure to wet soil; removal of dead leaves is recommended. **PROPAGATION**: seed. **LANDSCAPE USE**: specimen, accent in dry locations. **HARDINESS ZONE**: 9A–12B. In subtropical and warm temperate climates.

COMMENTS *Xanthorrhoea* plants are uncommon but available in some areas of the United States such as southern California. A few other species are known to exist, such as *X. australis*, *X. glauca*, *X. media*, *X. preissii*, and probably others.

Xanthorrhoea johnsonii

ADDITIONAL GENERA

Asphodelus albus (white-flowered asphodel), herbaceous perennial native to the
Mediterranean region but also in Libyan territory. In ancient Greek, the species was
associated with mourning and death.

Bulbine frutescens (bulbine lily), native to Southern Africa, a drought-tolerant plant that grows
in poor soils and blooms in spring and fall. Contrary to the name, it does not have a bulb.

AMARYLLIDACEAE

AMARYLLIS FAMILY

80 GENERA AND 2165 SPECIES

GEOGRAPHY Primarily warm temperate, tropical, and subtropical regions, extending into temperate areas.

GROWTH HABIT Predominantly terrestrial deciduous (dormant bulbs) or evergreen, herbaceous perennials, mostly bulbous plants. **LEAVES**: basal, linear or rarely elliptic, usually sessile but in few cases petiolate (e. g., *Eucharis*), 2-ranked or spirally arranged, few to many atop bulbs or rarely rhizomes. **FLOWERS**: bisexual, actinomorphic or rarely zygomorphic, solitary or in groups on scapose umbellate inflorescences subtended by a bract; perianth of 6 tepals in 2 whorls, free or connate, often showy; sometimes with a corona (in *Narcissus*) or staminal cup (in *Eucharis*), stamens 6, usually free; carpels usually 3, fused, inferior in majority of the species but superior in *Agapanthus* and the *Allium* group. **FRUIT**: loculicidaly dehiscent 3-locular capsule or fleshy and berry-like.

ECONOMIC USES A large number of horticulturally important ornamentals and a few vegetables such as onion (*Allium cepa*), chive (*Allium schoenoprasum*), Chinese scallion (*Allium chinense*), garlic (*Allium sativum*), leek (*Allium ampeloprasum* **Leek Group**), and shallot (*Allium cepa* **var.** *aggregatum*), and a few species are used for medicinal purposes.

CULTIVATED GENERA The family in its current broad sense includes species of families previously placed in Alliaceae (now subfam. **ALLOIDEAE**), to include *Allium*, *Nothoscordum*, *Tulbaghia*, and others; Agapanthaceae (with the single genus *Agapanthus*), now subfam. **AGAPANTHOIDEA**. Majority of the genera, however, are assigned to subfamily **AMARILLIDEAE** and are represented in trade by: *Amaryllis* (see comments below), *Clivia*, *Crinum*, *Cyrtanthus*, *Eucharis*, *Eucrosia*, *Haemanthus*, *Hippeastrum*, *Hymenocallis*, *Lycoris*, *Narcissus*, *Nerine*, *Pamianthe*, *Sprekelia*, *Stenomesson*, *Sternbergia*, and *Zephyranthes*, among others.

COMMENTS The genus *Amaryllis* is often confused with *Hippeastrum*, the common flowers of many colors sold under the vernacular name amaryllis. The genus *Amaryllis*, in a restricted sense, is monotypic and includes only *A. belladonna* (naked lady), which is native to the western region of South Africa's Cape Province. In fact, the family name Amarylllidaceae was originally based on this one species by Linnaeus. Other genera in **AMARILLIDEAE** include: *Hippeastrum* with about 90 species and 600 hybrid cultivars, and is of South American origin. Other common genera are exemplified by *Ammocharis coranica* (Karoo lily), *Boophone disticha* and *B. haemanthoides* (interesting leaf arrangement), *Brunsvigia radulosa* (candelabra flower), *Crossyne flava*, *Eucrosia bicolor* (uncommonly long stamens and style), *Galanthus nivalis* (common snowdrop); *Hebranthus* (= *Zephyranthella*, *Haylockia*; copper lily), *H. robustus*, *H. tubispathus*, *H. vittatus*, and others; *Griffinia rochae*, *Ipheion recuvifolium* (white flowers) and *I. uniflorum* (single, bright blue flower), *Ixiolirion tataricum*, *Leucojum aestivum* (snowflakes), *Nothoscordum bivalve*, *Phaedranassa* **spp.** (particularly *P. cinerea*, *P. dubia*, *P. carmiolii*, *P. chloracra*); *Phycella australis*, *P. bicolor*, *P. brevituba*; *Placea ornata*, *Pyrolirion* (= *Leucothauma;* fire lilies), *P. albicans*, *P. arvense*, *P. flavum*, *P. tubiflorum*, and others), *Rhodophiala bagnoldii*, *Scadoxus membranaceus*, *sperekelia formosissima*, *S. hawardii; Sternbergia lutea* (lily of the field), *Traubia modesta; Ungernia sewerzowii* (uncommon dark magenta flower), among others. Species of most genera are mostly in private collections. (see Garcia et al. 2019). Tribes are not used in this book.

Agapanthus praecox

AGAPANTHUS; LILY OF THE NILE

SYNONYMS *Agapanthus orientalis*, *Tulbaghia praecox*

ETYMOLOGY [From Greek *agape*, love; *anthos*, flower] [From Latin, very early, probably in reference to early flowering].

GROWTH HABIT Evergreen, clumping perennial herb with thick fleshy rhizomes, low growing, lacking aerial stem; variable with cultivar but generally 2 ft. (5 cm) or more; fine texture; fast growth rate. **STEM/BARK**: acaulescent, forms clumps from thick, fleshy, light brown to yellow rhizomes. **LEAVES**: simple, basal, often 2-ranked, arching, strap like, leathery, linear-lanceolate to 2 ft. (60 cm) long, 2 in. (5 cm) wide, entire margins, rounded apices, usually about 10 leaves per plant forming a mound-shaped clump. **FLOWERS**: in terminal umbels, to 4 in. (10 cm) across and containing 40–100 flowers, on a solid, smooth, green, leafless stalk (scape) about 2 ft. (60 cm) tall; bisexual, blue, violet, white, some multicolored, corolla tubular to campanulate, $1\frac{1}{2}$ in. (4 cm) across, 6 exerted stamens; Flowering in summer to early fall. **FRUIT**: dehiscent, 3-chambered capsule. **NATIVE HABITAT**: South Africa, in Natal and Cape of Good Hope, but has naturalized in a number of other regions.

CULTURE Full sun to partial shade, in fertile, well-drained soils; moderately salt-tolerant; chewing insects, maggots, borers, and botrytis may occasionally pose problems. **PROPAGATION**: seed or clump division. **LANDSCAPE USES**: edging, bedding plant, pot plant, cut flowers.

HARDINESS ZONE: 8A–12A. Foliage is usually damaged by frost, but plants recover with onset of warmer weather.

COMMENTS Three subspecies of *Agapanthus* have been recognized: *praecox*, *orientalis*, and *minimus* (≡ *A. comptonii*). These differ essentially in flower size and leaf numbers. Many cultivars, hybrids, and six related species (two evergreen and four deciduous). Dwarf and white ('**Albus**') or lilac-flowered and dwarf ('**Nanus**') cultivars are available. *Agapanthus africanus* and its cultivar '**Peter Pan**' are gaining popularity. A large number of cultivars of *A. campanulatus*, *A. inapertus* as well as subsp. *hollandii* '**Sky**', subsp. *intermedius*, and subsp. *pendulus*, and cultivars '**Major**', '**Hinrkop**', '**Nigrescens**', '**Pendulus grascoop**', and *A. praecox*, of various growth features, flower colors (various blues and whites) and inflorescence sizes (number of flowers) have been introduced. A few examples include: '**Arctic Star**' (large inflorescence with many large white flowers), '**Blue Triumphator**' (large umbellate inflorescences with many light blue flowers), '**Dwarf White**', '**Full Moon**' (bluish-white tepals with darker blue line), '**Northern Star**' (compact inflorescence of flowers with more than six tepals), '**Plenus**' (double flowers), '**Variegata**' (variegated leaves), '**Weaver**' (smaller blue flowers), among others. Making a distinction between cultivars of *A. Praecox* and *A africanus* is difficult.

Agapanthus praecox '**Albiflorus**' (may also be listed as 'Alba,' or 'Getty White')

Agapanthus africanus (African lily), above, *A. africanus* 'Stockholm' (below left), *A. africanus* 'Black Buddha' (below right)

Allium spp.

ONIONS
ETYMOLOGY [Classical Latin name for garlic]

GROWTH HABIT Aromatic evergreen or deciduous perennial herbs; fine texture; rapid growth rate. STEM/BARK: plants with tunicate bulbs or rhizomes (sect. *Rhizirideum*) and possess leafless flower stalks (= scapose). LEAVES: basal, sessile, linear, usually spirally arranged but may be 2-ranked. FLOWERS: umbellate inflorescences which are subtended by spathaceous (large sheathing), membranous, more or less imbricate bracts; actinomorphic, or occasionally zygomorphic, with perianth of 6 tepals in two whorls, 6 stamens that are fused to base of the tepals (= epitepalous); few to many flowers; ovary superior (in contrast to inferior ovary of *Hippeastrum* and related genera). FRUIT: trilocular fleshy capsule with numerous round black seeds. NATIVE HABITAT: predominantly in temperate climates of Northern Hemisphere in a wide range of habitats and soil types of varying moisture content, but a few in the American and African tropics.

CULTURE Full sun, in well-drained organic fertile soils. PROPAGATION: seed or bulblets that form at the base of old tunicate bulbs or from individual leaf-like segments of non-tunicate bulbs, or from bulbils at the end of stolons, or in a few species within the inflorescence and bulblet-like turions of rhizomes in a few cases. LANDSCAPE USES: in borders, mixed plantings, and some as vegetable crops. HARDINESS ZONES: 6A–10B. Usually not suitable for hot humid climates but thrive in warm Mediterranean regions.

COMMENTS Species of this genus have been taxonomically difficult and variously included in Alliaceae and Liliaceae, but currently included in the Amaryllidaceae by APG IV (2016) system. It constitutes the largest number of species in the family but often with uncertain delimitations. The group as a whole was prized by ancient civilizations as possessing medicinal properties or as aphrodisiacs. In addition to ornamental species, the genus includes onion and shallot (*A. cepa*), garlic (*A. sativum*), chives (*A. schoenoprasum*, *A. tuberosum*), scallions (*Allium* **spp**.), and leeks (*A. ampeloprasum*), all of which were also used for flavoring. It is important to note that so-called wild onions are often highly poisonous. Also, important to note that domestic animals are highly susceptible to poison-ing with these plants, and cows eating these plants will have onion-smelling milk. Ornamental species in the trade include, but are not limited to: *A. acuminatum*, *A. caeruleum*, *A. cari-natum*, *A. cyaneum*, *A. cyathophorum*, *A. flavum*, *A. gigan-teum*, *A. lusitanicum*, *A. narcissiflorum*, *A. paniculatum*, *A. sikkimense*, *A. splendens*, and *A. thunbergii*, among others, as well many hybrids and cultivars.

A collection of alliums
Allium giganteum (red, front), *Allium* **'Mount Everest'** (white, back),
Nectaroscordum siculum (above left)
(photographed at Chelsea Flower Show)

Allium nigrum   *Allium schubertii*

Allium moly (golden garlic)   *Allium* **'Purple Sensation'**

Allium schoenoprasum (chive)

Allium giganteum
(Gladiator allium, giant allium)

Allium ursinum
(wild garlic, wild cowleek)

Clivia miniata

KAFFIR LILY; NATAL LILY

ETYMOLOGY [Named in honor of Lady Charlotte Florentina Clive, Duchess of Northumberland (1787–1866), grand-daughter of Robert Clive, and the first to flower the type specimen in England] [miniate, painted red or vermilion, in reference to the color of flowers].

GROWTH HABIT Clump-forming evergreen herbaceous perennial to 18 in (45 cm) tall and 12 in (30 cm) or wider with age. **STEM/BARK**: bulb-like rhizomes. **LEAVES**: sword- or strap-shaped bright green leaves in 2-ranked arrangement. **FLOWERS**: slightly fragrant, red, orange, or yellow trumpet- or funnel-shaped flowers on acaulescent umbellate inflorescences, primarily in spring. **FRUIT**: attractive red berries when mature. **NATIVE HABITAT**: South Africa and Swaziland, in damp subtropical coastal woodlands to higher elevations.

CULTURE Bright filtered light or partial shade, in well-drained organic-based soil mix with the neck of the bulb above the soil level; except for cold, excess heat, and direct sun damage, clivias are essentially trouble free. Occasional mealybug, slug, or mite infestations, however, is possible in areas with poor air circulation. Plants should be kept on the dry side during winter months when grown outdoors. **PROPAGATION**: seed of fresh fruit, or by division (actually separation) after flowering. **LANDSCAPE USES**: borders and mixed plantings; often used as container plants for indoors and it is a favorite of conservatories and greenhouses. **HARDINESS ZONE**: 10A–13A. Plants are damaged below 41 °F (5 °C)

COMMENTS *Clivia miniata* **var.** *citrina* has yellow flowers and fruit. Other species that have often be used in hybridization attempts include: *C. nobilis*, *C. gardenii*, and *C. caulescens*. A number of cultivars with various color traits are in the trade, including: ‘**American Yellow**’, ‘**Belgian Hybrid Orange**’, ‘**Belgian Hybrid Yellow**’, ‘**French Hybrid**’, ‘**Golden Dragon**’, ‘**Variegata**’, and ‘**Victorian Peach**’, among others. All parts of the plant are reportedly toxic.

Clivia miniata (above),
C. minata '**Solomon Pink**' (below)

Clivia miniata '**Solomon Yellow**'

Clivia nobilis

Crinum spp.

CRINUM LILY, SPIDER LILY
ETYMOLOGY [From Greek *crinon*, a lily].

GROWTH HABIT Large perennial herbs with bulky tunicate bulbs, often with long necks, stalkless, often clump-forming, to 5 ft. (1 $\frac{3}{4}$ m) or more; coarse texture specimens; growth rate variable with species and cultivar. **STEM/BARK**: in some species such as *C. asiaticum*, a seemingly short stem is formed by leaf bases as an extension of the bulb's neck. **LEAVES**: simple, basal or from the neck, mostly spirally arranged, sword to strap-shaped, to 5 ft. (1 $\frac{3}{4}$ m) long, 4–5 in. (10–12 cm) wide, light green, fleshy, thick, midrib often keeled beneath, entire and sometimes with undulate margins. **FLOWERS**: bisexual, white, pinks, reds, some striped or multicolored, few or many-flowered on umbellate acaulescent inflorescences, corolla funnelform to 6 in. (15 cm) across with a long slender tube that is straight or curved, lobes nearly equal, linear or lanceolate, in umbels on a 2 ft. ($\frac{2}{3}$ m) long solid scape subtended by 2 bracts; flowering in summer-fall, some fragrant, especially at night. **FRUIT**: capsule, green or brownish, spherical, seeds globose and often germinate while still attached to the inflorescence (viviparous). **NATIVE HABITAT**: tropics and subtropics worldwide, in seasonally wet areas and along streams and marshes. Four species are native to Florida.

CULTURE Full sun, various well-drained but moist soils; moderately salt tolerant; chewing insects, caterpillars, *Botrytis*, leaf spots (especially in south Florida), and nematodes may cause problems. **PROPAGATION**: seed, division. **LANDSCAPE USES**: border, planter box, pot plant, bedding plant, in proximity of ponds (some species in water). In cooler climates, bulbs should be planted in containers that can be mostly buried in the ground and removed before onset of cold weather. **HARDINESS ZONE**: 8A–13B. Foliage is damaged by frost but plants recover.

COMMENTS The genus *Crinum* comprises about 180 species and includes many cultivars and hybrids, including US native taxa, and of which many are in cultivation. *Crinum asiaticum*, *C. asiaticum* '**Alan's White**', and *C. procerum* **var**. *splendens* seem to be the most popular landscape plants. *Crinum* × *augustum* '**Queen Emma**' (Purple Leaf crinum) (= *C. amabile* var. *augustum*), *C. americanum* (string lily, Florida swamp lily), *C. bulbispermum*, *C. gowenii*, *C.* × *herbertii* '**Schreck**', *C. jagus* (St. Christopher lily), *C. moorei*, *C.* × *powelii* (Cape lily), *C. zeylanicum* (milk and wine lily) which has naturalized in Florida, and a number of cultivars of uncertain origin are among the taxa in cultivation. Despite frequent use of the vernacular name "lily," crinums are not members of lily family and differ in having inferior ovary as opposed to that of superior ovary of Liliaceae.

Crinum asiaticum

Crinum powelii 'Album'

Crinum × *augustum* **'Queen Emma'** (Purple Leaf crinum)

Crinum × *digweedii* **'Marmaid'** (= *Crinum scabrum* × *C. americanum*; Mahon crinum lily)

Crinum zylanicum (Ceylon swamp lily)

Crinum augustum (Queen Emma crinum)

Crinum jagus

Crinum moori 'Album'

Cyrtanthus elatus

GEORGE LILY; SCARBOROUGH LILY
SYNONYMS *Vallota speciosa*; *Cyrtanthus purpureus*

ETYMOLOGY [From Greek *kyrtos*, arched *anthos*, flower, in reference to the curved flower tube] [From Latin, tall, in reference to the scapose flower stalk]

GROWTH HABIT Clumping bulbous herb. **STEM/ BARK**: relatively small tunicate bulbs. **LEAVES**: linear, mostly glaucous; may be produced before or after flowering depending on species. **FLOWERS**: the largest flower of the genus on umbellate scapes of 1–4 red to scarlet (as well as pink and white cultivars), funnel-form, straight flowers, each 2–3 in. (5–7 cm) across; scapes hallow, subtended by two bracts. Usually bloom in late summer but is known to do so erratically. **FRUIT**: loculicidal 3-lobed capsule; seeds flattened, winged, black. **NATIVE HABITAT**: eastern Cape Province of South Africa, in a wide range of habitats including stream edges.

CULTURE Full sun but tolerates partial shade, in well-drained soil kept moderately wet when plant actively growing but allow drying out after; to encourage flowering bulbs should be stored in refrigerator vegetable drawing for at least 6 weeks or longer if necessary (not is necessary in colder climates); mealybug and spider mites are two reported problems. **PROPAGATION**: seeds lose their via-

bility in a relatively short time and should be planted as soon as possible after collection; bulbs of this species tend to proliferate rapidly, hence bulbs should be divided periodically for producing additional plants, but also to encourage flowering. **LANDSCAPE USES**: borders and mixed planting but it is an ideal plant for container plantings. **HARDINESS ZONE**: 6A–12A. Despite statements to the contrary, this species has surprising tolerance to cold. The photograph below was taken in St. Petersburg, Russia in late September, and does not appear to be a recent planting.

COMMENTS *Cyrtanthus elatus* is widely grown for cut flower. The genus as a whole consists of about 60 species, of which some are evergreen, while others are deciduous. They also vary with regard to their growth cycle; some are summer growing, but others grow in winter in areas of milder climate. It is notable that a few species flower only following fire. Several species have tubular flowers: *Cyrtanthus brachyscyphus* (bright red flowers, tolerates wet rocky sites), *C. epiphyticus*, *C. falcatus*, *C. fergusoniae* (a deciduous species), *C. flavus* (yellow flowers), *C. helices* (white flowers), *C. herrei* (evergreen species with profusion of flowers), *C. mackenii* (has both evergreen and deciduous forms and prefers wet sites, '**Cream**', '**Himalayan Pink**', '**Peach**' flower colors, **var**. *cooperi* has long narrow yellow flowers), *C. obliquus* (long pinkish-red tubular flowers that terminate greenish-yellow), *C. sanguineus* (bright red solitary flowers), *C. spiralis* (long tubular flowers on umbellate inflorescence), and others.

Cyrtanthus elatus (photographed in St. Petersburg Botanical Garden, Russia)

Eucharis amazonica

AMAZON LILY; EUCHARIST LILY
SYNONYM *Eucharis × grandiflora*.

ETYMOLOGY [Greek word meaning pleasing, charming] [From the Amazon River region].

GROWTH HABIT Semievergreen, clumping, bulbous perennial herb, to 1 ft (30 cm) in foliage, 2–3 ft. (60–90 cm) high with flower scape. **STEM/BARK**: tunicate bulb, to 4 in. (10 cm), in diameter. **Leaves**: ovate to ovate-lanceolate, blade to 6 in. (15 cm) wide and about twice as long, born on long-petiolate succulent petiole, glossy dark green above, lighter green below, prominent midrib. **FLOWERS**: bisexual, white, 4 in. (10 cm) wide, nodding, 4–6 in. (10–15 cm) atop an umbellate 1–2 ft (30–60 cm) scape, 6 recurved tepals surround the staminal cup (connate stamens), which is tinged green inside, fragrant, long-lasting, appearing in spring and intermittently throughout the year. **FRUIT**: a 3-chambered, leathery capsule; $\frac{1}{2}$–1 in. (15–25 mm) in diameter, green or yellow; seeds are black, seldom formed. **NATIVE HABITAT**: Andes of northeastern Peru, but naturalized in several countries.

CULTURE Shade (bright sun burns leaves), in well-drained rich organic moist soil; plant bulbs 2–3 in. (5–8 cm) below soil level, or grow in pots; keep on dry side in winter to stimulate flowering; caterpillars, spider mite, bulb mites, virus, and bulb rot may become problems. **PROPAGATION**: division, tissue culture. **LANDSCAPE USES**: highlight for perennial bed in shade, container or house plant, interiorscapes, cut flower. **HARDINESS ZONE**: 10A–13B. Northern regions only with maximum protection or in container as house plant anywhere.

COMMENTS Although several related species are in cultivation, including *E. astrophiala*, *E. formosa*, *E. fosteri*, and *E. plicata*, among others, none have become commonly available. *Eucharis × grandiflora* is a natural hybrid, often sold as *E. amazonica* or *E. amazonica* 'Christine'. This plant is mostly sterile and may be of clonal origin. (Meerow and Dehgan, 1984)

Eucharis amazonica (According to A. Meerow et al. (2020), *Eucharis, Caliphruria,* and *Eucrosia dodsonii* have been transferred into *Urceolina,* hence the current accepted name is *U. amazonica*).

Hippeastrum Hybrids and Cultivars

"AMARYLLIS"

ETYMOLOGY [From Greek *hippos*, horse or *hippeus*, a rider, and *equestre*, belonging to a horseman. An apparent reference to the pair of spathes on the inflorescence that resemble horse's ears. However, *Hippeastrum* also translates to "Knight's Star," with no obvious intent].

GROWTH HABIT Terrestrial and a few epiphytic perennial herbs with tunicate bulbs, often clump-forming; sizes vary with cultivar, 1–2 ft. (30–60 cm) tall; medium texture; rapid growth. **STEM/BARK**: tunicate globular fleshy bulbs, to 6 in. (15 cm) in diameter. **LEAVES**: simple, basal, 2-ranked, linear to strap-shaped, to 2 ft. (30 cm) long, light green, glabrous, entire margins, arising with or after (not to be confused with *Amaryllis belladonna*) flower stalk emergence. **FLOWERS**: bisexual, actinomorphic, funnel-form, with 6 tepals, 6 stamens, and 3-lobed stigma; ovary inferior; predominantly in reds, pinks, whites, often lined or striped, but also yellow and orange; to 10 in. (25 cm) across, on umbels of 1–10 flowers (commonly 3–4) on a hollow leafless scape to 2 ft. (30 cm) tall; flowers in spring. **FRUIT**: trilocular globular capsule; seeds disk-shaped, winged, black, and numerous. **NATIVE HABITAT**: species native to Tropical America, primarily in eastern Brazil and southern Andes of Peru, in a wide range of habitats.

CULTURE Full sun to partial shade, various well-drained soils; moderately salt tolerant; chewing insects, caterpillars, red blotch virus mosaic, *Botrytis*. **PROPAGATION**: seeds, division, and tissue culture. **LANDSCAPE USES**: Bedding plant, border, mixed plantings, potted plants, planter box. **HARDINESS ZONE**: 8A–13B. Foliage is damaged by frost, but bulbs survive and plants recover.

COMMENTS There are about 90 species and 600 cultivars, of which many are in the trade. The true *Amaryllis belladonna* has white to pinkish flowers and is native to South Africa. Most *Hippeastrum* in trade are cultivars of hybrid origin, too numerous to mention, suffice it to note that they offer a wide range of sizes, colors, singles, and doubles.

For a detailed discussion of *Hippeastrum* see http://en.wikipedia.org/wiki/Hippeastrum

Photographed in Orchid show, Dijon (France)

***Hippastrum* 'Red Lion'** (above left), ***H.* 'Athene Pure White'** (above right). ***H.* 'Nymph'** and **'Nymph Double'** (middle), and a typical *Hippastrum* inflorescence with bracts and fruit with seed (below)

Hippeastrum puniceum (= *Hippeastrum equestre*), single (left), double hybrid (right)

Hippeastrum hybrid

×***Amarcrinum memoria-corsii*** (= *Crinum moorei* × *amaryllis belladonna*)

Lycoris radiata

SPIDER LILY; HURRICANE LILY
SYNONYMS *Amaryllis radiata*, *Lycoris terracianii*, *Nerine Japonica*, *N. radiata*

ETYMOLOGY [Named for the mistress of Mark Antony, a Roman beauty famed for intrigues] [With radiating rays, in reference to the long stamens of the species]

GROWTH HABIT Bulbous perennials that are dormant during summer months then begin growth and bloom in fall; fine texture; rapid growth rate. **STEM/BARK**: tunicate bulbs. **LEAVES**: long strap-shaped bluish-green, 1–2 ft. (30–60 cm) long and less than 1 in. (2 $\frac{1}{2}$ cm) wide. **FLOWERS**: on umbellate scapes of 4 8 flowers with long exerted filamentous stamens. In contrast to Amaryllis and *Hippeastrum* that bloom before leaves appear, *Lycoris* usually flower after the leaves have died and commonly after summer rains cease. **FRUIT**: trilocular capsule with many round hard black seeds. **NATIVE HABITAT**: eastern Asia from Japan to southern Korea, China, Vietnam, Thailand, but also in Pakistan and Iran and other countries of the region; naturalized in North Carolina.

CULTURE Full to partial sunny location, in well-drained soil; bulbs should be allowed to dry out after flowering; no major insect or disease problems reported.

PROPAGATION: seeds best scarified and planted soon after harvest and kept in a warm environment, large clumps of bulbs can also be separated, and tissue culture. **LANDSCAPE USES**: extensively cultivated in warm temperate regions in mixed borders and in containers, may also be used in mass plantings for late summer bloom. **HARDINESS ZONE**: 6A–10B.

COMMENTS There are two distinct group of species based on the length of their stamens: 2–3 times longer than the tepals (such as *L. radiata* **var**. *radiata* and *L. radiata* **var**. *pumila*) and those with about the same length as the tepals. A number of species that are probably of hybrid origin are sterile, although only one actual hybrid (*Lycoris* × *chejuensis*) is recognized. There are also 13–20 species and over 230 cultivars. A few of the more commonly available species include *L. albiflora* (with white flowers), *L. aurea* (golden spider lilies, with dark yellow flowers), *L. caldwellii* (sterile species with profuse pale yellow flowers), *L. chinensis* (golden surprise lily, with orange flowers), *L. incarnata* (peppermint lycoris, with red and white strips and short stamens) , *L. longituba* (white spider lily, with uncommonly long scape), *L. sanguinea* (orange surprise lily, is a small species with apricot color flowers and short stamens), *L. sprengeri* (electric blue surprise lily, is a small species with pinkish-blue flowers), *L. squamigera* (sterile plant of hybrid origin with large pink flowers and blooms before leaves), *Lycoris* bulbs are toxic.

Lycoris radiata

Lycoris aurea

Narcissus Species, Hybrids, and Cultivars

DAFFODIL, JONQUIL

ETYMOLOGY [From Greek mythology, in honor of the beautiful youth Narcissus, who became so entranced with his own reflection in water (narcissism) that Gods turned him into a flower].

GROWTH HABIT Low growing, stemless bulbous perennial herbs; with heights variable with cultivars, ranging from 3–24 in. (7 $\frac{1}{2}$ –60 cm), fine texture; rapid growth rate. **STEM/BARK**: contractile (pull down deeper), tunicate bulbs with membranous tunic (brown skin) and a prominent neck, 1–2 in. (2 $\frac{1}{2}$ –5 cm) in diameter. **LEAVES**: simple, basal, entire, linear (strap-shaped), rush-like or semi-terete in jonquil, nearly flat in daffodil, to 24 in. (60 cm) long, may be spirally twisted, generally dark green, leathery texture. **FLOWERS**: one or several usually fragrant flowers on a tall leafless mostly hallow umbellate scape; individual flowers mostly pedicellate but sometimes sessile, and covered by a spathe (a sheath) prior to opening; perianth segments 6 (petals + sepals); stamens 6, hidden in corona; central corona long and tubular or reduced to a shallow cup, flowers often nodding; prevailing colors are white, yellows, oranges, various shades and particolored, and often with complementary colors between the corona and the tepals; ovary is inferior; flowering in early spring. **FRUIT**: loculicidal capsule; seeds globose, hard, black. **NATIVE HABITAT**: unlike most species of Amaryllidaceae which are tropical or subtropical, species of *Narcissus* are native to cooler Mediterranean climates of southern Europe, western Mediterranean Region, and North Africa. Also, naturalized populations of horticultural origin in some areas.

CULTURE Full sun to partial shade, in various soils; moderately salt-tolerant. Bulbs remain dormant after flowers and leaves death, but will grow once again in spring and flower each year in cooler climates; while in warmer climates they will not bloom unless they are dug and stored in a cold storage when dormant – 6 weeks of cold period is required for flower initiation. It is advisable that in subtropical climates new bulbs be planted each year. Nematodes, basal rot, leaf spot, mosaic virus, and chewing insects may become troublesome. **PROPAGATION**: seeds, separation of bulblets. **LANDSCAPE USES**: bedding plant, border, planter box, pot, or dish plant. **HARDINESS ZONE**: varies with cultivar, generally zones 4A to 11.

COMMENTS Eleven basic types and thousands of named cultivars; much confusion over names due to extensive interbreeding. *Narcissus pseudonarcissus* is the original wild daffodil with a large number of synonyms and several subspecies, including *N. pseudonarcissus* **subsp.** *bicolor*, *N. P. calcicarpetanus*, *N. P.* **subsp.** *eugeniae*, *N. P.* **subsp.** *major*, and several other subspecies and numerous varieties. The Royal Horticultural Dictionary recognizes 13 Divisions to accommodate the 50 or so known wild species and numerous hybrids and cultivars of *Narcissus* known from cultivation. Any detailed discussion of Narcissus divisions is beyond the scope of this book but provided by The Royal Horticultural Society (RHS) and on the web (Wikipedia – List of Narcissus horticultural divisions), and noted in several encyclopedic plant books. https://en.wikipedia.org/wiki/List_of_Narcissus_horticultural_divisions and https://en.wikipedia.org/wiki/List_of_Narcissus_species

Examples of some *Narcissus* types, including singles and doubles (see Wikipedia and Royal Horticultural Society for grouping and differences between narcissus, daffodils, and jonquils). Two of the above photographs taken at Chelsea annual flower show

Narcissus bulbocodium. With exceptionally large corona and small perianth; Native to southern and western France, Portugal, and Spain

Scadoxus multiflorus

AFRICAN BLOOD LILY, FIREBALL LILY
SYNONYMS *Amaryllis multiflora*, *Haemanthus multiflorus* and several other names, *Nerissa multiflorus*

ETYMOLOGY [From Greek *skiadion*, parasol or umbel, and *doxa*, glory, in reference to the inflorescense] [Many-flowered]

GROWTH HABIT Evergreen perennial, rhizomatous bulb, coarse texture; rapid growth rate. **STEM/BARK**: white rhizomes with red spots, flattened, each produces 5–9 leaves, tubular bases of which produce a pseudostem to 1 in. (2 ½ cm) in diameter, and sometimes spotted. **LEAVES**: simple; spirally arranged, 12–15 in. (30–45 cm) long and about 5 in. (12 cm) wide or more, thin, with sheathing petiolate; midrib distinct; margins undulating. **FLOWERS**: bisexual; numerous small orange-red flowers (long narrow tepals) on acaulescent dense spherical multi-flowered umbellate head, to 6–8 in. (15–20 cm) in diameter; producing as many as 200 single blooms in mid- to late summer; stamens excerted, anthers yellow. **FRUIT**: showy small scarlet berries that last 6–8 weeks. **NATIVE HABITAT**: moist tropical and southern Africa, Yemen

CULTURE Full sun to part shade, in well-drained organic moist fertile soils; no serious pest or diseases reported, but bulbs should not be disturbed or waterlogged. **PROPAGATION**: seed, separation of bulblets, tissue culture. **LANDSCAPE USE**: primarily as container plant and suitable for greenhouse and conservatory. **HARDINESS ZONE**: 9B–12B. May be grown in cooler climates only if bulbs are dug and stored over winter months.

COMMENTS *Scadoxus* is a small genus of nine species, with *Scadoxus multiflorus* subsp. *katharinae* (= *Haemanthus katharinae*, named after Katharine Saunders, the well-known botanical artist) being the most common in cultivation. *Scadoxus multiflorus* subsp. *longitubus* (= *Haemanthus longitubus*) is found only in tropical West Africa and is also cultivated, but is not nearly as showy as subsp. *katharinae*. Other species in cultivation include *S. cinnabarinus*, *S. longifolius*, *S. nutans*, *S. puniceus*, and others. There are also about 25 species of *Haemanthus*, of which several are in cultivation (e.g., *H. albiflos*, *H. carneus*, and others). Bulbs of both genera are reportedly toxic.

Scadoxus multiflorus (above) and subsp. *katarinae* (below)

Tulbaghia violacea

SOCIETY GARLIC
SYNONYMS *Omentaria violacea*

ETYMOLOGY [Named in honor of Rijk Tulbagh (1699–1771), Dutch Governor of the Cape of Good Hope] [Violet colored].

GROWTH HABIT Aromatic (garlic smell) evergreen, clump-forming herbaceous tuberous perennial, reaches a maximum height of about 25 in. (60 cm) with variable spread; acaulescent; fine-textured; medium growth rate. **LEAVES**: simple; 4–8, in basal rosettes; linear, grass-like, to 1 ft. (30 cm) long, with an acute apex, and channeled at the base; distinct garlic odor when crushed. **FLOWERS**: bisexual, bright lilac to violet, salverform, $\frac{3}{4}$ in. (2 cm) long, in terminal umbels of 7–20 pedicellate flowers atop a long upright 20 in. (50 cm) scapes. Blooms in warm months of year. **FRUIT**: Insignificant capsules. **NATIVE HABITAT**: grassland areas of southern Africa.

CULTURE Flowers best in full sun, but can be grown in part shade, in light, well-drained sandy soils; not salt-tolerant; for the most part pest-free. **PROPAGATION**: seeds, clump division. **LANDSCAPE USES**: a pleasant ground cover, but avoid planting in entryways; works well massed in a naturalized area and in rock gardens; some people may find the garlic odor objectionable, particularly if forced to smell it for a long period of time; nevertheless, plants are suitable for container growth for patios, steps, etc. **HARDINESS ZONE**: 8A–11B.

COMMENTS *Tulbaghia* includes about 26 species and was included in the Alliaceae before that family was designated a subfamily (Allioideae) of the Amaryllidaceae. *Tulbaghia violacea* has been extensively used for medicinal purposes. The cultivar '**Alba**' (has white flowers), '**Tricolor**' and '**Silver Lace**' (large flowers), and '**Variegata**' (may be the same as 'Silver Lace', leaves striped cream-yellow) are offered. Related species *T. capensis*, *T. cominsii*, and *T. simmleri*, and perhaps other species are also offered.

Tulbaghia violacea

Zephyranthes spp.

RAIN LILY, ZEPHYR LILY
ETYMOLOGY [From Greek *zephyros*, God of the west wind; and *anthos*, a flower].

GROWTH HABIT Small herbaceous perennial with tunicate bulb, low growing, stemless; size varies with species, ranges from 4–8 in. (10–20 cm) tall, fine texture; rapid growth rate. **STEM/BARK**: Tunicate bulb, $\frac{3}{4}$ to 1 $\frac{1}{2}$ in. (2 to 4 cm) in diameter. **LEAVES**: simple, basal, narrow, more or less grass-like, evergreen in some species, ephemeral in others, and appearing after flowering in certain species, narrowly linear to 2 ft. (65 cm) long, dark to light green, glaucous, somewhat leathery texture, entire margins. **FLOWERS**: bisexual; solitary, sessile, or pedicellate, on a hollow scape to 8 in. (20 cm) tall, corolla erect, funnelform, segments 6, to 5 in. (12 cm) across, white, yellow, pink, rose, red, orange, some particolored; stamens 6; flowering in spring, summer, fall, and mostly short-lived; fragrant is some species. **FRUIT**: trilocular capsule; seeds many. **NATIVE HABITAT**: tropical America, a few species in Florida, and naturalized elsewhere, adaptable to diverse habitats but often observed in periodically wet areas such as roadside ditches.

CULTURE Full sun to partial shade, various soils; moderately salt-tolerant. Essentially trouble free but maggots, chewing insects, and botrytis may become problems; seeds often spread in greenhouses and because of their bulbous nature difficult to control. **PROPAGATION**: seed, separation of bulblets. **LANDSCAPE USES**: bedding plant, edging, planter box, potted plant. **HARDINESS ZONE**: 7A–13.

COMMENTS About 70 species and many cultivars and hybrids. *Zephyranthes atamasca* (rain lily, white, turning pink with age, somewhat weedy in greenhouses), *Z. candida* (zephyr flower, profuse bloomer with white flowers), *Z. citrina* (yellow flowers), *Z. grandiflora* (large pink flowers), *Z. rosea* (dark pink to magenta flowers), *Z. robusta* (larger light to dark pink flowers), *Z. simpsonii* (white, often with pink margins), and *Z. atamasca* var. *treatiae* (white to pinkish) are native to Florida. The common name "rain lily" refers to the flowering habit of some taxa which bloom after rain, if under dry conditions for a period of time. There are also a number of cultivars of hybrid origin, including 'Grandjax' (large pink flowers), 'Rose Perfection' (large pink tepals with a white line in the middle), and 'Krakatau' (bicolor, dark pink on upper and yellow on the lower part of tepals). Nearly all parts of the plants are said to be toxic. Genus *Hebranthus* is included in the now much larger genus *Zephyranthes*.

Zephyranthes atamasca (above and below left), ***Z. simpsonii*** (below left), and ***Z. atamasca* var.**
treatiae (below right)

Zephyranthes grandiflora

ADDITIONAL GENERA

Galanthus elwesii (giant snowdrop), with 2-3 narrow basal leaves and leafless scape inflorescence with a terminal single nodding flower; native to Balkans and western Turkey.

Ipheion uniflorum (= *Triteleia uniflora*; spring starflower), grass-like foliage, solitary star-shaped flowers, with each bulb producing multiple mildly fragrant nearly white to violet flowers; native to Uruguay, Argentina, and Chile

Ismene narcissiflora (= *Hymenocallis calathina, Pancratium littorale;* spider lily)

Leucojum aestivum (summer snowflake), long dark grassy leaves and white nodding bell-shaped flowers on naked hollow scapes, the six tepals are spotted green at tips, blooms in April (not summer); native to Britain and Ireland but naturalized in North America.

Nectaroscordum siculum

Nerine bowdenii, (Cornish lily, Cape flower, Bowden lily), herbaceous bulbous perennial with strap-shaped leaves and umbellate inflorescence of lily-like pink flowers that appear before leaves; includes several cultivars; native to South Africa.

Sternbergia lutea (winter daffodil, autumn daffodil, lily of the field, etc.), bulbous plant, with flowers appearing after the leaves, typical six actinomorphic yellow tepals, dormant during summer months; native to the Mediterranean region.

ASPARAGACEAE

ASPARAGUS FAMILY

143 GENERA AND ±3632 SPECIES

GEOGRAPHY Widespread in temperate, subtropical, and tropical predominantly arid regions of the Old World, but also Central America.

GROWTH HABIT A diverse group of monoecious, dioecious, or monoclinous (bisexual-flowered) shrubs or herbs, sometimes with cladodes (modified stems), robust rosette plants, and sometimes climbing. **STEM/BARK**: not woody (secondary thickening absent), often rhizomatous or bulbous. **LEAVES**: alternate, more or less sheathing, narrow, sharp, stiff, scaly, fleshy, and usually crowded (clustered) at the end of the stems, and sometimes spiny. **FLOWERS**: solitary or aggregated into a determinate pendulous or erect inflorescence; usually bisexual, sometimes unisexual; regular or only slightly irregular; 3-merous, tepals (sepals + petals) united and sometimes tubular; stamens 6, free; ovary of 3 fused carpels, superior, styles capitate or with 3 stigmatic branches. **FRUIT**: trilocular capsule or berry, often colorful; seeds one or numerous.

ECONOMIC USES Important source of fibers; fermented sap of Mexican *Agave* species is the source of pulque (a national drink of Mexico) and mescal, the distilled fermented sap. Tequila, a type of mescal, is made exclusively from the *Agave tequilana* plant. Agave syrup (maguey syrup) is made from several species of Agave. *Asparagus officinalis* shoots are eaten as vegetable. Species of most genera are used as ornamentals.

CULTIVATED GENERA *Agave*, *Anthericum*, *Asparagus*, *Aspidistra*, *Beaucarnea*, *Chlorophytum*, *Chrysodracon*, *Convallaria*, *Cordyline*, *Dasylirion*, *Dracaena*, *Eucomis*, *Hosta*, *Hyacinthus*, *Lachenalia*, *Liriope*, *Furcraea*, *Muscari*, *Nolina*, *Ophiopogon*, *Ruscus*, *Sansevieria*, *Scilla*, *Yucca*, etc. There is some disagreement as to placement of Agavaceae and Hyacinthaceae in this family and some suggest their segregation as distinct families. The genera listed here are only those considered common; there are many others in cultivation.

COMMENTS Earlier treatments of Asparagaceae included one or three genera and about 300 primarily Old-World species: *Asparagus*, *Myrsiphyllum*, and *Protasparagus* that were often treated as subgenera of *Asparagus*. In its current circumscription, however, **Asparagaceae** is a large and extremely diverse family, and perhaps fundamentally morphologically difficult for students to characterize. According to the APG III system (2009), it is a combination of several previously well-established families that are now for the most part recognized as seven subfamilies. The subsequent publication of the APG IV system (2016–2021) recognizes **Asphodelaceae**, and it is treated in this work as a distinct family. *Chrysodracon halapepe* (= *Dracaena halapepe*,

Peomele halapepe; O'ahu hala pepe), with its golden yellow flowers, is a unique plant, endemic to Hawaii. This book uses a practical horticultural approach, and specific detailed phylogenetic taxonomic discussions are beyond its scope.

Agave americana

CENTURY PLANT

SYNONYMS *Agave altissima*, *A. communis*, *A. complicata*, *A. ingens*, *A. ornata*, *A. picta*, *A. ramosa*, *A. spectabilis*, *A. variegata*, *A. zonata*, *Aloe americana*, etc.

ETYMOLOGY [Noble, in reference to the tall inflorescence of *A. americana*] [From America].

GROWTH HABIT Xerophytic, monocarpic (dies after blooming) evergreen short and thick-stemmed shrub, rarely to 9 ft. (3 m) tall and with about 4–5 ft. ($1\frac{1}{4}$–$1\frac{1}{2}$ m) spread; coarse texture; fairly slow growth rate. **STEM/BARK**: short, stout stem. **LEAVES**: in basal rosettes; sword-shaped, to 3 ft. (1 m) long and 10 in. (25 cm) wide; blue-green, thick, stiff, and fibrous; with a sharp terminal black spine; held straight or recurved; often have sharp marginal teeth. **FLOWERS**: bisexual; pale greenish-yellow, $2\frac{1}{2}$ in. (6 cm) long; in horizontal panicles borne on a stout terminal inflorescence up to 20 ft. (7 m) tall. **FRUIT**: capsules. **SEEDS**: viviparous, germinating while still on the flower stalk, and grow into small plantlets which drop to the ground and form roots. **NATIVE HABITAT**: arid regions of Mexico but naturalized in many areas of the world.

CULTURE Full sun, in a wide range of well-drained soils; tolerates salt and poor sandy soil; extremely drought-tolerant; no problem of major consequence; requires plenty of room to spread and develop; sharp spines can be hazardous and plantlets that arise after the mother plant dies may become troublesome if not removed early. **PROPAGATION**: seed or aerial plantlets, and lateral off-shoots. **LANDSCAPE USES**: distinctive, almost artificial-looking specimen; keep out of areas frequented by pedestrians. **HARDINESS ZONE**: 8B–13B. May be damaged in prolonged freezes.

COMMENTS A number of cultivars of *Agave americana* are in the trade, including: '**Blue Glow**' (blue foliage), '**Desert Diamond**' (wide variegated leaves, yellow margins), '**Marginata**' (yellow-white leaf margins), '**Gainesville Blue**' (bright blue foliage), '**Mediopicta**' (cream central stripe), '**Mediopicta Alba**' (with a central white band), '**Mediopicta Aurea**' (has a central yellow band), '**Snow Glow**' (spherical habit with narrow variegated leaves, yellow margins), and '**Variegata**' (with white leaf margins). Several varieties have also been described.

The common name of the genus comes from the plant seeming to take 100 years to bloom, but in fact it blooms and dies in as little as 6–10 years in warmer areas, but may take 15–20 years. *Agave* species are monocarpic plants, dying after the seed matures, but new plants often arise from their base. With nearly 200 species, *Agave* is a large genus of which many are in cultivation. A few examples include: *Agave albopilosa* (thick nearly cylindrical leaves with tufts of white hair at their apex), *A. attenuata* (fox tail agave, large recurved inflorescence), *A. flexispina*, *A. lophantha* '**Splendida**' (short triangular foliage with narrow yellow center and perfectly uniformly arranged marginal spines), *A. murpheyi* '**Engard**' (long narrow variegated foliage), *A. pelona* (mescal pelon, striking plant with short triangular leaves, its younger leaves with brown and older with red-orange upper half), *A. stricta* f. *rubra* (dense narrow needle-shaped orange-red leaves), *A. shawii* ('**Goldmaniana**', *A. shawii* × *attenuata*, coastal agave), *A. univittata* '**Quadricolor**' (wide triangular variegated leaves), among many other unique species and cultivars. *Agave* species are a somewhat unfriendly but magnificent presentation of nature at its best. Several *Agave* species are listed by nurseries: *A. attenuata* (spineless century Plant), *A. decipiens*, *A. desmettiana* (dwarf century plant, '**Variegata**'), *A. deserti*, *A. ferdinandi-regis*, *A. filifera*, *A. huachucensis*, *A. miradorensis* (= *A. desmettiana*), *A. parryi*, *A. sisalana* (sisal, sisal hemp), *A. stricta*, and *A. victoriae-reginae* ('**Bustamante**', '**Compacta**'; a relatively small, compact, choice species).

Agave Americana that apparently has never bloomed or produced suckers, in St. Petersburg Botanical Garden Conservatory, Russia (above left), inflorescence (above right), close-up of the inflorescence with well-developed capsules.

Agave tuberosa

TUBEROSE
SYNONYMS *Agave polyanthes*, *Crinum angustifolium*, *Polyanthes gracilis*, *P. amica*, *Polianthes tuberosa*

ETYMOLOGY [From Greek *poly*, many and *anthos*, flower, in reference to the many flowers on the inflorescence] [From Latin tuberosa, swollen, tuberous in reference to the bulb]

GROWTH HABIT Perennial rhizomatous plant with a bulb-like top (the turion), 2–2 $\frac{1}{2}$ ft. (24–30 cm). **STEM/BARK**: rhizomatous. **LEAVES**: cluster of long linear leaves atop the bulb-like turion. **FLOWERS**: numerous fragrant waxy white funnel-shaped flowers on elongated determinate (flowers open in succession, from base to top) bracteate racemes. **FRUIT**: loculicidal capsule. **NATIVE HABITAT**: although not known to exist naturally in the wild today, it is thought to have originated in Mexico and noted as having been grown there in pre-Columbian times.

CULTURE Full sun, in well-drained but moist fertile soil; planted 2 in. (5 cm) deep in spring and bloom during summer; rhizomes should be dug at the end of the growing season and kept in dry peat moss or vermiculite over winter; a viral infection has been reported. **PROPAGATION**: similar to *Gladiolus*, the plant is essentially monocarpic, where the original rhizome dies after flowering but gives rise to lateral rhizomes which are then separated and grown to mature size; also tissue culture. **LANDSCAPE USES**: mixed borders or in containers, however, it is a commercial crop and extensively planted for cut flowers. **HARDINESS ZONES**: 8B–12B.

COMMENTS flowers are used in perfumery, sold as cut flowers and often used in weddings, religious, and other events. A relatively small genus of 17 species of which *A. tuberosa* is the only one of interest. Cultivars usually refer to '**Mexican Single**', '**Double**', and '**Pearl Excelsior**'.

Agave tuberosa (= *Polyanthes tuberosa*)

Agave vivipara

CARIBBEAN AGAVE

SYNONYMS *Agave aboriginum*, *A. angustifolia* (an illegitimate name), *A. bergeri*, *A. breedlovei*, *A. costaricana*, *A. elongata*, *A. rigida*, *A. serrulata*, *A. textile*, *A. wightii*, etc.

ETYMOLOGY [Noble, in reference to the tall inflorescence of *A. americana*] [Narrow leaves].

GROWTH HABIT Xerophytic evergreen rosulate (rosette form) shrub; round or globose; rarely exceeds 5 ft. (1 $\frac{2}{3}$ m) in height or 6 ft. (2 m) in spread; medium texture; fairly slow growth rate. **STEM/BARK**: short-stemmed plant. **LEAVES**: in tight spirals; thick and stiff, narrow, sword-like, to 3 ft. (1 m) long and 4 in. (10 cm) wide; gray-green, leaves terminate in a sharp, conical, prominent black spine; margins have black teeth every $\frac{3}{4}$ in. (2 cm). **FLOWERS**: greenish, 2 in. (5 cm) long; in panicles on a stout, terminal, 9 ft. (3 m) tall flower stalk. This plant is monocarpic (dies after flowering), and the seeds are viviparous. The status of this condition is unclear since there seems to be some confusion with regard to correct identity of the species (*cf. A. vivipara*). **FRUIT**: dry dehiscent capsules. **NATIVE HABITAT**: origin unknown but widespread in tropical climates of Central America.

CULTURE Full or partial sun, in well-drained soil; tolerates salt and poor sandy soils; it is generally pest-free; sharp terminal leaf spine can be hazardous. **PROPAGATION**: seed, basal division of suckers when present. **LANDSCAPE USES**: same as *Agave americana*, but this plant is more rigid and is only one-third to one-half as large. Variegated forms are more commonly used. **HARDINESS ZONE**: 10A–13B.

COMMENTS In trade *Agave vivipara* is known as *A. angustifolia*, which is noted in synonyms as apparently an illegitimate name. There are a number of cultivars and related species: '**Marginata**' has a bold white stripe along the leaf margins and '**Variegata**' is essentially the typical species. Several *Agave* species are listed by nurseries: *A. attenuata* (spineless century Plant), *A. decipiens*, *A. desmettiana* (dwarf century plant), *A. deserti*, *A. fernandi-regis*, *A. filifera*, *A. parry* var. *huachucensis*, *A. miradorensis* (= *A. desmettiana*), *A. parryi*, *A. sisalana* (sisal, sisal hemp), and *A. stricta*.

Agave vivipara 'Marginata'

Agave vilmoriniana

OCTOPUS AGAVE
SYNONYMS *Agave eduardii, A. houghii, A. mayoensis*

ETYMOLOGY [Noble, in reference to the tall inflorescence of *A. americana*] [Named in honor of the firm of Vilmorin-Andrieux of Paris, celebrated French nurserymen and seedsmen, notably Maurice de Vilmorin and his wife]

GROWTH HABIT Medium size rosulate monocarpic succulent, with 3–4 ft. (1–1 $\frac{1}{4}$ m) tall and 5–6 ft. (1 $\frac{1}{2}$ –2 m) wide; medium texture; slow growth rate. **STEM/BARK**: short, compact stem. **LEAVES**: spirally arranged, gracefully arching, pointed, glaucous dark green, to about 3 ft. (1 m) long and 3 in. (7 $\frac{1}{2}$ cm) wide, margins entire but sometimes rough to touch with irregularly spaced small chestnut-brown teeth. **FLOWERS**: dense panicles on tall spike inflorescence 10–12 ft. (3 $\frac{1}{2}$ –4 m) long; stamens yellow, numerous, and showy. **FRUIT**: capsule; seeds viviparous with plantlets clinging to the flower stalk after the plant dies but ultimately fall to the ground and produce roots. **NATIVE HABITAT**: endemic to Sinaloa regions of Mexico on cliffs, at 1800–5100 ft. (600–1700 m) elevation.

CULTURE Sunny locations but tolerates some shade, in well-drained soil but benefits from an occasional irrigation; it is salt-tolerant; sensitive to cold despite its origin from higher elevations. **PROPAGATION**: plantlets on flower stalk, no suckers are produced. **LANDSCAPE USES**: specimen plant or in containers. Although margins are entire, the plant has a relatively sharp point and should not be too close to the public. **HARDINESS ZONE**: 10A–13B.

COMMENTS It is a somewhat variable species with no reported use other than as an ornamental.

Agave vilmoriniana (photographed in Kampong, National Tropical Botanical Garden in Florida

Other *Agave* **spp**.: *Agave colorata* (above left), *A. lophantha* (above right), *A. palmeri* (middle left), *A. lurida* (middle right), *A. parryi* **var.** *truncata* (below left), and *A. victoriae-reginae* (below right)

Asparagus spp.

ASPARAGUS FERN

SYNONYMS *Asparagopsis densiflora, Asparagus myriocladus, Protasparagus densiflorus*

ETYMOLOGY [Ancient Greek name *asparasso*, to tear or scratch, in reference to spiny leaves]

GROWTH HABIT Monoecious, evergreen; perennial herbs; ground cover; liana, rhizomatous or tuberous, spreading or climbing, fine texture; rapid growth rate. **STEM/BARK**: green, somewhat woody and spiny. **LEAVES**: cladophylls; true leaves scale-like; subtending narrow, green, leaf-like branchlets called cladophylls, linear, 1-nerved, usually flattened, and slightly curved; solitary or sometimes 3 or more at a node. **FLOWERS**: bisexual or unisexual; white or pale pink; usually borne in axillary or terminal racemes; usually rotate with 6 tepals; most relatively attractive when in full bloom. **FRUIT**: globose berry (baccate); red to nearly black; ovoid and of various lengths; seeds black, hard, 1–6 per fruit. **NATIVE HABITAT**: South Africa.

CULTURE Full sun to partial shade, in various well-drained soils; no salt tolerance; generally, pest free. **PROPAGATION**: seeds of most species germinate in 4–6 weeks, cuttings, or division of clumps. **LANDSCAPE USES**: border, ground cover, bedding plant, hanging basket, or in pots and urns. **HARDINESS ZONE**: 9A–12B. May be grown in warmer areas but frost kills the foliage, plants usually plants recover.

COMMENTS About 211 *Asparagus* species are listed in The Plant List. Two popular cultivars that are commonly grown include '**Myers**' (stems stiffly erect, very densely short branched, forming narrow plumes to 2 ft. ($\frac{2}{3}$ m) tall and 2 $\frac{1}{2}$ in. (7 $\frac{1}{2}$ cm) wide, tapering to the apex, slow growing); '**Sprengeri**' (stems drooping, loosely branched to 3 ft. (1 m) tall, branches to 5 in. (13 cm) long. *Asparagus officinalis* is a common edible vegetable, noted for its presumed diuretic and aphrodisiac properties.

Asparagus setaceus

Asparagus densiflorus 'Myers'

***Asparagus aethiopicus* 'Sprengeri'** (listed as an invasive exotic pest plant in Florida). It was included in *Asparagus densiflorus* but now considered a distinct species.

Aspidistra elatior

CAST IRON PLANT
SYNONYMS *Aspidistra variegata*, *Plectogyne variegata*

ETYMOLOGY [From Greek *aspidion*, a small, round shield; in reference to the shape of the stigma] [taller].

GROWTH HABIT Evergreen rhizomatous perennial herb, clumping, reaches a height of 2 ft. (50 cm) long, with a wider spread; coarse texture; slow growth rate. **Stem/Bark**: acaulescent, spreads slowly by a creeping rhizome. **LEAVES**: simple, oblong-elliptic, to 2 $\frac{1}{2}$ ft. (8 cm) and 4 in. (10 cm) wide, entire; the dark green, leathery, glossy leaves are borne on thin, erect, long-channeled petioles, 8–12 in. (20–30 cm) long, arising from rhizomes. **FLOWERS**: brown-purple, campanulate, 1" (2 $\frac{1}{2}$ cm) wide, inconspicuous, borne near the soil surface below the leaves. **FRUIT**: berries, small, 1-seeded, and inconspicuous. **NATIVE HABITAT**: eastern Asia, Japan, and Taiwan as understory plant.

CULTURE Partial to very dense shade; if grown in full sun, foliage becomes yellowed and burned; tolerates a wide range of soils (except heavily compacted) and marginal amounts of salt; good drought tolerance; withstands a wide range of conditions and abuse; old leaves should be removed periodically; may be attacked by a leaf-spotting fungus and mites when in low humidity situations, and scale insects; generally, pest-free; foliage injured by prolonged exposure to below-freezing temperatures and direct sunlight. **PROPAGATION**: clump division. **LANDSCAPE USES**: groundcover, planters, edging, massing, interiorscapes, textural contrast. As the common name implies, this is a tough plant that persists as a ground cover in areas where lack of light, water, or ventilation causes other plants to die. **HARDINESS ZONE**: 8A–11B.

COMMENTS It is commonly known as cast iron plant because it can withstand neglect including its high tolerance to shade. *Aspidistra lurida* and a number of its cultivars as well as *A. minutiflora* ('**Leopard**') are also in the trade. Cultivars include '**Akebono**', '**Asahi**' (leaves open chocolate brown, turn green from base upwards), '**Bubba**', '**Fugi No Mine**', '**Gold Spike**', '**Hoshi Zora**' (large, speckled leaves), '**Lennon's Green**', '**Lennon's Song**' (extended narrow apex and a paler stripe at center), '**Mary Sizemore**', '**Milky Way**', '**Okame**' (with creamy white striped leaves; tends to lose variegation in high fertility situations), '**Sekko Kan**', '**Snow Cap**', '**Starry Sky**', '**Stars and Stripes**', among others.

Aspidistra elatior, plants on left grown in shady location while that on the right in a mostly sunny location, hence the difference in color and some damage.

Beaucarnea recurvata

PONYTAIL PALM; BOTTLE PALM
SYNONYMS *Beaucarnea inermis*, *B. tuberculata*, *Dasylirion inerme*, *D. recurvatum*, *Nolina recurvata*, *Pincenectitia tuberculata*

ETYMOLOGY [Meaning obscure but perhaps a combination of Latin (and French) words *beau*, beautiful and *carne*, fleshy, in reference to the enlarged fleshy base of the plant] [Curved backward or downward]

GROWTH HABIT Caudiciform evergreen tree; round, irregular; single-stemmed when young, branching from the top of the caudex on older specimens; can reach 30 ft. (10 m) in height with a variable spread of 6–15 ft. (2–5 m) in older specimens; most plants do not exceed 10 ft. (3 $\frac{1}{2}$ m) in height; fine texture; slow growth rate. **STEM/BARK**: trunk woody, light gray-brown; tapering from a large caudex to the leafy portions; remains unbranched until the plant becomes fairly old. **LEAVES**: linear and recurved, sessile; to 5 ft. (1 $\frac{2}{3}$ m) long, $\frac{3}{4}$ in. (2 cm) wide at the base and tapering gradually to the long thin apex; margins have minute teeth, giving them a rough, bumpy feel; clustered at the tips of branches, resembling a pony's tail (hence the common name). **FLOWERS**: initially golden yellow turning whitish, small, in large erect panicles of 2–3 ft. ($\frac{2}{3}$–1 m) held above the leaves. **FRUIT**: small, long-pedicelled 3-winged capsules; rarely produced in cultivation. **NATIVE HABITAT**: dry regions of the states of Tamaulipas, Veracruz, and San Potosi in Mexico.

CULTURE Full sun or partial shade, in a wide range of well-drained soils; marginal salt tolerance; may be forced to branch by removal of top or by application of chemical growth regulators; chewing insects can disfigure the leaves; tends to develop root rot on poorly drained soils; micronutrient deficiencies common. **PROPAGATION**: seed which is usually imported from its natural habitat in Mexico. **LANDSCAPE USES**: distinctive specimen for the unusual trunk and the arching, drooping leaves; mainly grown in containers on patios, decks, and in rock gardens; occasionally used as a lawn specimen; interiorscape plant under high light intensities or in botanical conservatories. **HARDINES ZONE**: 10A–13B. Known to survive short freezing temperatures in zones 9A–9B, if under a tree or near interior windows.

COMMENTS The common name of the plant is deceiving; although it is a monocotyledonous plant, as are palms, there is no relationship between the two. No known cultivars but a few related species, such as *Beaucarnea compacta*, *B. goldmanii*, *B. gracilis*, *B. stricta*, and probably few others.

Beaucarnea recurvata

Chlorophytum comosum

SPIDER PLANT
SYNONYMS *Anthericum comosum*, *A. longituberosum*, *A. williamsii*, *Caesia comosa*, *Chlorophytum beniense*, *C. burchellii*, *C. elatum*, *C. elgonense*, *C. longum*, *C. magnum*, *C. miserum*, *C. nemerosum*, *Cordline vivipara*, *Hartwegia comosa*, *Phalangium comosum*, *P. viviparum*

ETYMOLOGY [From Greek *chloros*, green; and *phyton*, a plant] [Latin word for tufted].

GROWTH HABIT Evergreen herbaceous perennial ground cover, spreading, very short-stemmed, with fleshy tuberous roots; to 1 ft. (30 cm) tall; medium-fine texture; rapid growth rate. **STEM/BARK**: stoloniferous, small plantlets with roots are formed and persist on the stolon terminus. **LEAVES**: simple; alternate appearing whorled; linear-lanceolate to 1 $\frac{1}{2}$ ft. (40 cm) long and $\frac{3}{4}$ in. (2 cm) wide; in basal rosettes; sessile, soft, green; entire margins. **FLOWERS**: white, small; perianth rotate, segments and stamens 6; borne in a loose bracteate panicle to 30 in. (75 cm) long, becoming pendent on a cylindrical stolon. **FRUIT**: loculicidal capsule. **NATIVE HABITAT**: warm temperate southern Africa, naturalized elsewhere.

CULTURE Partial to dense shade, in interiors will tolerate low light level and low humidity; prefers rich organic well-drained soils; no salt tolerance; attractiveness of this plant is in part due to the number of plantlets, which will be reduced if overfertilized; spider mites, mealy bugs, root rot may be problems; it is sensitive to fluoride and boron. **PROPAGATION**: seeds, plantlet offsets, or division. **LANDSCAPE USES**: potted plant, hanging basket, or ground cover. It seems to be the favorite plant of restaurants probably because of low maintenance and its cascading habit. **HARDINESS ZONE**: 9B–12B.

COMMENTS Taxonomically the species consists of three described varieties that because of their geographical distance from one another should have been called subspecies. Nevertheless, the varieties include the typical *C. comosum* **var**. *comosum* with narrow strap-shaped leaves; *C. comosum* **var**. *bipindense* with petiolate leaves and long inflorescences; and *C. comosum* **var**. *spariflorum*, with broad leaves that narrow at the base, and are striped on the underside, and long inflorescences. Several cultivars are also available: '**Variegatum**' (leaves darker green, margined with white) and '**Vittatum**' (leaves recurved, with a central white stripe; slower growing). Other *Chlorophytum* species in the trade include: *C. laxum*, *C. orchidastrum* '**Fire Flash**', and *C. saundersiae*.

Chlorophytum comosum

Chlorophytum comosum '**Vittatum**'

Cordyline fruticosa

TI PLANT
SYNONYM *Dracaena terminalis; Cordyline terminalis*, as well as unusually large number of other synonyms.

ETYMOLOGY [From Greek *kŏrdyle*, a club, alluding to the shape of the root] [Latin shrubby, bushy]

GROWTH HABIT Evergreen shrub to 10 ft. (3 m) tall and 3–6 ft. (1–2 m) wide; medium texture; moderately rapid growth rate. **STEM/BARK**: scarcely branched trunk but sometimes clumping from base with suckers. **LEAVES**: ovate, 12–30 in. (30–75 cm) long and 4–6 in. (10–15 cm) wide; mostly crowded at the stem tips; glossy green to reddish-purple, or variously marked with white or yellow. **FLOWERS**: fragrant small yellowish to reddish on a ±12 in. (30 cm) paniculate inflorescence. **FRUIT**: showy red berry. **NATIVE HABITAT**: from tropical Southeast Asia to Papua New Guinea and northeastern Australia; naturalized in Hawaii and New Zealand.

CULTURE Full sun to partial shade, indoors for best color it should be placed in a bright location but not directly exposed to the sun; prefers well-drained but moist organic fertile soil; no serious disease or insect problems if outdoors but may be infested with spider mites and mealybugs indoors. **PROPAGATION**: stem cuttings which may be placed horizontally to produce stems at each node; air layering is also possible. **LANDSCAPE USES**: indoors or outdoors in the landscape but only in warmer climates; may be grown in containers and placed on a patio or moved indoors when temperature below 45 °F. **HARDINESS ZONE**: 10B–13B.

COMMENTS Starchy rhizomes have been used for food. A large number of cultivars have been selected primarily for color variation, including: '**Amabilis**' (wide ovate leaves spotted with pink and white), '**Auntie Lou**', '**Baby Doll**', '**Baby Ti**' (dwarf with copper color leaf margins), '**Baptisii**' (recurved leaves with pink and yellow stripes), '**Black Magic**', '**Black Prince**', '**Bolero**', '**Bronze**', '**Calypso Queen**', '**Colorama**', '**Compacta**', '**Exotica**', '**Fairchild Garden**', '**Firebrand**', '**Hawaiian Bonsai**' (dark crimson leaves), '**Hybrida**' (with pink margins), '**Imperialis**' (variegated red and pink leaves), '**Kilimanjaro**', '**Kiwi**', '**Madame Eugene Andre**', '**Magenta**', '**Nagi**', '**Peter Buck**', '**Pink Integrity**', '**Purple Prince**', '**Red**', '**Red Sister**', '**Ruby**', '**Tricolor**' (streaked with green, pink, and yellow), '**Turkey Tail**', '**Xerox**', and several others.

Cordyline fruticosa (above) and 'Rosa' (below)

Dracaena draco

CANARY ISLANDS DRAGON TREE
SYNONYMS *Asparagus draco*, *Dracaena resinifera*, *Draco*, *D. dragonalis*, *Drakaina draco*, *Palma draco*, *Shoerkia draco*, *Yucca draco*

ETYMOLOGY [From Greek *drakaina*, a dragon, or it may have been named after Sir Francis Drake (1540–1596), the English sea captain who circumnavigated the world in a single expedition] [Latin, dragon]

GROWTH HABIT Evergreen umbrella-like arborescent (tree-like) plant; 25 ft. (8 $\frac{1}{2}$ m) tall and as wide in cultivation; medium texture (leaves); growth rate slow. **STEM/BARK**: smooth single succulent trunk when young but profuse branching occurs soon after the first bloom in about 10–15 years and the process is repeated again thereafter. **LEAVES**: sword-shape blue-green, about 2 ft. (75 cm) long and 1 $\frac{1}{2}$ in. (3 $\frac{1}{2}$ cm) wide in clusters atop succulent branches. **FLOWERS**: much-branched spike with fragrant greenish-white lily-like bloom. **FRUIT**: pinkish-red berries. **NATIVE HABITAT**: Canary Islands (in Tenerife), western Morocco, and Cape Verde, in dry forests of coastal cliffs of warm Mediterranean-type climate without freezing temperatures.

CULTURE Full sun but tolerates some shade in well-drained soil that should be kept on the dry side –deep irrigation followed by allowing soil to dry out. **PROPAGATION**: seeds germinate readily, when available. **LANDSCAPE USES**: drought and salt-tolerant plant best used in dry gardening. It can also be used in containers but in time (several years) it may become unwieldy and difficult to handle. **HARDINESS ZONE**: 10B–13B.

COMMENTS The mythical origin of *Dracaena draco* is said to be resulted from Hercules having killed the dragon whose blood flowed over the land and resulted in dragon tree. The reddish resin (dragon's blood) from the trunk is used for staining wood and for medicinal purposes. Dragon tree is considered vulnerable by IUCN Red List and endangered in Cape Verde. Loss of habitat and several biological reasons are mentioned as the cause of the tree's endangerment. There are about 113 species of *Dracaena* of which several are in the trade and sold predominantly as house plants. However, none is as interesting and strange as that of *D. draco*. It is said to be "the ultimate in architectural planting."

Dracaena draco, old plant in Tenerife (left) and younger plants in Gran Canarias Park, Canary Islands (right)

Dracaena marginata

MADAGASCAR DRAGON TREE

ETYMOLOGY [From Greek *drakaina*, a dragon, or it may have been named after Sir Francis Drake (1540–1596), the English sea captain who circumnavigated the world in a single expedition] [Margined, in reference to the leaf margin colors].

GROWTH HABIT Evergreen small tree or shrub; much branched; 15–20 ft. (5–7 m) tall and 3–10 ft. (1–3 m) wide when grown outdoors; fine texture; relatively slow growth rate. **STEM/BARK**: gray slender upright stems ringed with diamond-shaped leaf scars and a dense rounded canopy. **LEAVES**: simple, clustered in close spirals at the ends of branches; narrow-oblong to $2\frac{1}{2}$ ft. (75 cm) long and $\frac{1}{2}$ in. ($1\frac{1}{2}$ cm) wide; entire, smooth, flexible, ranging in color from variegated light greens and pinks to reds to nearly black; petioles deeply channeled, 2–6 in. ($2\frac{1}{2}$–15 cm) long, clasping the stem. **FLOWERS**: yellowish, white, or reddish, inconspicuous; $\frac{1}{3}$ in. (8 mm) long; borne in 12 in. (30 cm) long panicles; blooms in spring. **FRUIT**: berry; color changes from yellow to red; $\frac{1}{4}$ in. (6 mm) wide; rarely produced in cultivation. **NATIVE HABITAT**: Madagascar

CULTURE Grows well in partial to dense shade, in well-drained but moist fertile soils; no salt tolerance; best foliage color occurs in shade; discoloration results with exposure to full sun; mealy bugs, mites, nematodes, and leaf spot may be problems; plant is sensitive to fluoride damage, appearing as marginal necrosis. **PROPAGATION**: tip and cane cuttings, air layering, and occasionally seed. **LANDSCAPE USES**: outdoors as a color accent for shaded areas; indoors in tubs or planters as a specimen plant. **HARDINESS ZONE**: 10B–13B.

COMMENTS A few cultivars with foliage color variations are available, and commonly used as indoor foliage plants, including: '**Colorama**', '**Magenta**', and '**Tricolor**'. There are, however, a large number of other *Dracaena* species and cultivars that are commonly grown as foliage plants. A few examples include: *Dracaena arborea*, *D. dermensis* (with several variously green with white stripes cultivars), *D. fragrans* (also all green with white bands but shorter leaves and more stout stems), *D. reflexa* (with predominantly shorter green leaves and more compact), *D. sanderiana* (green with yellowish wide margins), *D. surculosa* (distinct from other species, with ovate often spotted leaves and red berry-like fruit), and *D. terminalis* (see *Cordyline terminalis*), among others. There is significant confusion between these related genera, species, and cultivars: although authorities of plant names are not mentioned in this book, an exception is necessary in this case: *Dracaena marginata* Aiton is a synonym of *Aloe purpurea* but *Dracaena marginata* hort. is the accepted name and the plant in cultivation and discussed here; it has no synonym. To further complicate the issue, *D. marginata* Lam. is a synonym of *D. refelxa* var. *angustifolia*. Further examples and discussions are left to botanical authorities.

Dracaena marginata propagation stock plants (above), several cultivars, including the two cultivars (below). Photographed in Indonesia.

Dracaena reflexa

SONG OF INDIA
SYNONYM *Draco reflexa*, *Lomatophyllum erflexum*, *Pleomele reflexa*

ETYMOLOGY [From Greek *drakaina*, a dragon, or it may have been named after Sir Francis Drake (1540–1596), the English sea captain who circumnavigated the world in a single expedition] [Reflexed, in reference to position of the leaves].

GROWTH HABIT Evergreen shrub or small tree with an erect branched stem and an irregular open crown; to 15 ft. (15 m) tall and with 3–8 ft. (1–3 m) spread; medium texture; slow growth rate. **STEM/BARK**: mature stems may become thick, rigid, irregular, and ringed with leaf scars. **LEAVES**: simple, spirally arranged in whorls; lanceolate to elliptic, $1\frac{1}{2}$ ft. (18 cm) long and 2 in (5 cm) wide at base; sessile; margins entire, leaf color ranges from green to creamy white or yellow margins; plants grown in the landscape often terminate with leaves that have yellow margins; leaves tend to persist longer than other *Dracaena* species. **FLOWERS**: white petals but purple sepals, $\frac{1}{2}$ in. ($1\frac{1}{3}$ cm) in diameter; borne in 12 in. (30 cm) long panicles; rarely produced and not of any particular interest. **FRUIT**: small orange-red berry, insignificant. **NATIVE HABITAT**: Madagascar and other islands of the Indian Ocean.

CULTURE Partial to dense shade, bright light but no direct afternoon sun, in a wide variety of soils but prefers a well-drained and well-aerated soil containing organic matter; drought-tolerant; roots tolerant of some salinity, but foliage has no salt spray tolerance; mites, thrips, chewing insects, leaf spot; sensitive to fluoride. **PROPAGATION**: tip cuttings or layering of older stems. **LANDSCAPE USES**: outdoors, it can be used as a specimen or accent plant in shaded areas and sometimes used as a pruned or unpruned hedge; a good, low-maintenance specimen for public interiorscapes and homes. **HARDINESS ZONE**: 10B–13B.

COMMENTS *Dracaena reflexa* is locally used for medicinal purposes. A few cultivars are in the trade, including: '**Anita**', '**Honoriae**', '**Riki**' (with narrower and longer leaves and yellow and green bands), '**Song of India**' (with wide yellow margins and perhaps to most commonly sold cultivar), '**Song of Jamaica**', and '**Variegata**'. There seems to be some confusion as to the species and related cultivars in the trade with those of *D. reflexa*.

Draecina reflexa used as hedge (above) and in-ground nursery production (below)

Dracaena fragrans

CORN PLANT

SYNONYMS *Aletris fragrans*, *Cordyline fragrans*, *Dracaena albanensis*, *D. broomfieldii*, *D. deremensis*, *D. lindenii*, *D. masseffana D. smithii*, *D. victoria*, *Draco fragrans*, *Pleomele dremensis*, *P. fragrans*, *Sansevieria fragrans*

ETYMOLOGY [From Greek *drakaina*, a dragon, or it may have been named after Sir Francis Drake (1540–1596), the English sea captain who circumnavigated the world in a single expedition] [fragrant].

GROWTH HABIT Evergreen large shrub to small tree, with an unbranched stem (unless top is removed); columnar form, 25 ft. (8 $\frac{1}{3}$ m) tall with a 15 ft. (5 m) spread; medium-coarse texture; slow growth rate. **STEM/BARK**: light brown stem ringed with leaf scars. **LEAVES**: simple, whorled in terminal clusters; linear to 3 ft. (1 m) long, 4 in. (10 cm) wide; sessile; dark, emerald green with a wide yellow-green center stripe; sun brings out center stripe; prominent midrib on the lower surface light green or yellowish. **FLOWERS**: yellowish, extremely fragrant, nocturnal flowers (hence the specific epithet); borne in terminal panicles to 1 $\frac{1}{2}$ ft. (45 cm) long; not commonly produced. **FRUIT**: berry, also uncommon. **NATIVE HABITAT**: Tropical Africa.

CULTURE Best in partial to dense shade; leaf tips and margins will scorch in full sun; tolerates a wide variety of soils, but prefers an organic soil; excellent for low light interiors; the foliage is not salt-tolerant, but the roots will tolerate some salinity; mites, chewing insects, leaf spot, and fluoride toxicity may be problems. **PROPAGATION**: tip and cane cuttings. **LANDSCAPE USES**: outdoors as an accent plant because of its colorful foliage; excellent interior specimen because of its durability. **HARDINESS ZONE**: 10B–13B.

COMMENTS The common name derives from the yellow variegated, corn-like, arching leaves and trunk of the cultivar 'Massangeana'. Other available cultivars include: 'Lindenii'. Several cultivars are listed under the *Dracaena deremensis*, which is actually a synonym of *D. fragrans*, including: 'Janet Craig' (has strap-like, dark green leaves with a prominent midrib); 'Warneckii' (striped dracaena; has linear leaves with two white stripes), 'Baueri', 'Golden King', 'Green Stripe', 'Compacta', 'Lemon and Lime', and 'Warneckii Compacta', and probably others. *Dracaena fragrans* is produced in mass quantities in Central America and shipped to United States and other countries, where, once "canes" stuck is soil, they develop roots and produce leafy stems.

Dracaena fragrans 'Janet Craig' (above left), *D. fragrans* 'Massangeana' (above right), and in production field (below left) and rootless and leafless stems ready for shipment (below right) in Costa Rica.

Dracaenia trifasciata

SNAKE PLANT; MOTHER-IN-LAW'S TONGUE

SYNONYMS *Sansevieria craigii*, *S. jacquinii*, *S. laurentii*, *S. trifasciata*

ETYMOLOGY [After Raimond de Sangro, prince of Sanseviero, eighteenth-century Italian patron of horticulture] [In three bundles].

GROWTH HABIT Hermaphroditic, evergreen, herbaceous, perennial subshrub; upright, spreading; acaulescent, with stiff, erect leaves arising from a short thick rhizome; to 5 ft. ($1\frac{2}{3}$ m) tall, spreads and forms thick stands by means of the underground rhizome; medium texture; moderately fast growth rate. **STEM/BARK**: rhizomatous. **LEAVES**: simple, basal, erect; linear-lanceolate, to 5 ft. ($1\frac{2}{3}$ m) long and $2\frac{3}{4}$ in. (7 cm) wide; conspicuously cross-banded with light and dark green bands; fibrous, flat, stiff, and fairly thick and pointed. **FLOWERS**: bisexual, greenish-white, tubular, and fragrant; to $\frac{1}{2}$ in. ($1\frac{1}{4}$ cm) long; borne on a racemose inflorescence to $2\frac{1}{2}$ ft. (30 cm) tall. **FRUIT**: red berry. **NATIVE HABITAT**: tropical west Africa, from Nigeria east to Congo.

CULTURE Grows in light conditions ranging from full sun to dense shade; indoors will tolerate low light intensities, grows on a wide range of well-drained soils; drought-tolerant; moderate salt tolerance; mites, thrips, and chewing insects occasionally cause problems, but generally this plant has few insect or disease problems; overwatering induces root rot. **PROPAGATION**: division of rhizomes or rooting of leaf segment cuttings. It is noteworthy that once rooted, leaf cuttings of the cultivars 'Laurentii' and 'Silver Hahnii' revert to their original green forms. **LANDSCAPE USES**: upright forms can be used as a vertical accent, while rosette forms make a good ground cover for very shady or dry areas in the landscape. Because of its tolerance to low light levels, *D. trifasciata* makes an excellent interior specimen. **HARDINESS ZONE**: 10B–13B.

COMMENTS There are two commonly available cultivars of *Draecenia* (*Sansevieria*) *trifasciata* 'Hahnii' (bird's nest sansevieria) is a dwarf cultivar to 8 in. (20 cm) tall; the short leaves form a funnel-shaped rosette; *D. trifasciata* 'Laurentii', 'Silver Laurentii', and 'Silver Hahnii' are chimeras that resemble the species except leaves bear golden yellow and silver margins, respectively. Other cultivars include 'Black Coral', 'Black Gold', 'Compacta', 'Futura', 'Futura Superba', 'Gold Flame', 'Goldiana', 'Moonglow', 'Moonshine', 'Sibersee', 'Twisted Sister', and probably others. *D. cylindrica*, *D. ehrenbergii*, *D. grandis*, *D. zeylanica* (Ceylanese bowstring hemp), and other species are also offered. *Dracaena trifasciata* is considered invasive in Florida and probably in similar regions.

Dracaena trifasciata (= Sansevieria trifasciata; rooted and grown leaf segment cuttings (left), and *D. trifasciata* **'Laurentii'** (right), *D. trifasciata* **'Black Gold'** (below)

Inflorescence of *Dracaena trifasciata* 'Laurentii'

Dracaena trifasciata field production

Dracaena hanningtonii
(= *Sansevieria ehrenbergii;*
sword sansevieria)

Draceana sanderiana

Hosta spp.

PLANTAIN LILIES; HOSTAS
ETYMOLOGY [Named in honor of Nicolaus Tomas Host (1761–1834), physician to the Emperor of Austria]

GROWTH HABIT Perennial scapose herbs with a range of sizes from less than 4 in. (10 cm) across and 3 in. (8 cm) high to 6 ft. (2 m) across and 4 ft. ($1\frac{1}{3}$ m) high. **STEM/BARK**: modified stems, rhizomes and stolons. **LEAVES**: significance of hostas in botanical descriptions and particularly in horticulture lies in their leaf shapes, sizes, and color variegations; from rounded to oval, cordate, and lanceolate; surfaces range from flat to furrowed; and colors vary from green and glossy to white-cream, yellowish, or with a bluish cast (see *H. sieboldiana* below). **FLOWERS**: white, lavender, or violet pendulous funnelform or bell-shaped, and arise on scapose racemes, but very few are fragrant. **FRUIT**: a 3-locular capsule. **NATIVE HABITAT**: northeastern Asia, including China, Japan, Korea, and Russia.

CULTURE Hostas are shade-tolerant plants but grow and bloom well in the direct or partial sun; intense leaf colors are best developed in shady but bright sites; they prefer well-drained but moist organic fertile soils and should be irrigated during hot summer months; although for the most part hostas are pest and trouble free, apparently nematodes may pose a serious problem where they exist, and other than soil fumigation there does not appear to be a pesticide specifically approved for their control; virus-infected plants must be destroyed. **PROPAGATION**: division of offsets in early spring or seeds. **LANDSCAPE USES**: with the wide range of color, texture, and leaf sizes, hostas may well be one of the most useful plants for shade gardens and can be used in borders, background (larger cultivars), or ground covers, and even specimen-accent plants. **HARDINESS ZONE**: 4A–11B, species- and cultivar-dependent, generally larger numbers do best in cooler climates and few if any suitable for tropical humid zones.

COMMENTS Although the genus as a whole is not considered a large taxon, it constitutes one of the largest numbers of cultivars that originated as sports or by hybridization. Thousands of such cultivars have been reported and registered by the American and British Hosta societies. Careful selection of cultivars for specific localities is highly recommended. Accordingly, enumerating cultivars in this treatment would be a pointless exercise. There are not many hostas that do well in hot humid climates but are perfectly suitable for such places as cool or warm temperate Mediterranean climate regions. It is also important to note that hostas contain saponins and are toxic to domesticated and possibly also wild animals. It is noteworthy that *Hosta* cultivars are generally difficult to identify.

Hosta plantaginea (above) and *H.* '**Patriot**' (below)

Hosta seiboldiana (above) and *H. seiboldiana* **'Elegans'** (below)

Hosta 'Designer Genes' (?)

Hosta plantaginea (plantain lily), native to China, with nocturnal fragrant flowers, it is listed as suitable for zones 3-9, these photographs, however, were taken in Indonesia which is essentially a tropical zone.

Hyacinthus orientalis

HYACINTH
SYNONYMS *Hyacinthus albulus*, *H. brumalis*, *H. modestus*, *H. praecox*, *H. provincialis*, *H. rigidulus*

ETYMOLOGY [From Greek mythology, name used by Homer, the flowers said to have sprung from the blood of *Hyakinthos*, a youth of great beauty accidentally killed by Apollo when teaching him to thorough the discus] [From the Orient]

GROWTH HABIT An acaulescent bulbous plant. **STEM/ BARK**: modified stem, tunicate bulb. **LEAVES**: strap-shaped, in a whorl atop the bulb; 5–10 in. (15–30 cm) long and $\frac{1}{2}$ to 1 in. (1–2 $\frac{1}{2}$ cm) broad. **FLOWERS**: tubular on a spike, 7–14 in. (20–35 cm) long, with few blue bell-shaped flowers in its original natural form, but numerous flowers of many colors tightly organized along the leafless rachis in cultivars; flowers are generally fragrant but much less intense in commercially produced cultivars. **FRUIT**: fleshy spherical 3-locular capsules with many black seeds. **NATIVE HABITAT**: southwestern Asia and parts of the Middle East.

CULTURE Best in Mediterranean-type climates in direct sun or part shade, in well-drained soil that should BE kept on the dry side. **PROPAGATION**: Bulbs of cultivars are shallowly cross-cut at their base and planted; bulblets are produced and separated. It is also produced by seed, but the development is much too slow for common use; there are several reports of tissue culture as well. **LANDSCAPE USES**: as landscape plants in cooler climates, but bulbs and potted plants are sold; flowers do not last for long in warmer climates and bloom only the first year, but if dug and stored in a cool storage, they probably will bloom again. Plant 4–5 in. (10–12 cm) deep, in fall. **HARDINESS ZONES**: 6B–10A.

COMMENTS An early bloomer, bulbs are forced to flowers and used at Christmas, also used on Persian New year (Nowruz, March 21st), at its normal blooming time, where it is collected in lower mountain elevations. Popularity of the plant has resulted in selection of hundreds of cultivars primarily for various flower colors. Mention of cultivar names serves no useful purpose at this time particularly since hyacinths are not appropriate plants for commercial production in warm winter climates. New bulbs are annually imported from the Netherlands. There are three accepted *Hyacinthus* species names: *H. litwinowii*, *H. orientalis*, and *H. transcaspicus*.

Hyacinthus orientalis bulbs (left) and original wild species (right)

Hyacinthus orientalis cultivars (Photographed at Keukenhof Gardens, Amsterdam)

Lachenalia aloides

OPAL FLOWER; CAPE COWSLIP
ETYMOLOGY [Named after Werner de la Chenal (1736–1800), professor of botany at Basel, Switzerland] [Aloe-like]

GROWTH HABIT Bulbous perennial. **STEM/BARK**: fleshy, modified stem, tunicate bulb. **LEAVES**: strap-shaped, solid green or spotted, usually few in number. **FLOWERS**: yellow tubular, pendent with reddish orange lip; to $1\frac{1}{2}$ in. (4 cm) long, often reddish in bud; on raceme inflorescences of 5–10 in. (15–30 cm) tall. **FRUIT**: loculicidal 3-locular capsule. **NATIVE HABITAT**: western Cape Province of South Africa, and in Namibia, on granite and sandstone outcrops.

CULTURE Full sun, in well-drained sandy-loam soil; as in most other bulbous plants this species has a dormancy period and should be treated accordingly; mealybugs and slugs may be problems. **PROPAGATION**: bulbs multiply freely and can be separated. **LANDSCAPE USES**: mixed plantings with succulents or in borders, and as potted plants; blooms in early spring. **HARDINESS ZONE**: 9B–11B. More suitable for Mediterranean climate zones.

COMMENTS There ae 110 listed species in the genus of which the largest number are native to South Africa. *Lachenalia aloides* and its varieties and cultivars are the most commonly grown taxa. Cultivar '**Pearsonii**' (vigorous growth and large flowers), '**Nelsonii**' (golden yellow below and light green above, pendent flowers); *L. aloides* **var**. *aloides* (syn. *L. tricolor*) has yellow flowers with reddish-orange tip; *L. aloides* **var**. *aurea* has a pair of plain green or spotted leaves and golden yellow flowers *L. aloides* **var**. *quadricolor* has four color flowers, reddish-orange at base, then yellow and purplish-maroon at tips; and *L. aloides* **var**. *vanzyliae* often with purplish-brown marked leaves and flowers pale blue on the outer side and yellow-green with white margins on the inside. Other reported species include: *L. bulbifera*, *L. mutabilis*, *L. pendula*, and others (for information on other species see "Lachenalia Species One" on: http://www.pacificbulbsociety.org/pbswiki/index.php/LachenaliaSpeciesOne)

Lachenalia aloides

Ledebouria socialis

SILVER SQUILL; WOOD HYACINTH; LEOPARD LILY
SYNONYMS *Ledebouria violacea*, *Scilla laxa*, *S. paucifolia*, *S. socialis*, *S. violacea*

ETYMOLOGY [After Carl Fredrich von Ledebour (1785–1851), German botanist] [Social, in reference to prolific rapid reproduction and formation of a group of bulbs]

GROWTH HABIT Clump-forming evergreen bulbous perennial plants; to about 10 in. (25 cm); fine texture; rapid growth rate (bulb production). **STEM/BARK**: A cluster of purple tunicate bulbs, 1–2 in. ($2\frac{1}{2}$–5 cm) in diameter. **LEAVES**: spirally arranged atop bulbs; lanceolate, 4–6 in. (10–15 cm) long and about $\frac{3}{4}$ in. (2 cm) wide, with numerous purple blotches on the entire surface. **FLOWERS**: small green bell-shaped with purple exerted stamens, on a single acaulescent racemose inflorescence from each bulb. **FRUIT**: loculicidal capsule; seeds two per capsule, obovoid, reddish or yellowish-brown, or black. **NATIVE HABITAT**: South Africa, Eastern Cape Province, in seasonally dry savanna.

CULTURE Partial shade but tolerates full sun, in well-drained humus-rich sandy soil; should be kept on the dry side except when growing actively in spring; bulbs should be above soil level. **PROPAGATION**: separation of the rapidly developing bulbs. **LANDSCAPE USES**: ground cover or rock garden in warmer climates otherwise as potted plants; prolific production of bulblets soon results in attractive potted plants with spotted leaves and small delicate but prolific flower stocks, usually twice per year. **HARDINESS ZONES**: 10B–12B.

COMMENTS *Ledebouria socialis* is treated as a succulent and favored by succulent enthusiasts. It is a variable species with respect to leaf spots, which may range from mottled black to purple and various flower color variations, such as pink pedicel versus green. Although not common in the trade, a few cultivars have been noted on the web: '**Juda**' (with variegated leaves), '**Laxifolia**' (globose bulbs and pale green leaves mottled with darker green), '**Miner**' (a dwarf selection), '**Pauciflora**' (ovoid bulbs predominantly with two leaves mottled with green spots), and '**Violacea**' (with silver-colored leaves and mottled light purple above and maroon violet below). Of the reported 55 species, *Ledebouria cooperi* (narrow leaves with dark purple marking and pink flowers) is the only other species reported from cultivation.

Ledebouria socialis '**Violacea**' illustrating a cluster of bulbs (above, left), profuse inflorescences (above, right), cymose inflorescences (below, left), and close-up of flowers (below right).

Liriope muscari

LILYTURF, MONKEY GRASS, LIRIOPE
SYNONYMS *Liriope exiliflora*, *L. gigantea*, *L. platy-phylla*, *L. yingdeensis*, *Ophiopogon muscari*

ETYMOLOGY [Named in honor of a Greek woodland nymph, Liriope, the mother of Narcissus] [Greek musk, in reference to the sweet scent of the flowers].

GROWTH HABIT Perennial evergreen clumping herb, which forms dense, grass-like mats, reaches a height of 8–24 in. (60 cm), depending on cultivar, with an indeterminate spread; fine texture; moderate growth rate. **STEM/BARK**: acaulescent, spreads by rhizomes, forms tubers. **LEAVES**: simple, linear, arching, 8–24 in. (20–60 cm) long and $\frac{1}{2}-\frac{3}{4}$ in. (12–20 cm) wide, dark green, leathery, grass-like, with parallel veins, tuft-forming. **FLOWERS**: bisexual; purple, small $\frac{1}{4}$ in. (7 mm) across, dense, hyacinth-like, in short spikes held on stalks several inches taller than the leaves; flowers on some cultivars can be showy in full bloom in late spring to early summer. **FRUIT**: fleshy, berry-like, $\frac{1}{2}$ in. (12 mm) wide, black, in fall and winter, held above the foliage. **NATIVE HABITAT**: eastern Asia.

CULTURE Partial to deep shade, some cultivars can tolerate full sun; variegated cultivars revert to the green form in dense shade; tolerates a wide range of soils and moderate amount of salt; somewhat drought-tolerant, but may develop tip burn in severely hot, dry locations; can be heavily sheared (actually mowed) every few years to remove ragged old foliage; highly competitive against weeds once established; no serious pest problems; fruiting stalks may be unsightly and require removal. **PROPAGATION**: seeds, clump division of cultivars. **LANDSCAPE USES**: ground cover for shady locations, especially useful where grass will not grow, and for edging woodland walkways. It will not withstand foot traffic. **HARDINESS ZONE**: 6B–11A.

COMMENTS Numerous available cultivars: '**Aztec Grass**' (variegated silvery, bright white and green foliage), '**Aztec Gold**' (white flower spikes and green leaves with pale yellow bands), '**Big Blue**' (18–20 in. (40–50 cm) tall, bold blue-green foliage), '**Evergreen Giant**' (18–24 in. (40–60 cm) tall, white flowers), '**Lilac Beauty**' (prolific flowering, lilac; does well in sun or shade), '**Majestic**' (to 24 in. (60 cm) tall, rich lavender flowers), '**Monroe White**' (white flowers; leaves tend to bleach in full sun), '**Royal Purple**' (tall inflorescences of purple flowers, arching green foliage), '**Silver Midget**' (8 in./20 cm) tall; variegated with white banding, slightly twisted foliage), '**Samantha**' (profuse bloomer, pink flowers, clumping-does not spread), '**Silvery Sunproof**' (lavender flowers; white variegation; does well in sun or shade; leaves lose some of variegation in dense shade), '**Variegata**' (variegated with yellow striping); and others. *Liriope spicata* has narrower leaves, $\frac{1}{2}$ in (1 $\frac{1}{3}$ cm) wide, 8–16 in. (20–40 cm) tall, similar to *Ophiopogon japonicus*, but flowers are borne above leaves; spreads vigorously and is difficult to confine as an edging plant.

Liriope muscari

Ornithogalum dubium

SUN STAR, STAR OF BETHLEHAM
SYNONYMS *Eliokarmos dubius*, *Myogalum flavescens*, *Ornithogalum alticola*, *O. aureum*, *O. citrinum*, *O. flavescens*, *O. miniatum*, *O, pillansii*, *O. vandermerwei*, etc.

ETYMOLOGY [From Greek *ornis*, a bird and *gala*, milk, in reference to predominantly white flowers of the species] [Dubious, doubtful]

GROWTH HABIT Hermaphroditic bulbous perennial, to 20 in. (50 cm) tall; fine texture; bulbs grow slowly but sprout rapidly. **STEM/BARK**: deciduous bulbs. **LEAVES**: simple; basal, linear or strap-shaped, 3–8 per bulb, erect or spreading; margins minimally pubescent. **FLOWERS**: bisexual; in racemose inflorescenes, yellowish-orange or rarely white, cup- or star-shaped, often with a darker brown center; tepals 6, stamens 6; style short; to about 1 in. (2 $\frac{1}{2}$ cm) across; blooms from September to December. **FRUIT**: globose capsule, deeply 3-angled, ribbed; seeds discoid, black and shiny.

NATIVE HABITAT: mountain slopes and flats of western and eastern Cape Province of South Africa.

CULTURE Full sun to part shade, in well-drained sandy soil; dormancy begins in mid-spring through summer, bulbs should be planted about 1 in. (2 $\frac{1}{2}$ cm) deep, in fall, keep on dry side after leaves sprout; mealy bugs and aphids could be problems, and the latter spread mosaic virus. **PROPAGATION**: seed, bulb scoring, or offsets. **LANDSCAPE USE**: border or mass planting but commonly used as container plant. **HARDINESS ZONE**: 8A–11B. Best in Mediterranean climate zones.

COMMENTS A large genus of 190 species native to Africa, Europe, and western Asia. Plants are used for medicinal purposes, but all parts are known to be poisonous. In addition to *Ornithogalum dubium*, other species in cultivation include *O. arabicum* (white flowers), *O. caudatum*, *O. fimbrimarginatum*, *O. longibracteatum*, *O. magnum*, *O. nutans*, *O. saundersiae*, *O. thyrsoides* (white flowers), *O. umbellatum* (white flowers), and probably others. A few cultivars have been released by the U.S. National Arboretum. Holland is the principal producer of bulbs.

Ornithogalum dubium

Ornithogallum orthophyllum, native from central Europe to Meditteranean, and Iran
(photographed at Denver Botanical Garden, Colorado)

Maianthemum racemosum

FALSE SPIKENARD; FEATHERY FALSE SOLLOMON'S SEAL

SYNONYMS *Convallaria ciliata*, *C. racemosa*, *Polygonastrum racemosum*, *Sigillaria multiflora*, *Smilacina ciliata*, *S. racemosa*, *Tovaria racemosa*, *Unifolium racemosum*, *Vagnera australis*, etc.

ETYMOLOGY [From Greek *Maios*, May; *anthemum*, bloom, in reference to blooming season] [Raceme, in reference to the inflorescence]

GROWTH HABIT Clump-forming herbaceous perennial, to 3 ft. (1 m) tall and equal spread, but forms large colonies in some areas. **STEM/BARK**: elongated thick branched cylindrical rhizome to 3 ft. (1 m) long; stems erect. **LEAVES**: simple alternate, elliptic-ovate, 3–5 in. (7–15 cm) long and 1–2 ½ in. (3–6 cm) wide; base rounded; margins entire; apex acute; sessile and clasping or sometimes petiolate.

FLOWERS: small fragrant flowers with 6 creamy white tepals on many-flowered racemose panicles, 4–7 in. (10–15 cm) showy inflorescences; ovary globose; stigma obscure. **FRUIT**: berry-like translucent red in showy clusters when mature; seeds 1–4, globose. **NATIVE HABITAT**: North America, including continental United States and Canada, but excluding Mexico, in predominantly moist shady woodland locations.

CULTURE Partial to full shade, in well-drained organic moist soil; although it is native to northern climates and does not tolerate hot humid summers, it is native to parts of southern U. S., including a small area of the Florida Panhandle. No serious problems have been reported. **PROPAGATION**: cold moist stratified seeds or careful separation of clumps with rhizome. **LANDSCAPE USES**: ideal for naturalized woodland landscapes and native plant gardens; mixed plantings with smaller ferns such as *Adiantum* spp., or other shade-tolerant plants near ponds or streams; also used to prevent soil erosion. **HARDINESS ZONE**: 4A–9A.

COMMENTS Only young shoots without leaves are edible. Rhizomes have laxative properties and the plant is said to have been used as cough suppressant. The western form with longer inflorescence and shorter leaves is referred to as *M. racemosum* **subsp**. *amlexicaule*. The smaller eastern U. S. species, *M. stellatum*, has longer star-shaped flowers and larger dark red berries. Other listed species include *M. bifolium* and *M. flexuosum*.

Maianthemun racemosum, Photographed at Kew Gardens, England

Ophiopogon japonicus

MONDO GRASS, DWARF LILYTURF
SYNONYMS *Convallaria graminifolia*, *C. japonica*, *Flueggea argulata*, *F. japonica*, *Liriope gracilis*, *Mondo gracile*, *M. japonicum*, *M. longifolium*, *M. stolonifera*, *Ophiopogon argyi*, *O. gracilis*, *O. stolonifer*, etc.

ETYMOLOGY [From Greek *ophis* a snake and *pogon*, a beard, allusion obscure, but perhaps in reference to the growth habit of the plant] [From Japan].

GROWTH HABIT Evergreen, clumping perennial herb which forms dense, grass-like mats, to 10 in. (25 cm) high and with an indeterminate spread; fine-textured; medium growth rate. **STEM/BARK**: acaulescent, spreads by long stolons. **LEAVES**: simple, linear, grass-like, 6–15 in. (15–45 cm) long, $\frac{1}{2}$ in. (13 mm) wide, dark green and curving toward the ground, occurring in tufts. **FLOWERS**: lilac or white, $\frac{1}{2}$ in. (13 mm) wide, inconspicuous in short, loose, few-flowered racemes below the leaves; blooms early summer. **FRUIT**: capsules, blue, pea-sized, fleshy. **NATIVE HABITAT**: Japan and Korea.

CULTURE Grows in partial to full shade on a wide range of soils; good salt and drought tolerance; requirements very similar to those of *Liriope muscari* but generally slower growing; pest-free. **PROPAGATION**: seeds, but commonly division of clumps. **LANDSCAPE USES**: along walkway borders and in planters and foundation plantings; makes an excellent, low-maintenance ground cover in areas too shady for grass. It will not withstand foot traffic, but can be mowed infrequently. **HARDINESS ZONE**: 7A–11B.

COMMENTS Only a few of 67 listed species and cultivars are available: "Gyoku-Ryu' (dwarf clump only 1–2 in. (2 $\frac{1}{2}$ –5 cm) tall, 'Nana' (dwarf form, 2–3 in. (5–7 $\frac{1}{2}$ cm) tall, small white flowers), 'Kyoto Super Dwarf' (tightly clumped, to 4 in. (10 cm), 'Silver Dragon' (variegated leaves), 'Variegatus' (foliage with white and green striping); the related *O. jaburan* 'Vittatus' (leaves symmetrically arranged, pale green stripes and white margins), and 'White Dragon' (wide white stripes), *O. planiscapus* 'Nigrescens' (black mondo, deep green, nearly black basal leaves), and 'Little Tabby' and 'Silver Ribbon' (both green with white borders), 'Black Beard' (large black mondograss).

Ophiopogon japonicus

Ophiopogon jaburan **'Vittatus'** (giant lilyturf)

Ruscus aculeatus

BATCHLOR'S BROOM
SYNONYMS *Oxymyrsine pungens*, *Ruscus flexuosus*, *R. laxus*, *R. parasiticus*, *R. ponticus*

ETYMOLOGY (Latin name, bachelor's broom, in reference to growth habit; said to have been used by bachelors to sweep) [Sharply pointed, in reference to "leaves"]

GROWTH HABIT Plants hermaphroditic or dioecious erect low-growing shrub or subshrub, to 3 ft. (1 m) tall and with equal spread, medium texture; slow growth rate. **STEM/BARK**: flattened leaf-like shoots, known as cladophylls, where flower and fruit are produced. **LEAVES**: reduced to small membranous scales with flattened, leaf-like, cladodes in their axils, with alternate, opposite or whorled arrangement. **FLOWERS**: unisexual (but sterile staminal column may be present in pistillate flowers) or bisexual, small and inconspicuous, on upper surface (in *Ruscus*) (but on lower surface or margins of cladodes or rarely in short terminal racemes on normal branches (in *Danae*); perianth of 6 tepals, free or more or less connate, usually inconspicuous; stamens 3 or 6, filaments connate forming a short tube or column, carpels 3, fused, unilocular or trilocular, ovary superior, ovules 2 or rarely 1 per locule; style short, stigma sessile or subsessile. **Fruit**: bright red berry; seeds 1–4, pale (not black), embryo straight, endosperm copious. **NATIVE HABITAT**: northern Africa, western Asia, Europe.

CULTURE Bright shade but tolerates full shade, in well-drained moist soil; tolerates drought and heavy shade; does not tolerate wet soils; leaves (cladodes) are stiff and have a sharp point and should not be planted near the public; no serious pests or diseases reported. **PROPAGATION**: seed (requires cold stratification), tissue culture. **LANDSCAPE USE**: woodland gardens, foundation, stems used in floral arrangements (fresh or dried). **HARDINESS ZONE**: 7B–10A.

COMMENTS *Ruscus aculeatus* (butcher's-broom), *R. hypoglossum* and *R. hypophyllum*, and four other species are evergreen, dioecious shrubs with small leaves and flowers on leaf-like cladophylls, native primarily to the Mediterranean region, sometimes grown in the landscape and used in dried arrangements. Leaves and young stems of *R. aculeatus* can be used as a vegetable. A few cultivars of *R. aculeatus* have been introduced: '**Christmas Berry**' (low-growing, displays bright red berries in winter) '**Elizabeth Lawrence**' (hermaphroditic, compact, thick stems, and numerous berries), '**John Redmond**' (compact carpet-like growth), '**Landeolatus**' (long narrow 'leaves'), and '**Sparkler**' (good growd cover with orange-red berries). Formerly known as Ruscaceae, with cladode (leaf-like stems), was a small family of 3 genera and 8 (currently 14) primarily Mediterranean species, all cultivated as ornamentals: *Danae racemosa* (Alexandrian laurel) is an evergreen shrub with cladodes, scale-like leaves and attractive red berries, native from Syria to Iran, used in the landscape or as cut foliage for flower arrangements. *Semele androgyna* (climbing butcher's-broom), and two other species are shrubby, dioecious vines with umbellate inflorescences, but otherwise similar to *Ruscus*, native to the Canary Islands, sometimes grown in greenhouses or outdoors in milder climates.

Ruscus aculeatus (butcher's broom), evergreen shrub with flat spine-tipped shoots (cladodes, resemble leaves), with small whitish-green small flowers at center of each cladode that become red berries; native to Eurasia.

Scilla luciliae

**GLORY-OF-THE-SNOW, LUCILES'S
GLORY-OF-THE-SNOW**
SYNONYMS *Chionodoxa gigantea*, *C. grandiflora*, *C. luciliae*

ETYMOLOGY [Greek word *scilla*, used by Hippocrates, meaning to wound or to harm, in allusion to toxic properties of some species] [Named in honor of his wife Lucile Boissier, by the Swiss botanist Edmond Boissier (1810–1885)] [Common names are in reference to early blooming of the species)

GROWTH HABIT Bulbous perennial, 4–8 in. (10–20 cm) tall and as wide; fine texture; rapid sprout from bulb but relatively slow growth from seed. **STEM/BARK**: deciduous ovoid bulb. **LEAVES**: simple, basal, 2–4, often recurved, narrow linear, 3–4 in. (7–10 cm) long and about 1 in. (2 $\frac{1}{2}$ cm) wide. **FLOWERS**: bisexual; on solitary scapes with 1 or 2 flowers; perianth blue with white center; tepals $\frac{1}{3}$ – $\frac{1}{2}$ in. (12–15 mm); filaments white; blooming early to mid-spring in lawns.
 FRUIT: globose capsule; seeds with white appendages. **NATIVE HABITAT**: on mountain slopes of western Turkey; naturalized in some areas of United States.

CULTURE Partial shade but tolerates morning sun, in well-drained organic soils, about 3 in. (7 cm) deep; plants self-seed and spread, forming a carpet but go dormant by late spring; no serious insect or diseases reported, but nematodes may be troublesome. **PROPAGATION**: seed, division of dormant bulb offsets. **LANDSCAPE USE**: woodland naturalized settings. **HARDINESS ZONE**: 3B–9B. More suitable for cooler northern or Mediterranean climate zones.

COMMENTS There is some disagreement as to the correct name for this plant. It is listed as **Chionodoxa luciliae** (e.g., in Flora of North America, Plant Finder, etc.) while The Plant List recognizes **Scilla luciliae**. To the extent that except for arrangement of tepals, species of the two genera are morphologically similar some authorities have combined the two into a single genus. No judgment made as to status of the two genera. Other species of **Scilla** in cultivation include *S. autumnalis* (lilac pink flowers from August to September), *S. forbesii* (taller scapes with more and larger flowers than *S. luciliae*), *S. hispanica* (flowers in clusters of 12–15, in 12 in. (30 cm) erect stems. *S. nonscripta* (English bluebell, often found in lawns), *S. peruviana* (error by Linnaeus, actually native to Mediterranean region, with star-shaped pink flowers), *S. siberica* (Siberian squill, drooping, bell-shaped deep blue flowers on 3–4 scapes), and probably others.

Scilla luciliae (= *Chionodoxa luciliae*)

Yucca aloifolia

SPANISH BAYONET
SYNONYMS *Dracaena lenneana*, *Sarcoyucca aloifolia*, *Yucca arcuata*, *Y. atkinsii*, *Y. purpurea*, *Y. serrulata*, *Y. tenuifolia*, *Y. tricolor*, *Y. yucatana*, etc.

ETYMOLOGY [Caribbean name for cassava (*Manihot esculenta*), based on error by Linnaeus] [Aloe-like leaves].

GROWTH HABIT Evergreen shrub; upright, round; can reach a height of 15–20 ft. (5–7 m) with the spread depending on the degree of clumping; coarse texture; fairly rapid

growth rate. **STEM/BARK**: trunks erect, about 4 in. (10 cm) wide; usually covered with a skirt of brown dead leaves; frequently bend outward or arch with age; rarely branch, instead sucker from the base to form clumps. **LEAVES**: spirally arranged, dagger-like, to 30 in. ($\frac{3}{4}$ m) long and 2 $\frac{1}{2}$ in. (5 cm) wide, narrowing to a very sharp, conical terminal brown spine; leaf margins are denticulate. **FLOWERS**: bisexual, creamy-white, pendant, and bell-shaped; 1 in. (2 $\frac{1}{2}$ cm) long, in large showy panicles held above the leaves; blooms May through November. **FRUIT**: capsules, turning black at maturity; oblong, 2 in. (5 cm) long. **NATIVE HABITAT**: coastal and dry areas of the southern United States and West Indies.

CULTURE Full sun or partial shade, in well-drained soils; tolerates salt and hot dry locations; scale and leaf spot can be a problem in areas with poor air circulation; larvae of the yucca moth may bore through, weaken, and damage the terminal shoots. **PROPAGATION**: usually by sucker division; less commonly by seeds or large stem tip cuttings. **LANDSCAPE USES**: durable, maintenance-free plant for coastal areas and parking lots; sometimes seen as a foundation plant or rock garden specimen. Wicked leaf spines preclude its use in areas with pedestrian traffic. **HARDINESS ZONE**: 6B–13B.

COMMENTS There are about 50 *Yucca* species, but very few are actually in cultivation, perhaps because of dangerously sharp spines. Examples include *Y, angustissima* (and its varieties), *Y. baccata*, *Y. brevifolia*, *Y. filamentosa*, *Y. flaccida*, *Y. gloriosa*, *Y. recurvifolia*, and others, all are polycarpic but *Y. whipplei* is monocarpic (dies after flowering). Yucca leaves are a source of fiber used in manufacture of ropes. There are a few cultivars in the trade, including: '**Marginata**' (with yellow leaf margins), '**Purpurea**' (with purplish leaves), '**Vittorio Emanuele II**' (= *Y. aloifolia* × *Y. recurvifolia*), '**Blue Boy**' is the same as '**Purpurea**'. For additional information see: http://succulent-plant.com/families/agavaceae/yucca.html

Yucca aloifolia

Yucca filifera

TREE YUCCA; ST. PETER'S PALM
ETYMOLOGY [Caribbean name of cassava (*Manihot esculenta*)] [Thread-bearing]

GROWTH HABIT Monoecious evergreen caudiciform, comparatively massive tree 46 ft. (14 m) high and caudex diameter of about 15 ft. (5 m) or more. **STEM/BARK**: straight trunk, branched at and near the top once they reach 10–12 ft. (3–4 m) in height. **LEAVES**: blue-green rigid enciform (sword-shaped) with a terminal spine, in rosettes at stem ends but also alternate along the stem; these usually die but remain attached; leaves are interspersed with white filaments as well as on their margins. **FLOWERS**: numerous creamy white, on large pendulous paniculate inflorescence of 4–6 ft. (1 $\frac{1}{4}$–1 $\frac{1}{7}$ m) at stem terminus. **FRUIT**: 3-lobed capsule. **NATIVE HABITAT**: Chihuahua Desert of northeastern Mexico.

CULTURE Full sun, may not be appropriate for wetter climates but thrives in most well-drained, preferably sandy soils does not tolerate wet soils, and despite some reports that it tolerates freezing temperature, perhaps it should not be grown below zone 9; generally, it is a trouble-free plant. **PROPAGATION**: seed, which may reduce germination time by soaking in warm water for 24 h prior to planting; root cuttings have been mentioned as a possibility. **LANDSCAPE USES**: accent plant in dry gardens, but may also be used in containers for a few years but eventually will have to be transplanted into the ground. The sharp terminal leaf spine should be taken into consideration when planting this species. **HARDINESS ZONE**: 9B–12B.

COMMENTS Cooked or fresh fruit is said to be edible; flower petals are also edible but stamens and carpels have a bitter taste and have to be removed. Roots contain saponin, which is toxic to people.

Yucca filifera. Photographed at Huntington Botanical Gardens, California

Yucca gigantea

SPINELESS YUCCA

SYNONYMS *Dracaena lennei* (= *D. hanningtoniai*), *D. yuccoides*, *Sarcoyucca elephantipes*, *Yucca eleana*, *Y. elephantipes*; *Y. guatemalensis*, *Y. lenneana*, *Y. mazelii*, *Y. mooreana*, *Y. roezeii*

ETYMOLOGY [Caribbean name for cassava (*Manihot esculenta*)] [Large (as an elephant's foot) in reference to the caudex].

GROWTH HABIT Caudiciform evergreen tree; usually columnar but sometimes irregular; can reach a height of 30 ft. (9 m) with a 15 ft. (5 m) spread, depending on degree of clumping, caudex size usually stays under 20 ft. (6 $\frac{1}{2}$ m) in height; coarse texture; fairly rapid growth rate. **STEM/BARK**: solitary, thick, rough trunk with a swollen base and usually a few short branches; there are multi-trunked specimens (see left side of figure). **LEAVES**: simple, spirally arranged; shiny, strap-like, to 4 ft. (1 $\frac{1}{3}$ m) long and 3 in. (7 cm) wide; margins rough to the touch; leaves lack a sharp terminal spine, hence the vernacular name. **FLOWERS**: bisexual, creamy-white, bell-shaped and pendant; in upright large showy panicles held above the leaves; waxy petals are edible; blooms May through November. **FRUIT**: capsule; brown, oblong, to 1 in. (2 $\frac{1}{2}$ cm) long, pulpy. **NATIVE HABITAT**: Central and northern South America, from southeastern and southern Mexico to Belize.

CULTURE Full sun to dense shade, in a wide range of well-drained soils; moderate salt tolerance; yucca moth borers and scale may be problems; it does not tolerate wet soil. **PROPAGATION**: seed; suckers, cuttings of any size. **LANDSCAPE USES**: tallest of all yuccas, it is used as a framing plant for large buildings, in indoor malls and planters, and as a specimen, especially with other succulents; lack of spines allows its use where people frequent. **HARDINESS ZONE**: 9B–13B. May require some protection in lower zones when young.

COMMENTS This species is national flower of El Salvador. Two cultivars noted as *Yucca elephantipes* '**Silver Star**' and *Y. elephantipes* '**Variegata**' should be changed to *Y. gigantea* '**Silver Star**' and *Y. gigantea* '**Variegata**', respectively. However, the two cultivars may be the same or very similar. Other listed cultivars include '**Artola Gold**', '**Jewel**', '**Jewel Gold**', and '**Pluck**' all listed under *Y. elephantipes*.

Yucca gigantea, photographed at Fairchild Tropical Botanical Gardens, Miami

Yucca faxoniana (faxon yucca), native to Mexico and Chihuahua Desert of West Texas

ADDITIONAL GENERA

Convallaria majalis (lily-of-the-valley), rhizomatous herbaceous perennial spreading groundcover useful for shady areas; native to Europe but naturalized in North America.

Dasylirion wheeleri (dessert spoon, spoon flower, common sotol), native to Chihuahua, Sonora, Mexico, and southwestern United States.

Furcraea foetida **var.** *mediopicta*, (variegated false agave), native to the Caribbean and northern South America, naturized in several countries and in Florida, evergreen stemless subshrub, closely related to *Agave*.

Polygonatum biflorum (Solomon's seal), Eastern U.S. and southcentral Canada, upright arching rhizomatous perennial, with parallel veins, small bell-shaped flowers in pairs, and blue-black berries. Rhizomes are edible. Photographed at Scotney Castle, England

***Polygonatum odoratum*'Variegatum'**(widespread in temperate Europe and much of Asia; pink
flowers are *Aquilegia*)

Eucomis autumnalis (autumns pineapple flower, pineapple lily), native to Malawi, Zimbabwe,
and southern Africa

Muscari botryoides (*Hyacinthus botryoides*; grape hyacinth), perennial bulbous plant with round blue flowers, the lower ones fertile and point downwards, upper ones paler, sterile, and point upwards; native to southeastern Europe.

Scilla siberica (Siberian squill), strap-like leaves and bell-like deep blue flowers with blue anthers on thin drooping scapes; native to southern Russia.

ARECACEAE (= PALMAE)

PALM FAMILY

195 GENERA AND ±2400 SPECIES

GEOGRAPHY Chiefly tropical (especially eastern Asia and S. America), with some subtropical, and very few temperate, in a wide range of habitats, from rainforests to desert.

GROWTH HABIT Monoecious, dioecious, or polygamous; evergreen small to moderate and large to massive tree-like, shrub-like, or rarely climbing. **STEM/BARK**: usually tough and fibrous, monopodial predominantly unbranched stems; sometimes with a short or rarely long rhizome (as in *Rhapis excelsa*) or prostrate stems (as in *Serenoa repens*), sympodially branching (from lateral buds in leaf axils) to form clustered stems (as in *Phoenix reclinata*), and/or basal leaves, or with thin reed-like stems and long internodes (as in the climbing *Calamus* and *Desmoncus*); smooth or often covered with old leaf bases, leaf scars of various forms and shapes, or simple or branched spines. **LEAVES**: large, coriaceous, palmately dissected, pinnately or rarely bipinnately compound (as in *Caryota*), costapalmate (as in *Sabal*), or rarely entire (as in some species of *Chamaedorea*); occurring usually in a crown atop the stem; petiole with entire or spinose margins and with a firm, persistent, often conspicuous sheath around the stem; a structure known as the hastula is often present on one or rarely both sides of the junction of the petiole and base of the blade in palmate and costapalmate leaves; blades entire or bifid in juvenile plants and plicate (folded, crumpled) at least in bud; leaflets of pinnate leaves may be 2–3-ranked along the rachis, folded inwardly or V-shaped (induplicate), or folded outwardly or inverted

V-shaped (reduplicate), and bifid at the apex. **FLOWERS**: unisexual or bisexual, small, actinomorphic, hypogynous, in usually axillary or rarely terminal (as in *Corypha*, and some species of *Metroxylon*, where the plants are often monocarpic – die after flowering) compound inflorescences with racemose branches, usually enclosed within and later subtended by a cymbiform (boat-shaped), tubular and two-keeled prophyll at the base of the peduncle; one to several persistent or caducous "peduncular bracts" and "rachis bracts" may also be present above the prophyll; perianth of 6 similar segments, uniseriate or biseriate, free or connate, valvate (touching but not fused) or imbricate (overlapping); stamens usually 6, or sometimes 3, 9 or numerous, filaments mostly free or sometimes connate with each other or adnate to the perianth, anthers longitudinally or rarely poricidally (as in some species of *Areca*) dehiscent; carpels usually 3, fused and uni- or trilocular, or free; ovary superior, ovules 1 or 3 per locule; stigmata vary from short, recurved and tapering to bilabiate or relatively undeveloped. **FRUIT**: berry or drupe with endocarp attached to the seed and the exocarp smooth or variously ornamented with scales, prickles, warts or hairs, and of various colors; seeds usually 1 or less frequently 2–10, usually ovoid, ellipsoidal or globose, from pea-size to the large single coconut (*Cocos*) and double coconut (*Lodoicea*), the largest known seed; embryo small, cylindrical to conical, endosperm copious.

ECONOMIC USES Palms are among the most important economic and horticultural plants. Some are valuable sources of vegetable fats, margarine, and soap, as well as waxes. Many have edible fruit or seeds, while the stem of several contain much starch (sago). Sugar-containing fluid is obtained by tapping the stem and may be utilized directly or fermented into alcoholic beverages. The apical bud of some species is used for salad or "heart-of-palm," which results in the eventual death of solitary-stemmed plants. Hearts are sustainably harvested from several species of clustering palms. Stems are used in building, leaves for thatching and basket-making and for other items. Palms are also among the most popular ornamental plants in warm temperate, subtropical, and tropical regions. It is, therefore, not surprising that the majority of known palms are in cultivation to a greater or lesser extent.

CULTIVATED GENERA *Acanthophoenix, Acoelorrhaphe, Adonidia Acrocomia, Aiphanes, Archontophoenix, Areca, Arenga, Asterogyne, Astrocaryum, Attalea, Bactris, Bismarckia, Borassus Brahea, Brassiophoenix, Butia, Calyptrocalyx, Calyptronoma, Carpentaria, Caryota, Chamaedorea, Chamaerops, Chambeyronia, Coccothrinax, Cocos, Copernicia, Corypha, Cryosophila, Dictyosperma, Drymophloeus, Dypsis, Elaeis, Euterpe, Gaussia, Geonoma, Hedyscepe, Heterospathe, Howea, Hyophorbe, Hyphaene, Johannesteijsmannia Jubaea, Latania, Lepidorrhachis, Licuala, Livistona, Lodoicea, Lytocaryum Mauritia Nannorrhops, Normanbya, Nypa, Phoenix, Pinanga, Polyandrococos, Pritchardia, Pseudophoenix, Ptychosperma, Ravenea, Reinhardtia, Rhapidophyllum, Rhapis, Rhopalostylis, Roystonea, Sabal, Salacca, Satakentia, Serenoa, Syagrus, Synechanthus, Thrinax, Trachycarpus, Veitchia, Wallichia, Washingtonia, Wodyetia, Zombia*, among many others.

(For additional list of palm genera in cultivation see Riffle, Craft and Zona, 2012, also see references for palms, as well as Wikipedia.)

IMPORTANT PALM CHARACTERISTICS

Costapalmate leaf: petiole becomes a *costa* (L. "rib") and continues through part or all of the blade

Crown Shaft is an elongated clear green area between the trunck and leaf bases in some pinnate-leaved palms (left); **Hastula** is located at the point of petiole attachment to the blade and usually appears as a ring (middle); **Inflorescence** consists of a boat-shaped **spathe** that initially encloses the **branched spadix**.

COMMENTS The most recent taxonomic treatment of the Arecaceae (APG IV, 2016–2921) recognizes 5 subfamilies, 15 tribes, and 35 subtribes. Representative examples of cultivated palm species, in addition to those described in detail, include: For genera not listed in the subfamilies below see the full descriptions. Numbers in parenthesis after descriptions indicate subfamily assignments.

1. **SUBFAMILY CALAMOIDEAE**: Lianas, climbing by more or less recurved spines, plants often dioecious, flowers perfect; calyx valvate, stamens basally connate; style branched; fruit covered with reflexed scales.

Calamus, with about 370 species is the largest genus in the family of which very few are cultivated: *C. formosanus*, *C. ornatus*, *C. rotang* (rattan cane), and others, are climbing or sometimes shrub-like, or acaulescent spiny palms with the lower part of the stems becoming bare; leaves usually pinnate or rarely bifid with a terminal spine; petiole absent or well-developed and variously armed; sheaths with scattered whorled spines and much indumentum, native to tropical regions of the Old world, the cane-like stems of the plants provide rattan canes used in furniture construction; *Eleiodoxa conferta* (kelubi, assam paya), leaves are used for thatching and the acrid sour fruit for cooking; *Eugeissona tristis* (moriche, buriti, or ita palm), and few others, are large, solitary, dioecious palms with erect, unarmed, smooth stem; leaves reduplicate, briefly costapalmate with blades divided partly to the base into 1-ribbed segments; petioles elongate, unarmed, sheaths splitting opposite the petiole, native to tropical America, they yield oil, starch, wine, timber, cork, fiber, and palm heart (hence the name "tree of life"), also planted in wetter areas as landscape plants; *Mauritiella aculeata* (buriti), is not reported from cultivation but is locally important for its edible fruit; *Metroxylon amicarum* (Caroline ivory nut palm), *M. sagu* (true sago palm), native to tropical southeastern Asia yields "vegetable ivory," used for buttons and other products; *Pigafetta elata*, is occasionally cultivated as a landscape plant; *Plectocomia elongata*, is a monocarpic palm only rarely seen in cultivation; *Raphia farinifera* (raffia palm), is the source of "raffia fiber," locally used and exported for a variety of uses, petioles of other species are used for furniture construction, a wine is made by tapping the stem apex, and plants are sometimes used for landscaping;

2. **SUBFAMILY NYPOIDEAE**: Rhizomatous, stems dichotomously branched; inflorescence racemose, staminate (branches occur on alternate sides of stem and inflorescence a spike, carpellate inflorescence a head, adnate to the internode above; perianth free; stamens 3, opposite outer perianth, connate; carpels 3.

Nypa fruticans (nypa, mangrove palm), is usually grown near brackish water as a landscape plant but leaves are used for thatching (palm shingles, "atap"), the inflorescences are taped for sugar which is fermented into alcoholic beverages, and the seeds are boiled with sugar and eaten;

3. **SUBFAMILY CORYPHOIDEAE**: Leaves palmate or costapalmate; inflorescences various (terminal or adnate to the internode above); flowers perfect, solitary or in cincinni (branches occur on alternate sides of the stem and inflorescence is bent to one side); corolla often valvate, connected to carpels 1, style well-developed.

Arenga engleri, *A. microcarpa*, *A. pinnata* (sugar palm, black fiber palm), *A. tremula*, and others, are a variable group of dwarf to large, solitary or clustered, usually unarmed, monoecious or exceptionally dioecious, mostly monocarpic (flowering from the axils of the upper leaves successively downward, then dies), palms with stems often densely covered by persistent black fibrous leaf bases and sheaths; leaves pinnate or rarely undivided, leaflets in several ranks, toothed or often lobed along the margins; petiole slender or well-developed, channeled at base and usually covered with indumentum, native from India to China and through the Southeast Asian Islands to Australia, an economically important group (particularly *A. pinnata*), providing sugar, fiber, thatch, wine, and starch (sago), and widely planted as ornamental plants; *Acoelorraphe wrightii; Borassus flabellifer* (palmyra palm), is used as a landscape plant but locally important in providing "toddy," a popular drink, and its long primary seedling root is considered a delicacy; *Brahea armata* (blue fan palm), *B. brandegeei* (San Jose fan palm), *B. calcarea*, *B. edulis* (Guadupe palm), *B. pimo*, and a few others, are small to moderate, armed or unarmed, monoecious palms with solitary or rarely clustered stems; leaves briefly costapalmate, with 1-ribbed, induplicate, bifid segments; petiole short or long, concave, flattened, or channeled on the upper side, margins smooth or spinose; sheaths fibrous, persistent, splitting basally, native to Mexico and Central America, often used as landscape plants; *Corypha elata* (Gebang palm), *C. umbraculifera* (talipot palm, producing the world's largest inflorescence), are used as landscape plants; *Daemonorops grandis*, *D. margaritae*, *D. ochrolepis*, *D. periacantha*, *D. pseudosepal*, among many others, are solitary or clustered, acaulescent or erect or climbing, spiny, dioecious palms with slender cane-like stems, leaves pinnate, well-developed, petiole variously armed, rachis terminating in a cirrus for climbing, sheath tubular, densely spined, native from India and southern China through Southeast Asian Islands to New Guinea, plants are often used for rattan furniture construction and utilization of plants as medicinal and food sources have also been noted; *Lodoicea maldivica* (coco-de-mer, double coconut), is a rarely cultivated plant but is notable for its unusual bilobed seeds, the largest in the plant kingdom; *Pritchardia pacifica* (Fiji fan palm) and *P. thurstonii*, are used as landscape plants and their large leaves are used as umbrellas and fans; *Wallichia caryotoides* and *W. disticha*, are grown as ornamentals and locally provide sago and thatch; *Zombia antil-*

larum, is used for landscaping but requires some caution because of its long sharp nodal spines.

Zombia antillarum trunk

4. **SUBFAMILY CEROXYLOIDEAE**: Plants usually dioecious; inflorescence racemose, spicate (arranged in spikes); flowers solitary, flowers perfect, calyx and corolla elongate, free; carpels 3–10, free; receptacle elongated; germination pores absent.

Ammandra decasperma (ivory palm), is used for vegetable ivory and thatch; *Phytelephas aequatorialis* (ivory palm, tagua), is rarely cultivated but yields vegetable ivory used in carving; *Ceroxylon alpinum* (wax palm), and other species, are tall to very tall, unarmed, dioecious palms with smooth, usually waxy stem with prominent leaf scars, leaves pinnate with reduplicate, acute leaflets; petioles leathery, channeled on the upper side; sheaths splitting opposite the petioles, native to the higher elevations of the Andes of South America, used as landscape plants but yield wax for candles and matches and their fruit is used for cattle feed, and *C. quindiuense*, the world's tallest monocot; *Phytelephas macrocarpa* (ivory nut palm, tagua), and other species are moderate, solitary or clustered, unarmed, dioecious palms with usually robust, erect or procumbent stem which is covered with a mass of fibers and leaf bases or when base marked with spiral, triangular leaf scars; leaves evenly pinnate, erect or arching; petiole short, elongate or lacking, shallowly channeled on the upper side, margins rounded or spinescent; sheaths tubular, sometimes with a large ligule opposite the petiole, fibrous, native to central and South America, used as ornamentals but have edible fruit and seed, mature endosperm is used as vegetable ivory for carving, and leaves are used as

thatch; *Ravenea* species are not reported from cultivation but *R. madagascariensis* is said to have hard flexible wood and is locally used for various purposes, *R. robustior* has starch-rich pith and is a source of sago, and all species have edible terminal bud;

5. **SUBFAMILY ARECOIDEAE**: Plants monoecious; inflorescences spicate; flowers in 3s, protandrous (stamens mature before the carpels), upper flower female, lateral flowers male, or in two vertical rows; corolla valvate; stamens 3, carpel 1, fertile; style single, branches separate, short to long; germination pores 9.

Acanthophoenix rubra (barbel palm), critically endangered in Mauritius and La Reunion islands, prized for palm heart, *A. crinita* (barbel palm), endemic to Reunion Island, and *A. rousselii*, are rare in the wild (Mauritius and Reunion Islands), and listed as critically endangered by IUCN, because of their edible cabbage (palm heart); *Acrocomia aculeata* (gru-gru palm, mucaja), is small to large, solitary, spiny, monoecious palm with very short and subterranean or erect stems which are covered with persistent leaf bases or rings of spines, soon becoming smooth and ringed with leaf scars; leaves pinnate, leaflets 1-ribbed; petiole short or leaves subsessile, channeled on the upper side, sheath fibrous, spiny and bristly, native to West Indies and from Mexico to Argentina, yield starch, fiber, oil (from seed endosperm), and have edible cabbage and fruit, as well as medicinal properties, *A. crispa* (= *Gasterococos crispa*) is occasionally grown as ornamental; *Adonidia*; *Aiphanes minima* (coyure), *A. horrida* (ruffle palm), *A. corallina*, *A. erosa*, *A. lindeniana*, *A. truncata*, and others, are a diverse group of small to moderate size, solitary, spiny, acaulescent or erect, monoecious palms with bare stems which are conspicuously ringed with leaf scars and horizontal rows or rings of black spines; leaves pinnate or entire-bifid, leaflets truncate and toothed at the apex, petiole short to long, channeled on the upper side, with golden-yellow to black spines, sheaths tubular at first but soon fibrous and shredded, usually densely spiny and/or tomentose, native to the West Indies and northern South America, used as ornamentals, and *A. horrida* has edible fruit; *Archontophoenix alexandrae* (Alexander palm, king palm), and *A. cunninghamiana* (= *Ponapia ledermanniana*, piccabean palm, bangalow palm), are fast growing moderate to tall, solitary, unarmed, monoecious palms with basally slightly swollen but slender, columnar stem; leaves pinnate, sometimes basally twisted about 90°, leaflets lanceolate, acute; petiole short, grooved on the lower side; sheath tubular, forming a prominent leathery, green, rusty brown or purplish-red crownshaft, native to eastern Australia, used in landscaping because of their rapid growth and stately habit; *Areca catechu* (Betel palm, Betel nut), *A. glanduliformis*, *A. triandra*, *A. vestiaria*, *A. wrightii* (= *Colpothrinax wrightii*, bottle palm), widely grown for its edible nuts, and others are small to moderate, solitary or clustered, acaulescent to erect, unarmed, monoecious palms with slender stem often conspicuously marked by leaf scars and sometimes stilt roots; leaves pinnate or unlobed and pinnately 1 to several-ribbed and apically toothed, petiole present or absent, glabrous or variously tomentose, sheaths tubular, usually forming a crownshaft, native from India and China through Malaysia to New Guinea and Solomon Islands, often cultivated as landscape plants; *Asterogyne martiana*, is used as an ornamental; *Astrocaryum aculeatum* (tucuma), *A. mexicanum*, and others are cultivated as ornamentals but also yield seed oil; *Attalea amygdalina* is a source of seed oil; *A. cohune* (Cohune palm), *A. leandroana* (= *Scheelea leandroana*), *A. macrocarpa* (= *Scheelea macrocarpa*, yagua palm), *A. maripa* (maripa palm, cucurite palm), is sometimes cultivated in the tropics, *A. speciosa* (Barbassu palm), *A. spectabilis*, and others, are moderate to large, solitary, erect or acaulescent, unarmed, usually dioecious palms with subterranean to tall, bare straight stems; leaves pinnate, large, leaflets many, acute, with prominent midrib and fibrous margins, petiole lacking or short or long, channeled on the upper side and tomentose when present; sheaths large, thick, with margins fibrous, native from Mexico to South America, in addition to their use as ornamentals the species yield oil from seeds, wine from the apical bud, and leaves and petioles are utilized for various purposes, from cooking oil to fiber, etc.; *Bactris*, with 79 currently recognized species, is a large genus of which relatively few species are cultivated: *B. gasipaes* (peach palm, pejibaye, pupunha), *B. guineensis* (Tobago cane), *B. major* (prickly palm), *B. mexicana*, *B. trichophylla*, and others, are small to large, solitary or clustered, spiny or rarely unarmed, monoecious palms with very short to erect, slender to moderate stems which have conspicuous nodal scars and are often scaly and frequently armed with short or long spines; leaves pinnate or entire-bifid, leaflets acute to acuminate, rarely oblique and toothed at the apex; petiole short to long, channeled on the upper side, flat or angled; sheaths usually splitting opposite the petiole, with smooth or fibrous margins, unarmed or densely spiny, scaly, hairy, or bristly, native from Mexico to Brazil and the West Indies, several species have edible fruit, yield seed oil and fiber, and are used for landscaping, the most widely cultivated and economically important species, *B. gasipaes* is not known from the wild; *Basselinia eriostachys*, is occasionally grown as an ornamental; *Brassiophoenix schumannii*, is grown as an ornamental; *Butia capitata* (jelly palm), among other species, is cold hardy and commonly grown as a landscape plant and has edible fruit; *Butia capitata* (jelly palm), among other species, is cold hardy and commonly grown as a landscape plant and has edible fruit; *Calyptronoma dulcis* (Cuban manac), and *C. occidentalis* (Jamaican manioc), are used as ornamentals; *Calyptrocalyx spicatus*, is also grown as an ornamental; *Calyptrogyne ghiesbreghtiana*, is sometimes used as an ornamental;

Carpentaria acuminata (carpentaria palm, not to be confused with *Carpentaria californica*, the bush anemone), is grown as an ornamental; *Deckenia nobilis*, is occasionally planted as an ornamental; *Ceratolobus kingianus*, is a rarely cultivated climbing palm; *Desmoncus orthocanthos*, *D. polycanthos*, and *D. quasillarius*, are the only species of this relatively large genus occasionally cultivated as landscape plants; *Dictyosperma album* (common princess palm, and its variety *conjugatum*) and *D. aureum* (yellow princess palm), are prized ornamental feather palms; *Drymophloeus beguinii* and *D. oliviformis*, are cultivated as ornamentals; *Dypsis decaryi* (triangle palm), *D. lastelliana*, *D. lutescens* (yellow butterfly palm), and others are attractive landscape plants, *D. fibrosa* is a dichotomously branched palm cultivated as an ornamental but is the source of "paissava" or vegetable fiber, and has edible fruit and cabbage; *Elaeis guineensis* (African oil palm), and *E. oleifera* (American oil palm), are moderate to large, solitary, armed, monoecious palms with procumbent or erect stem, persistent leaf bases and wide leaf scars, leaves pinnate, leaflets 1-ribbed, acute, 1 or several-ranked, petiole channeled on the upper side, with fiber spines, tomentose, sheath of an interwoven mass of spinose fibers at the base of the petiole, native to Africa and South America, respectively, economically among the most important palm genera, widely cultivated in the tropics for oil production, as well as a variety of other local uses; *Euterpe edulis* (juçara, açai-do-sul palm), locally used for making rural buildings, *E. oleracea* (açai palm), *E. precatoria*, and others, are small to medium, solitary or clustered, unarmed, monoecious palms with slender or relatively stout, sometimes basally swollen stems; leaves pinnate or undivided, leaflets 1-ribbed, acute and spreading or pendulous; petiole elongate or short, often slender, channeled on the upper side, densely tomentose; sheath tubular, splitting opposite the petiole, only exceptionally forming a prominent crownshaft, native to West Indies and from Central to South America, some of the species are the sources of commercial "heart of palm" which is exported primarily from Brazil, and several are used as ornamentals; *Gaussia attenuata*, is sometimes grown as an ornamental; *Geonoma deversa*, *G. interrupta*, and *G. schottiana*, among others, are small to moderate, solitary or clustered, unarmed, monoccious palms with very short, subterranean, erect or creeping, slender, cane-like stems; leaves pinnate or undivided, bifid at the apex and pinnately ribbed, leaflets acute or sometimes toothed; petiole short to long, slightly grooved or flattened on the upper side, glabrous or tomentose; sheaths short, splitting opposite the petiole and with fibrous margins, native from Mexico to South America, used as ornamentals but leaves are locally used as thatch and the cabbage is sometimes eaten; *Hedyscepe canterburyana* (umbrella palm), is often planted as a landscape plant or potted plant; *Heterospathe elata* (sagisi palm), and *H. glauca*, are attractive feather palms grown as landscape or potted plants, native to Guam; *Howea belmoreana* (Belmore sentry palm), and *H. forsteriana* (kentia or Forster's sentry palm), are medium, solitary, unarmed, monoecious palms with erect, slender, bare stems which are conspicuously marked with leaf bases, leaves pinnate, curved or flat, leaflets 1-ribbed, acute, erect or drooping, petiole short or somewhat elongated, flattened or slightly channeled on the upper side; sheath well-developed, splitting longitudinally and opposite the petiole and becoming fibrous, endemic to Lord Howe Island, among the most commonly grown palms, used as foliage or landscape plants; *Hydriastele beguinii* (= *Siphokentia beguinii*) is occasionally cultivated as a landscape plant, native to Obi Island, *H. microcarpa* (pinang salea) and *H. ramsayi*, are sometimes cultivated as landscape plants; *Iriartea deltoidea* (horn palm), has hard and durable trunk and is used in construction; *Iriartella setigera* (palma de cerpatana), leaf bases are used medicinally; *Juania australis* (juania palm, Robinson Crusoe palm), has edible terminal bud and has been used for cabinet work and carving; *Jubaea chilensis* (Chilean wine of honey palm), trunks tapped for wine and sugar but now threatened, commonly cultivated in Mediterranean type climates; *Jubaeopsis caffra* (Pondoland palm), is only rarely seen in cultivation; *Kentiopsis oliviformis*, sometimes used as an ornamental; *Laccospadix australasica*, is sometimes used as an ornamental; *Lepidorrhachis mooreana*, is cultivated as an ornamental; *Linospadix minor* and *L. monostachya* (walking-stick palm), are often grown as foliage plants; *Nephrosperma vanhoutteanum* is used as a landscape plant; *Normanbya normanbyi* (black palm); *Oncosperma fasciculatum* and *O. tigillarium* (Nibung palm) are used as landscape plants; *tigillarium*, is occasionally grown as a landscape plant but also has edible fruit and yield seed oil; *Phoenicophorium borsigianum*, is used as a landscape plant; *Pinanga*, with about 120 species is one of the larger genera of palms but few are reported from cultivation: *P. furfuracea*, *P. coronata*, *P. malaiana*, *P. patula*, *P. pectinata*, *P. tashiroi*, and others, are small to medium, solitary or clustered, erect or acaulescent, monoecious palms with conspicuous leaf scars and sometimes stilt roots, leaves pinnate or undivided and pinnately ribbed, often with bifid apex, petiole present or absent, rounded or channeled on the upper side; sheaths tubular, usually forming a well-defined crownshaft, native from Himalayas and southern China to New Guinea, very attractive plants often used as ornamentals; *Podococcus barteri* (buri palm), is occasionally cultivated in botanical gardens; *Prestoea montana* (mountain cabbage palm); *Ptychococcus paradoxus*, is sometimes grown as a landscape plant; *Reinhardtia gracilis* and *R. simplex*, are usually grown as foliage plants but are also used in the landscape, leaves with only two broad leaflets pr side that have distinctive windows at their base; *Rhopaloblaste augusta*, *R. ceramica*, and *R. singaporensis*, are grown as landscape plants; *Rhopalostylis baueri* and *R. sapida* (feather-duster palm), are also grown as landscape plants; *Roystonea borinquena* (Puerto Rican royal palm), *R. oleracea* (Caribbean royal palm, Caribee royal palm), *R. regia* (Cuban Royal Palm, Florida Royal Palm), and others, are tall, stout, solitary unarmed, monoecious palms with columnar, variously tapered or swollen, tan, gray or white stem which may be ringed by prominent leaf scars, leaves pinnate, leaflets acute, midrib prominent, petiole short, channeled on the upper side; sheath large, tubular, forming a prominent crownshaft, native to the West Indies and from Florida to northern Central America, among the most majestic palms, often used as street trees or for parks; *Satakentia liukiuensis* is sometimes used as a landscape plant; *Synechanthus fibrosus* and *S. warscewiczianus* are similar to species of *Chamaedorea* and are cultivated as ornamentals for their attractive foliage and yellow to bright orange fruit; *Verschaffeltia splendida* is sometimes cultivated as an ornamental; *Wendlandiella gracilis*, is also similar to *Chamaedorea* but only occasionally cultivated as an ornamental; *Wettinia praemorsa* (palma parapa), parts of trunk used in construction.

Acoelorrhaphe wrightii

PAUROTIS PALM, SAW CABBAGE PALM, EVERGLADE PALM

SYNONYM *Acanthosabal caespitosa*, *Acoelorrhaphe arborescens*, *A. pinetorum*, *Brahea psilocalyx*, *Copernicia wrightii*, *Paurotis arborescens*, *P. wrightii*, *Serenoa arborescens*

ETYMOLOGY [From Greek *a-*, without, *koilos*, hallow, and *rhaphis*, needle; in reference to the needle-like leaf segments] [After Charles Wright (1811–1855), the American plant taxonomist].

GROWTH HABIT Multi-trunked cluster palm; bushy, dense; basal sprouts assure that there is always foliage at the base of the trunks; up to 25 ft. (8 m) tall with variable spread; mature specimens can reach 40 ft. (13 m) in height; medium texture; moderate to slow growth rate. **STEM/BARK:** slender trunks is about 6 in. (15 cm), covered with old leaf bases and matted fiber, giving them a burlap-like appearance. **LEAVES:** palmate, orbicular in outline, 2–3 ft. ($\frac{3}{4}$–1 m) wide and long; leaves divided halfway toward center into pointed, stiff segments that are split at apex; lighter green undersides; petioles 3 ft. (1 m) with orange teeth along the margins that point upward and often occur in pairs. **FLOWERS:** yellow-green, small, on erect panicles that arch as fruit matures; flower stalks are usually around 3 ft. (1 m) long or somewhat longer, extend beyond the leaves, and are borne in late winter/early spring. **FRUIT:** orange at first, then black at maturity; round, $\frac{1}{4}$ in. (7 mm) in diameter;

mature in December. **NATIVE HABITAT**: low moist areas in the Everglades region of South Florida, the West Indies, and Central America. (2)

CULTURE Full sun to partial shade; any soil, but growth rate is slow in drier soils; thrives in low moist, reasonably fertile soils; slight salt tolerance; pest-free but micronutrient deficiencies; salt-tolerant and resistant to lethal yellowing. **PROPAGATION**: seed and lateral shoot separation.

LANDSCAPE USES: near pools or patios, as a screen plant, a specimen, or as a corner plant for houses. **HARDINESS ZONE**: 9A–13B, but zone 8B in protected locations.

COMMENTS A monotypic genus *Acoelorrhaphe wrightii* is currently protected by Florida State law but readily available in nurseries. Although slow-growing, the plant tends to extend beyond its designated location as it grows.

Acoelorrhaphe wrightii young plant (above left) and an older specimen (above right), leaf close-up (below left), and fiber and leaf base-covered trunk close-up (below right)

Adonidia merrillii

CHRISTMAS PALM, MANILA PALM
SYNONYM *Actinorhytis calapparia*, *Normanbya merrillii*, *Veitchia merrillii*

ETYMOLOGY [After Adonis, Greek god of beauty and desire] [Elmer Drew Merrill (1876–1956), an American botanist specializing in the flora of the Asia Pacific region].

GROWTH HABIT Monoecious, small, single-trunked erect palm (said to resemble a small version of the *Roystonia regia*, the royal palm); rigidly arched compact canopy; to 20 ft. (7 m), usually 6–10 ft. (2–3 $\frac{1}{3}$ m); medium-coarse texture; moderate growth rate. **STEM/BARK**: smooth, to 10 in. (25 cm) in diameter; faintly ringed; whitish-grey; tapering to green crownshaft, somewhat swollen at base. **LEAVES**: pinnate to 6 ft. (2 m) long; rigidly arched; bright green leaflets in 2-ranks, sword-shaped, reduplicate, to 30 in. (75 cm) long; petiole short, unarmed. **FLOWERS**: unisexual, yellow-green to white, small; inflorescence stalk short, arise below crownshaft. **FRUIT**: drupe; elliptic, to 2 in. (5 cm) long, pointed, green becoming red when ripe; glossy; very showy. **NATIVE HABITAT**: coastal areas of Philippines (Palawan and Dongguan Islands), and Malaysia (Sabah). (5)

CULTURE Full sun to partial shade, in various well-drained but moist soils; will tolerate short drought periods; moderate salt tolerance; susceptible to Lethal Yellowing disease and scale. **PROPAGATION**: seed germinate in 1–3 months. **LANDSCAPE USES**: specimen, framing tree, patio/terrace in containers, street plantings, in group plantings. **HARDINESS ZONE**: 10B–13B.

COMMENTS Listed by IUCN Red List as threatened due to habitat destruction and other factors. The species was formerly part of *Veitchia*, a genus of several cultivated species. *Adonidia* is a monotypic genus with *A. merrillii* as the only species.

Adonidia merrillii growth habit in the landscape (above left), closeup of the distinct crownshaft (above right), attractive fruit immediately below the crownshaft (middle left), and closeup of the inflorescence with fruit (middle right), individual fruit (below)

Allagoptera arenaria

SEASHORE PALM
SYNONYMS *Allagoptera pumila, Cocos arenaria, Diplothemium arenarium, D. littorale, D. maritimum*

ETYMOLOGY [From Greek *allage*, change and *petron*, wing, in reference to changing direction of the feathery leaves] [From Latin *arenaria*, in reference to the sandy habitat of the species]

GROWTH HABIT Monoecious shrubby plant to about 6 ft. (2 m) tall. **STEM/BARK**: solitary main trunk subterranean but commonly branched at ground level. **LEAVES**: pinnately compound leaves arise spirally from the ground level atop the essentially invisible trunk but spread in different directions; 3–6 ft. (1–2 m) long and leaflets to about 2 ft. ($\frac{2}{3}$ m) long, turning in various directions.

FLOWERS: greenish-yellow; male and female flowers arranged spirally on the same spike-like erect inflorescence. **FRUIT**: yellowish orange, oval, similar to a coconut; about 1 in. (2 $\frac{1}{2}$ cm) in diameter. **NATIVE HABITAT**: southeastern Brazil, along the sandy Atlantic coastal strand. (5)

CULTURE Full sun, in very well-drained sandy soil. **PROPAGATION**: seed. **Landscape Uses**: highly salt and drought-tolerant and cultivated as ornamental and for its edible fruit along coastal areas; may also be grown as a container plant. **HARDINESS ZONE**: 10B–12B.

COMMENT A highly endangered species due to land clearing and construction. Fruit is edible and is used for making jellies and jams. Related species include *A. brevicalyx*, *A. campestris*, and *A. leucocalyx*.

Allagoptera arenaria plant in the landscape (left), leaf close-up (middle), immature fruit (right)

Bismarckia nobilis

BISMARK PALM
SYNONYMS *Medemia nobilis*

ETYMOLOGY [Named in honor of Prince Otto von Bismarck (1815–1898), first chancellor of the German Empire] [From Latin, noble]

GROWTH HABIT Dioecious palm, to 75 ft. (25 m) tall (usually much smaller in cultivation) and about 24 ft. (8 m) spread; coarse texture; slow growth rate. **STEM/BARK**: gray to brown solitary trunk with leaf scars on older segments, 15–16 in. (35–40 cm) in diameter; slightly swollen at base. **LEAVES**: palmate but with the petiole extending slightly into the blade, hence interpreted by some as Costa palmate, rounded, with 20 or more segments, grayish to silvery blue in color, about 9 ft. (3 m) across; petiole 6–8 ft. (2–2 $\frac{1}{2}$ m) long and covered with a whitish wax.

FLOWERS: small, brownish in color and more numerous in male inflorescences. **FRUIT**: brown drupe with a single seed. **NATIVE HABITAT**: endemic to Madagascar, in open savanna grasslands.

CULTURE Full sun, in well-drained but moist fertile soils. **PROPAGATION**: seed. **LANDSCAPE USES**: notable as specimen plants that should be given adequate room; also used as street trees and along wide walkways. It is seen as a container plant in colder botanical gardens where it can be moved indoors. **HARDINESS ZONE**: 10A–13B. Silvery-blue form is said to tolerate 28 °F, but the green form is damaged by temperatures below 35 °F.

COMMENT Despite the large number of palm species and their enormous diversity, no other palm can match the uniqueness and beauty of the monotypic Bismarck palm.

Bismarckia nobilis in Nong Nooch Botanical Garden, Thailand (above left), as street tree in a Florida nursery (above right), a relatively young specimen (below left), and close-up of leaves and immature fruit (below right)

Butia odorata

PINDO PALM, JELLY PALM
SYNONYMS *Butia capitata*, *B. nehrlingiana*, *B. pulposa*, *Cocos elegantissima*, *C. erythrospatha*, etc.

ETYMOLOGY [The Brazilian name for the species] [dense-headed].

GROWTH HABIT Monoecious, medium-sized single-trunked erect palm, stiff, strongly recurving canopy; to 30 ft. (10 m) tall, commonly seen 10–20 ft. (3–6 m), medium texture; slow growth rate. **STEM/BARK**: stout trunk, 12–24 in. (30–60 cm) in diameter, persistent leaf bases along entire trunk. **Leaves**: pinnate, 8–10 ft. (2 2/3 –3 m) long, strongly recurving toward trunk, leaflets stiff, standing upward from rachis in 1 rank, blue-green, reduplicate, petiole armed, spines pointing toward leaf tip. **FLOWERS**: unisexual, in groups of 3, 2 male and 1 female, small, stalked inflorescences, to 5 ft. (2 1/2 m) long. **FRUIT**: drupe, yellow to red, oblong-ovoid to 1 in. (2 1/2 cm) long and wide, dense clusters, pulpy, fibrous, and edible, primarily used for jellies and jams. **NATIVE HABITAT**: dry grasslands and savannas of South America: Argentina, Brazil, and Nicaragua. (5)

CULTURE Full sun, various soils; moderately salt-tolerant; tolerates hot windy conditions along asphalt and concrete areas; palm leaf skeletonizer, scale, micronutrient deficiencies may cause problems; fruit can be messy on sidewalks. **PROPAGATION**: seed, but germination difficult, cracking of the seed coat has been shown to enhance germination. **LANDSCAPE USES**: specimen, accent, street plantings, and planters. **HARDINESS ZONE**: 7B–12B.

COMMENTS According to Dr. Larry Noblick of Montgomery Botanical Center (personal communication), what has long been referred to as *Butia capitata* in the trade and landscape is actually *B. odorata*. *Butia odorata* **var**. *nehrlingiana* (smaller bright red fruit) and **var**. *strictior* (strongly ascending, bluish leaves) are available. Fruit is used in making jellies and jams. Nearly all 20 currently

known species are reported from cultivation, including *B. archeri*, *B. bonnetii*, *B. campicola*, *B. eriospatha*, *B. microspadix*, *B. noblickii*, *B. paraguayensis*, *B. purpurascens*, and *B. yatay*. Apparently, the species readily hybridize making specific identification difficult. Two nothospecies (= artificial intergeneric hybrids), ×*Jubautia splendens* (*Jubautia chilensis* × *Butia odorata*) and ×*Butiagrus nabonnandii* (*Butia odorata* × *Syagrus romanzoffiana*) are also known from cultivation. Selections from the latter intergeneric hybrid include cultivars '**Dick Douglas**' and '**Don Nelson**'.

Butia odorata in landscape (above left), close-up of pinnate foliage (above right), much branched inflorescence with ripened fruit (middle left), close-up of the edible fruit (middle right), and leaf close-up to illustrate induplicate leaflet characteristic (below)

Caryota mitis

BURMESE FISHTAIL PALM, CLUSTERING FISHTAIL PALM

SYNONYMS *Caryota furfuracea*, *C. griffithii*, *C. javanica*, *C. nana*, *C. speciosa*, *Drymoroeus zippellii*, *Thuessainkia speciosa*

ETYMOLOGY [From Greek *karyon*, a nut] [From Latin gentle, without spines, soft].

GROWTH HABIT Monocarpic; monoecious, large, multi-trunked, cluster palm; height ranges from 25– 40 ft. (8–14 m), with variable spread; coarse texture; medium to rapid growth rate. **STEM/BARK**: the slender trunks, 4 in. (10 cm) in diameter, are covered with gray leaf-bases and black matted fibers. **LEAVES**: dark green bipinnately-compound, 4–9 ft. (1 $\frac{1}{3}$ –3 m) long; leaflets 6 in (15 cm) long by 5 in. (12 cm) wide, triangular, like a fish's tail; leaflets have a jagged, toothed apex, with numerous parallel veins and are slightly induplicate; petioles unarmed and round in cross-section. **FLOWERS**: each trunk blooms successively from top leaf axils to bottom axils and then dies, replaced by younger trunks in 5–7 years; many-branched, hanging flower stalks resemble a horsetail, usually 2 ft. ($\frac{2}{3}$ m) long with whitish flowers on separate male and female inflorescences; blooms all year. **FRUIT**: black, globular, $\frac{1}{2}$ in. (1 $\frac{1}{4}$ cm), ripening all year; stinging crystals in the outer pulp. **NATIVE HABITAT**: tropical rainforests but also grows in open areas in the Philippines, Thailand, southern China, and Southeast Asia. (3)

CULTURE Full sun to partial shade, in well-drained organic soils with reasonable fertility; no salt tolerance; red spider mites when plants used in interiors; lethal yellowing disease. **PROPAGATION**: seeds and division of clumps. **LANDSCAPE USES**: useful as an interior/patio tub specimen or as a specimen or screen planting and in borders and under larger trees. **HARDINESS ZONE**: 10B–13B.

COMMENTS *Caryota mitis* is reportedly an invasive species in some areas, including south Florida. The genus includes 12–13 species of which *Caryota urens* (wine palm) differs in having a thick, solitary, gray, smooth trunk to 80 ft. (27 m) and is larger overall than other taxa. Its flowers are used to make sugar and wine which is known as toddy. Other species in cultivation include *Caryota maxima* and its cultivar 'Himalaya', which is reportedly colder-tolerant, fast-growing, and not monocarpic, *Caryota no* (endemic to the island of Bornea, *C. obtusa*, *C. ophiopelis*, *C. rumphiana*, *C. urens*, *C. zebrina*, and probably others are native from tropical Asia to Australia, and the Solomon Islands. Among the most commonly grown ornamental palms, often used as foliage plants, but *C. urens* is particularly economically important providing sago, fibers, wine, sugar, timber, etc.

Caryota mitis

Caryota rumphiana flowering and fruiting habit

Chamaedorea elegans

NEANTHE BELLA, PARLOR PALM
SYNONYMS *Chamaedoria deppeana*, *C. helleriana*, *C. humilis*, *C. pulchella*, *Collinia deppeana*, *C. elegans*, *C. humiolis*, *Kunthia deppei*, *Neanthe bella*, *N. elegans*, *N. neesiana*, *Nunnezhria elegans*, *N. humilis*, *N. pulchella*

ETYMOLOGY [From Greek *chamai*, on the ground, dwarf, low growing; *dorea*, a gift, in reference to low growth habit and bright-colored fruit] [From Latin elegant].

GROWTH HABIT Dioecious, small, single-trunked palm; usually 4 ft (1 $\frac{1}{3}$ m) high, occasionally reaching 6–10 ft. (2–3 $\frac{1}{3}$ m), with a spread of 2–3 ft. ($\frac{3}{4}$ –1 m);

medium-fine texture; rapid growth rate. **STEM/BARK**: slender light green trunks, 1–1 $\frac{1}{2}$ in. (2 $\frac{1}{2}$ –3 $\frac{1}{2}$ cm) wide; leaf scar rings dark green. **LEAVES**: dark green, pinnately compound, 18–36 in. (40–100 cm) long; leaflets 20–40, lanceolate, to 1 in. (2 $\frac{1}{2}$ cm) wide and 8 in. (20 cm) long; petioles short, unarmed. **FLOWERS**: unisexual; whitish-yellow, small; on branched, orange flower stalk to 2 ft. ($\frac{3}{4}$ m) long, in or below the leaves; flower stalks borne all year. **FRUIT**: black, globose, $\frac{1}{4}$ in. (6 mm) wide; maturing throughout the year. **NATIVE HABITAT**: understory plant of dense rain forests of Mexico, Guatemala, and Belize. (5)

CULTURE Partial to full shade, in well-drained fertile organic soil; no salt tolerance; spider mites in interior situations may become problematic. **PROPAGATION**: seeds but air layering of stems results in adventitious roots. **LANDSCAPE USES**: excellent house and interiorscape plant; also useful for tubs and planters in shady locations; can be used as an accent plant on north side locations and in shady underplantings. **HARDINESS ZONE**: 10A–13B.

COMMENTS *Chamaedorea elegans* is said to be exceeding rare in its natural habitat because of overcollecting. It is however, an easy plant to grow and the most widely used indoor palm. With about 100 understory species *Chamaedorea* is one of the larger genera and with several species popular as indoor foliage plants. A few examples include *C. arenbergiana*, *C. brachypoda*, *C. cataractarum*, *C. costaricana*, *C. elatior*, *C. ernesti-angusti*, *C. erumpens* (bamboo palm), *C. glaucifolia*, *C. metallica*, *C. microspadix* (bamboo palm), *C. oblongata*, *C. radicalis*, *C. seifrizii* (bamboo palm), *C. tepejilote*, among others. These are a variable group of small or sometimes moderate, acaulescent or trunked or rarely climbing, solitary or clustered, unarmed palms with erect or procumbent, smooth and green stems that are prominently ringed with leaf scars and fibers; leaves bifid or variously pinnate, reduplicate; petiole short to elongate, sometimes with a pale green or yellow stripe on the lower side; sheath closed or split, short or elongate. Native to Mexico, Central and South America.

Chamaedorea elegans in a hotel lobby

Chamaedorea seifrizii

REED PALM
SYNONYM *Chamaedorea donneli-smithii*, *C. erumpens*, *C. campechana*

GROWTH HABIT Dioecious, small, delicate, multiple-trunked (clumping) palm; up to 12 ft. (4 m) tall with variable spread but usually about 9 ft. (3 m); medium-fine texture; rapid growth rate. **STEM/BARK**: slender green trunks are $\frac{3}{4}$ in. (2 cm) thick. **LEAVES**: even-pinnately compound, to 2 ft. ($\frac{2}{3}$ m) long; leaflets 24–36, 10 in. (30 cm) long by $\frac{3}{4}$ in. (2 cm) wide, narrowly lanceolate with ribbed margins; petioles short, smooth. **FLOWERS**: unisexual; whitish, small; on once-branched, 1 ft. (30 cm) long, orange flower stalks, in or below the leaves; flower stalks borne all year. **FRUIT**: globose green to black when mature, $\frac{1}{2}$ in. (1 cm); fruit mature all year. **NATIVE HABITAT**: dense rainforests of Yucatan, Mexico and Central America. (5)

CULTURE This species will endure more light than most *Chamaedorea* spp. but still needs some shade part of the day; in fertile well-drained soil; no salt tolerance; susceptible to nematodes and spider mites when used in interiorscapes. **PROPAGATION**: seeds and division. **LANDSCAPE USES**: often used as a screen in shade locations; also used as a tub plant for patios and interiors. **HARDINESS ZONE**: 11A–13B. Not as cold hardy as other related species.

COMMENTS *Chamaedorea seifrizii* 'Florida Hybrid' is presumably an interspecific hybrid between *C. erumpens* × *C. seifrizii*. Since the two parental species are now synonymous, 'Florida Hybrid' should be considered a cultivar selection of *C. seifrizii*, a variable species.

Chamaedorea seifrizii in the landscape (above left), close-up of leaves with unripe fruit (above right), stem showing leaf scars (below left), and close-up of mature fruit (below right)

Chamaedorea costaricana

COSTA RICAN BAMBOO PALM
SYNONYMS *Chamaedorea biolleyi*, *C. linearia*, *C. quetzaltica*, *C. seibertii*, *Legnea lacinata*, *Nunnezharia costaricana*, *Omanthe costaricana*

ETYMOLOGY [From Greek *chamai*, on the ground, dwarf, low growing; *dorea*, a gift, in reference to low growth habit and bright-colored fruit] [From Costa Rica, where the species was originally discovered]

GROWTH HABIT Dioecious bamboo-like palm in dense clusters of up to about 18 ft. (6 m) tall and 6–10 ft. (2–3 m) wide, but usually shorter in cultivation. **STEM/BARK**: stems to 1–2 in. (2 ½ –5 cm) thick, with numerous equidistant rings. **LEAVES**: pinnately compound, 3 ft. (1 m) or longer, arching, with glossy green lanceolate opposite leaflets; petioles to about 2 ft. (⅔ m) long. **FLOWERS**: inflorescences infrafoliar (below the leaves), pendulous, branched; flowers unisexual, yellow-green, small. **FRUIT**: globose to subglobose, about ⅓ in. (1 cm) in diameter, becoming purplish-black as they ripen on bright red branched peduncle and rachis. **NATIVE HABITAT**: from Mexico through Central America to Costa Rica, in rainforests in the Pacific slopes up to 7500 ft. (2300 m) elevation.

CULTURE Partial to bright full shade, in well-drained but fertile moist soil; should be given enough space to grow. **PROPAGATION**: seed or division. **LANDSCAPE USES**: excellent container candidate or as specimen in protected corners. **HARDINESS ZONE**: 10A–13B. Grows well in warmer Mediterranean climate regions.

COMMENT This is an attractive and useful and easy to grow plant, with bamboo-like stems that should be sold and used more often. In the photograph below, the height of the plant is three times that of the man standing beneath it.

Chamaedorea costaricana in the landscape (left), and the attractive and unusual inflorescence (right). Photographed at Canary Islands Botanical Garden in Tenerife

Chamaerops humilis

EUROPEAN FAN PALM; DWARF FAN PALM
SYNONYMS *Chamaerops arborescens*, *C. bilaminata*, *C. conduplicata*, *C. depressa*, *C. elegans*, *C. macrocarpa*, *Phoenix humilis*

ETYMOLOGY [From Greek *chamai*, dwarf and *rhops*, a bush, in reference to low growing habit of the species] [Low-growing, dwarf].

GROWTH HABIT Dioecious or polygamodioecious (bisexual and male flowers on same plant), bushy, medium-sized, low growing palm, forming a clump of several trunks, often reclining; to 20 ft. (6 ½ m), usually shorter in cultivation with crown spread of 6–9 ft. (2–3 m) or sometimes greater; coarse texture; slow growth rate. **STEM/BARK**: one to several, to 10 in. (25 cm) in diameter, persistent leaf bases burlap-like. **LEAVES**: palmately divided to near base, segments sword-shaped, split at ends, to 3 ft. (1 m) across, green, grey-green or glaucous blue, often silvery beneath, very stiff (not drooping); petiole with short spines similar to a hacksaw, pointing toward leaf apex. **FLOWERS**: unisexual or occasionally bisexual, small, yellow, hidden among leaves; inflorescence short and stiff. **FRUIT**: drupe; brown or yellow, globose, 3-sided near base, to 1 in. (2 ½ cm) in diameter, in short dense clusters. **NATIVE HABITAT**: Mediterranean southwestern region of Europe and N. Africa; northernmost species of palm in the world. (3)

CULTURE Full sun to bright shade, various soils; slightly salt-tolerant; slow growth makes it expensive to produce in nursery; scales may become problematic. **PROPAGATION**: seed, division. **LANDSCAPE USES**: specimen, framing, urn or planter box, terrace, patio, giving a tropical effect; an excellent small palm for homes and smaller yards. **HARDINESS ZONE**: 8B–12A. Despite statements to the

contrary, it is known to be damaged with sudden prolonged freezing temperatures.

COMMENTS *Chamaerops* is a monotypic species with few taxa of lower ranks, such as *C. humilis* **var**. *argentea* (= var. *cerifera*) with glaucous bluish-silvery leaves, *C. humilis* **var**. *elatior* has a non-suckering solitary trunk (sometimes erroneously referred to as var. *arborea*), *C. humilis* **var**. *epondraes*, also with glaucous leaves, and *C. humilis* '**Blue**', and *C. humilis* '**Vulcano**' is a compact spineless cultivar. A number of uses have been reported for this species, including use of its young leaves and buds as vegetable.

Chamaerops humilis specimen at San Diego Zoo, California (above), inflorescence with flowers (below left), and immature fruit (below right)

Chambeyronia macrocarpa

**FLAME THROWER PALM; RED FEATHER PALM,
BLUSHING PALM**

SYNONYMS *Chambeyronia hookeri, Cyphokentia macrocarpa, Kentia lindenii, K. lucianii, K. macrocarpa, K. rubicaulis, Kentiopsis lucianii, K. macrocarpa*

ETYMOLOGY [Named after Charles Marie-Léon Chambeyron (1827–1891), the French naval officer who published the nautical instructions for New Caledonia in 1881, et seq.] [Latin for large fruit]

GROWTH HABIT Monoecious, single-trunked palm, to 45 ft. (15 m) or taller. **STEM/BARK**: trunk solitary, with widely spaced rings, and distinct green crownshaft 3–4 ft. (1–1 $\frac{1}{3}$ m) tall. **LEAVES**: newly emerging leaves orange to dark red or nearly purple, to 12 ft. (4 m) long gracefully turning up towards the trunk; leaflets alternate, arranged in a single plane along the grooved rachis, linear-oblong, with a distinct central vein, to 4 ft. (1 $\frac{1}{2}$ m) long. **FLOWERS**: inflorescences pendent, arising at the base of the crownshaft; flowers greenish-white. **FRUIT**: globose or subglobose, about 2 in. (5 cm) wide, red when mature. **NATIVE HABITAT**: endemic to rainforests of New Caledonia. (5)

CULTURE Partial shade or full sun in milder climates, in well-drained moist fertile soils; although generally trouble free, scales and whiteflies may become problems. **PROPAGATION**: seed. **LANDSCAPE USES**: focal point specimen because of its very handsome colorful new leaves but may be grown in large containers or as conservatory subjects. **HARDINESS ZONE**: 10B–12B. Plants grow well in warm Mediterranean climates but have no tolerance to freezing.

COMMENT *Chambeyronia macrocarpa* is the national tree of New Caledonia. Riffle and Craft (2003) have pointed out significant diversity in growth habit of this species. Apparently, some specimens in the wild grow as tall as 90 ft. (30 m) or more and show leaf and stem color variations as well as pubescent and glabrous crownshafts. Others recognize four variants in the cultivated specimens: the "**green form**" (with dark green crownshaft), the "**yellow form**" (= *C. macrocarpa* var. *hookeri*, with yellow crownshaft and most red leaves), the "**watermelon form**" (with yellow and green streaked crownshaft), and *C.* **macrocarpa** **var.** **houailou** (with unsegmented leaves and apparently lacking the red new leaves). The only other known species in the genus is *Chambeyronia lepidota*, which is a solitary palm that reaches a height of 45 ft. (15 m), has a slender trunk, and slightly arching leaves. It is also endemic to higher elevations in rainforests of New Caledonia and rare in cultivation.

Chambeyronia macrocarpa (above) with typical dark red leaf and solid green crown shaft and the **"watermelon form"** with striped crown shaft and also stem rings (below).

Coccothrinax argentata

FLORIDA SILVER PALM

SYNONYMS *Cocothrinax garberi*, *C. jacunda*, *Palma argentata*, *Thrinax altissima*, *T. garberi*

ETYMOLOGY [From Greek *kŏkkŏs*, a berry and *Thrinax*, a related genus, in reference to the berry-like appearance of the fruit] [From Latin, silvery].

GROWTH HABIT Hermaphroditic small, single-trunked palm; medium-coarse texture; can reach 20 ft. (6 m) tall, but 4–8 ft. (1 $\frac{1}{2}$ –2 $\frac{1}{2}$ m) is the normal range; 6 ft. (2 m) spread; slow growth rate. **STEM/BARK**: trunk is usually 6 in. (15 cm) wide; smooth and gray or sometimes covered on the upper part with woven-appearing fiber. **LEAVES**: palmate, silver-green, 2–3 ft. ($\frac{2}{3}$ –1 m) long; divided nearly to the base, 30–40 slender, drooping segments; dark green on the upper surface, silvery below; petioles 2 $\frac{1}{2}$ ft. ($\frac{3}{4}$ m) long, unarmed. **FLOWERS**: bisexual, small, whitish-green, numerous; borne on 2 ft. (60 cm) long flower stalks among the leaves in summer. **FRUIT**: purple to black, globose; $\frac{1}{4}$ – $\frac{1}{2}$ in. (1–1 $\frac{1}{2}$ cm) wide; ripening in late summer and fall. **NATIVE HABITAT**: limestone pinelands of south Florida and the Bahamas. (3)

CULTURE Full sun to part shade, tolerates any well-drained soil; moderate salt tolerance; no problem of major importance. **PROPAGATION**: seed. **LANDSCAPE USES**: prized as a diminutive specimen for the silver flashing of leaves in the wind; especially good for near coastal locations. **HARDINESS ZONE**: 10A–12B.

COMMENTS Related species in the landscapes include: *Coccothrinax alta*, *C. argentea* (silver thatch palm), *C. crinita* (thatch palm), *C. barbadensis* (= *C. dussiana*), *C. miraguama*, (including *C.* **subsp.** *miraguama*, *C. miraguama* **subsp.** *havanensis*, and *C. miraguama* **subsp.** *roseocarpa*), *C. spissa*, and others, are small to moderate, solitary or clustered, unarmed or partly armed, monoecious palms with slender stems, covered with closely ringed narrow leaf scars; leaves palmate, leaflets induplicate; petiole long and slender, flat to ridged on the upper side, densely tomentose; sheaths fibrous or more or less spiny, persistent. Native primarily to the West Indies. Often used as landscape plants and leaves ocally utilized in weaving baskets and hats. The genus contains more than 50 species.

Coccothrinax argentata, photographed in south Florida habitat

Coccothrinax crinita

OLDMAN PALM; MAT PALM; THATCH PALM
SYNONYMS *Antia crinata, Thrinax crinata*

ETYMOLOGY [From Greek *kŏkkŏs*, a berry and *Thrinax*, a related genus, in reference to the berry-like appearance of the fruit] [From Latin, with long-hair]

GROWTH HABIT Dioecious, single-stemmed palm, 6–30 ft. (2–10 m) tall; coarse texture; slow growth rate. **STEM/BARK**: solitary trunk 3–8 in. (8–20 cm) in diameter, although it appears visually wider because of the dense wool-like fibers. **LEAVES**: palmate, nearly rounded, green above, dull gray below; to about 5 ft. (1 $\frac{3}{4}$ m) in diameter, segments many, lobed to nearly $\frac{2}{3}$ or more down and with split tips; petiole to about 4 ft. (1 $\frac{1}{2}$ m) long. **FLOWERS**: small, yellowish, on drooping long stalks from among the canopy leaves. **FRUIT**: globose, fleshy, purplish-black, less than 1 in. (2 $\frac{1}{2}$ cm) in diameter. **NATIVE HABITAT**: endemic to seasonally wet savannas of Cuba. (3)

CULTURE Partial to full sun, in well-drained but moist, low nutrient soils; it does tolerate some drought and has some salt tolerance. **PROPAGATION**: seed. **LANDSCAPE USES**: it is a useful landscape or potted specimen of interest because of its wooly trunk, but may also be used in conservatories because of its small stature and interesting appearance. **HARDINESS ZONE**: 9B–12B. Reportedly tolerates temperatures of 20 °F, most likely none, or for short periods.

COMMENT *Coccothrinax crinita* is a critically endangered species as a result of habitat destruction. The *Coccothrinax crinata* **subsp**. *brevicornis* has shorter hairs than the typical subspecies.

Coccothrinax crinita

Cocos nucifera

COCONUT PALM
SYNONYMS *Callapa nucifera, Cocos indica, Cocos nana, Palma cocos*

ETYMOLOGY [From the Spanish and Portuguese *coco*, a smiling or a grinning face, in reference to marks of the seed] [nut-bearing].

GROWTH HABIT Monoecious, large, single-trunked palm; often arched, gracefully curving canopy, to 90 ft. (30 m), tall, commonly seen to about 40 ft. (13 m), with a 21 ft. (7 m) spread; coarse texture; rapid growth rate. **STEM/BARK**: enlarged at base to 3 ft. (1 m) in diameter, upper portion to 1 ft. (30 cm) in diameter; irregularly ringed; vertical cracks present. **LEAVES**: pinnate to 15 ft. long, 60–80 leaflets; leaflets stiff and leathery, in 2 ranks, green to yellow-green, reduplicate; petiole unarmed, channeled above, with coarse woven fiber wrapped around base. **FLOWERS**: unisexual; inconspicuous but in groups of three consisting of two smaller males and one larger female; inflorescence stalk pendant, to 6 ft. (2 m) long. **FRUIT**: technically a drupe (the husk consist of exocarp and mesocarp, endocarp is the seed or "nut"); obovoid to 1 ft. (30 cm) long, somewhat 3-sided; fibrous husk; albumen lining the endocarp; fluid filling the cavity is known as "coconut water." The large seed has three distinct depressions, hence the generic name. **NATIVE HABITAT**: unknown, but probably South Pacific; floats with ocean currents, now widespread in tropics particularly in coastal areas. (5)

CULTURE Full sun, various soils; high salt tolerance; susceptible to lethal yellowing disease; falling coconuts may be a hazard. **PROPAGATION**: seed. **LANDSCAPE USES**: parks, large areas, street (?) and seaside plantings; specimen plants. **HARDINESS ZONE**: warmest areas of 10B–13B.

COMMENTS A monotypic genus with *Cocos nucifera* as the only species but with some 1200 cultivars worldwide; '**Green Malayan Dwarf**', '**Golden Malayan Dwarf**', and the '**Maypan**' cultivars are resistant to lethal yellowing disease and the most widely cultivated. Coconut palm is the source of many products and of various uses: the white flesh, milk, oil, fermented drinks from the seed and coir (the fiber) from the husk which is used for matting and soil compost, timber from the trunk known as "porcupine wood" for building construction, sugar, known as "jaggery" is harvested from the inflorescence, and the apical bud is known as "heart of palm" or "palm cabbage" is eaten, oil and soft drinks, and coconut milk.

Cocos nucifera on Honolulu beach in Hawaii (above), pinnate leaves (middle, left) and coconut fruit (middle, right), a typical coconut with the husk removed (below, left). A double coconut, coco de mer (*Lodoicea maldivica*), the largest known seed, is added to point out the generic difference and avoid any potential confusion (below, right).

Copernicia baileyana

BAILEY FAN PALM
SYNONYM *Copernicia fallaensis*

ETYMOLOGY [Named in honor of the polish Renaissance mathematician and astronomer Mikolaj Kopernic, best known as Nicolaus Copernicus (1473–1543)] [Named in honor of Liberty H. Bailey (1858–1954), Horticulturist/botanist at Cornell University]

GROWTH HABIT Monoecious, tall columnar palm 50–60 ft. (18–20 m) tall with a dense crown of leaves; coarse texture; slow growth rate. **STEM/BARK**: whitish-gray, concrete-like massive trunk to 24 in. (60 cm) in diameter, somewhat swollen near the top. **LEAVES**: Costapalmate, circular in outline, rigid ascending, about 5 ft. (1 $\frac{2}{3}$ m) across; darker green on the upper and gray-green on the lower surfaces; segments erect, lanceolate, divided to about one-third towards the middle of the blade; petiole stout, to 4 ft. (1 $\frac{1}{3}$ m) long. **FLOWERS**: whitish, on much-branched paniculate inflorescences, to 7 ft. (2 $\frac{1}{7}$ m) long, from center of the crown. **FRUIT**: globose, about 1 in. (2 $\frac{1}{2}$ cm) in diameter; brownish-black. **NATIVE HABITAT**: endemic to eastern and central seasonally dry savanna and open woodlands in Cuba.

CULTURE Full sun and well-drained soil; said to be drought-tolerant but benefits from regular irrigation and fertilization, particularly when young. **PROPAGATION**: seed. **LANDSCAPE USES**: it is best as specimen plant on larger properties but sometimes seen in groups of few; much too large and inappropriate for container or conservatory plantings. **HARDINESS ZONE**: 10B–13B.

COMMENT Although *Copernicia baileyana* is perhaps one of the most "majestic" palms, at least some of the 26 other species in the genus are also in the trade and landscapes, including *Copernicia alba*, *C. berteroana* (palma de cana), *C. ekmanii*, *C. glabrescens*, *C. hospital* (and its blue form), *C. macroglossa* (Cuban petticoat palm), *C. prunifera* (carnauba palm), *C. rigida*, and *C. tectorum*, and others, are moderate to tall, solitary, slow-growing, armed, monoecious palms with persistent sheaths covering the stem; leaves palmate, segments induplicate, shallowly divided, with single-ribbed segments; petiole elongate or short or rarely absent, with stout teeth on the margins; sheaths fibrous, persistent. Native to the West Indies and South America. Usually used as landscape plants, and *C. prunifera* is the source of carnauba wax.

Copernicia baileyana in the landscape (above left), a group of plants showing location of inflorescences (above right), and close-up of leaf crown (below). Photographed at Fairchild Tropical Gardens.

Copernicia macroglossa

CUBAN PETTICOAT PALM
SYNONYMS *Copernicia leoniana, C. torreana*

ETYMOLOGY [Named in honor of the Polish Renaissance mathematician and astronomer Mikolaj Kopernic, best known as Nicolaus Copernicus (1473–1543)] [A medical term (macroglossia), meaning large tongued, in this case referring to the large hastula of the palmate leaf]

GROWTH HABIT Hermaphroditic solitary stemmed palm, to 15 ft. (5 m) tall; coarse texture; slow growth. **STEM/BARK**: solitary trunk to 8 in. (2 $\frac{3}{4}$ cm) in diameter covered to near the apex with dried leaf remains (sometimes removed), hence the name petticoat. **LEAVES**: spirally arranged palmate (nearly costapalmate) leaves at and near the stem apex; 5–7 ft. (1 $\frac{3}{4}$–2 $\frac{1}{4}$ m) wide consisting of many lanceolate segments divided to about one-half of the blade; there is a distinct and relatively large hastula at the point of blade connection to the short petiole. **FLOWERS**: brownish-yellow, bisexual, on long paniculate inflorescence to 6 ft. (2 m) long. **FRUIT**: oval, black when mature, to about 1 in (2 $\frac{1}{2}$ cm). **NATIVE HABITAT**: endemic to savanna and coastal salt marshes of Cuba. (3)

CULTURE Full sun, in well-drained soil, as it is said to be drought-tolerant. **PROPAGATION**: seed. **LANDSCAPE USES**: the unique growth habit of this species provides an opportunity for locating it in the landscape as in-ground accent specimen. It is not suitable as a container plant and ultimately overwhelming as a conservatory specimen. **HARDINESS ZONE**: 10B–13B. No tolerance to frost and may show signs of damage at about 40 °F.

COMMENT There seems to be some disagreement as to sexuality of the plant as it is noted as being dioecious or monoecious, and having unisexual or bisexual flowers. However, if the plant does in fact have bisexual flowers, then it is neither monoecious nor dioecious, but hermaphroditic. There is also some question as to whether the plant is drought-tolerant. Without personal experience with this plant, it is inappropriate to make a judgment but the top attached photograph was taken in a moist site partially shaded by oak trees in a personal property.

Copernicia macroglossa (Cuban petticoat palm) with full skirt. (Photographed at a private residence in Sarasota, Florida)

Copernicia fallaensis, photographed in a botanical palm collection site in Canary Islands.

Cyrtostachys renda

SEALING WAX PALM; LIPSTICK PALM
SYNONYMS *Areca erythrocarpa*, *A. erythropoda*, *Bentinckia renda*, *Cyrtostachys lakka*, *Pinanga purpurea*, *P. rubicaulis*, *Ptychosperma coccinea*

ETYMOLOGY [From Greek *kyrtos*, arched and *stachys*, a spike, in reference to the curved inflorescence] [A Malayan aboriginal word for palm]

GROWTH HABIT Monoecious clumping palm with brilliant scarlet or red crownshaft which is slightly swollen at base; to 60 ft. (20 m), but much smaller in cultivation; fine texture; slow to moderate growth rate. **STEM/BARK**: stems slender, smooth, 2–3 in. (5–7 $\frac{1}{2}$ cm) in diameter, arising from base; distinctly equidistantly ringed and sometimes reddish or brownish. **LEAVES**: bright green pinnately compound, about 2 ft. ($\frac{2}{3}$ m) long, with many pairs of opposite elliptic leaflets, each 18 in. (50 cm) long; green glabrous above waxy white beneath; petioles short and together with the rachis the same color as the crownshaft. **FLOWERS**: greenish, on long branched inflorescences with green or purplish-red peduncle from below the crownshaft. **FRUIT**: ovoid, turning shiny bluish-black at maturity. **NATIVE HABITAT**: lowland peat swamps, tidal coastal areas, riverbanks, and rainforests of Thailand, Malaysia, Sumatra, and Borneo. (5)

CULTURE Should be grown in a protected area to lessen the effect of wind and perhaps not suitable for areas with high stormy winds such as Florida; it requires rich moist soil and does not tolerate drought. **PROPAGATION**: seeds are apparently slow to germinate. **LANDSCAPE USES**: best in greenhouses and conservatories but as specimen plants in less windy tropical areas; often grown in ponds. **HARDINESS ZONE**: 10B–13B.

COMMENT Listed as threatened by IUCN in 1995 but removed in 2000. Meristematic apical areas are locally eaten as heart of palm and stems are used in flooring. A few cultivars have been introduced, such as *Cyrtostachys renda* 'Apple', *C. renda* 'Orange', and *C. renda* 'Theodora Buhler'.

Cyrtostachys renda in Fairchild Tropical Garden Conservatory (above) and older plants at a
landscape in Indonesia (below)

Dypsis lutescens

YELLOW BUTTERFLY PALM; GOLDEN CANE PALM
SYNONYMS *Areca flavescens*, *Chrysalidocarpus glauce-scens*, *C. lutescens*

ETYMOLOGY [Meaning of the word dypsis is said to be obscure but perhaps it originated from the Greek word *dyptein*, to dip, in reference to the original description of *Dypsis noronha* in which inflorescence rises then dips down with the curved peduncle] [yellowish, becoming yellow, in reference to stem color].

GROWTH HABIT Dioecious, multiple-trunked clump-ing palm, reaches a height of 20 ft. (6 m), but usually smaller in cultivation, with a variable spread; medium tex-ture; rapid growth rate. **STEM/BARK**: yellow-green, prominently ringed; 4–6 in. (10–15 cm) wide; surmounted by the crownshaft. **LEAVES**: pinnately compound, to 6 ft. (2 m) long; arching; leaflets linear, 80–120, narrowly indu-plicate, to 20 in. (50 cm) long and $1\frac{1}{2}$ in. (4 cm) wide, with tapering apices; petioles smooth, to 2 ft. (60 cm), channeled on the upper side, expanded at the base, golden yellow with black streaks at the base; light green.

FLOWERS: unisexual; whitish, small; on many-branched, yellowish inflorescence among the leaves, to 3 ft. (1 m) long; borne all year. **FRUIT**: yellow turning violet-black when matures; $\frac{3}{4}$ in. (2 cm) long, oblong; ripens all year. **NATIVE HABITAT**: endemic to wet forests of eastern Madagascar. (5)

CULTURE Full sun to partial shade, in well-drained fertile moist soil; no salt tolerance; pest-free in outdoor use. **PROPAGATION**: seed and division of off-shoots. **LANDSCAPE USES**: valued as a specimen, screen, corner, foundation plantings, and as a patio tub plant; can be used in interior location if acclimated properly, but quickly grows too large. Some references note the over use of the species. **HARDINESS ZONE**: 10B–13B.

COMMENTS *Dypsis lutescens* is considered by IUCN as critically endangered. It is often incorrectly called *Areca* palm. Of the related 145 species several are reported from cultivation including: *D. bejofo*, *D. cabadae*, *D. crinita*, *D. decaryi*, *D. decipiens*, *D. lanceolata*, *D. lutescens*, *D. lastel-liana*, *D. madagascariensis* (including **var.** *lucubensis*), *D. mananjarensis*, *D. onilahensis*, *D. pinnatifrons*, *D. tsarav-otsira*, *D. utilis*, and probably others.

Dypsis lutescens (yellow butterfly palm)

Dypsis decaryi (triangle palm)

Hemithrinax ekmaniana

JUMAGUA PALM, LOLLIPOP PALM
SYNONYM *Thrinax ekmaniana*

ETYMOLOGY [From Greek *hemi*, half and *thrinax*, a trident (a three-pronged spear), in reference to the long spines on the trunk] [Named in honor of Erik Leonard Ekman (1883–1931), Swedish botanist and explorer]

GROWTH HABIT Hermaphroditic, medium size solitary palm; fine texture; slow growth rate. **STEM/BARK**: slender gray trunk to about 2 in. (5 cm) in diameter, covered with spines; said to take 6 years before a trunk develops. **LEAVES**: rigid, closely grouped in dense spiny spherical crown; petioles are practically nonexistent. **FLOWERS**: bisexual, in tightly clustered inflorescence. **FRUIT**: globose, whitish, mature to nearly black. **NATIVE HABITAT**: A single wild population endemic to steep cliffs of seasonal scrub of Las Villas in northern Cuba. (3)

CULTURE Direct sun and well-drained but moist soil; salt and wind-tolerant. **PROPAGATION**: seed, germinates in 3 months to 1 year. **LANDSCAPE USES**: specimen and perhaps container grown. **HARDINESS ZONE**: 9B–13B. May require protection in lower zones.

COMMENT *Hemithrinax ekmaniana* is listed by IUCN in the 2006 Red List of Threatened Species as critically endangered. It is noted as being one of the rarest palms in the world. Apparently, fewer than 100 mature individuals remain and regeneration is poor. The plant is actually uncommon in cultivation but is becoming increasingly available. It is discussed here because of its unique characteristics. *Hemithrinax compacta, H. ekmaniana,* and *H. rivularis* are small to moderate size palms with solitary, erect, naked or leaf base-covered, tan or gray stem, and usually with a mass of basal fibrous roots.

Hemithrinax ekmaniana, young plants

Hyophorbe verschaffeltii

SPINDLE PALM
SYNONYMS *Areca verschaffeltii, Mascarena verschaffeltii*

ETYMOLOGY [From Greek *hys*, a pig and *phorbe*, food alluding to the fleshy fruit eaten by pigs] [Named in honor of Ambrose Colletto Alexander Verschaffelt (1825–1886) of Ghent, Belgian nurseryman and author of book on camellias]

GROWTH HABIT Monoecious solitary trunk palm to 20–25 ft. (6–8 m), slow growth, coarse texture. **STEM/BARK**: gray trunk swollen about the middle below the gray-green crownshaft and narrows above and middle the swollen area so as to resemble a spindle. **LEAVES**: pinnately compound, 6–10 stiff leaves that are slightly ascending and have induplicate leaflets. **FLOWERS**: fragrant, yellowish-orange on much branched inflorescences that arise below the crownshaft with an upward spathe that resembles a horn. **FRUIT**: ovate, green then brownish-red at maturity, $\frac{3}{4}$ in. (2 cm) long. **NATIVE HABITAT**: endemic to Rodrigues Island, Mauritius. (5)

CULTURE Full sun in well-drained moist soil; slightly susceptible to lethal yellowing and micronutrient deficiencies; somewhat salt-tolerant. **PROPAGATION**: seeds that germinate in 3–6 months. **LANDSCAPE Uses**: accent plants in warmer climates and suitable for near coastal areas; may be grown as indoor or outdoor container plants and small enough to be used in conservatories. **HARDINESS ZONE**: 10B–12B. They defoliate and not recover if exposed to freezing cold.

COMMENT Critically endangered in its **NATIVE HABITAT** with fewer than 50 plants remaining. The genus consists of five species of which in addition to *Hyophorbe verschaffeltii*, the other four known species are also reported from cultivation: *H. amaricaulis, H. indica, H. lagenicaulis* (bottle palm), and, *H. vaughanii,* are moderate, solitary, unarmed, monoecious palms with basally swollen (spindle-shape) or more or less uniform diameter stem that are ringed with narrow leaf scars; leaves pinnate, leaflets acute; petiole short, robust, channeled on the upper side; sheaths forming a prominent crownshaft. Endemic to Mascarene Islands. These are valuable landscape plants but considered endangered in the wild.

Hyophorbe verschaffeltii in the landscape (above left), specimen showing a pair of large inflorescences with fruit (above right), and below close-up of one of the inflorescences (below left), with immature fruit (below middle), and mature fruit (below right)

Hyphaene thebaica

DOUM PALM; GINGERBREAD TREE
SYNONYMS *Chamaeriphes crinita, C. thebaica, Corypha thebaica, Chamaeriphes crinita, Hyphaene crinita, H. occidentalis, Palma thebaica*

ETYMOLOGY [From Greek *hyphaino*, to entwine, in allusion to the fibers of the fruit] [From Thebes, a city in ancient Egypt, near Nile River]

GROWTH HABIT Dioecious, to about 60 ft. (20 m) high, but usually shorter in cultivation; coarse texture; medium growth rate. **STEM/BARK**: dark gray smooth but closely ringed trunk of uniform diameter of about 35 in. (90 cm), branches in perfect dichotomy and are topped with masses of leaves; at least part of the trunk is often covered with leaf bases. **LEAVES**: strongly costapalmate, spirally arranged in dense crown, 3 ft. (1 m) long; petioles sheathing at the base and with sharp upward facing spines; blades are 4 ft. (1 $\frac{1}{3}$ m) wide and 6 ft. (2 m) long and curve distinctly downward; there are up to 40 segments with a single fiber in between them. **FLOWERS**: unisexual, creamy white, on about 4 ft. (1 $\frac{1}{3}$ m) long inflorescences arise from the crown between the leaves and branch only once. **FRUIT**: large ovate to globose woody drupe, variable in shape and size; yellowish-brown when mature. **NATIVE HABITAT**: widely distributed in wetter localities of hot arid regions of the Sahel in northern Africa, from Mauritania to Senegal and Egypt, Kenya, Tanzania, and elsewhere. (3)

CULTURE Full sun, in well-drained soil; annual fertilizer application containing micronutrients is recommended. **PROPAGATION**: seed which should be planted deep and germinates in 5–6 months; also by suckers. **LANDSCAPE USES**: the uniform dichotomous branching of this palm is unique and when mature it provides the eye-catching feature no other plant can match. Its use is essentially limited to landscapes. **HARDINESS ZONE**: 10B–13B.

COMMENTS The tree was cultivated in ancient Egypt and in many areas considered sacred. Indeed, existence of this plant is a blessing in the hot dry regions where it grows by providing shade. Leaves are utilized in weaving baskets, mats, brooms, and used as thatch. The timber is used for building construction and furniture making, and all parts utilized for a whole host of purposes, including medicine and food. Although eight species are known *H*. *coriacea* and *H*. *dichotoma* are the only other species reportedly available in the trade.

Hyphaene thebaica, older plants with distinct dichotomous branching at Foster Botanical Garden in Honolulu, Hawaii (above) and younger recently transplanted plant that is beginning to branch dichotomously at Fairchild Tropical Botanical Garden (below)

Johannesteijsmannia altifrons

JOEY PALM; DIAMOND JOEY
SYNONYMS *Teysmannia altifrons*

ETYMOLOGY [After Johannes Elias Teijsmann (1808–1882), the "hortulanus" – chief gardener and nurseryman of Bogor (Buitenzorg) Bogor Botanical Garden in Indonesia.

GROWTH HABIT Hermaphroditic, without an aerial trunk, 10–20 ft. (3–6 $\frac{1}{2}$ m) tall and 10–15 ft. (3 to 5 m) wide; coarse texture; slow growth rate. **STEM/BARK**: trunkless. **LEAVES**: large, simple, coriaceous, undivided, triangular, with serrate margins and pleated surface, arising directly from the subterranean rootstock; petioles 6–10 ft. (2–3 $\frac{1}{4}$ m) long and armed with small saw tooth-like spines. **FLOWERS**: bisexual, cream colored, in short inflorescences that arise from amongst the leaves. **FRUIT**: globose, brown. **NATIVE HABITAT**: montane rainforests of southern Thailand, Malaysia, Sumatra, and Borneo at an elevation of 3600 ft. (1200 m). (3)

CULTURE Prefers shady protected (from wind) location, in well-drained moist fertile soil; requires application of fertilizers containing micronutrients. **PROPAGATION**: seed. **LANDSCAPE USES**: accent plant in the landscape or in containers for indoor or outdoor use; most appropriate for conservatory use in cooler climates. **HARDINESS ZONE**: 10B–12B. Said to have cold-intolerant roots.

COMMENT Considered endangered by IUCN. Three other species in addition to *Johannesteijsmannia altifrons* have been described, including *J. lanceolata*, *J. magnifica*, and *J. perakensis*, all endemic to Malaysia but uncommon in the trade.

Kerriodoxa elegans

WHITE ELEPHANT PALM; KING THAI PALM
ETYMOLOGY [Named in honor of Arthur Francis George Kerr (1877–1942), a physician turned botanist, who is recognized as the "founding father" of Thai botany] [Latin for elegant]

GROWTH HABIT Dioecious, to 15 ft. (5 m) tall and 10–12 ft. (3–4 m) wide when mature. **STEM/BARK**: light brown, 6–8 in. (15–20 cm), in diameter, smooth but with closely set rings on older parts while younger parts are covered with leaf bases. **LEAVES**: palmate, circular, 6–8 ft. (2–2 $\frac{3}{4}$ m) across, dark green above, silvery white below; segments lanceolate, shallowly divided; petioles unarmed, dark colored, to 3 ft. (1 m). **FLOWERS**: small yellowish; male inflorescences much branched, short to about 1 ft. (25 cm) long, female inflorescences with fewer branches to about 3 ft. (1 m) long. **FRUIT**: globose, yellow to orange, 2 in. (5 cm) in diameter. **NATIVE HABITAT**: understory palm endemic to the west coast lowland rainforest of Thailand.

CULTURE Sheltered (from wind) shady, in well-drained but moist organic fertile soil. **Propagation**: seed. **LANDSCAPE USES**: although mentioned as a "stunning house plant" it may in fact be difficult or inappropriate for this purpose because of its large leaf dimensions and indoor low humidity. It may, however, be an excellent choice for large conservatories, patios, courtyards, and as accent plant in landscapes. **HARDINESS ZONE**: 10A–13B. Although doubtful, said to withstand temperature of 34 °F.

COMMENT *Kerriodoxa* is a monotypic genus only recently introduced into cultivation and of increasing interest. Apparently, the plant does not set seed in cultivation probably because mature plants of both sexes are usually not present in the same location.

Johannesteijsmannia altifrons

Kerriodoxa elegans

Latania loddigesii

LATAN PALMS

SYNONYMS *Chamaerops excelsior*, *Cleophora dendriformis*, *C. loddigesii*, *Latania glaucophylla*

ETYMOLOGY [From Mauritius vernacular name] [In honor of Joachim Conrad Loddiges (1738–1826), a British nurseryman]

GROWTH HABIT Dioecious, single-trunked palms; extremely coarse texture; to 35 ft. (12 m) tall with a spread of 15 ft. (5 m); moderate growth rate. **LEAVES**: costapalmate; very thick and stiff, to 8 ft. ($2\frac{3}{4}$ m) in diameter; divided into 3-in. ($7\frac{1}{2}$ cm) wide segments with finely toothed margins, undivided at base. Petioles 5 ft. ($1\frac{3}{4}$ m) long, sometimes with basal marginal teeth, as the palms mature; whitish, glaucous, or waxy pubescence covers the undersides of leaves and petioles. **STEM/BARK**: all three latania palms have 10 in. (25 cm) wide trunks with swollen bases; ringed with leaf scars. **FLOWERS**: unisexual; whitish, small; on 3–6-foot-long flower stalks with short branches; borne within the leaves all year. **FRUIT**: brown, glossy, oblong; to 2 in. (5 cm) wide; 1–3-seeded, ripening all year. **NATIVE HABITAT**: Mascarene Islands in the Indian Ocean. (3)

CULTURE Full sun, but tolerates partial shade, in fertile, well-drained soil; moderate salt tolerance; susceptible to lethal yellowing disease. **PROPAGATION**: seed. **LANDSCAPE USES**: specimen for its leaves; especially useful in coastal situations. **HARDINESS ZONE**: 10B–13B.

COMMENTS Species may be distinguished by leaf color; use only young leaves to make a determination as differences in color between the palms fade with age: *L. loddigesii* (blue latan) entire leaf blue-gray; *L. lontaroides* (red latan) petiole, leaf margins, veins reddish; and *L. verschaffeltii* (yellow latan) petiole, leaf margins, veins deep orange-yellow. These are moderate to tall, solitary, dioecious palms with spirally marked elliptic leaf scars; leaves induplicate, costapalmate; petioles long and relatively wide, reddish-brown when young but changing to densely whitish tomentose with age and with spinose teeth; sheaths narrow, horizontally split at the base and smooth or densely tomentose. Native to Mascarene Islands. Among the most attractive landscape plants of the tropics and subtropics.

Latania loddigesii

Licuala grandis

RUFFLED FAN PALM; VANUATU FAN PALM
SYNONYM *Pritchardia grandis*

ETYMOLOGY [From the Moluccan vernacular name, *leko wala*] [Latin for majestic, impressive, in reference to shape of the leaves]

GROWTH HABIT Hermaphroditic solitary palm, to 9 ft. (3 m) in height; slow growing; coarse texture. **STEM/ BARK**: trunks $1\frac{1}{2}$ –2 in. (5–6 cm) in diameter; covered with brown fibers from triangular leaf bases. **LEAVES**: palmate, large, undivided, and mostly orbicular or somewhat wedge-shaped in outline; segments are fused appearing pleated and serrate on the entire blade margins; blades seemingly crowded in the crown, green on both sides, 3 ft. (1 m) wide; petioles also about 3 ft. (1 m) long. **FLOWERS**: bisexual, yellowish-white, on 6 in. (15 cm) long inflorescences that rise from amongst the leaves. **FRUIT**: globose, bright red when mature, about $\frac{1}{2}$ in. (1 $\frac{1}{4}$ cm) in diameter. **NATIVE HABITAT**: native to lowland rainforests of Solomon Islands and Vanuatu. (3)

CULTURE In sheltered (from wind) partial shady location, in well-drained but moist organic soil, and should be given regular fertilization to maintain the rich leaf color. **PROPAGATION**: seeds germinate readily. **LANDSCAPE USES**: excellent candidate for container plant for indoor use or in conservatories, but it is also superb candidate as specimen or accent plant in tropical landscapes. **HARDINESS ZONE**: 11A–13B.

COMMENTS *Licuala* is a large genus of ±150 species of various growth habits including trunkless species and undivided and divided leaves. Species reported in cultivation include but are not limited to *L. glabra*, *L. lauterbachii*, *L. orbicularis*, *L. paludosa*, *L. peltata*, *L. pumila*, *L. ramsayi*, *L. rumphii*, *L. spinosa*, and *L. triphylla*, among others. These are small to moderate, solitary or clustered, acaulescent to shrubby or rarely arborescent palms with short or subterranean, creeping or erect stems which are covered with remains of disintegrating leaf sheaths; leaves palmate or shortly costapalmate, undivided or deeply divided; petiole armed or less often unarmed. Native from India to Southern China through Southeast Asia to New Guinea and Australia. Commonly grown in subtropical and tropical landscapes or as foliage plants.

Licuala grandis (photographed in Bogor, Indonesia)

Livistona chinensis

CHINESE FAN PALM

SYNONYMS *Chamaerops biro*, *Latania chinensis*, *L. japonica*, *L. mauritiana*, *L. oliviformis*, *L. sinensis*, *L. subglobosa*, *Saribus chinensis*, *S. oliviformis*, *S. subglobosus*

ETYMOLOGY [After Patrick Murray, Baron of Livingston (1632–1671), premiere botanist and physician, whose plant collection helped establish Edinburgh Botanical Gardens in 1670] [From China].

GROWTH HABIT Hermaphroditic single-trunked palm; up to 40 ft. (13 m) high and 15 ft. (5 m) wide; coarse texture; moderately slow growth rate. **STEM/BARK**: trunk grayish, ringed when young, becomes smooth when mature, to 12 in. (25 cm) wide. **LEAVES**: light green, palmate, 6 ft. (2 m) long and $4\frac{1}{2}$ ft. ($1\frac{3}{4}$ m) wide; mature leaves divided one-third of the way toward the petiole into 60–100, 2 in. (5 cm) wide segments; segment tips are cleft for 8 in. (20 cm), forming two slender sections which droop to form a fringe around the leaf; juvenile leaves palmate-orbicular with drooping, solid segment tips, giving a "footstool" appearance; petiole 6 ft. (2 m) long, curving toward the trunk, armed with stout spines at the basal end. **FLOWERS**: bisexual, whitish, small; borne on yellowish, 6 ft. (2 m) flower stalks with many short branches; flower stalks appear all year from among the leaves. **FRUIT**: blue-green, ellipsoidal, $\frac{3}{4}$ in. (2 cm) long; ripening all year. **NATIVE HABITAT**: southern Japan, Central China. (3)

CULTURE Full sun but tolerates partial shade; young plants thrive in dense shade; fertile, well-drained soil; no salt tolerance; susceptible to lethal yellowing. **PROPAGATION**: seed. **LANDSCAPE USES**: excellent freestanding specimen; young plants are very popular as large interior/patio tub plants, but tend to become too large with age. **HARDINESS ZONE**: 8B–12B.

COMMENTS *Livistona* encompasses 28 species, of which several are in cultivation, including: *L. australis* (Australian fan palm, Gippsland palm), *L. benthamii*, *L. decora*, *L. drudei*, and *L. fulva*, *L. humilis*, *L. jenkinsiana*, *L. mariae*, *L. muelleri*, *L. nitida*, *L. oliviformis*, *L. rotundifolia*, *L. saribus*, and perhaps others. These are robust or rarely slender, solitary shrubby or arborescent palms with stems bare or covered with persistent petiole bases; leaves palmate or costapalmate, deeply divided, 1-ribbed, shallowly or deeply bifid; petioles usually with spinose marginal teeth, well-developed upper hastula; sheaths conspicuous, interwoven reddish-brown fibrous. Native to tropical and subtropical Asia, Africa to Malaysia and Australia. Among the more popular landscape plants of the tropics and subtropics. Apparently drooping leaves provide shelter for fruit-eating bats.

Livistona chinensis in the landscape (above left), close-up of leaves to illustrate the divided cleft segment tips (above right), and close-up of leaf from a young plant (below)

Phoenix canariensis

CANARY ISLAND DATE PALM
SYNONYMS *Phoenix cycadifolia*, *P. erecta*, *P. jubae*, *P. macrocarpa*, *P. tenuis*, *P. vigieri*

ETYMOLOGY [Greek name for date palm] [From Canary Islands].

GROWTH HABIT Dioecious, large single-trunked upright palm; stiff globular canopy; to 60 ft. (20 m) high; coarse texture; slow growth rate. **STEM/BARK**: to 4 ft. (1 $\frac{1}{3}$ m) in diameter, dense diamond-shaped, persistent leaf bases; often large fibrous root mass at its base. **LEAVES**: pinnate, to 20 ft. (7 m) long, leaflets short, narrow, stiff, 2–3-ranked, green, induplicate, reduced to spines at base of the petiole. **FLOWERS**: unisexual, inconspicuous, inflorescence to 6 ft. (2 m) long, much branched. **FRUIT**: berry, with membranous endocarp between seed and flesh; elongate to 1 in. (2 $\frac{1}{2}$ cm)

in diameter, in heavy clusters, orange. **NATIVE HABITAT**: drier often rocky hillsides in the Canary Islands. (3)

CULTURE Full sun, various well-drained soils, moderately salt-tolerant; palm weevil, leaf spot, macro- and micronutrient element deficiencies, moderately susceptible to Lethal Yellowing Disease. **PROPAGATION**: seed. **LANDSCAPE USES**: specimen, large areas, street plantings; probably too large for most residential sites. Often purchased and grown as moderate size specimens. **HARDINESS ZONE**: 8B–12B. It is a common sight in most mild climate areas.

COMMENTS *Phoenix dactylifera* (date palm), and its cultivars are economically important plants cultivated for their fruit in warm, arid regions since ancient times; other species, such as *P. acaulis*, *P. loureiroi* and var. *humilis*), *P. paludosa*, *P. reclinata* (Senegal date palm), *P. roebelinii* (miniature date palm), *P. rupicola* (Indian date palm), *P. sylvestris* (wild date palm), *P* (Ceylon date palm), and few others, are commonly grown as landscape and/or foliage plants. These are dwarf or creeping to large, solitary or clustered, armed, dioecious palms with stems covered with spirally arranged leaf bases; leaflets induplicate, pinnate, lower leaflets modified as spines; petiole short to well-developed, channeled, flattened or ridged on the upper side; sheath form a fibrous network. Native to tropical and subtropical Africa and Asia.

Phoenix canariensis in natural habitat (above left), as street planting (above right), mature fruit (middle left), plant in a Florida landscape (middle right), inflorescences with young fruit (below left), and trunk sowing diamond-shaped leaf scars and adventitious roots at base (below right)

Phoenix reclinata

SENEGAL DATE PALM

SYNONYMS *Fulchironia senegalensis, Phoenix abyssinica, P. baoulensis, P. comorensis, P. djalonesis, P. dybowskii, P. equinoxialis, P. leoensis, P. spinosa*

ETYMOLOGY [Greek name of date palm] [Latin bent backward].

GROWTH HABIT Dioecious, medium-sized clustering palm, often with reclining trunks, stiff recurving canopy; to 30 ft. (10 m), medium texture; slow growth rate. **STEM/ BARK**: branches 4–7 in. (10–21 cm) thick, mostly free of leaf bases, covered with fibers, often reclining. **LEAVES**: pinnate, 7–15 ft. (2 $\frac{1}{2}$–4 $\frac{1}{2}$ m) long, recurving; leaflets short, narrow, stiff, 2–3-ranked, green, induplicate, reduced to spines at base of petiole. **FLOWERS**: unisexual, inconspicuous, stalk to 3 ft. (1m) long, much branched. **FRUIT**: berry, with membranous endocarp between seed and flesh; ovoid to $\frac{3}{4}$ in. (2 cm) long, brown or orange. **NATIVE HABITAT**: Tropical Africa, Arabian Peninsula, Madagascar, and Comoro Islands. (3)

CULTURE Full sun, in various well-drained soils; moderately salt-tolerant; palm weevil, macro- and micronutrient deficiencies, and somewhat susceptible to Lethal Yellowing Disease. **PROPAGATION**: seed, division. **LANDSCAPE USES**: specimen, accent, large areas for full development. **HARDINESS ZONE**: 9B–12B; known to survive in protected spots in zone 8b but is damaged with prolonged severe colds.

COMMENTS Many hybrids with other *Phoenix* spp. resulting in some variations. Several uses have been reported for this species, including edible fruit, wine making, carpet, tannins from its roots, etc. Three varieties, which should have been designated as subspecies because of their geographical distance, are recognized: *P. reclinata* **var**. *comorensis*, *P. reclinata* **var**. *madagascariensis*, and *P. reclinata* **var**. *somalensis*.

Phoenix reclinata (above), Base of a plant showing branching at soil level (middle), inflorescence with developing fruit (below, left), and with mature fruit (below, right).

Phoenix roebelenii

PYGMY DATE PALM, MINIATURE DATE PALM
ETYMOLOGY [Greek name of date palm] [After M. Roebelin who collected in Southeast Asia and China].

GROWTH FORM Dioecious, diminutive, single-trunked palm (often used as multiple trunked); up to 9 ft. (3 m) tall and 5 ft. (1 $\frac{2}{3}$ m) wide; medium-fine texture; slow growth rate. **STEM/BARK**: trunk slender, 5 in. (12 cm) wide, studded with knobs of old leaf bases. **LEAVES**: light green pinnately compound, to 4 ft. (3 $\frac{1}{3}$ m) long; graceful, arching; leaflets 90–100, induplicate, linear, 10 in. (25 cm) long; modified near the trunk to form long, straight, sharp spines. **FLOWERS**: unisexual, whitish, small; on numerous short branches at the end of a 1 ft. (30 cm) stalk; flower stalks appear among the leaves all year. **FRUIT**: berry, with membranous endocarp between seed and flesh; dark red when mature, ovoid, $\frac{1}{2}$ in. (1 $\frac{1}{4}$ cm) long; ripening all year. **NATIVE HABITAT**: northern Laos, northern Vietnam, and southwestern China. (3)

CULTURE Full to partial sun, in fertile, well-drained soil; no salt tolerance; chewing insects, leaf spot, and bud rot may become troublesome. **PROPAGATION**: seed. **LANDSCAPE USES**: excellent as a lawn specimen or group of three; very popular as a planter, tub, and pot plant; suitable for interiorscapes. **HARDINESS ZONE**: 9A–12B. In zone 8B it is usually planted in protected locations.

COMMENTS As is the case with other *Phoenix* species, *P. roebelenii* also freely hybridizes with other species, such as *P. rupicola* (see Figs).

Phoenix roebelenii (male inflorescences)

Phoenix dactylifera (date palm)

Phoenix robelenii × *P. rupicola* (Photographed at Harry P. Leu Gardens, Orlando, Florida)

Pritchardia hillebrandii

MOLOKAI FAN PALM; HAWAIAN FAN PALM; LOULU LELO
SYNONYMS *Euprichardia hillebrandii*, *Pitchardia insignis*, *Styloma hillebrandii*, *S. insignis*, *Washingtonia hillebrandii*

ETYMOLOGY [Named in honor of William Thomas Pritchard (1829–1907), the first British Council stationed at Fiji Island] [Named in honor of William Hillebrand (1821–1886), a Prussian physician and plant collector]

GROWTH HABIT Hermaphroditic single-trunked medium size palm, 25–60 ft. (8–20 m) tall, course texture; slow growth rate. **STEM/BARK**: solitary, to 8 in. (25 cm) in diameter, fibrous when young becoming smooth with age, with vertical fissures on dark-gray background; leaf scars not clearly visible. **LEAVES**: costapalmate, waxy-glaucous gray-green, 4–6 ft. (1 $\frac{1}{3}$ –2 m) across, pleated, divided to about one-half the length; segments stiff or only slightly drooping. **FLOWERS**: bisexual, pale white to yellowish, on paniculate inflorescences that are shorter or equal to the length of the petioles. **FRUIT**: ovoid-globose, small, $\frac{3}{4}$ to 1 in. (2–2 $\frac{1}{2}$ cm) in diameter, yellowish to reddish brown but become bluish to nearly black when mature. **NATIVE HABITAT**: endemic to rocky islets of north coast of Molokai Island of Hawaii. (3)

CULTURE Full sun, in well-drained moist soils; probably susceptible to lethal yellowing disease. **PROPAGATION**: seed. **LANDSCAPE USES**: probably best as specimen plant but may be used in rock gardens as long as it can be occasionally irrigated. **HARDINESS ZONE**: 10B–13B.

COMMENT There are 27 known species of *Pritchardia*, of which 23 are endemic to the Hawaiian Islands, although they are mostly endangered, threatened, or vulnerable. Other *Pritchardia* species reported from cultivation include *P. affinis*, *P. beccariana*, *P. pacifica*, and *P. thurstonii*, among others.

Pritchardia hillebrandii

Pseudophoenix vinifera

CHERRY PALM, WINDE PALM
SYNONYMS *Aeria vinifera*, *Cocos vinifera*, *Euterpe vinifera*, *Gaussia vinifera*, *Pseudophoenix insignia*

ETYMOLOGY [From Greek *Pseudo*, false; *phoenix*, the date palm which these palms resemble] [Latin for wine-bearing, in reference to use of the stem sap from the pith in wine production]

GROWTH HABIT Polygamous solitary trunked palm, to 70 ft. (23 m), but 15–40 ft. (5–15 m) tall in cultivation and a crown of 15 ft. (5 m) wide; slow growing. **STEM/BARK**: light brown to gray, strongly ventricose (bottle-shaped with swollen or inflated belly) in about the middle, narrow above the swelling and with a tapering relatively short crownshaft; leaf scars prominent when young but indistinct on older trunks. **LEAVES**; pinnate, 10–12 ft. (3–4 m) long; petiole to about 2 ft. ($\frac{2}{3}$ m) long, silvery green to nearly white; leaflets are arranged more or less spirally along the rachis. **FLOWERS**: yellowish bisexual as well as unisexual, on pendent few-branched inflorescences that arise from leaf crown (as opposed to below the crownshaft). **FRUIT**: globose, scarlet, $\frac{3}{4}$–1 in (2–2 $\frac{1}{2}$ cm) in diameter. **NATIVE HABITAT**: endemic to seasonally dry coastal hillside forests in the Caribbean islands of Cuba, Hispaniola, Dominican

Republic, Haiti, and Puerto Rico as well as Yucatan Peninsula in Mexico. (4)

CULTURE Full sun, in well-drained soil; no serious problems has been reported. **PROPAGATION**: seed. **LANDSCAPE USES**: with its curious growth habit it may be used as accent or specimen plant in the landscape or in containers; it may also be a good candidate for conservatory succulent sections or as a container plant. **HARDINESS ZONE**: 10A–13B.

COMMENT The species is considered endangered as a result of its use in the past for wine making. All four known species of *Pseudophoenix* are reported from cultivation: *Pseudophoenix eckmanii*, *P. lediniana*, *P. sargentii* (buccaneer palm), and *P. vinifera* (cherry palm), are moderate solitary, monoecious or polygamous palms with erect, gray-green, waxy, swollen trunk prominently ringed with wide leaf scars; pinnate, leaflets reduplicate, arranged in several ranks; petiole channeled on the upper side; sheaths splitting distally opposite the petiole. Native from Florida to the West Indies and Mexico to Belize. Among the most attractive plants for landscaping because of its large graceful leaves and cherry-like fruit, leaves are locally used for thatch, fruit for animal feed, and a fermented beverage had been made from the sweet juice of the trunk in Haiti.

Pseudophoenix vinifera, at Montgomery
Botanical Center, Coral Gables, Florida

Ptychosperma macarthurii

MACARTHUR PALM, MACURTHUR FEATHER PALM, CLUSTER PALM

SYNONYM *Actinophloeus bleeseri*, *A. hospitus*, *A. macarthurii*, *Carpentaria bleeseri*, *Kentia macarthuri*, *Ptychosperma bleeseri*, *P. hospitum*, *P. julianettii*, *Saguaster macarthurii*

ETYMOLOGY [From Greek, *Ptyche*, a fold; *sperma*, a seed, in allusion to a characteristic of the seed; a folded seed] [After Sir William Macarthur (1800–1882), an Australian botanist and horticulturist].

GROWTH HABIT Monoecious, medium size, multiple-trunked (clumping) palm; can reach a height of 25 ft. (8 m), but more commonly seen in the 12 ft. (4 m) range with variable spread; medium texture; rapid growth rate. **STEM/BARK**: slender, widely spaced leaf scars, gray-green trunks; 3–4 in. (7–10 cm) wide; surmounted by a crownshaft. **LEAVES**: pinnately compound, to 6 ft. (2 m) long; leaflets 40–70, reduplicate, 12 in. (25 cm) long and 2 in. (5 cm) wide, with jagged, square-ended apex and a torn appearance; petioles short, smooth, and expanded at the base to form a crownshaft. **FLOWERS**: unisexual, whitish, small, on 24 in. (60 cm) long branched, bushy flower stalk; flower stalks appear all year below the crownshaft. **FRUIT**: red, ovoid, $\frac{1}{2}$ in. (1 $\frac{1}{4}$ cm) long, ripening all year; seeds grooved. **NATIVE HABITAT**: rainforests of Fiji, New Guinea, and Queensland Australia. (5)

CULTURE Partial shade best, but tolerates full sun and dense shade; grows in any soil of reasonable drainage and fertility; no salt tolerance; relatively pest-free in outdoor uses. **PROPAGATION**: seed. **LANDSCAPE USES**: valued as a specimen or in corner foundation plantings; makes a good indoor/patio tub plant, but becomes too large quickly. **HARDINESS ZONE**: 10B–13B.

COMMENTS *Ptychosperma macarthuri* is listed by IUCN as endangered in its **Native Habitat**s. Other species known in cultivation include: *P. burretianum*, *P. caryotoides*, *P. elegans* (solitaire or Alexander palm, has single trunk to 20 ft., 7 m tall), *P. ledermanniana*, *P. microcarpum*, *P. propinquum*, *P. salomonense*, *P. sanderianum*, *P. schefferi*, and *P. waitianum*, among others, are small to medium, solitary or clustered, unarmed, monoecious palms with erect, usually slender, smooth stems which are sometimes conspicuously marked with leaf scars; leaves pinnate, relatively short, few in the crown. leaflets cuneate to linear, obliquely toothed at apex; petiole short or elongate, channeled on the upper side, tomentose or with scales; sheaths elongate and tubular, forming a prominent crownshaft. Native primarily to New Guinea but extending into Australia. Commonly used as landscape plants.

Ptychosperma macarthurii (Photographed at Fairchild Tropical Botanical Garden

Rhapidophyllum hystrix

NEEDLE PALM
SYNONYMS *Chamaerops hystrix*, *Corypha hystrix*, *Rhapis caroliniana*, *Sabal hystrix*

ETYMOLOGY [From Greek *raphis*, needle; and *phyllon*, leaf, in reference to the needle-like spines] [From Greek, *hysterix*, porcupine, also alluding to the needle like spines].

GROWTH HABIT Polygamodioecious (bisexual and male flowers on one plant and bisexual and female flowers on another plant), low bushy palm with single or multiple trunks and medium texture; reaches a maximum height of 9 ft. (3 m) with a variable spread; grows very slowly. **STEM/BARK**: trunk is very short and thick; may be solitary when young, but sucker with age and are covered with brown matting and long, slender, sharp, black, 6–8 in (15–20 cm) long, erect spines or needles, arising from the leaf bases. **LEAVES**: palmate, to 3 ft. (1 m) wide, separated almost to the base into 7–20 spreading, stiff, 3-ribbed segments. The 1 $\frac{1}{2}$ in. (4 cm)

wide linear segments are toothed and 2-cleft at the apex, and are powdery below; petioles 2–3 ft. ($\frac{3}{4}$ –1 m) long, slender, and unarmed. **FLOWERS**: unisexual or bisexual, reddish, small, on short flower stalks, hidden among the leaf bases and spines. **FRUIT**: drupe, brown, egg-shaped to 1 in. (2 $\frac{1}{2}$ cm) long, woolly. **NATIVE HABITAT**: endemic to low, moist areas of the Southeastern United States; uncommon. (3)

CULTURE Partial to full shade, but tolerates full sun, in poorly drained soils, but will grow in moderately moist soils of reasonable fertility; not salt-tolerant; easily transplanted; pest-free. **PROPAGATION**: seed. **LANDSCAPE USES**: usually grown as a specimen if available. **HARDINESS ZONE**: 7A–10A. It is the hardiest palm species, surviving temperatures of −6 °F. (= −21 °C).

COMMENTS *Rhapidophyllum hystrix* is an endangered monotypic taxon because it is slow growing and often collected from wild populations. It is commonly planted as clumping palm distinguished by its long, sharp needles of the leaf sheaths.

Rhapidophyllum hystrix growth habit (above left), close-up of leaves (above right), and close-up of the base with spines and old flowers (below)

Rhapis excelsa

LADY PALM

SYNONYMS *Chamaerops excelsa; C. kwanwortsik, Rhapis cordata, R. divaricata, R. flabelliformis, R. kwanwortsik, R. major, Trachycarpus excelsa*

ETYMOLOGY [From Greek *Raphis*, needle, in reference to slender leaf segments] [Latin for tall, in reference to cane-like stems].

GROWTH HABIT Dioecious, small delicate palm, forming dense clusters of cane-like trunks, to 15 ft. (5 m) tall, commonly 6–10 ft. (1–3 m), medium texture; moderate growth rate. **STEM/BARK**: slender, rhizomatous, covered with leaf bases and fiber, smooth, cane-like near base, green-ringed. **LEAVES**: palmate, 5–10 segments, divided nearly to base, to 1 ft. (25 cm) across; segments with 2–4 strong ribs at tip; petiole unarmed. **FLOWERS**: unisexual, inconspicuous, fragrant stalks to 12 in. (25 cm), branched, hidden by leaves. **FRUIT**: drupe; oblong, white when mature, to 3 in. (7 $\frac{1}{2}$ cm) long, sometimes several-seeded. **NATIVE HABITAT**: probably southern China. It is not known from the wild. (3)

CULTURE Partial to deep shade, in fertile organic soils; not salt-tolerant; scale, sooty mold, and palm aphids may become troublesome; rhizomes may grow beyond intended boundaries similar to non-clumping bamboos and require periodic control – known to grow through paved roads and break containers. **PROPAGATION**: seed, division. **LANDSCAPE USES**: barrier, foundation for large buildings, accent, urn, planter box, as indoor or outdoor potted plants. **HARDINESS ZONE**: 8A–12B. May require protection in zone 8A when severely cold.

COMMENTS *Rhapis excelsa* includes dwarf ('**Gyokuho**', '**Kodaruma**') and variegated ('**Zuikonishiki**', '**Zuiko-Lutino**', '**Chiyodazuru**', and '**Kotobuki**') cultivars, as well as leaf forms (such as '**Koban**', '**Daruma**', and '**Tenzan**'), especially in Japan where it is referred to as 'Rolls Royce of Palms' (see Okita and Hollenberg, 1981). However, *R. humilis* (fine-leaved lady palm, dwarf lady Palm), *R. multifida*, and *R. subtilis* (Thailand lady palm, Thai dwarf *Rhapis*) are also in cultivation. Related species such as *R. gracilis*, *R. micrantha*, *R. siamensis*, *R. subtilis*, and probably few others, are small, clustering and spreading palms with slender, erect, cane-like stems which are covered with persistent black or gray-brown fibrous leaf sheaths when young but bare when mature; leaves palmate, induplicate, divided nearly to the base into 2-ribbed segments which are also divided shallowly between the folds; petiole elongate, slender, margins smooth; sheaths consist of gray-brown interwoven fibers. Native to southern China and southeastern Asia. Much valued landscape and foliage plants of warmer regions.

Rhapis excelsa growth habit, illustrating spread of the plant by rhizomes (above), leaf close-up (middle left), stem close-up showing petiole remains and fibers (middle right), inflorescences with male flowers (below left), and inflorescence with fruit (below right).

Roystonea regia

ROYAL PALM, CARIBBEAN ROYAL PALM; FLORIDA ROYAL PALM

SYNONYMS *Euterpe jenmanii, E. ventricosa, Oenocarpus regius, Oreodoxa regia, Palma elata, Roystonea elata, R. floridana, R. jenmanii, R. ventricosa*

ETYMOLOGY [After General Roy Stone (1836–1905), American army engineer in Puerto Rico] [Rigid, stiff, in reference to its stature].

GROWTH HABIT Monoecious, large, solitary, erect palm; majestic, gently drooping canopy; to 100 ft. (35 m) tall, commonly seen 50 ft. (18 m); coarse texture; moderate growth rate. **STEM/BARK**: smooth, greyish, irregularly bulged particularly at base; large, glossy green crownshaft; leaf scars when young. **Leaves**: pinnate, to 12 ft. (4 m) long; leaflets to 3 ft. (1 m) long, stoutly nerved on either side of midrib, 3–4-ranked, bright green, reduplicate; petiole unarmed. **FLOWERS**: unisexual, born in spadices below crownshaft; flower stalks to 3 ft. (1 m) long, much branched. **FRUIT**: oblong globose drupe; purplish-black, $\frac{1}{2}$ in. ($1\frac{1}{4}$ cm) long. **NATIVE HABITAT**: Caribbean, South America, one in Florida. (5)

CULTURE Full sun, fertile organic soils; moderate salt tolerance; palm leaf skeletonizer, royal palm bug, and micronutrient deficiencies may become problems. **PROPAGATION**: seed. **LANDSCAPE Uses**: parks, large areas, street plantings. **HARDINESS ZONE**: 10B–13B.

COMMENTS *Roystonea regia* (royal palm) and *R. elata* (Florida royal palm) are the most commonly cultivated species. Two other species, *R. borinquena* (Puerto Rican royal palm) *R. oleracea* (Caribbean royal palm), and the uncommon *R. violacea*, and probably all 10 known species are used in landscapes. In all cases the common name "Royal Palm" is most fitting as the majestic form of these palms is unparalleled by any other.

Roystonea regia used as street trees (above left), palm nursery specimens showing the enlarged base of young plants (above right), developing inflorescences below the distinct crownshaft (below left); immature fruit (below right).

Roystonea borinquena in Puerto Rico natural habitat

Sabal palmetto

CABBAGE PALM

SYNONYMS *Chamaerops palmetto*, *Corypha palmetto*, *Inodes palmetto*, *Sabal bahamensis*, *S. jamesiana*, *S. parviflora*, *S. schwarzii*, *S. viatoris*, etc.

ETYMOLOGY [From South American vernacular name] [*Palmito*, Small palm].

GROWTH HABIT Hermaphroditic, erect, medium-sized, solitary palm; dense, tight globular canopy; to 90 ft. (30 m), commonly 20–40 ft. (7–13 m), coarse texture; moderate growth rate. **STEM/BARK**: to 18 in. (45 cm) in diameter but quite variable, fibrous, rough; sometimes covered with old split leaf bases but often clean and smooth; very rarely known to branch from the apex. **LEAVES**: costapalmate, to 6 ft. (2 m) long, 3 ft. (1 m) wide, divided onethird of the way to base; segments long, tapering, pointed, split at apex, many threads in sinuses, green or grey-green; petiole unarmed but somewhat sharp. **Flowers**: bisexual; inflorescence stalk to 4 ft. ($1\frac{1}{3}$ m), much branched. **FRUIT**: globose drupe; brown-black, to $\frac{1}{4}$ in. (1 cm) in diameter, shiny. **NATIVE HABITAT**: predominantly in swampy freshwater areas and woodlands of eastern United States (Carolinas to Florida), the Bahamas, and Cuba. (3)

CULTURE Full sun to partial shade, various soils; highly salt-tolerant; easily transplanted (leaf removal recommended); palm weevil and palm leaf skeletonizer may become problems. **PROPAGATION**: seed, often germinate readily where they fall. **LANDSCAPE USES**: specimen, framing tree, street plantings. **HARDINESS ZONE**: 7A–13B and possibly warmer areas of zone 6.

COMMENTS Most of the 15 or so known species of *Sabal* are known from cultivation, including *S. bermudana*, *S. causiarum* (Puerto Rican hat palm), *S. domingensis*, *S. etonia* (scrub palmetto), *S. maritima*, *S. mauritiiformis*, *S. mexicana* (Mexican or Texan palmetto), *S. minor* (dwarf palmetto), *S. rosei*, *S. uresana*, and *S. yapa*, and others. These are dwarf to moderately tall palms with solitary, acaulescent on erect stems that are covered with leaf bases becoming smooth with age; leaves costapalmate, segments induplicate, partly divided into 1-ribbed segments or a pair of segments; petiole long, unarmed, channeled on the upper surface; sheaths with a conspicuous cleft below the petiole, margins fibrous. Native from southern US to the West Indies and northern South America. Commonly planted as ornamentals but leaves are used for thatching and the terminal bud (heart of palm, palm cabbage) of some species are eaten as salad.

Sabal palmetto in the landscape showing general growth habit (above left), close-up of a costapalmate leaf (above middle), a typical prolific inflorescence (above right), branched specimens (middle left and right), group planting of three *S. palmetto* (left) and *S. causiarum* (Puerto Rican hat palm) at center surrounded by *S. palmetto* (right)

Sabal mexicana (Mexican palmetto, Texas sabal palm), native to northern Mexico and southwestern North America, reaches a height of about 50 ft. (18 m), typical costapalmate leaves and spineless petioles. Photographed in Nong Nooch Botanical Garden, Thailand.

Salacca magnifica

SALAK PALM
ETYMOLOGY [Latinized form of the aboriginal name in Malaysia] [Latin for magnificent]

GROWTH HABIT Dioecious acaulescent (trunkless) spiny palm; coarse texture; slow growth rate. **STEM/BARK**: None visible-subterranean. **LEAVES**: undivided, obovate, bifid at apex, to 15 ft. (5 m) long and about 30 in ($\frac{3}{4}$ cm) wide; margins coarsely toothed; green above but whitish-gray below; petiole densely spiny. **FLOWERS**: inflorescences arise from within the leaf bases. **FRUIT**: in cluster at base of the plant, reddish-brown with scaly snake-like skin covering a white fleshy edible tasty pulp, about 2 in. (5 cm) long, with one and sometimes two seeds. **NATIVE HABITAT**: endemic to mountainous rainforests of Borneo. (1)

CULTURE Full sun to partial shade, well-drained but moist organic soil. **PROPAGATION**: seed. **LANDSCAPE USES**: this is an essentially beautiful plant that should be handled with some caution because of its sharp spiny petioles. It is otherwise suitable as accent or specimen plant and may be grown for its edible fruit. **HARDINESS ZONE**: 11A–13B.

COMMENT This is one of the species in the genus *Salacca* with undivided leaves. It is currently uncommon in the trade and landscape but presented here for its unique features, particularly the edible fruit characteristics. A number of other species such as *S. flabellata*, *S. wallichiana*, and *S. zalacca* are noted as being in cultivation, all with short stems, petiolar spines, and fleshy edible fruit. *Salacca zalacca* (salak), is cultivated for its edible fruit, and petioles of *S. wallichiana*, are used for house construction.

Salacca magnifica growth habit showing undivided leaves (left), densely covered petiolar spines (middle), and the edible fruit (right). Photographed in a private residence in Indonesia.

Serenoa repens

SAW PALMETTO
SYNONYMS *Brahea serrulata*, *Chamaerops serrulata*, *Corypha obliqua*, *C. repens*, *Diglossophyllum serrulatum*, *Sabal serrulata*, *Serenoa serrulata*

ETYMOLOGY [Honoring Sereno Watson (1826–1892), distinguished American botanist at Harvard] [Latin for creeping].

GROWTH HABIT Monoecious, low, clumping palm, bushy with multiple trunks; can reach heights of 20 ft. (7 m), but normally in the 4–5 ft. (1 $\frac{1}{4}$–1 $\frac{1}{2}$ m) range; both height and spread increase with age. It has a medium-coarse texture; slow to moderate growth rate. **STEM/BARK**: trunks are initially subterranean but later normally creep along the ground and become somewhat erect with age. They sucker in contact with the ground; 9–12 in. (25–30 cm) wide and covered with brown fiber and old leaf bases. **LEAVES**: palmate, to 3 $\frac{1}{2}$ ft. (1 $\frac{1}{4}$ m) wide, deeply divided into 25–30 stiff,

tapering segments with cleft tips; blades are normally green, but there are bluish-silver varieties. The 3–4 ft. (1–1 $\frac{1}{4}$ m) petioles have small, sharp sawtooth (hence the vernacular name), covering the margins of their basal half. **FLOWERS**: whitish, small, on a 3 $\frac{1}{2}$ ft. (1 $\frac{1}{4}$ m) long flower stalk with numerous short branches that appear among the leaves in spring; flowers are visited by honey bees and the source of a high-grade honey. **FRUIT**: ellipsoidal drupe; to 1 in. (2 $\frac{1}{2}$ cm) long, yellowish, turning black at maturity, ripening from August through October. **NATIVE HABITAT**: sandy areas and pinelands, coastal locations throughout the southeast United States mostly as an understory plant. It often forms extensive dense colonies. (3)

CULTURE Needs full sun to partial shade, will grow even in poor soils as long as they have good drainage. It is highly salt-tolerant; pest-free. **PROPAGATION**: seed; seedlings grow very slowly; mature specimens transplant poorly from wild. **LANDSCAPE USES**: leave when clearing land to give a naturalistic effect, as accent. **HARDINESS ZONE**: 7B–12B. May survive in protected spots of zone 7A.

COMMENTS *Serenoa repens* is monotypic but there is a cultivar with bluish-gray foliage ('**Cinerea**') in central and south Florida. Fruit of *Serenoa* is used as medication for benign prostate problems (BPH). The ripe fruit is edible but of little interest, it is, however, of considerable importance to wildlife.

Serenoa repens growth habit inland (above left), in a coastal area (above right), close-up of older specimen of inland plants with well-developed trunks (middle left), deeply divided stiff leaves (middle right), inflorescence arising from amongst the leaves (below left), close-up of a prolific flowering inflorescence (below middle), and typical fruit (below right).

Syagrus cearensis

COCO-BABÃO; CATOLÉ

ETYMOLOGY [Latin name for a species of palm used by Pliny the Elder, Roman author and naturalist] [Latinized name of Ceará state in Brazil where the species is native]

GROWTH HABIT Dioecious, mostly twin trunks, 12–30 ft. (4–10 m) tall; medium growth rate. **STEM/BARK**: trunks unarmed, 2–4 in a single plane, tan to gray, 4–6 in. (10–18 cm) in diameter; leaf scars prominent, spaced wider at base than near the top. **LEAVES**: 10–15 in crown, 30–40 (90–100 cm) long, sheathing at base; true petiole nearly absent; leaflets 100–130, irregularly distributed along rachis. **Flowers**: unisexual, yellow, inflorescence interfoliar, 18–34 in. (45–85 cm) long. **FRUIT**: light orange when mature, $\frac{1}{4} - \frac{1}{3}$ in. (7–10 mm), as long as wide. **NATIVE HABITAT**: mountainous areas and seasonal forests in state of Ceará of northeastern Brazil. (5)

CULTURE Full sun to partial shade, adaptable to most soils; somewhat drought-tolerant but best with periodic irri-gation. **PROPAGATION**: seed. **LANDSCAPE USES**: with its mostly double trunk it is a superior accent plant but may be used in most landscape situations. **HARDINESS ZONE**: 10B–13B.

COMMENT Although relatively common in cultivation, this species was erroneously identified as *S. flexuosa* and *S. comosa* but more recently recognized as a distinct species. Several other species of *Syagrus* are in cultivation such as *S. campestris*, *S. cocoides*, *S. comosa*, *S. coronata* (licury palm), *S. flexuosa*, *S. orinocensis*, *S. romanzoffiana* (queen palm), among others, are a variable group of small to large, solitary or clustered, armed or unarmed, monoecious palms, with short subterranean to erect and tall, sometimes basally swollen, straight stems; leaves pinnate; leaflets reduplicate, acute, shortly bifid or obtuse at the apex; petiole short to long, channeled or flattened on the upper side, margins smooth; sheath fibrous, usually slender to flattened. Native to South America, primarily in Brazil. Planted mostly as ornamentals but some have edible fruit and yield seed oil and wax, leaves are used for thatch.

Syagrus cearensis twin-trunk specimens (left) and close-up of leaves (right). Photographed at Montgomery Botanical Center, Coral Gables, Florida

Syagrus romanzoffiana

QUEEN PALM

SYNONYMS *Arecastrum romanzoffianum*, *Calappa acrocomioides*, *C. plumosa*, *C. romanzoffiana*, *Cocos acrocomioides*, *C. australis*, *C. plumosa*, *C. romanzoffiana*

ETYMOLOGY [Latin name for a species of palm used by Pliny the Elder, Roman author and naturalist] [After Count Nicholas De Romanzoff (1754–1826) chancellor of the Russian Empire and great patron of science].

GROWTH HABIT Monoecious, large single-trunked erect palm, graceful drooping canopy; to 40 ft. (33 m) or more, medium texture; rapid growth rate. **STEM/BARK**: 1–2 ft. ($\frac{1}{3}$–$\frac{3}{4}$ m) in diameter, slightly and irregularly bulging, grey, smooth except for ring-scars; crownshaft absent. **LEAVES**: pinnate, 8–15 ft. (3–5 m) long, glabrous leaflets to 3 ft. (1 m) long, soft, drooping, 2–3-ranked, dark green on both sides, reduplicate; petiole unarmed. **FLOWERS**: unisexual; cream colored, arising in lower leaf axils, on stalk to 6 ft. (2 m) long, at first covered by long woody spathe. **FRUIT**: broad ovoid, short-beaked drupe; yellow, to 1 in. (2$\frac{1}{2}$ cm) long, fleshy fibrous exterior.

NATIVE HABITAT: southern Brazil to northern Argentina, including Bolivia. (5)

CULTURE Full sun, various well-drained soils (well adapted to sandy soils); moderately salt-tolerant; relatively short-lived (35–40 years); palm leaf skeletonizer, scale, minor nutrient deficiency (Mn deficiency resulting in frizzle top symptoms); leaves are persistent after death and require pruning; resistant to Lethal Yellowing disease. **PROPAGATION**: seed, which germinate readily and seedlings show considerable variation; plant has become invasive in some areas, including Florida. **LANDSCAPE USES**: accent, specimen, street plantings, commonly used to line avenues. It can substitute for *Roystonea* spp. (royal palms) in areas that are too cold for the latter. **HARDINESS ZONE**: 9A–12B and warmer areas of 8b. Can be cultivated in protected areas.

COMMENTS *Butia odorata* × *Syagrus romanzoffiana*, is a widely known and used hybrid. *Syagrus* is a relatively large genus of 65 species of which several are in cultivation, including *S. amara*, *S. botryophora*, *S. campylospatha*, *S. comosa*, *S. coronata*, *S. flexuosa*, *S. harleyi*, *S. macrocarpa*, *S. oleracea*, *S. orinocensis*, *S. pseudococos*, *S. ruschiana*, *S. schizophylla*, among others.

Syagrus romanzoffiana growth habit with dead laves remaining on trunk (above left), plants used in a street landscape (above right), prolifically blooming inflorescence (below left), prolific immature fruit on a pair of inflorescences (below middle), and mature fruit (below right).

Branching is uncommon in *Syagrus* but on very rare occasions the apical meristematic region seems to multiply and create few stems.

Trachycarpus fortunei

WINDMILL PALM; CHUSAN PALM
SYNONYMS *Chamaerops fortunei, Trachycarpus caespitosus, T. wagnerianus*

ETYMOLOGY [From Greek *trachys*, rough and *karpos*, fruit; rough fruit] [After Robert Fortune (1812–1880), Scottish horticulturist and plant hunter].

GROWTH HABIT Dioecious and sometimes polygamodioecious, erect, small to medium-sized, single trunked palm; compact globose canopy; to 60 ft. (30 m), but usually seen 5–15 ft. (2–5 m) in cultivation; coarse texture; slow growth rate. **STEM/BARK**: about 1 ft. (30 cm) in diameter, usually much more slender; densely covered with brown fibers and old leaf bases. **LEAVES**: palmate; to 3 ft. (1 m) across, divided almost to base, segments sometimes drooping near tips, dark green above, glaucous beneath; petiole to about 40 in. (1 m) long, unarmed but bumpy, rough. **FLOWERS**: unisexual or sometimes together with bisexual flowers, small, yellow male and greenish female, fragrant; inflorescence paniculate, stalk very short, branched; not showy. **FRUIT**: reniform (kidney-shaped) drupe; blue-black when ripe, three-lobed, to $\frac{1}{2}$ in. (1 $\frac{1}{4}$ cm) long. **NATIVE HABITAT**: central China to northern Myanmar (Burma) and India, growing from about 300 to 7800 ft. (100–2400 m) altitudes. (3)

CULTURE Full sun to partial shade, fertile well-drained soils; moderately salt, drought-tolerant, and wind-tolerant; may suffer from scale, palm aphids, root rot, and it is somewhat susceptible to Lethal Yellowing Disease. **PROPAGATION**: seed. **LANDSCAPE USES**: framing tree, accent, specimen, urn, does well in confined areas. **HARDINESS ZONE**: 9A–12B, warmer areas of 8b, recent consecutive freezing temperatures did not harm plants if under trees or close to a building.

COMMENTS A few related species are also in cultivation, such as *Trachycarpus latisectus*, *T. martianus*, *T. nanus*, *T. oreophilus*, *T. princeps*, *T. takil*, and *T. wagnerianus* (the latter species is not known from the wild and is treated as semi-dwarf cultivar '**Wagnerianus**' by some). These are dwarf or moderate, solitary or clustering palms with stems decumbent or erect, bare and conspicuously marked with oblique leaf scars or covered with persistent petiole bases and fibrous sheaths; leaves palmate, induplicate; petiole elongate, armed with fine teeth or unarmed. Native from Himalayas to northern China. Commonly used as landscape plants in warm temperate climates and stems are used locally as posts, fibers for brushes and other items, and seeds for medicinal purposes.

Trachycarpus fortunei growth habit (above left), close-up of the fiber-covered trunk (above right), and close-up of leaves (below).

Washingtonia robusta

MEXICAN WASHINGTON PALM, MEXICAN FAN PALM

SYNONYMS *Brahia robusta*, *Neowashingtonia robusta*, *N. sonorae*, *Pritchardia robusta*, *Washintonia gracilis*, *W. sonorae*

ETYMOLOGY [Named after George Washington (1732–1799), first president of the United States] [Stout; strong].

GROWTH HABIT Monoecious, large single-trunked erect palm; loose globose canopy; to 80 ft. (27 m) tall, commonly seen 40–50 ft. (13–17 m); course texture; rapid growth rate. **STEM/BARK**: trunk to 2 ft. ($\frac{3}{4}$ m) in diameter, wider at base, angled rings, vertical cracks present; sometimes large root mass at base. **LEAVES**: palmate; to 4 ft. (1 $\frac{1}{3}$ m) across, divided half way to base, many threads in segment sinuses when young, disappearing with age; segments bright green; petiole reddish-brown, armed, spines pointing in both directions; leaves often persist forming a dense skirt. **FLOWERS**: unisexual, small, numerous; white; inflorescence stalk 9–12 ft. (3–4 m) long, erect at first, then hanging; not showy. **FRUIT**: ovoid-globose drupe; black, to $\frac{1}{3}$ in (8 mm) long. **NATIVE HABITAT**: western Sonora, Baja California, in northwestern Mexico. (3)

CULTURE Full sun, various well-drained soils; moderately salt-tolerant; palm weevils, root rot, persistent leaves and armed petioles on falling leaves are dangerous; apparently resistant to Lethal Yellowing disease; more prone to uprooting than other palms during severe storms. **PROPAGATION**: seed. **LANDSCAPE USES**: specimen, accent for tall buildings, street plantings (if dead leaves are removed). **HARDINESS ZONE**: 8B–12B. Said to be cold hardy to about 20 °F (7 °C).

COMMENTS *Washingtonia filifera* (desert Washington palm), and *W. robusta* (Mexican Washington palm), are tall, solitary palms with erect stems which are at least partially covered with a dense skirt of dead leaves, ringed with close leaf scars, and with loose globose canopy; leaves costapalmate, induplicate, divided half way to the base, with fibrous threads in sinuses when young, petiole reddish-brown, armed, thorns pointing in both directions, sheaths with a conspicuous cleft on the lower side. Native to Southwestern US and Mexico. These are among the most common palms used as street and landscape trees in dryer warm-temperate-subtropical regions. The hybrid between the two species *W. filifera* and *W. robusta* (*Washingtonia filifera* × *filibusta*) is common in the trade.

Washingtonia robusta, growth habit with clear trunk skirt removed (above left), specimen with a full skirt (above middle), close-up of the petiole showing spines' up and down direction (above right), and close-up of laves (below).

Wodyetia bifurcata

FOXTAIL PALM

ETYMOLOGY [Wodyeti, an aboriginal name for a gentleman known as Johnny Flindor, the last surviving male of the tribe with knowledge of the region] [*in* reference to the equally forked leaflets].

GROWTH HABIT Monoecious, solitary or sometimes with two or more planted together, smooth, with a canopy of 8–10 leaves; to 30 ft. (10 m) tall; fine texture; fast growth rate. **STEM/BARK**: trunk slender, gray, swollen at base, ringed with leaf scars; crownshaft narrow, green with whitish waxy scales; leaf sheaths with dark brown scales at top. **LEAVES**: pinnately compound, reduplicate, arching, 8–10 ft. ($2 \frac{3}{4}$–$3 \frac{1}{4}$ m) long; several hundred leaflets attached in several ranks forming a plumose appearance (hence the name foxtail), to 6 in. (15 cm) long by 2 in. (5 cm) wide, often divided into two or more linear segments; deep green above, silvery on underside. **FLOWERS**: unisexual, white, both sexes on same much branched inflorescence, borne below the crownshaft. **FRUIT**: ovoid, orange-red when mature, 2 in. (5 cm) long. **NATIVE HABITAT**: endemic to gravelly hills of Cape Melville National Park in Queensland, Australia. (5)

CULTURE Widely adaptable, requires moderate to high light, in well-drained soil; moderate drought and salt tolerance; may develop fungal leaf spots if irrigated with overhead sprinklers. **PROPAGATION**: seed, germinating in 2–3 months. **LANDSCAPE USES**: popular specimen plant for the landscape; although described in 1978 and introduced relatively recently, the plant has taken the palm world by storm. **HARDINESS ZONE**: 10A–12B.

COMMENTS *Wodyetia bifurcata* is an elegant palm that may be planted individually or with two or more plants in a clustered group. It is a monotypic genus with *W. bifucata* as its sole species. It is considered endangered by IUCN Red List.

Wodyetia bifurcata growth habit in a landscape (above left), close-up of leaves showing the very attractive plumose leaves (above right), close-up of the trunk showing the rings (middle left), mature and immature fruit-bearing inflorescences below the crownshaft (middle right), multiple and paired trunks showing potential uses of the plants in landscapes (below).

ADDITIONAL GENERA

Bactris gasipaes (peach palm, pupunha), native to Central and South America, a naturally common and cultivated for its edible fruit, heart of palm, and timber. Photographed in Herinnering Ann Johannes Flias Park, in Kebun Raya Bogor, Indonesia (5)

Borassodendron borneenes, native to rainforests of Borneo, tall solitary palm with costapalmate leaves, rarely seen in cultivation. It is one of only two species, *B. machadonis*, is the other. Photographed in Bogor Botanical Garden, Indonesia. (3)

Veitchia spiralis (Kajewskia palm), native to Vanuatu Islands, nearly extinct in the wild. Photographed at Fairchild Tropical Garden, Florida

COMMELINACEAE

SPIDERWORT FAMILY

41 GENERA And ±730 SPECIES

GEOGRAPHY Moist locations in tropical, subtropical, and warm temperate regions.

GROWTH HABIT Andromonoecious, erect or ascending, more or less succulent annual or perennial herbs, stemless or with often swollen jointed stems; stoloniferous, rhizomatous, and sometimes tuberous and with fibrous adventitious roots. **LEAVES**: alternate, distichous, or spirally arranged, entire, sheathing at base. **FLOWERS**: short-lived bisexual and male; in terminal or lateral cymes; actinomorphic or rarely zygomorphic, 3-merous with distinct green sepals and free or rarely fused petals; stamens 6, in two whorls, free, often with colorful hairs; ovary superior. **FRUIT**: trilocular capsule.

ECONOMIC USES Ornamentals and latex of a few species are locally used for medicinal purposes.

CULTIVATED GENERA *Callisia*, *Cochliostema*, *Commelina*, *Cyanotis*, *Dichorisandra*, *Palisota*, *Tradescantia* (including *Setcreasea* and *Zebrina*), etc.

COMMENTS Other than interest in some species for foliage or interesting fruits, members of this family are not of interest for their flowers since usually they do not last for more than a few hours or a day and are not profusely produced. There is also the problem of recent name changes of the more familiar plants as noted in list of synonyms – no doubt necessary, but annoying to commercial growers.

Tradescantia fluminensis 'Variegata'

Callisia ornata

Palisota barteri

PALISOTA
SYNONYMS *Palisota ambrophila*, *P. staudtii*

ETYMOLOGY [*Ambroise Marie François Joseph Palisot, Baron de Beauvois* (1752–1820), French botanist and explorer, author of *Flore d'Oware et de Benin* (1804–1821) and worked on grasses and mosses] [Charles Barter (1821–1859), British botanist who trained at Kew Gardens from 1849 to 1851 and joined Balfour Niger expedition from 1857 to 1859]

GROWTH HABIT Rhizomatous perennial herbs, 2–5 ft. (25–60 cm) tall; coarse texture; moderate growth rate. **STEM/BARK**: except for the rhizome, essentially acaulous (stemless). **LEAVES**: basal; oblong to obovate-lanceolate, 10–25 in. (25–60 cm) long, margins entire; apex acuminate, arranged in rosette; petiole to 8 in. (20 cm) long. **FLOWERS**: bisexual, white, strongly scented, 3-merous, open late pm; on terminal cymose inflorescence with a short peduncle; ovary superior. **FRUIT**: indehiscent berry-like, initially green but bright red when mature and attractive. **NATIVE HABITAT**: tropical rainforests of western and central Africa.

CULTURE Partial or filtered shade, in well-drained moist organic fertile soil; inflorescence and leaf galls are reported problems. **PROPAGATION**: seed and division. **LANDSCAPE USES**: primarily as a container plant, but quite suitable for borders and mixed plantings, and in greenhouses and conservatories. **HARDINESS ZONE**: 10B–12B.

COMMENTS All parts of this plant are reportedly toxic, containing calcium oxalate. The plant is not readily available in the nursery trade in the United States, but is recommended for tropical and subtropical areas.

Palisota barteri (photographed at Missouri Botanical Gardens)

Tradescantia pallida

PURPLE SECRETIA; PURPLE HEART, PURPLE QUEEN
SYNONYMS *Setcreasea jaumavensis*, *S. lanceolata*, *S. pallida*, *S. purpurea*, *Tradescantia purpurea*

ETYMOLOGY [John Tradescant the Elder (1570–1638) and John Tradescant the Younger (1608–1662), father and son, both naturalists, plant collectors, and gardeners to King Charles I] [Pale in color]

GROWTH HABIT Evergreen perennial herb; sprawling, mat-forming, succulent ground cover; reaches a height of 15 in (38 cm) and spreads to cover large areas; medium-coarse texture; fairly rapid growth rate. **STEM/BARK**: succulent, reclining stems; break easily; root at the nodes in contact with the ground; have a watery sap that may irritate skin. **LEAVES**: simple; alternate; lanceolate and trough-shaped, to 7 in (18 cm) long, clasping the stem; color varies from green to intense purple. **FLOWERS**: pink; petals 3, to 1 in. (2 $\frac{1}{2}$ cm) wide; usually terminal and subtended by two large, leaf-like bracts; last only one morning. **FRUIT**: small, inconspicuous capsules. **NATIVE HABITAT**: gulf coast of eastern Mexico.

CULTURE Full sun or partial shade, in a wide range of soils; marginal salt tolerance; does well under trees; generally pest-free, although chewing insects may attack. **PROPAGATION**: cuttings root easily. **LANDSCAPE USES**: ground cover, occasionally used as a pot plant or hanging basket. **HARDINESS ZONE**: 9A–12A. Often damaged by cold in zone 8 but grows back in spring.

COMMENTS *Tradescanthia pallida* '**Blue Sue**', '**Pale Puma**', '**Purpurea**', and '**Tricolor**' are among the listed cultivars. Latex may be irritating to some people.

Tradescantia pallida

Tradescantia zebrina var. *zebrina*

WANDERING JEW
SYNONYMS *Commelina zebrina*, *Cyanotis vittata*, *Tradescantia pendula*, *T. tricolor*, *Zebrina pendula*, *Z. purpusii*

ETYMOLOGY [John Tradescant the Elder (1570–1638) and John Tradescant the Younger (1608–1662), father and son, both naturalists, plant collectors, and gardeners to King Charles I] [Striped].

GROWTH HABIT Evergreen perennial herb; creeping, succulent ground cover; reaches a height of 10 in. (25 cm) with an indefinite spread; medium-fine texture; rapid growth rate. **STEM/BARK**: fleshy stems break easily and root at the nodes when in contact with the ground. **LEAVES**: simple; alternate; ovate to 2 in. (5 cm) long; sheathing leaf bases; the typical form has purple-green leaves with two longitudinal silvery stripes and purple undersides; cultivars have different coloring and variegations. **FLOWERS**: pink; small, inconspicuous; in sessile clusters subtended by 2 unequal, spathe-like bracts. **FRUIT**: small capsules; 3-valved. **NATIVE HABITAT**: tropical America.

CULTURE Partial shade, in a wide range of well-drained soils; marginal salt tolerance; sometimes mites, if grown indoors; may become weedy. **PROPAGATION**: stem cuttings root easily. **LANDSCAPE USES**: ground cover for shady areas; it covers quickly, but tends to spread beyond bed boundaries; also grown as an indoor pot plant, hanging basket, or in shady planters. **HARDINESS ZONE**: 9A–12B. May be grown in zone 8B with some protection.

COMMENTS Cultivars '**Purpusii**' (mat-forming, with purple leaves and small pink flowers), and '**Quadricolor**' (leaves with pink, red, and white stripes).

Tradescantia zebrina var. *zebrina*

Tradescantia spathacea

OYSTER PLANT, MOSES IN THE CRADLE

SYNONYM *Ephemerum bicolor*, *E. discolor*, *Rhoeo discolor*, *R. spathacea*, *Tradescantia discolor*, *T. versicolor*

TRANSLATION [Named after John Tradescant the Elder (1570–1638) and John Tradescant the Younger (1608–1662), father and son, both were naturalists, plant collectors, and gardeners to King Charles I] [Spathe-like].

GROWTH HABIT Evergreen perennial herb; clumping succulent ground cover; to 18 in. (45 cm) tall, with an indeterminate spread; medium texture; moderate growth rate. **STEM/BARK**: purple, succulent, erect stems fall over with age and form roots in contact with the ground, new erect shoots follow, fleshy rhizome. **LEAVES**: simple, alternate, lanceolate, and dagger-like, to 12 in (30 cm) long, in tight rosettes; stiff, fleshy, upright, green on the upper surface, purple below. **FLOWERS**: white, small, and inconspicuous; petals 3, in scorpioid cymes, enclosed by small, clam-like purple bracts. **FRUIT**: small, inconspicuous capsules hidden within the bracts. **NATIVE HABITAT**: tropical America: Belize, Guatemala, and southern Mexico.

CULTURE Grows under any light condition, in a wide range of soils, including bare rock; marginal salt tolerance; caterpillars and mites are sometimes problems. **PROPAGATION**: seed, stem cuttings, or division. **LANDSCAPE USES**: low-maintenance ground cover, edging, or planter plant that spreads, but can be kept under control without difficulty. **HARDINESS ZONE**: 9A–12B. May be grown in protected locations in zone 8.

COMMENTS Variegated and dwarf cultivars are available, including *T. spathacea* 'Tricolor', 'Variegata', 'Vittata'. Among a number of other *Tradescantia* species in cultivation is *T.× andersoniana* (including a group of cultivars with various growth habits and flower colors), *T. cerinthoides*, *T. fluminensis*, *T. gigantea*, *T. hirsuticaulis*, *T. microfolia*, *T. occidentalis*, *T. ohiensis*, *T. virginiana*, and others. *Tradescantia spathacea* is considered an invasive species in some areas such as Florida and Hawaii.

Tradesantia spathacea in the landscape (above), close up of growth habit (below left), and close up of flower (below right)

ADDITIONAL GENERA

Cochliostema velutinum, epiphytic unbranched rosette herbs with somewhat succulent leaves, native from southern Nicaragua to southern Ecuador

Callisia fragrans   *Tradescantia virginiana*

STRELITZIACEAE

BIRD OF PARADISE FAMILY
GEOGRAPHY Endemic to wet swampy sites in tropical America (*Phenakospermum*), South Africa (*Strelitzia*), and Madagascar (*Ravenala*).

GROWTH HABIT Hermaphroditic, medium to very large rhizomatous perennials, stemless (acaulescent) and shrubby or with pseudostems from sheathing leaf bases, and tree-like. **LEAVES**: medium to large, simple, alternate, distichous, entire; with or without a petiole; penniparallel lateral veins with a well-defined midrib. **FLOWERS**: bisexual, zygomorphic, on terminal or lateral, bracteate inflorescences, with each bract subtending a cluster of flowers; tepals 3, two noticeably lager than the other; stamens 5 or 6; ovary inferior, with 3 fused carpels. **FRUIT**: loculicidally dehiscent 3-locular woody capsule with numerous arillate seeds.

ECONOMIC USES Widely cultivated in tropical and subtropical regions as landscape subjects and *Strelitzia reginae* is also used for cut flower production.

CULTIVATED GENERA *Phenakospermum* (1 sp.), *Ravenala* (1 sp.), and *Strelitzia* (5 spp.).

COMMENTS Strelitziaceae was formerly included in Musaceae, but is recognized as distinct in the current molecular-based classification. Various past authors had also recognized the affinity of Strelitziaceae with Musaceae (*Ensete* and *Musa*), Lowiaceae (*Orchidantha*), and Heliconiaceae (*Heliconia*).

Ravenala madagascariensis

MADAGASCAR TRAVELLER'S TREE; TRAVELLER'S PALM
SYNONYMS *Heliconia ravenala*, *Urania madagascariensis*, *U. revenala*, *U. speciosa*

ETYMOLOGY [Malagasy vernacular *ravinala*, forest leaves] [From Madagascar]

GROWTH HABIT Large evergreen tree-like herb, 30–40 ft. (10–13 m) tall and 18 ft. (6 m) spread; coarse texture; moderate growth rate. **STEM/BARK**: unbranched woody stout ringed trunk, gray, with numerous suckers arising at base. **LEAVES**: the 2-ranked large banana-like leaves are arranged in a fan-shaped single plain arrangement atop the stem; petioles long and with an enlarged base that can retain rainwater. **FLOWERS**: small white flowers are born in large axillary inflorescences that arise at leaf bases with an erect series of large green boat-shaped bracts. **FRUIT**: woody dehiscent capsule to 3 in. (8 cm) long, with numerous seeds covered with fleshy bluish-indigo arils. **NATIVE HABITAT**: rainforests of Madagascar.

CULTURE Full sun when fully developed, in well-drained but moist organic soil; does not tolerate cold and when possible (at least when young) should be protected from high winds; suckers are often removed to allow only a single specimen; may be planted in coastal areas but not on the actual coast. **PROPAGATION**: seed and separation of suckers. **LANDSCAPE USES**: excellent specimen or accent plant for its unique appearance and tropical effect. **HARDINESS ZONE**: 10B–13B.

COMMENTS The rainwater that is stored in the leaf bases may enable it to withstand dry periods and may be used by travelers in emergency conditions (hence the common name).

Ravenala madagascariensis, a group of mature plants in the landscape (above and below left) and upper part of the plant with dried inflorescences (below right). Photographed in Java, Indonesia.

Strelitzia nicolai

GIANT BIRD OF PARADISE; WHITE BIRD OF PARADISE
SYNONYMS *Strelitzia alba* subsp. *nicolai*, *S. quensonii*

ETYMOLOGY [Sophia Charlotte of the family Mecklenburg-Strelitz (1744–1818), wife of King George III and a patron of botany] [Czar Nicholas I (1796–1855) of Russia]

GROWTH HABIT Evergreen upright palm-like clump-forming herb; with large leaves; reaches a height of 20 ft. (6 $\frac{2}{3}$ m) tall and a spread of 10 ft. (3 $\frac{1}{3}$ m), commonly much smaller; coarse texture; medium growth rate. **STEM/BARK**: solitary trunk (pseudostem) formed by overlapping tightly packed sheathing leaf bases; gray, stout; suckers at the base. **LEAVES**: banana-like (musoid); 2-ranked; to 5 ft. (1 $\frac{2}{3}$ m) long and 2 ft. ($\frac{2}{3}$ m) wide; dark green; less drooping and tattered than *Ravenala;* midribs channeled on the upper side. **FLOWERS**: bisexual, with 3 white sepals and 1 purple-blue and 2 white petals; stamens and carpels are held within the blue petal; borne in a compound inflorescence of several reddish-brown, boat-shaped bracts which often function as a nectar container. **FRUIT**: trilocular dehiscent capsule; many seeds with filamentous arils. **NATIVE HABITAT**: evergreen coastal forests of eastern South Africa.

CULTURE Prefers full sun but with partial daily shade, on fertile, well-drained but moist acid soil; no salt tolerance; scales can be a problem in poorly ventilated areas; shredded leaves make the plant unattractive. **PROPAGATION**: division of suckers; seeds germinate slowly. **LANDSCAPE USES**: specimen plant for an exotic, tropical effect and sometimes used in interiorscapes or in large containers. **HARDINESS ZONES**: 9A–13B. Should be grown in protected locations to minimize wind damage and tolerate short frost periods.

COMMENTS No cultivars or taxa of lower ranks have been reported for this species.

Strelitzia Nicolai specimen (left) and with suckers allowed to grow (right)

Strelitzia reginae

BIRD OF PARADISE; CRANE FLOWER
SYNONYMS *Heliconia bihai*, *H. strelizia*, *Strelitzia angustifolia*, *S. cucullata*, *S. farinosa*, *S. gigantea*, *S. glauca*, *S. humilis*, *S. ovata*, *S. parvifolia*, *S. prolifera*, *S. pumila*, *S. rutilans*, *S. spathulata*

ETYMOLOGY [Sophia Charlotte of the family Mecklenburg-Strelitz (1744–1818), wife of King George III and a patron of botany] [The queen, in reference to Queen Marie Hennette (1836–1902) of Austria, wife of King Leopold II].

GROWTH HABIT Evergreen herbaceous shrubby perennial; clump-forming; reaches a height of 4–6 ft. (1 $\frac{1}{3}$ –2 m) tall and with increasing spread with age, old clumps may measure 10 ft. (3 m) across; slow growth rate; coarse texture. **STEM/BARK**: acaulescent, trunkless. **LEAVES**: basal, simple; oblong to lanceolate; somewhat 2-ranked; thick, waxy; blue-green above and grayish glaucous below; not separated into segments; basal undulate margins; petioles long and channeled; blades to 1 $\frac{1}{2}$ ft. (1 $\frac{1}{2}$ m) long and 6 in. (15 cm) broad. **FLOWERS**: blue, fused petals, long orange sepals; borne laterally on solitary scapose inflorescence, with 3 yellow to red 8 in. (20 cm) sepals of which 2 are fused and 3 blue petals, on stout scapes; bracts subtend flowers like flying birds (hence the common name), with up to 6 flowers per bract; one opening daily. Blooms periodically throughout the year. **FRUIT**: trilocular dehiscent capsules containing many $\frac{1}{4}$ in. ($\frac{3}{4}$ cm) black seeds with orange filamentous arils; ripen all year. **NATIVE HABITAT**: South Africa, but naturalized in a number of other countries.

CULTURE Light, drifting shade, on fertile acid, moisture-retentive soils; no salt tolerance; scales may become problem. **PROPAGATION**: division of clumps; seeds germinate slowly. **LANDSCAPE USES**: specimen for the attractiveness and curiosity of its flowers and contrasting gray leaves; also grown for cut flowers. **HARDINESS ZONE**: 9B–13A. Should be planted in protected locations.

COMMENTS Several cultivars are listed: '**Farinosa**' (oblong glaucous leaves with long petiole), '**Humilis**' (= '**Pygmaea**') (dwarf, clump-forming), '**Kirstenbosch Gold**' (has orange-gold flowers, '**Mandela's Gold**' (yellow flowers), and '**Rutilans**' (leaves with red-purple midrib). *Strelitzia reginae* is the official city flower of Los Angeles, California.

Strelitzia reginae in a landscape (above left), flowers close-up (above right), and a display (below)

Strelitzia juncea (leafless bird of paradise). Photographed at Harry P. Leu Gardens, Orlando, Florida

HELICONIACEAE

HELECONIA FAMILY

1 GENUS AND ±200 SPECIES

GEOGRAPHY American tropics and a few species in Southeast Asia.

GROWTH HABIT Hermaphroditic, terrestrial erect perennial herbs with rhizomes, to about 21 ft. (7 m) tall; stems mostly formed by overlapping petiole sheaths or in some cases true aerial shoots. **LEAVES**: simple, alternate, distichous, elliptic to oblong, long petiolate, sheathing around the stem, without a ligule or petioles absent; margins entire. **FLOWERS**: bisexual, conspicuously colored bracts subtend the flower clusters that form on the terminal erect or pendulous inflorescence; zygomorphic, 3-merous, with 5 fertile stamens and one staminode; ovary inferior, 3-locular. **FRUIT**: drupaceous, blue to purple, with 3 non-arillate seeds.

ECONOMIC USES Landscape and cut flowers.

CULTIVATED GENERA *Heliconia*.

COMMENTS There is significant species and cultivar diversity, most having originated from interspecific crosses. Heliconias are ideal hummingbird plants. Heliconiaceae is closely allied to Costaceae, Musaceae, Marantaceae, Strelitziaceae, Zingiberaceae, and others.

Heliconia cut flowers being prepared for shipment to flower market

Heliconia angusta

CHRISTMAS HELICONIA
SYNONYMS *Bihai angusta, B. angustifolia, Heleconia angustifolia, H. aurorae, H. bicolor, H. bidentata, H. citrina, H. fluminensis, H. lacletteana, H. laneana*

ETYMOLOGY [Mount Helicon in Greece, the home of the Muses, reflecting the closeness of this genus to Musa – nine daughters of Zeus] [Latin narrow, probably in reference to the bracts]

GROWTH HABIT Evergreen erect rhizomatous herb, 3–4 ft. (1–1 $\frac{1}{3}$ m) tall; course texture, rapid growth rate. **STEM/BARK**: rhizomes with short internodes, resulting in relatively compact plants; stems are formed by leaf sheaths. **LEAVES**: dark green banana-like, somewhat coriaceous. **FLOWERS**: bisexual; 4–8 red bracts and white flowers; rachis red to pink; sepals white; pedicel red; flowers usually emerge in October-December. **Fruit**: blue drupe, 3-seeded. **NATIVE HABITAT**: moist forest understory of Atlantic Southeastern Brazil.

CULTURE Full sun to partial shade, in moist but well-drained rich organic soils; requires large spaces in the garden or sizable containers. It does not tolerate cold or drought. **PROPAGATION**: seed, division of the rhizomes, and tissue culture. **LANDSCAPE USES**: landscape or container specimens for patios or conservatories, where it may have to be kept under control to prevent spreading too far. **HARDINESS ZONE**: 10B–13. Do not tolerate temperatures below 35 °F.

COMMENTS IUCN considers *Heliconia angusta* populations as vulnerable due to agriculture but genetic diversity is maintained through *ex situ* conservation. There are cultivars of varying bract colors, including 'Orange Christmas', 'Red Christmas', 'Yellow Christmas', and 'Yellow Waltz'.

Heliconia angusta 'Holiday'

Heliconia caribaea

CARIBBEAN HELECONIA; HELECONIA LOBSTER CLAW
SYNONYMS *Bihai borinquena, B. conferta, B. luteofusca, Heliconia borinquena, H. conferta, H. luteofusca*

ETYMOLOGY [Mount Helicon in Greece, the home of the Muses, reflecting the closeness of this genus to Musa – nine daughters of Zeus] [of the Caribbean area]

GROWTH HABIT Large erect rhizomatous herb, 12–18 ft. (4–6 m) tall. **STEM/BARK**: true stems lacking. **LEAVES**: simple, large, banana-like, coriaceous, dark green, with a waxy white coating on the lower midvein; long petioles arise directly from the ground, 4–5 ft. (3 $\frac{1}{3}$–3 $\frac{2}{3}$ m) long and with white waxy coating. **FLOWERS**: deep red bracts are shaped like a lobster (hence the common name), 9–15 in. (3–6 m) long, terminal on pedunculate racemes. **FRUIT**: blue-purple drupe, 3-seeded. **NATIVE HABITAT**: humid lowland tropical forest of the West Indies, usually in wet areas such as rivers; different inflorescence colors are often specific to the island of their origin.

CULTURE Full sun to partial shade, in well-drained organically rich moist soil; no major problems reported. **PROPAGATION**: division of the rhizomes and tissue culture. **LANDSCAPE USES**: landscape specimens or in large containers for patio or conservatory use; inflorescences are used as cut flowers. **HARDINESS ZONE**: 10B–13B.

COMMENTS Cultivars mostly based on color of the bracts include 'Barbados Flat', 'Big Red', 'Black Knight', 'Black Magic', 'Bonnie Kline', 'Chartreuse', 'Cream', 'Dominator', 'Dominican Lime', 'Dominique', 'Emerald Gold', 'Fire Ball', 'Flash', 'Gold', 'Green Tips', 'Prince of Darkness', 'Purpurea', 'Ruby Royale', 'Springfield Estates' and 'Vivian Rosy Cheek'. There is also interspecific hybrid between *Heliconia caribaea × H. bihai* with several selected cultivars, including 'Carib Flame', 'Criswick', 'Grand Etang', 'Green Thumb', 'Grenadier', 'Hot Rio Nights', 'Jacquinii', 'Kawauchi', 'Pyrotechnic', 'Rauliniana', 'Richmond Red', 'St. Rose', 'Tropical Fruit', 'Vermillion Lake', and 'Yellow Dolly'. The extensive list of cultivars indicates the significance of these plants in the floriculture business.

Heliconia caribaea growth habit (above left), inflorescence close up (above right), *H. caribaea* 'Yellow' (below left), and *H. caribaea* 'Burgundy' (below, right)

Heliconia chartacea

PINK FLAMINGO HELECONIA
SYNONYMS *Heliconia chartacea* var. *chartacea*

ETYMOLOGY [Mount Helicon in Greece, the home of the Muses, reflecting the closeness of this genus to Musa – nine daughters of Zeus] [papery, in reference to the thin leaves]

GROWTH HABIT Evergreen rhizomatous clumping herb, reportedly 6–24 ft. (2–8 m) tall, fast growth; coarse texture. **STEM/BARK**: stems waxy. **LEAVES**: large, banana-like (= musoid), with waxy coating; older blades often torn into lateral segments. **FLOWERS**: green, on pendent inflorescence, rachis pinkish-red and zig-zagged, bracts bright pink with green margins, 4–28 flowers per inflorescence. There is some variation in the length of the bracts and color of flowers.

FRUIT: blue-black drupe with three seeds. **NATIVE HABITAT**: upland disturbed tropical forests of Venezuela, French Giana, Ecuador, and Peru.

CULTURE Similar to other species, in direct sun or partial shade, in well-drained but moist organic rich soil. **PROPAGATION**: separation of clumps or division of the rhizomes, tissue culture. **LANDSCAPE USES**: screening, border plantings, as landscape or container specimens, and cut flowers. **HARDINESS ZONE**: 10B–13B. Sensitive to cold.

COMMENTS Inflorescences with pink bracts are rare in *Heliconia*. The only reported cultivars include the '**Sexy Pink**', '**Meeana**' (yellowish-white), and '**Sexy Scarlet**' (scarlet rachis and along base and dark green lip). The hybrid cultivar *H. chartacea × platystachys* '**Temptress**' has red narrow bracts.

Heliconia chartacea '**Sexy Pink**', green flowers (left) and yellow flowers (right)

Heliconia lingulate

FALSE BIRD OF PARADISE, YELLOW FAN
SYNONYMS *Bihai lingulata, Heleconia weberbaueri*

ETYMOLOGY [Mount Helicon in Greece, the mountain home of the Muses, reflecting the closeness of this genus to Musa – nine daughters of Zeus] [tongue-shaped, in reference to the bracts]

GROWTH HABIT Evergreen erect herbaceous perennial to 15 ft. (5 m) tall; coarse texture, moderately fast growth rate. **STEM/BARK**: short pseudostem formed by sheathing bases of petioles. **LEAVES**: simple; entire, dark green, broadly oblong-lanceolate, musoid (resembling banana leaves); long petioles from the base, sheathing at base, prominent primary vein; margins entire; apex acute. **FLOWERS**: bisexual; on erect inflorescences that rise above the foliage; zygomorphic tubular flowers clustered at base of 11–23 alternately arranged coriaceous bracts, often red-tinged near the pointed apex, flowers greenish-yellow, smaller near the apex but become increasingly larger down the rachis to the base of inflorescence; flowers 3-merous, with 5 fertile stamens and a single sterile staminode. **Fruit**: globose blue fruit with usually 3 seeds. **NATIVE HABITAT**: lowland humid forests of Peru and Bolivia

CULTURE Full sun to bright shade, in well-drained fertile organic moist soils; intolerant of dry soils and must be irrigated during summer months; no serious insect or disease problems reported. **PROPAGATION**: division, tissue culture. **LANDSCAPE USE**: specimen, hedge, fresh or dried cut flowers. **HARDINESS ZONE**: 10A–13B. Sensitive to cold but if damaged it may be covered with leaves and will regenerate in spring.

COMMENTS Inflorescence is sold as cut flower. The reported examples of cultivar include 'Fan', (seems to have narrower more pointed bracts), 'Red Tip Fan' (more intense red bract tips), 'Southern Cross' (with numerous and larger flowers), 'Spiral Fan' (bracts arranged in spiral around rachis).

Heliconia lingulata

Heliconia pogonantha

POGONANTHA
SYNONYMS *Heliconia pogonantha* **var**. *pogonantha*

ETYMOLOGY [Mount Helicon in Greece, the home of the Muses, reflecting the closeness of this genus to Musa – nine daughters of Zeus] [Latin "bearded flowers"]

GROWTH HABIT Erect rhizomatous herb, 12–25 ft. (4–8 m) tall. **STEM/BARK**: pseudostems formed by basal sheathing of the petioles. **LEAVES**: musoid, ovate, with long petiole. **FLOWERS**: bright pink bracts with light yellow base, rachis twisted (zig-zagged), golden yellow; inflorescence terminal, pendent. **FRUIT**: drupe with 3 seeds, blue-purple when ripe. **NATIVE HABITAT**: Honduras to Costa Rica

CULTURE Full sun but prefers partial shade in well-drained fertile organic soil. **PROPAGATION**: division of rhizomes, tissue culture. **LANDSCAPE USES**: screening, border plantings, as landscape or container specimens, and cut flowers. **HARDINESS ZONE**: 10B–13B.

COMMENTS There is disagreement as to the height of this species. Most references note its height as 6–8 ft. (2–2 $\frac{3}{4}$ m). *Heliconia pogonantha* **var**. *holerythra* (native to Panama; rachis red), *H. pogonantha* **var**. *pubescens* (pubescent rachis), *H. pogonantha* **var**. *veraguasensis* (native to Panama). *H. pogonantha* × *mariae* 'Dinosaur'.

Heliconia pogonantha

Heliconia psittacorum

PARAKEET FLOWER; PARROT'S FLOWER
SYNONYM *Heliconia cannoidea*

ETYMOLOGY [Mount Helicon in Greece, the home of the Muses, reflecting the closeness of this genus to Musa – nine daughters of Zeus] [of parrots]

GROWTH HABIT Rhizomatous upright perennial herb, relatively compact and smaller than most species, 3–9 ft. (1–3 m) tall; medium-course texture; fast growth rate. **STEM/BARK**: pseudostems originate from the rhizome and are formed by the sheathing petiole bases. **LEAVES**: musoid, simple, alternate, elliptic-lanceolate, glaucous green; apex acuminate; petioles longer at basal leaves and short and nearly sessile on the upper leaves. **FLOWERS**: few (usually 4–6), commonly yellow or cream-colored, with pinkish-red to mauve bracts that resemble a parrot's beak (hence the common name), on terminal upright inflorescences; prolific bloomer all year. **FRUIT**: globose fleshy dark blue. **NATIVE HABITAT**: rainforests of the Caribbean and northern South America, but naturalized in a number of other areas, including Hawaii.

CULTURE Full sun but prefers partial bright shade, warm humid air, and well-drained rich organic moist to wet soil. **PROPAGATION**: division of rhizomes. **LANDSCAPE USES**: screening, border plantings, as landscape or container specimens, and inflorescences with long vase life used as cut flowers. **HARDINESS ZONE**: 10A–12B. Apparently, it tolerates some cooler temperatures in zone 9B.

COMMENTS *Heliconia psittacorum* is one of the most commonly cultivated and extensively selected species for bract color and plant height. Some of the cultivars include 'Adrian', 'Andromeda', 'Bella Donna', 'Black Cherry'. 'Blush', 'Borinquen Sunrise', 'Brazil', 'Celeste', 'Choconiana', 'Double B Red', 'Flamingo', 'Fuchsia', 'Green Rubra', 'Johanna', 'Kathy', 'Keanea', 'Lady Di', 'Lavidia', 'Lena', 'Lisa', 'Lizzete', 'Nadja', 'Parakeet', 'Peter Bacon', 'Petra', 'Pinky', 'Rosi', 'Red Vee', 'Rosi', 'Ruby', 'Sassy', 'Shamrock', 'Sherbet', 'Strawberries and Cream', 'St. Vincent Red', 'Tangerine', and 'Tropical Night', and others. There are also a number of hybrid cultivars including: *H. psittacorum* × *spathocircatinata:* 'Alan Carle', 'Baby Flame', 'Golden Torch', 'Golden Torch Adrian', 'Golden Torch Sunshine', 'Guyana', 'Johnson Beharry V. C'., 'Keanae Red', 'Lucille Gibson', 'Pink Golden Torch', 'Tropics', and 'Yellow Parrot'.

Heliconia psittacorum with container in water (left), close-up of inflorescence (middle), and cultivar with yellow inflorescence (right)

Heliconia psittacorum × ***H. tropiflora*** (photographed in
Golden Gate Park Conservatory, San Francisco, California)

Heliconia rostrata

HANGING LOBSTER CLAW; FALSE BIRD OF PARADISE
SYNONYMS *Bihai poeppigiana*, *B. rostrata*, *Heliconia poeppigiana*

ETYMOLOGY [Mount Helicon in Greece, the home of the Muses, reflecting the closeness of this genus to Musa – nine daughters of Zeus] [beaked, in reference to the bract apex]

GROWTH HABIT Large evergreen rhizomatous perennial herb, 5–20 ft, (1 ½ –6 ½ m) tall; rapid growth; coarse texture. **STEM/BARK**: pseudostems are 3–6 ft. (1–2 m) tall, formed by leaf sheaths; on rhizomes. **LEAVES**: musoid (banana-like), simple, alternate, ovate-lanceolate, entire, dark green, coriaceous, with long petiolate leaves, blades are readily shredded by wind. **FLOWERS**: small yellow flowers, subtended by a conspicuous scarlet red, edged with yellow and green claw-shaped bracts, on long twisted (zig-zagged rachis) pendent inflorescences that can be up to 3 ft. (1 m) long. **FRUIT**: violet colored drupe with three hard seeds. **NATIVE HABITAT**: rainforests of Peru, Bolivia, Colombia, and Ecuador.

CULTURE Full sun to partial shade, in well-drained but moist fertile organic soils; eEnough room should be allowed for the plant spread; plants grown in higher than pH 6.5 may show iron deficiency. **PROPAGATION**: division of rhizomes, tissue culture. **LANDSCAPE USES**: screening, border plantings, as landscape or container specimens on patio or greenhouse/conservatory, and for long vase life cut flowers. **HARDINESS ZONE**: 10B–13B.

COMMENTS *Heliconia rostrata* is the most readily recognized and popular species. It is the national flower of Bolivia. Several cultivars are in the trade, including 'Arborescent', 'Bucky', 'Carli's Sharonii', 'Dimples', 'Dorado Gold', 'Dwarf', 'Dwarf Jamaican', 'Fire Bird', 'Green Tip', 'Iris Bannochie', 'Long John', 'Misahualli', 'Orange', 'Orange Marmalade', 'Peru', 'Pink Peru', 'Spiral', 'Sun-kissed Orange', 'Tagami', and 'Variegata', as well as *H. rostrata* × *H. spathocircinata*.

Heliconia rostrata

Heliconia rostrata **'Dwarf'** (photographed in Bolivia)

Heliconia vellerigera

FURRY HELICONIA; KING KONG HELICONIA
SYNONYM *Bihai vellerigera*

ETYMOLOGY [Mount Helicon in Greece, the home of the Muses, reflecting the closeness of this genus to Musa – nine daughters of Zeus] [fleecy, with a woolen coat, in reference to the wooly bracts]

GROWTH HABIT Large erect rhizomatous evergreen herb, to about 18 ft. (6 m) tall; fast growth; medium texture. **STEM/BARK**: pseudostems formed by the leaf sheaths. **LEAVES**: musoid, ovate-lanceolate; with distinct penniparallel venation; petioles shorter than the blades. **FLOWERS**: yellow sepals subtended by chocolate brown bracts densely covered with cinnamon colored (or white) pubescence, on long pendent inflorescence with long peduncle, rachis twisted (zig-zagged), red and with wooly hairs. **FRUIT**: yellow berry-like drupe that turns blue at maturity. **NATIVE HABITAT**: tropical forests of western South America, from Colombia, Ecuador to Peru, and also in Panama.

CULTURE Full sun to partial bright shade, in well-drained rich organic moist soil; as with other heliconias, it is sensitive to cold and drought. **PROPAGATION**: division of rhizomes. **LANDSCAPE USES**: as one of the larger species, it is more suitable for landscape specimen or screening, but may be grown in large containers as well; long vase life as cut flower. **HARDINESS ZONE**: 10B–13B.

COMMENTS Only a few cultivars are listed, including *Heliconia vellerigera* '**King Kong**', '**Red Velvet**', '**Root Beer**', '**She Kong**', '**Tena**', and '**Zamora Giant**'.

Heliconia vellerigera, foliage removed to intensify color of the bracts (left), and typical inflorescence (right). Photographed at a nursery in Indonesia.

Heleconia bractiaca

Heleconia stricta

MUSACEAE

BANANA FAMILY

3 GENERA AND ±80 SPECIES

GEOGRAPHY From wet lowlands of West Africa to the Asian Pacific.

GROWTH HABIT Monoecious, large, tree-like monocarpic suckering perennial herbs, with (in *Musa*) or without (in *Ensete*) laticifers; with the sheathing overlapping leaf bases forming a pseudotrunk. **LEAVES:** very large, simple, alternate, spirally arranged, and with a distinct midrib and penniparallel venation. **FLOWERS:** zygomorphic, usually unisexual (females on the lower and males on the upper portions of the terminal inflorescence); tepals in 2 whorls; stamens 5 + one staminode; ovary inferior and consist of 3 fused carpels; inflorescence arises from ground level and grows through the pseudotrunk to a terminal pendent position where flower clusters are subtended by a series of large overlapping bracts. **FRUIT:** fleshy berry with numerous seeds, although commercially grown banana is a parthenocarpic (seedless) fruit.

ECONOMIC USES Banana and plantain are among the most important commercial tropical fruit crops; some species and cultivars are used as ornamentals.

CULTIVATED GENERA *Ensete* and *Musa*.

COMMENTS In its current circumscription Musaceae includes only the two cultivated genera. The genera *Heliconia* and *Strelitzia* that were previously included in Musaceae are currently recognized as distinct families Heliconiaceae and Strelitziaceae, respectively.

Ensete spp.

ABYSSINIAN BANANA; FALSE BANANA, ENSET

SYNONYMS *Musa aroldiana*, *Musa ventricosa*, *Musa ensete*.

ETYMOLOGY [Local African name for banana]

GROWTH HABIT Large monoecious monocarpic evergreen non-suckering perennial herbs, to 20 ft. (6 m) tall; rapid growth; course texture. **STEM/BARK:** pseudostem results from overlapping sheathing leaf bases. **LEAVES:** very large, musoid (banana-like) blades with a distinctly colored pinkish-red or brown midrib and penniparallel laterals. **FLOWERS:** blooms only once before they die in a few years, producing a large pendant spherical inflorescence with clusters of flowers subtended by large pinkish bracts. **FRUIT:** berries, similar to the common banana but with hard spherical black seeds. **NATIVE HABITAT:** rainforests and near streams in Central Africa, across Southeast Asia, to New Guinea and Indonesia.

CULTURE Full sun, in well-drained but moist soil (wet during active growth period); benefits from fertilizer applications. **PROPAGATION:** seeds (see comments). **LANDSCAPE USES:** specimen plants in landscapes or in containers. **HARDINESS ZONE:** 9B–13A. In cooler climates some protection may be necessary.

COMMENTS *Ensete ventricosa* is cultivated for fiber and its stem tissue and young shoots are cooked as food. Fruit is edible but usually has a tart insipid taste but roots are a major diet in Ethiopia. Although wild and seed grown plants usually do not produce suckers, according to the article in Wikipedia, apparently domesticated plants produce as many as 400 suckers which are used in plantations.

***Ensete ventricosum* 'Maurelii'**
(red false banana)

Ensete superbum
(cliff banana)

Musa spp.

BANANA; PLANTAIN

ETYMOLOGY [From Arabic or Farsi *mouz* or *moz*, but may be named after Antonio Musa (63–14 B.C.), physician to Octavius Augustus, the first Roman Emperor].

GROWTH HABIT Monoecious, evergreen clumping, monocarpic perennial, tree-like herb; with very large leaves clustered at the top of each pseudostem to7 ft. (2 $\frac{1}{2}$ m) for dwarf species to 30 feet (10 m) tall for the largest types; rapid growth; coarse texture. **STEM/BARK**: succulent pseudostems made of concentric layers of expanded petiole bases; suckers continuously at pseudostem bases. **LEAVES**: spirally arranged; arching or drooping; oblong, very large, to 9 ft. (3 m) long and 2 ft. ($\frac{3}{4}$ m) wide; entire at first, but become separated into many lateral, tattered segments with age; prominent midrib, deeply channeled on the upper surface which continues into the short, stout, fleshy petioles. **FLOWERS**: terminal, on a peduncle which grows out of the top of the pseudostem; borne in a spicate or paniculate inflorescence often terminated by a large purple bud that peels away one bract at a time; flowers are borne in the axils of the bracts; lower flowers female or rarely bisexual, upper flowers male. **FRUIT**: berry; elongate, fleshy, edible, usually yellow when ripe (in banana) or remaining green (in plantain). **NATIVE HABITAT**: originated in tropical Asia, through southern China to the Philippines, and northern Australia, but now cultivated in tropics worldwide.

CULTURE Full sun or partial shade, in fertile moist soil; require wind and cold protection and heavy fertilization; no salt tolerance; Sigatoka leaf-spot disease may be a problem. **PROPAGATION**: division of thick, narrow-leaved suckers is best, commercially tissue culture. **LANDSCAPE USES**: different species and cultivars are grown for the flowers, variegated foliage, or fruit; also, as landscape specimens to give a tropical effect. **HARDINESS ZONE**: 8B–13A, but need a protected location in lower zones and usually does not produce fruit. Also, it is often damaged in winter months but regenerates in spring. With the recent consistent higher winter temperatures some fruiting has been observed in cooler climates.

COMMENTS *Musa acuminata*, 'Dwarf Cavendish', a triploid form, is one of the best fruit cultivars. In addition, several species and hybrids are commonly planted in southern landscapes, including: *M. acuminata*, *M. coccinea*, *M. ornata* (Flowering Banana), *Musa × paradisiaca* (the edible banana, plantain, platano), *M. textilis*, *M. velutina* (seeded banana), among others in private and tropical fruit collections. In several countries banana leaves are used in cooking, usually as wrap. There is a large number of species and cultivars for fruit and landscape use in the trade.

***Musa* spp.** growth habit (above left), close-up of the inflorescence in early developmental stage
showing protective bracts (above right), partially developped inflorescence with developing fruit (below
left), and fully developed but unripe fruit stock (below right)

Musa **'Orinoco'**, a dwarf cultivar (above) and *Musa* **'Dwarf Green'** (below)

Musa acuminata **'Lady Finger' (much smaller plant and with small fruit and thinner skin)**

Musa coccinea (scarlet banana), native to tropical China and Vietnam

CANNACEAE

CANNA FAMILY

1 GENUS (*Canna*), 12 (10) SPECIES

GEOGRAPHY In tropical America, West Indies (including Florida), along riverbanks.

GROWTH HABIT Rhizomatous perennial herbs with aerial stems. **LEAVES**: large, broad, with a strong central and penniparallel lateral veins, sheathing at base. **FLOWERS**: showy, bisexual, 3-merous, on racemose inflorescences and subtended by a bract; sepals usually free, green; petals 3 with one smaller than the others and fused at base and to the staminal column; stamens 4–6, usually brightly colored; ovary inferior and consist of 3 fused carpels. **FRUIT**: warty, 3-locular capsule containing several spherical black (when ripe) seeds.

ECONOMIC USES Primarily as ornamentals. The rhizome of the South American/ West Indian *C. edulis* is the source of canna starch (tous les mois).

CULTIVATED GENERA *Canna*

COMMENTS Cannaceae is a monogeneric family; all known species of *Canna* are in cultivation and used primarily in hybridization.

Canna × *generalis*

CANNA

TRANSLATION [From Latin *canna*, cane, reed, in reference to growth habit of the plant] [Common].

GROWTH HABIT Monoecious, tall, erect perennial herb, thick branching rhizomes, mostly solitary or single main stems, upright, heavily foliated, vigorous; size varies with cultivar, from 1 $\frac{1}{2}$ to 5 ft. ($\frac{1}{2}$–1 $\frac{2}{3}$ m) tall; coarse texture; rapid growth rate. **STEM/BARK**: Unbranched, green or somewhat colored. **LEAVES**: simple, alternate to spirally arranged, ovate to elliptic-lanceolate, to 2 ft. (60 cm) long, prominent veins, usually glaucous, entire margins, sheathing petioles; color varies with cultivar from green to reddish-purple to bronze to variegate. **FLOWERS**: red, pink, yellow, white, striped or splashed; in spikes, terminal to 18 in. (45 cm) tall, flowers 3-merous, irregular, reflexed; to 6 in. (15 cm) across; flowering in summer and fall. **FRUIT**: capsule; 3-valved, warty surface. **NATIVE HABITAT**: original species along river banks and swamps of tropical America but those in cultivation are of horticultural origin.

CULTURE Full sun or bright shade, in various well-drained, fertile soils; some cultivars moderately salt-tolerant; withstands summer heat if moisture is adequate; canna leaf roller (a major problem), borers, chewing insects. **PROPAGATION**: seeds, division of rhizome. **LANDSCAPE USES**: bedding plant, border, pot plant, planter box, mass-plantings, formal or informal gardens. **HARDINESS ZONE**: rhizomes may be left in the ground year-round in zone 7B-12B; farther north, rhizomes must be dug up each fall.

COMMENTS A large group of complex hybrids from which many cultivars have been selected for flower and foliage color and size. *Canna indica* which was introduced from the West Indies is actually the origin of the common red cultivars. *Canna coccinea* was introduced from South America. *Canna glauca*, from Mexico, was introduced to Europe and have been used as a parental species in hybridizations. A few related species are also cultivated, such as the yellow flowering *C. flaccida*, which is native to the wetlands of the southeastern US There are more than 1000 recorded cultivars.

Canna 'Bengal Tiger'

Canna × *generalis* cultivars (above) and *C. indica* and fruit (below)

Canna × generalis, used as bedding plant in Bogor Botanical Garden, Indonesia

Canna × generalis 'King Humbert'   *Canna × generalis* 'Australia'

MARANTACEAE

ARROWROOT FAMILY; PRAYER-PLANT FAMILY

27 (29) GENERA AND ABOUT 555 (627) SPECIES

GEOGRAPHY Predominantly tropical America, with a few species in Africa and Asia, absent from Australia.

GROWTH HABIT Hermaphroditic rhizomatous or tuberous perennial herbs. **LEAVES**: often colorful and showy; distichously arranged; the blades variously narrow or broad, with penniparallel venation; petiolate, with a swollen pulvinus at the point of attachment to the blade; sheathing at base. **FLOWERS**: bisexual, mostly small and not showy, zygomorphic, each subtended and partially enclosed by a bract; perianth with a distinct series each of free sepals and tubular 3-lobed petals; stamens epipetalous, with usually only one fertile and functions as a trigger when touched by pollinating insects; ovary inferior and consist of 3 carpels. **FRUIT**: fleshy or loculicidal capsule.

ECONOMIC USES Used as ornamentals (particularly as foliage plants), flowers and tubers of some are eaten (arrow root), and leaves are used in roofing and basket making.

CULTIVATED GENERA *Calathea*, *Ctenanthe*, *Ischnosiphon*, *Maranta*, *Marantochloa*, *Schumannianthus*, *Stachyphrynium*, *Stromanthe*, *Thalia*, *Thaumatococcus*, and probably others.

COMMENTS Flowers of Marantaceae represent one of the most interesting pollination phenomena. The single stamen functions as a trigger, where once it is touched by insects it bends and forcefully hits the stigmata and releases its pollen. This can be readily demonstrated by using a pencil or pen. Maranthaceae encompasses 28 genera and 555 or more species. Most likely representative taxa of all genera are in cultivation used predominantly as foliage plants.

Maranta leuconeura, illustrating foliar color patterns

Calathea crocata

ETERNAL FLAME
SYNONYM *Phyllodes crocata*

ETYMOLOGY [From Greek *kalathos*, a basket, in allusion to the flower cluster which looks like a flower in a basket] [Latin for saffron-colored, yellow-orange].

GROWTH HABIT Rhizomatous evergreen perennial herb; compact clump-forming; 10–20 in. (25–50 cm) tall; coarse texture. **STEM/BARK**: stemless; leaves and inflorescence arise directly from the subterranean rhizome. **LEAVES**: simple; elliptic to ovate; upper surface puckered ribbed, with green markings along the major lateral veins; lower surface purplish-maroon; petioles reddish-purple, sheathing. **FLOWERS**: bisexual, on short spike inflores-

cences, with bright yellowish-orange (hence the common name) bracts. **FRUIT**: capsule; usually 3-seeded. **NATIVE HABITAT**: tropical forests of Brazil.

CULTURE Partial to dense shade, at higher light intensities foliage discolors, in well-drained but moisture-retentive soils with high organic matter content; high humidity; no salt tolerance; very sensitive to dissolved salts in the growing medium often showing as marginal burn; nematodes, mites, and root rot. **PROPAGATION**: division of rhizomes, separation of clumps, and tissue culture. **LANDSCAPE USES**: primarily used as a container-grown specimen for interiors; can be used as an accent or specimen for its unusual foliage. May be grown as a specimen or ground cover outdoors in warmer climates. **HARDINESS ZONE** 10B–13B. Protect from cold and direct sun; injured below 40 °F.

COMMENTS The showy flowers (bracts) of this species are somewhat unusual for the family. *Calathea crocata* '**Tasmania**' (more upright, darker leaves, and taller flower stalks), '**Candela**' (also leaves upright, compact, but shorter flower stalks). *Calathea* is a large genus of 287 tropical species of which several are reported from culti- vation: *C. congesta*, *C. dodsonii*, *C. ecuadoriana*, *C. lan- cifolia*, *C. louisae*, *C. makoyana*, *C. orbifolia*, *C. ornata*, *C. roseopicta*, *C. vitichiana*, and *C. warscewiczii*, among others. Calathea currently contains about 60 species, the remaining 140 species have been transferred to *Goeppertia*.

Calathea crocata

Calathea crotalifera

RATTLESNAKE PLANT
SYNONYM *Calathea insignis*, *C. quadratispica*, *C. sclerobractea*, *Phyllodes insignis*

ETYMOLOGY [From Greek *kalathos*, a basket, in allusion to the flower cluster which looks like a flower in a basket] [From Greek *krotalon*, a rattle or clapper, in this case with reference to the arrangement of the bracts that resemble rattlesnake tail]

GROWTH HABIT Evergreen spreading perennial herb, 3–5 ft. (1–1 $\frac{1}{2}$ m) tall and wide. **STEM/BARK**: stemless, leaves and inflorescences arise directly from the rhizomes.

LEAVES: ovate, large, dark green above and paler below, with distinct penniparallel, seemingly ribbed venation; peti- oles shorter than the blades and sheathing at base. **FLOWERS**: small tubular yellow-orange flowers are sub- tended by yellow leathery bracts that are arranged on an upright terminal pedunculate, 10 in. (25 cm) long inflores- cences that resemble a rattlesnake's tail (hence the common name). **FRUIT**: oval capsules are about $\frac{1}{2}$ in. (1 $\frac{1}{3}$ cm) long and contain dark blue seeds surrounded by white fleshy endosperm. **NATIVE HABITAT**: wet tropical forests from southern Mexico to Guatemala.

CULTURE Prefers partial but bright shade, in well-drained, but with high moisture holding capacity, organic soil. It does not tolerate cold or low humidity and root rot may result if rhizomes are planted deep or kept in standing water. **PROPAGATION**: division of rhizomes or separation of clumps. **LANDSCAPE USES**: attractive foliage plant in containers for indoor or outdoor use and as accent plant, hedge, mass planting, and cut flowers. **HARDINESS ZONE**: 10B–13A.

COMMENTS *Calathea crotalifera* is naturalized in Hawaii. Some species are listed by IUCN as threatened due to habitat destruction. *Calathea* includes several cultivars that are primarily selected for leaf colorations. A few examples include '**Cora**', '**Corona**', '**Jungle Velvet**', '**Kopper Krome**', '**Maria**', '**Medallion**', '**Pink Aurora**', '**Red Rattle**' (reddish-brown inflorescence), and '**Velvet touch**'. Some of the well-known related taxa include: *C. burle-marxii* ('**Blue ice**', '**Green Ice**', and '**White Ice**'), *C. casapito* ('**Bronze Pagoda**'), *C. elliptica*, *C. fasciata*, *C. louisae*, *C. loeseneri*, *C. lutea*, *C. majestic* (including '**Princeps**'), *C. makoyana*, *C. picta*, *C. roseopicta*, *C. warscewiczii*, and *C. zebrina*.

Calathea crotalifera (photographed in the Philippines by F. Almira)

Calathea ornata

PIN-STRIPED PLANT
SYNONYMS *Calathea arecta*, *Maranta albolineata*, *M. cordifolia*, *M. ornata*, *M roseolineata*, *Phrynium ornatum*, *Phyllodes arrecta*, *P. ornata*

ETYMOLOGY [From Greek *kalathos*, a basket, in allusion to the flower cluster which looks like a flower in a basket] [Latin for ornate, adorned, embellished, in reference to the leaf surface ornamentation]

GROWTH HABIT Evergreen clump-forming perennial herb, to about 3 ft. (1 m) tall; medium growth rate, coarse texture. **STEM/BARK**: stemless, leaves and inflorescences arise directly from the rhizomes. **LEAVES**: ovate-elliptic, somewhat coriaceous, dark olive green, veins penniparallel with distinctly white or pinkish horizontally parallel laterals (hence the common name); maroon to reddish beneath; peti-oles green, longer than the blades and sheathing at base; as in other members of Marantaceae leaves fold up at night. **FLOWERS**: bisexual, small tubular white or mauve with spirally arranged orange bracts; of little significance and uncommon in cultivation. **FRUIT**: capsule. **NATIVE HABITAT**: South America, Columbia and Venezuela.

CULTURE Partial but bright light, in well-drained moist organic media; no tolerance to dryness, cold, high winds, or salt, either on the foliage or in the media, and it is susceptible to spider mites. **PROPAGATION**: division and tissue culture. **LANDSCAPE USES**: best used as container plant for indoor or covered patio, but may be included in mixed planters or for mass planting under trees. **HARDINESS ZONE**: 10B–13A.

COMMENTS *Calathea ornata* '**Sanderiana**' and '**Roseolineata**' are the two known cultivars, with the latter having shorter petioles and smaller leaves.

Calathea ornata (photographed in Busch Gardens, Tampa. Florida)

Calathea roseopicta

Maranta leuconeura

PRAYER PLANT

SYNONYMS *Calathea leuconeura*, *C. massangeana*, *Maranta kerchoveana*, *M. messangeana*

ETYMOLOGY [After Bartolomeo Maranta (1500–1571), Venetian physician and botanist] [Latin for white-veined].

GROWTH HABIT Evergreen; rhizomatous perennial herb; ground cover; low growing, forms clumps with spreading or pendent branching stems; to 2 ft. (50 cm) tall, generally seen to 12 in. (30 cm) tall; medium-coarse texture; medium fast growth rate. **STEM/BARK**: stems short and cylindrical with zigzag branching pattern. **LEAVES**: simple; alternate; elliptic to ovate, to 6 in. (15 cm) long and 3 $\frac{1}{2}$ in. (9 cm) wide; velvety; colorful, variegated; upper surfaces marked with dark green, light green, or brown; main and lateral veins sometimes gray or red; lower surfaces may be purple or reddish-purple; margins entire; petioles sheathing, flat during the day and folded with an upright position at night (hence the common name). **FLOWERS**: white marked with purple; small, irregular; 3 sepals, 3 petals; borne in loose racemes. **FRUIT**: capsule; 1-seeded. **NATIVE HABITAT**: rainforest of Brazil.

CULTURE Dense to partial shade; at higher light intensities the foliage will bleach; well-drained but moisture-retentive organic soils; high humidity; no salt tolerance; nematodes, mites, slugs, and root rots. **PROPAGATION**: one- or two-node Cuttings. **LANDSCAPE USES**: primarily in interiors for its unusual foliage coloration; also, in cascading hanging baskets, containers, ground beds, or even on short totems. **HARDINESS ZONE**: 10–13B.

COMMENTS Several varieties are available: **var.** *erythro-neura* (Herringbone plant) with rose-red main lateral veins and a reddish-purple lower surface; var. *kerchoveana* (rabbit's-tracks, with a row of dark brown or dark green blotches aligned on each side of the midrib on a light-green leaf); 'Massangeana' – leaves tinted blue, dull rusty-brown toward center. A relatively large number of *Maranta* species have been relegated to other genera.

Maranta leuconeura **var.** *leuconeura*

Maranta leuconeura
var. *erythroneura*

M. leuconeura
var. *kerchoveana*

ADDITIONAL GENUS

Stromonthe sanguinea 'Tricolor', rhizomatous perennial, native to Brazilian rainforests

COSTACEAE

COSTUS FAMILY

7 GENERA AND 140 SPECIES
GEOGRAPHY Pantropical: topical Asia, Africa, and Central and South America, from Mexico to Brazil and Paraguay, and also in the Caribbean.

GROWTH HABIT Hermaphroditic evergreen rhizomatous perennial herbs, with tuberous roots; stems with an acrid juice. **LEAVES**: simple, ovate to elliptic, entire, spirally arranged around bamboo-like stems, closed sheathing leaf bases with a ligule (a projection) on top. **FLOWERS**: solitary or aggregated zygomorphic flowers in terminal head or spike inflorescences on a leafy stem or sometimes on leafless basal shoots; with five fused infertile stamens from a large petal-like spreading labellum and a single fertile stamen; individual herbaceous or mostly coriaceous bracts subtend each flower; petals fused; ovary inferior.

FRUIT: a berry or 2- or 3-locular capsule crowned with persistent calyx; seeds numerous, black or brown, ovoid, with an aril.

ECONOMIC USES Primarily as ornamentals but locally used for medicinal purposes.

CULTIVATED GENERA All eight genera are reported from cultivation: *Chamaecostus*, *Cheilocostus*, *Costus*, *Dimerocostus*, *Helenia*, *Monocostus*, *Paracostus*, and *Tapeinochilos*

COMMENTS Until recently Costaceae was included in the Zingiberaceae and is currently placed in Zingiberales. Species of all listed 8 genera are reported from cultivation.

Costus comosus

SPIRAL GINGER; RED FLOWER GINGER
SYNONYMS *Alpinia comosa*; *Costus maritimus*.

ETYMOLOGY [Classical Latin name derived from the Arabic name *qust* or *kust*, an imported aromatic root] [Latin "furnished with a tuft or conspicuous bracts," as inflorescences in this case]

GROWTH HABIT Perennial rhizomatous herb, 3–5 ft. (1–1.75 m); coarse texture; medium rapid growth. **STEM/BARK**: somewhat fleshy cane-like upright stems that are sometimes slightly arching; also rhizome. **LEAVES**: elliptic to oblanceolate, spirally arranged around the cane-like stems; glabrous above but somewhat pubescent below. **FLOWERS**: single bright yellow in axils of bright red spirally arranged stiff bracts; pine cone-shaped inflorescence terminal on stems, usually with only 2 or 3 flowers apparent at a time. **FRUIT**: loculicidally dehiscent ellipsoid capsule. **NATIVE HABITAT**: endemic to a small wet forest area of Costa Rica in the Central Valley of San Jose which is currently a shopping center and apartments.

CULTURE Tolerant of a wide range of lights and soils as long as it is maintained in moist to wet condition.

PROPAGATION: seed, division, or tissue culture. **LANDSCAPE USES**: specimen plants in ground or in large containers; inflorescences are used as cut flowers. **HARDINESS ZONE**: 9B–13B. In lower zones the aerial parts of the plant may die if frozen but will grow back in spring, although it may not bloom.

COMMENTS IUCN Red List (2014) considers this plant as critically endangered due to loss of habitat. This species has often been sold and cultivated as *Costus barbatus* in the United States and in Australia it is known as *C. comosus* var. *barbatus*. **Costus barbatus**, however, is listed as an acceptable species and is commonly known as spiral or red tower ginger. Stem extracts are locally used for medicinal purposes. A total of 104 accepted species is listed, most of which are probably in cultivation in either botanical gardens or private collections. The genus **Costus** has recently been divided into four genera: **Costus**, **Paracostus**, **Hellenia** (= *Cheilocostus*), and **Chamaaecostu**s.

Costus comosus (above), *C. barbatus* (below)

Costus stenophyllus

BAMBOO COSTUS; BAMBOO GINGER; SNAKE GINGER
ETYMOLOGY [Classical Latin name derived from the Arabic name *qust* or *kust*, an imported aromatic root] [Latin *stenos*, narrow and *phyllus*, leaves, in reference to the narrow lanceolate leaves]

GROWTH HABIT Evergreen rhizomatous perennial herb to 12 ft. (4 m) tall, although in cultivation it is usually about one-half as tall; medium-fine texture; rapid growth rate. **STEM/BARK**: stems resemble that of bamboos or rattle snake, hence the common names, upright unbranched, with alternating brown and gray to white banding. **LEAVES**: linear-lanceolate, grass-like, about 1 in (2.5 cm) wide, spirally arranged and sheathing around the stems. **FLOWERS**: yellow, subtended by a brilliant red bract, on short leafless basal stems; inflorescences 4–12 in. (9–14 cm) long and 1.5–2 in (3.5–4.5 cm) in diameter. **FRUIT**: A dehiscent capsule. **NATIVE HABITAT**: endemic to lower elevation tropical forests of Osa Peninsula and adjacent areas of Costa Rica.

CULTURE Prefers direct sun or bright light, in well-drained but moist fertile organic soil. **PROPAGATION**: division of rhizomes, tissue culture. **LANDSCAPE USES**: most attractive around water ponds, as accent plant, or in containers; stems are used in floral arrangements. **HARDINESS ZONE**: 10B–13B.

COMMENTS According to IUCN Red List (2014) *Costus stenophyllus* is found in protected areas and is not threatened or endangered.

Costus stenophyllus (photographed in Indonesia)

Costus spectabilis (spiral ginger), is an unusual acaulescent species that goes completely dormant during dry season but new leaves and a single large brilliant yellow flower appears after rain. It is uncommon in cultivation. Photographed at Huntington Botanical Garden, California.

Costus woodsonii

RED BUTTON GINGER; INDIAN HEAD GINGER

ETYMOLOGY [Classical Latin name derived from the Arabic name *qust* or *kust*, an imported aromatic root] [Robert E. Woodson Jr. (1904–1963), curator of the herbarium at the Missouri Botanical Garden]

GROWTH HABIT Evergreen rhizomatous perennial herb, usually in dense clumps, 4 $\frac{1}{2}$ –6 ft. (1.5–2 m) high but usually in cultivation to about 3 ft. (1 m), somewhat variable in spread. **STEM/BARK**: erect or sometimes slightly leaning green stems arise from the rhizomes. **LEAVES**: broadly elliptic to ovate, about 6 in. (15 cm) long and 2 $\frac{1}{2}$ in. (8 cm) wide, dark green, glabrous, drooping, entire margins, spirally arranged around the stems. **FLOWERS**: tubular reddish orange and yellow flowers in dark red tightly overlapping waxy bracts that form an ovoid cone-shaped inflorescence resembling a male cycad cone, at terminus of the stems; flowers appear usually one at a time. **FRUIT**: oblong many-seeded dehiscent capsule. **NATIVE HABITAT**: lowlands and coastal areas of Central America (Nicaragua, Costa Rica, and Panama), and Colombia in South America; naturalized in Hawaii.

CULTURE Full sun to partial and full bright shade, in moist but well-drained fertile soils, has some salt tolerance; essentially requires minimal care. **PROPAGATION**: young plantlets that appear at the base of inflorescences, stem cuttings, or division of rhizomes. **LANDSCAPE USES**: specimen or accent plants in the landscape or in containers; inflorescences are used in floral arrangements. **HARDINESS ZONE**: 10B–13B. There are suggestions that it tolerates zones 8 or 9 but it will not bloom if temperature is below 50 °F (10 °C).

COMMENTS *Costus woodsonii* is used for treatment of several ailments, for flavoring, and as a fixative in cosmetics. The orange-yellow labellum is edible. There are a few dwarf cultivars, such as 'Dwarf Lipstick', 'French Kiss', and 'Red Button'.

Costus woodsonii growth habit which probably became a large clump by inflorescence plantlets as well as by branching of the rhizome (above left), close-up of the leaves and inflorescences (above right), and close-up of individual flowers subtended by a bract (below). Photographed at Fairchild Tropical Botanical Garden, Coral Gables, Florida.

Tapeinochilos ananassae

INDONESIAN WAX GINGER, RED WAX GINGER, PINEAPPLE GINGER

SYNONYMS *Costus ananassae*, *C. pungens*, *Tapeinochilos australis*, *T. pungens*, *T. queenslandiae*, *T. teysmannianus*

ETYMOLOGY [Greek *tapeinos* "low" and *cheilos* "lip," in allusion to the short flower labellum] [Latinized form of *nana*, the South American Tupi Indian name for pineapple, in reference to the inflorescence]

GROWTH HABIT Evergreen rhizomatous perennial herbs to 9 ft. (3 m) tall. **STEM/BARK**: bamboo-like with distinctly marked nodes, brown upright stiff stems, sometimes twisted at the apex. **LEAVES**: Spirally arranged, elliptic to oblanceolate, glabrous or sometimes slightly pubescent, sheathing around the stem, to about 15 in. (38 cm) long and 2 in. (5 cm) wide. **FLOWERS**: orange-yellow in deep red waxy bracts of 4–12 in. (10–30 cm), pineapple-shaped inflorescences on leafless stalks that arise directly from the rhizomes. **FRUIT**: dehiscent trilocular capsule, with fleshy white mass containing the arillate seeds. **NATIVE HABITAT**: an understory plant of lower elevation rainforests in Queensland Australia, New Guinea, and Indonesia.

CULTURE Prefers shady to partial sun in moist, rich, organic soils; no specific problems other than drought injury have been reported. **PROPAGATION**: division of clumps or rhizomes, stem cuttings, and tissue culture. **LANDSCAPE USES**: as specimen or background plants in the landscape or container plants on patios. Inflorescences are used in floral arrangements. **HARDINESS ZONE**: 9B–13B.

COMMENTS This species lacks the essential oils that are used for flavoring or medicinal purposes. The genus consists of 16 species with nearly all endemic to Papua New Guinea. *Tapeinochilos* species in cultivation include *T. dahli* 'Backscratcher' (dark purple-black bracts in cone-shaped inflorescence), *T. brassii* (bright red bracts in compact cone-shaped inflorescence), *T. pubescens* (orange bracts on tall narrow inflorescence), and probably others.

Tapeinochilos ananassae (photographed in vicinity of Darwin, Australia)

ADDITIONAL GENUS

Hellenia (= Cheilocostus) **sp.** (?)

Costus dubius (spiral flag, thebu), a densely grown, sometimes invasive plant, native to Sri Lanka and Ghana and Tanzania in Africa

ZINGIBERACEAE

GINGER FAMILY

50 GENERA AND 1548 SPECIES

GEOGRAPHY Pantropical, especially in Indomalaysia (Indonesia, Singapore, Brunei, the Philippines, and Papua New Guinea).

GROWTH HABIT Hermaphroditic, rhizomatous, aromatic perennial herbs, often with tuberous roots; cauline stems, when present, usually short and leafless or sometimes foliated; pseudostems sometimes created by sheathing petiole bases. **LEAVES**: often large, distichous, petiolate, sheathing and with a distinct ligule at base, parallel veins arising from a distinct midrib. **FLOWERS**: bisexual, zygomorphic, subtended by a bract, with a conspicuous 2- or 3-lobed labellum; stamens 1–3 but only 1 fertile; perianth segments fused into a tubular calyx and 3 more or less fused petal-like showy corolla segments; ovary inferior and consists of 3 fused carpels. **FRUIT**: brightly colored, often fleshy capsule; seeds frequently arillate.

ECONOMIC USES Ornamentals, condiments (e.g., ginger from *Zingiber officinale*, turmeric from *Curcuma angustifolia*, etc.), herbs (e.g., cardamom from *Elettaria cardamomum*), dyes (from several genera, including *Curcuma* and *Hedychium*), perfumes (e.g., abeer (= abir, gutal, colored powder used for Hindu rituals), from the rhizome of *Hedychium spicatum*), and medicinal (e.g., arrowroot from *Curcuma* spp.).

CULTIVATED GENERA *Alpinia*, *Curcuma*, *Elettaria*, *Etlingera*, *Globba*, *Hedychium*, *Kaempferia*, *Renealmia*, *Siphonochilus*, and *Zingiber*, among many others.

COMMENTS There is basic reversal in the APG IV (2016), where distinct families are once again recognized for Cannaceae, Heliconiaceae, Lowiaceae, Marantaceae, Musaceae, Streliziaceae, and Zingiberaceae. In this book species of these families are treated individually. Monogeneric Lowiaceae is not discussed.

Alpinia calcarata

SNAP GINGER

SYNONYMS *Alpinia alata*, *A. bracteata*, *A. cernua*, *A. erecta*, *A. roscoeana*, *A. simsii*, *A. spicata*, *Catimbium erectum*, *Globba erecta*, *Languas calcarata*, *Renealmia calcarata*, *R. erecta*, *R. minor*.

ETYMOLOGY [Prospero Alpino (1553–1616), Italian botanist, professor of botany at Padua, Italy, who wrote on plants of Egypt] [Spurred, with a spur]

GROWTH HABIT Aromatic rhizomatous clump-forming perennial herb without aerial stems, to 5 ft. ($1 \frac{2}{3}$ m) tall; medium-coarse texture; fast growth rate. **STEM/BARK**: thick subterranean rhizomes; slender pseudostems are produced by the overlapping of clasping leaf bases. **LEAVES**: aromatic when crushed, alternate, lanceolate, 10–12 in. (25–30 cm) long, acuminate, glabrous on both surfaces but lighter on the underside. **FLOWERS**: bisexual, fragrant, pinkish white and maroon, zygomorphic; on pedunculated unbranched scorpoid cymose inflorescence, usually with a pair of flowers in bract axils, the upper flower usually developing first, on short pedicels; calyx is 3-lobed, fused into a campanulate tube, staminodes seemingly absent; fertile stamens one; ovary inferior, densely pubescent. **FRUIT**: red trilocular capsule. **NATIVE HABITAT**: India, southern Malay Peninsula, Sri Lanka.

CULTURE Full sun to partial shade, well-drained but moist organic soil; no serious pests or diseases reported. **PROPAGATION**: usually by division of the rhizome but also tissue culture. **LANDSCAPE USES**: can be used as low screen or as accent plants. **HARDINESS ZONE**: 9B–13B.

COMMENTS The rhizome is sometimes used as an ingredient of curries and for medicinal purposes in treatment of rheumatoid arteritis and for a number of other ailments including inflammation and diabetic treatments, as well as snake bites. It is listed as one of the Indian Ayurvedic medicinal plants.

Alpinia calcarata

Alpinia purpurata

RED GINGER; OSTRICH PLUME
SYNONYMS *Alpinia grandis*, *Guillania novo-ebudica*, *G. purpurata*, *Languas purpurata*

ETYMOLOGY [After Prospero Alpino (1553–1616), Italian botanist, professor of botany at Padua, who wrote on plants of Egypt] [Purple]

GROWTH HABIT Tall, upright evergreen rhizomatous herb, 3–15 ft. (1–5 m) tall and 2–4 ft. ($\frac{3}{4}$ –1 $\frac{1}{3}$ m) wide; coarse texture; fast growth rate. **STEM/BARK**: thick aromatic fleshy rhizomes give rise to leafy cane-like pseud-ostems that terminate in an inflorescence. **LEAVES**: alternate, oblong, sessile, with sheathing bases, 12–30 in. (30–75 cm) long and 4–9 in. (10–22 cm) wide; base sheathing, apex acuminate. **FLOWERS**: small inconspicuous tubular white flowers on distinct spike inflorescence with long brightly colored red or pink ovate bracts, to about 12 in. (30 cm) tall. **FRUIT**: infrequently produced globose capsules 4–6 in. (10–15 cm) long containing black arillate seeds. **NATIVE HABITAT**: New Caledonia, New Hebrides, Solomon Islands, etc., and naturalized in Hawaii.

CULTURE Prefers partial shade but it is commercially grown in Central and South America in direct sunlight, in moist but well-drained organic soil; chlorosis appears in high pH and dry soils. **PROPAGATION**: plantlets that may form on bract axils can be rooted; division of rhizome. **LANDSCAPE USES**: grown as an accent plant in tropical landscapes, may be used as a hedge or screen, and background foundation planting, but also for cut flowers and as house plant. **HARDINESS ZONE**: 10A–13B.

COMMENTS *Alpinia purpurata* is the national flower of Samoa. It is widely cultivated in Central and South America for cut flowers. Cultivars '**Jungle King**' (dark red), and '**Jungle Queen**' (light pink) are best known and most common, a number of other cultivars, particularly dwarf ones, however are also available, such as '**Dwarf Pink**', '**Hot Pink**', '**Red Dwarf**', as well as other cultivars of varying inflorescence color or growth habits, such as '**Anne Hironaka**', '**Eileen McDonald**', '**Fireball**', '**Kazu**', '**Raspberry**', '**Rosy Dawn**', '**Tahitian**' (double), and '**Tomi Pink**', among others.

Alpinia purpurata (photographed in Bolivia)

Alpinia zerumbet

SHELL GINGER; SHELL FLOWER; VARIAGATED GINGER
SYNONYMS *Alpinia cristata*, *A. fimbriata*, *A. fluviatilis*, *A. schumanniana*, *A. speciosa*, *A. nutans*, *A. speciosa*, *Amomum nutans*, *Catimbium speciosum*, *Costus zerumbet*, *Languas schumanniana*, *L. speciosa*, *Renealmia nutans*, *R. spectabilis*, *Zerumbet speciosum*.

ETYMOLOGY [Prospero Alpino (1553–1616), Italian botanist, professor of botany at Padua, who wrote on plants of Egypt] [Aboriginal vernacular name].

GROWTH HABIT Evergreen; herbaceous clumping perennial herb; densely foliated, clump-forming; to 12 ft. (4 m) tall; spreading; coarse-textured; rapid growth rate. **STEMS/BARK**: rhizome; aerial stems arch gracefully, to about 10 ft. (3 m) tall. **LEAVES**: simple; alternate; 2-ranked; elliptic-oblong to 2 ft. (0.75 m) long and 5 in. (12 cm) wide; deep green; pinnately veined; smooth with hairy margins. **FLOWERS**: bisexual; funnelform, white tipped with red; yellow lip variegated with red and brown, broad, curved, crinkled, to 2 in. (5 cm) long; irregular, bell-shaped; borne in the axils of obtuse bracts; stamens 3 but only one with pollen; inflorescence terminal, pendulous, hairy; fragrant; blooms in summer and fall. **FRUIT**: capsule; red; globose; to $\frac{3}{4}$ in. (2 cm) in diameter. **NATIVE HABITAT**: Dense thickets in forests and wetlands of tropical and warm temperate Asia, particularly on stream banks and shady places.

CULTURE Full sun or partial shade on moist fertile soil; slight salt tolerance but not recommended for dune plantings; mites may become problematic. **PROPAGATION**: seed, division of clumps (rhizome), tissue culture; seeds several per fruit. **LANDSCAPE USES**: use as a specimen or accent plant for the unusual flowers and lush effect of the foliage; also used as indoor foliage plant. **HARDINESS ZONE**: 8B–13B. Plant may freeze back to the ground, but grows back from the rhizome in spring.

COMMENTS Sheathing leaf bases are a source of fiber for rope in southeastern Asia. Leaves are used in cooking, and leaves and rhizomes are used for medicinal purposes. Other *Alpinia* species in cultivation include *A. aquatica*, *A. formosana*, *A. galanga*, *A. japonica*, *A. mutica*, *A. purpurata* and its cultivars, and *A. vittata*, among others. *Alpinia zerumbet* 'Variegata' is common in the trade. The word "shell" in its vernacular name is based on the similarity of its flower buds to seashells.

Alpinia zerumbet (above) and *A. zerumbet* 'Variegata' (below)

Etlingera elatior

RED TORCH GINGER; PHILIPPINE WAX FLOWER
SYNONYMS *Alpinia elatior, A. magnifica, A. speciosa,*
Amomum magnificum, Cardamomum magnificum,
Dracodes javonica, Nicolaia elatior; N. speciosa;
Phaeomeria speciosa

ETYMOLOGY [Andreas Ernst Etlinger, (1730–1790),
German botanist who described several species of *Salvia*]
[Winged]

GROWTH HABIT Herbaceous clump-forming perennial,
9–18 ft. (3–6 m) high; coarse texture; fast growth rate.
STEM/BARK: stout rhizome; pseudostems that are created
by petiole sheaths closely grouped. **LEAVES**: lanceolate,
33 in. (85 cm) long and 7 in. (18 cm) wide, glabrous; petiole

$1\frac{1}{8}$ –1 $\frac{1}{3}$ in. (3–4 cm) long; sheathing at base. **FLOWERS**:
peduncles raised above ground, 2–3 ft. ($\frac{2}{3}$ –1 m) tall,
robust; involucral bracts bright red, corolla usually pink to
red; labellum deep red with yellow or white margins, inner
perianth segments pink; floral bracts similar to the involucral
bracts but smaller and pinkish; calyx $1\frac{1}{8}$ –1 $\frac{1}{3}$ in.
(3–4 cm), with the apex 3-toothed; anthers red. **FRUIT**: globose capsule, pubescent green then red, about 1 in. (2 $\frac{1}{2}$ cm)
in diameter; seeds small, many, black. **NATIVE HABITAT**:
widely distributed in Malaysia, Indonesia, and southern
Thailand and introduced into the Philippines.

CULTURE Partial to full shade, in organic moist soil but
tolerates most well-drained wet soils. **PROPAGATION**:
seed, division of rhizome, and tissue culture. **LANDSCAPE
USES**: commercially grown for its showy flowers that are

used in floral arrangements; can be used as accent or specimen plants. **HARDINESS ZONE**: 9B–13B.

COMMENTS Flower buds and inflorescence peduncles are used in cooking and for medicinal purposes, particularly as an antioxidant. There are about 100 species of **Etlingera** native to the Old-World tropics. A few examples reported in cultivation include **E. coccinea**, **E. corneri**, **E. hemisphaerica**, **E. littoralis**, and **E. venusta**, and probably others.

Etlingera elatior **'Red Torch'** in early development stage (above left), and fully opened (above right), *E. elatior* **'Porcelain Pink'** early development (below left) and fully open (below right). The white cultivar **'Thai Queen'** is not shown.

Globba globulifera

PURPLE GLOBE GINGER, DANCING LADY GINGER

ETYMOLOGY [From the Indonesian common name *galoba* for zingibers] [Latin *globulus*, small sphere, spherical, in reference to its bulbils]

GROWTH HABIT Semievergreen rhizomatous clumping herbaceous perennial, to 2 ft. (60 cm) tall and wide; coarse texture; moderate growth rate. **STEM/BARK**: rhizomes creeping, slender; pseudostems erect, formed by sheeted foliage bases, to 4 $\frac{1}{2}$ ft. (1 $\frac{1}{2}$ m). **LEAVES**: alternate (on rhizomes), sessile or shortly petiolate, blades oblong-elliptic or elliptic-lanceolate, to about 10 in. (24 cm). **FLOWERS**: bisexual; yellow, pendent, in terminal thyrse or raceme, each cincinnus (scorpioid cyme) of flowers (or bulbils) subtended by an ovate reddish-purple bract; calyx campanulate, 3-lobed, tubular; corolla tube slender; lateral staminodes petaloid, stamen filaments long, curved, anther with or without appendages on each side; ovary unilocular; blooming summer-fall. **FRUIT**: globose dehiscent capsule; seeds small, whitish arillate. **NATIVE**

HABITAT: forest understory in Southeast Asia, Thailand and Vietnam.

CULTURE Filtered shade or shade beneath trees, in well-drained organic moist fertile soils; moderately salt spray-tolerant, does not tolerate prolong drought; mealybugs, spider mites, and slugs may become problems. **PROPAGATION**: bulbils, or division of rhizomes. **LANDSCAPE USE**: excellent long-lasting inflorescence for use in flower arrangements. **HARDINESS ZONE**: 9B–12B. May be grown in colder zones if rhizomes are dug and stored until spring.

COMMENTS *Globba globulifera* may potentially become weedy by spreading bulbils. Nearly 100 species of *Globba* have been reported. Several species have been reported from cultivation: *G. campsophylla*, *G. clarkei*, *G. marantina* (Vietnamese globba), *G. paniculata*, *G. racemosa*, *G. schomburgkii* (dancing girl ginger), *G. sessiflora*, *G. winitii* (including the dark purple '**Thai Beauty**' (white), '**Purest Angel**', and '**White Dragon**' (bracts white). *Globba* × '**Ruby Queen**' is used in flower arrangements.

Globba globulifera

Globba schomburgkii (dancing girl ginger), native to India, Thailand, and China

Globba winitii **'Alba'** (dancing ladies, dancing jewels), native to Southeast Asia, Thailand. Bracts vary in color, from white to pinkish-red. Photographed at Filoli Gardens in California.

Hedychium **Species and Cultivars**

SYNONYM *Brachychilum*.

ETYMOLOGY [From Greek *hedys*, sweet and *chion*, snow, in reference to the color and fragrance of the *H. coronarium* flower]

GROWTH HABIT Rhizomatous perennial herbs, 4–6 ft. (1 $\frac{1}{3}$ –2 m) tall; mostly terrestrial but a few epiphytic. **STEM/BARK**: some have leafy shoots but pseudostems are often formed by sheathing of leaf bases. **LEAVES**: distichous, more or less sessile, and mostly lanceolate to elliptic. **FLOWERS**: mostly fragrant, in a range of colors (except blue), on showy terminal spike inflorescences; calyx tubular; corolla tube slender and with unequal lobes, petals long and narrow, reflexed, labellum bilobed; stamens petaloid, usually only one with pollen and excreted. **FRUIT**: spherical or oblong capsule, with usually red or orange interior arillate seeds. **NATIVE HABITAT**: originally native to wetter tropical forests of Southeast Asia, southern China, the Himalayas, and Madagascar, but some species naturalized in other tropical regions.

CULTURE Most taxa prefer partial to full shade but some can tolerate direct sun for short periods, in well-drained but moist organic soils; some species may require control every few years so that they do not spread too widely. **PROPAGATION**: readily propagated by division but a few species have been tissue cultured; seeds when produced or available are not often used, except in cases of hybridization. **LANDSCAPE USES**: specimen or container plants, sometimes used indoors. **HARDINESS ZONE**: 8B–13B. Species dependent, in cooler climates aerial portions may die but rhizomes give rise to aerial parts in spring.

COMMENTS There are about 50 species of *Hedychium* and many attractive cultivars. Some examples include: *H. coronarium*, *H. chrysoleucum*, *H. coccineum* ('**Disney**'), *H. densiflorum*, *H. ellipticum*, *H. flavescens*, *H. flavum*, *H. forrestii*, *H. gardnerianum* ('**Fiesta**', '**Tara**'), *H. greenii* , *H. horsfieldii*, *H. longicornutum*, *H. spicatum* ('**White Wings**'), *H. thyrsiforme*, *H. wardii*, and probably other species as well as cultivars, many of which are of hybrid origin or undocumented parentage. Rhizomes of some species are used for medicinal purposes. Some species are considered invasive.

Hedychium gardnerianum (above left), *H. gardnerianum* 'Tara' (above right), *Hedychium* 'Pink V' (below left), and *Hedychium* 'Elizabeth' (below right).

Hedychium coronarium

SYNONYMS *Amomum filiforme*, *Gandasulium coronarium*, *G. lingulatum*, *Hedychium chrysoleucum*, *H. gandasulium*, *H. lingulatum*, *H. maximum*, *H. prophetae*, *H. spicatum*, *H. sulphurium*, *Kaempferia hedychium*

ETYMOLOGY [Greek *hedys*, sweet and *chion*, snow, in reference to the color and fragrance of the flower] [Coronary, used in garlands].

GROWTH HABIT Clumping herbaceous perennial, 4–6 ft. (1 $\frac{1}{3}$ –3 m) tall; course texture; fast growth rate. **STEM/BARK**: stiff, erect, mostly unbranched stems to 6 ft. (2 m) tall, which droop with age after flowering. **LEAVES**: alternate, simple, entire margins, clasping the stem, to 24 in. (60 cm) long and 5 in. (12 $\frac{1}{2}$ cm) wide. **FLOWERS**: several white, zygomorphic, fragrant terminal flowers subtended by large stiff bracts, on a spike; corolla tube about 3 in. (8 cm) long; staminodial lip large, sometimes with a yellow tinge. **FRUIT**: a 3-locular capsule. **NATIVE HABITAT**: in shaded wet (but not submerged) forests of tropical Asia (the Himalayas, Nepal, and India), and naturalized in tropical America.

CULTURE Partial shade, in moist fertile soil; requires removal of older stems and may become too large if not periodically divided. **PROPAGATION**: division of the rhizome which should be done once every 3–4 years to keep the plant under control. **LANDSCAPE USES**: outdoors or in greenhouse as specimen or accent plant for its tropical effect and large, showy, fragrant flowers; no serious problem reported. **HARDINESS ZONE**: 8B–13B. (8A with some protection); aerial portion may die but will grow from rhizomes in spring.

COMMENTS *Hedychium coronarium* is considered a global invasive in some tropical regions such as Bolivia, Brazil, Hawaii and elsewhere, but it is the national flower of Cuba, where it has also become naturalized. Rhizomes are edible and used for medicinal purposes.

Hedychium coronarium growth habit (above left), fruit with seeds (above right), and flowers (below).
Top left photographed was taken in Bolivia

Hedychium longicornutum

HORNBILL GINGER
SYNONYMS *Hedychium crassifolium*, *H. longicornutum* var. *minor*

ETYMOLOGY [Greek *hedys*, sweet and *chion*, snow, in reference to the color and fragrance of the flower] [long garland or crown]

GROWTH HABIT Evergreen clump-forming epiphytic (on trees) or lithophytic (on rocks) perennial, 2–3 ft. ($\frac{2}{3}$ –1 m) high; coarse texture; slow growth rate. **STEM/BARK:** thick fleshy rhizome but also short, red same as stems. **LEAVES:** aromatic (when crushed) glossy deep green, lanceolate, sessile, sheathing at base. **FLOWERS:** striking multicolor bloom with long **fertile stamen**, white filament, and yellow anther more than twice the length of the lobed down curved red labellum; rachis not visible once the scarlet terminal inflorescence opens. The flower has been likened to Medusa's head. Unlike other hedychiums, it is a short-day plant that blooms early to late winter; not fragrant. **FRUIT:** brown, pubescent capsule with about 15 red seeds. **NATIVE HABITAT**: Malaysia in shady forests

CULTURE Often grown in organic orchid media in containers or hanging baskets in a bright but shady location and kept moist. May also be grown on rocks with addition of compost and bark mixture. **PROPAGATION:** seed, division, and perhaps a good candidate for tissue culture. **LANDSCAPE USES:** grown as an epiphyte in greenhouses and conservatories. **HARDINESS ZONE**: 10B–13B. May require protection in zone 10 or moved indoors if cold.

COMMENTS The fleshy roots are used for medicinal purposes.

Hedychium longicornutum

Hedichium flavum (yellow butterfly ginger, Nardo ginger lily), photographed in Mavis Bank, Jamaica. Although Flora of Jamaica (Adams and Proctor, 1972), lists the plant as *H. flavum*, specimens in the trade usually have creamy yellow flowers. The species is actually native to Tibet and China.

Renealmia alpinia

SWEET LEAF; HONEY LEAF

SYNONYMS *Alpinia bicalyculata, A. exaltata, A. macrantha, A. tubulata, Amomum alpinia, A. renealmia, A. repens, Ethanium bracteosum, E. exaltatum, E. macranthum, E. pacoseroca, Renealmia bracteosa, R. coelobractea, R. exaltata, R. foliosa, R. goyazensis, R. lativagina, R. macrantha, R. pacoseroca, R. raja, R. rubroflava*

ETYMOLOGY [Paul de Renaeulme (Latinized Renealmi) (1560–1624), French physician and author of *Specimen Historiae Plantarum* (1611)] [Prospero Alpino (1553–1616), Italian botanist, professor of botany at Padua, who wrote on plants of Egypt]

GROWTH HABIT Aromatic rhizomatous herb, 6–18 ft. (2–6 m) tall; coarse texture; rapid growth rate. **STEM/BARK**: thick rhizome 4–12 in. (10–30 cm); pseudostems from sheathing leaf bases. **LEAVES**: elliptic, 12–43 in. (30–110 cm) long and 2–6 in. (5–18 cm) wide; base sheathing; with ginger fragrance when crushed. **FLOWERS**: bisexual, inflorescence racemose, basal, to 22 in. (55 cm); tubular, zygomorphic, orange with yellow lips. **FRUIT**: ellipsoid capsules, $\frac{1}{2}$–1$\frac{1}{2}$ in. (1$\frac{1}{2}$–3$\frac{1}{2}$ cm); crowned by red-purple-black remains of the calyx. **NATIVE HABITAT**: wet thickets or dense forests or on-stream banks, from Mexico through Central and South America and the Caribbean, where it often forms large colonies.

CULTURE Prefers shady locations, in well-drained but moist organic soil; no problems reported. **PROPAGATION**: seed or division of rhizome, also tissue culture. **LANDSCAPE USES**: similar to *Alpinia* and *Hedychium*, but the inflorescences arise at or near ground level, similar to *Zingiber*, hence plants should be within visible view. **HARDINESS ZONE**: 10B–13B. Not as cold hardy as *Alpinia*.

COMMENTS Fruit of *Renealmia alpinia* is edible and highly valued in some localities. Fish are wrapped and roasted in leaves to add flavor. Rhizomes are used for medicinal purposes, including snake bites. Dye from ripe fruit is used for weaving fibers, tattoos, and as ink. Despite 86 known species, no cultivars are reported.

Renealmia alpinia (photographed in Amazonian Forest in Bolivia)

Zingiber zerumbet

PINE CONE LILY, WILD GINGER
SYNONYMS *Amomum silvestre*, *A. zerumbet*, *Zerumbet zingiber*

ETYMOLOGY [Classical Greek *zingiberi*, in turn said to derive from Malay *inchiver* or Indian *singivera*. The word *zingiber* is reportedly derived from a Sanskrit word that refers to bull's horn, in reference to the rhizome] [Aboriginal vernacular name].

GROWTH HABIT Deciduous herbaceous perennial herb; rhizomatous; heavily foliated; inclined stems form dense clumps; to 6 ft. (2 m) tall, spread is variable; coarse texture; fast growth rate. **STEM/BARK**: leafy stems arise from a tuberous, aromatic rhizome. **LEAVES**: simple; alternate; lanceolate, 2-ranked; long and narrow, to 12 in. (30 cm) long and 3 in. (8 cm) wide; thin; hairy beneath; sheathing bases. **FLOWERS**: bisexual; white or yellowish with a yellow lip; tubular, 3-lobed; inconspicuous; arise from overlapping bracts to $1\frac{1}{4}$ in. (3 cm) long; bracts showy, green when young, becoming red with age, 2–3 in. (5–$7\frac{1}{2}$ cm) long; dense, pine cone-like inflorescences borne on a peduncle to 12 in. (30 cm) long, separate from the leaves. The bracts develop maximum color in fall and winter. **FRUIT**: capsule; 3-valved; seeds are arillate. **NATIVE HABITAT**: India and the Malay Peninsula introduced to Hawaiian Islands by migrating Polynesians and has naturalized.

CULTURE Full sun or partial shade, in moist, fertile soil; moderate salt tolerance; mites may be troublesome. **PROPAGATION**: division of clumps (rhizome). **LANDSCAPE USES**: specimen for its unusual inflorescence, may be used as a ground cover; inflorescence is popular in floral arrangements. **HARDINESS ZONE**: 8B–12B. Plants die back to ground in response to the short daylengths of autumn, but grow from the shallow rhizome in spring.

COMMENTS *Zingiber officinale*, the ginger of commerce, is much like **Z. zerumbet**. About 140 species of ***Zingiber*** are known of which only a few reported from cultivation include **Z. citriodorum**, **Z. malaysianum**, **Z. mioga**, **Z. niveum**, **Z. spectabile**, and **Z. vinosum**, and probably others in private collections. In Japan, inflorescences of *Z. mioga* are dipped in tempura batter and deep-fried. The cultivar '**Darceyi**' (has white variegated leaves); '**Awapuhi**' (said to be one of the "Diety Kanes" in Hawaiian mythology; it has longer and narrower inflorescences). Rhizome extracts of *Z. zerumbet* is essentially used worldwide for various ailments. Its active pharmacological component is zerumbone.

Zingiber zerumbet

Zingiber macradenium

Zingiber spectabile 'Beehive'

NOTE: Associated green sword-shaped leaves are that of *Dracaena angolensis* (= *Sansevieria cylindrica*), not of *Zingiber*. Combination of the two plants is merely intended as design elements.

ADDITIONAL GENERA

Curcuma aeruginosa (turmeric, curcumin), perennial with unbranched leafy stems from a large subterranean rhizome that has been the source of turmeric, used for medicinal and flavoring for centuries; native to Southeast Asia, China, India, Guinea and Australia, naturalized elsewhere

Curcuma petiolata (jewel of Thailand, queen lily), native to Thailand and Malaysia, variable rhizomatous species with bracts ranging from white to purple. It is related to turmeric plant.

HAEMODORACEAE

BLOODWORT FAMILY

14 GENERA AND ±100 SPECIES

GEOGRAPHY Primarily Southern Hemisphere, Australia, South Africa, and New Guinea

GROWTH HABIT Hermaphroditic rhizomatous, tuberous, or bulbous, perennial herbs, with or without laticifers or latex. **LEAVES**: basally aggregate, linear or ensiform (sword-shaped), alternate, distichous, coriaceous, sessile, succulent; margins entire; base amplexicaul (partially closing the stem). **FLOWERS**: much branched terminal cymes, panicles or racemes; usually wooly-hairy; actinomorphic or zygomorphic; 3-merous, with 6 free or connate tepals; stamens fertile, 3 or 6, adnate to the perianth; carpels 3, inferior or superior, gynoecium 3-locular. **FRUIT**: dehiscent loculicidal capsule or indehiscent nut; seeds winged or wingless.

CULTIVATED GENERA *Anigozanthos*, *Conostylis*, *Haemodorum*, *Macropidia*, *Xiphidium*

COMMENTS Although very few species of other genera are reported from cultivation, *Anigozanthos* appears to be the most commonly noted, much hybridized, and commercially available.

Anigozanthos spp.

KANGAROO PAW; CATSPAW; AUSTRALIAN SWORD LILY

ETYMOLOGY [Greek *anigo* "open" and *anthos* "flower," in reference to open branching of the flower stems]

GROWTH HABIT Evergreen or deciduous grass-like rhizomatous perennials with or without colored latex. **STEM/BARK**: rhizomes short, horizontal, with red pigment in some species. **LEAVES**: simple, linear, alternate, distichous, sessile, sheathing, basal rosettes of linear green or grayish leaves; pubescent in some species. **FLOWERS**: bisexual, tubular green and yellow to dark red and sometimes bicolor zygomorphic 3-merous flowers with 6 equal size and all fertile stamens; ovary 3-locular, inferior or half inferior; flowers covered with variously colored pubescence; situated on much branched long leafless pubescent raceme or spike inflorescence stalks, to 6 ft. (2 m) tall, that arise from the center of the rosettes. **FRUIT**: dehiscent loculicidal capsule; seeds winged or wingless. **NATIVE HABITAT**: endemic primarily to dry sandy areas of southwest and western Australia.

CULTURE Full sun in well-drained soil, with preference for rich organic media; no serious problems reported. **PROPAGATION**: seeds and division of clumps, and in some cases tissue culture. **LANDSCAPE USES**: in addition, choice garden subjects, inflorescences are important cut flowers in flower arranging. **HARDINESS ZONE**: 8B–12B. Best in seasonally dry Mediterranean climates.

COMMENTS There are 11 species in the genus, several subspecies, and many hybrids and cultivars, such as *Anigozanthos* 'Amber Velvet' (orange), 'Autumn Mystrey', 'Autumn Sunrise', 'Baby Roo', 'Big Red' (red), 'Bush Ballad' (orange), 'Bush Dawn' (yellow), 'Bush Blaze' (orange-red), 'Bush Devil' (red), 'Bush Games' (red), 'Bush Garnet' (burgundy red), 'Bush Gold' (golden yellow), 'Bush Lantern' (yellow fuzzy), 'Bush Pearl' (pink, dwarf), 'Bush Pearl'(pink fuzzy, hybrid), 'Bush Ranger' (red, dwarf), 'Bush Sunset' (dark red), 'Bush Tango' (orange), 'Cape Aurora' (mustard yellow), 'Cape Magenta' (magenta), 'Coral Pink' (pink), 'Gold Velvet' (gold), 'Harmony' (yellow with red stem), 'Orange Cross' (orange), 'Pink Joey' (pink), 'Red Cross' (dark burgundy), 'Rambueleg' (red with red stems), 'Rambodiam' (white) 'Tequila Sunrise' (golden orange), and 'Yellow Gem' (golden yellow), among others. For an extensive list of species, hybrids, and additional cultivars see Wikipedia. It should be noted that recognition of specific taxa from online photographs is nearly an impossible task.

"Anigozanthos rufus"

Anigozanthos flavidus (kangaroo paw)

Anigozanthos humilis (catspaw). All photographed at Huntington Botanical Gardens,
California

Anigozanthos 'Yellow Gem'

BROMELIACEAE

PINEAPPLE FAMILY

51 GENERA AND 3475 SPECIES

GEOGRAPHY All but one species from tropical and sub-tropical America, ranging from dry deserts of Southwestern US to equatorial tropical rain forests, one species, *Pitcairnia feliciana*, native to central Guinea in West Africa (said to be long distance dispersal).

GROWTH HABIT Hermaphroditic, mostly epiphytic, saxicolous (= clinging on rocks), or less often terrestrial; often monocarpic (= primary plants die after flowering), rosette plants, usually with reduced stems and adventitious roots which primarily function as fasteners rather than in absorption of nutrients. **LEAVES**: basal, linear to ovate-lanceolate, spiral forming rosettes (frequently form a "tank" at their base from which water and nutrients from decaying insects and debris are absorbed by specialized hairs), or rarely distichous (in some *Tillandsia*), often fleshy and/or coriaceous; base broadly sheathing; surfaces typically covered with scales; margins often dentate or spinose; **FLOWERS**: bisexual, in terminal simple or compound spikes, racemes or panicles, often with showy bracts; actinomorphic or infrequently slightly zygomorphic, 3-merous in two series, stamens 6, in two series; carpels 3, united, trilocular; ovary superior or half-inferior, stigmata 3. **FRUIT**: berry, multiple fruit (as in pineapple – see *Ananas*), or septicidal or rarely loculicidal capsule; seeds small and numerous.

ECONOMIC USES Majority of known species are used as Ornamentals, fruit (*Ananas comosus*, pineapple), fibers (*Tillandsia usneoides*, Spanish moss).

CULTIVATED GENERA *Aechmea*, *Alcantarea*, *Ananas*, *Billbergia*, *Brocchinia*, *Bromelia*, *×Cryptabergia* (*Bilbergia* × *Cryptanthus*), *Cryptanthus*, *×Cryptbergia* (*Bilbergia nutans* × Cryptanthus bahiana), *Dyckia*, *Greigia*, *Guzmania*, *×Guzvriesia* (*Guzmania* × *Vriesia*), *Hechtia*, *Hohenbergia*, *Neoregelia*, *Nidularium*, *Pitcairnia*, *Portea*, *Puya*, *Tillandsia*, *Vriesea*, and *Wittrockia*, among others.

COMMENTS Similar to the orchids, the majority of known bromeliads are in cultivation as ornamentals and most genera are listed here. In earlier classifications bromeliads were grouped in only three subfamilies but the most recent phylogenetic treatment (APG IV, 2017), eight subfamilies are recognized, which include: **Brocchinioideae, Lindmanioideae, Tillandsioideae, Hechtioideae, Navioideae, Pitcairnioideae, Puyoideae,** and **Bromelioideae**. Any discussion of these subfamilies does not seem to serve any useful purpose in this book. Interested individuals should refer to the Angiosperm Phylogeny Website for details: http://www.mobot.org/MOBOT/research/APweb/

Neoregelia spectabilis: A "tank" is formed at the base of spirally arranged rosette leaves (left) from which water and nutrients resulted from decaying insects and debris are absorbed by the specialized hairs (right). The "tank" is a mini-ecosystem where potentially many small animals and plants may thrive.

Genera uncommon or not described in detail include: *Acanthostachys*, with 2 species; *Araeococcus*, with 9 species, including *A. calcicola*, *A. strobilacea* (pinecone bromeliad), *A. pitcairnioides* and others; *Ayensua ualpanensis*, monotypic; *Brocchinia*, with 21 species; *Bromelia*, with 62 species, relatively common in cultivation, such as *B. antiacantha*, *B. balansae* (heart of flame), *B. hemisphaerica*, *B. humilis*, *B. palmeri*, *B. pinguin*, *B. serra*, and others, are large terrestrial plants with basal rosettes of sharp-spiny usually green leaves (except bright red when grown in direct sun or in bloom), and flowers in white pedunculate heads, native to tropical America, in addition to attractiveness of the plants these have edible fruit which are also used for medicinal purposes and the leaves yield fiber; *Catopsis*, with 21 species, relatively common in cultivation primarily by collectors: *C. bertoroniana*, *C. compacta*, *C. floribunda*, *C. nitida*, *C. nutans*, *C. sessiflora*, *C. subulata*, and others, are epiphytic rosette plants with soft, entire, glaucous, green or less often banded or speckled leaves, and small white or yellow flowers on terminal erect or pendent inflorescences, native from Florida and the West Indies to northern South America; the monotypic *Cottendorfia florida*; *Deuterocohnia* (incl. *Abromeitiella*), 18 species, including *D. lotteae*, *D. absrtrusa*, *D. brevifolia*, *D. longipetala*, *D. lorentziana*, and others; *Dyckia* is a large genus of 159 species of which several are in cultivation, such as: *D. brevifolia* (pineapple dyckia), *D. encholirioides*, *D. fosteriana*, *D. frigida*, *D. leptostachya*, *D. marnier-lapostollei*, *D. microcalyx*, *D. minarum*, *D. remotiflora*, and *D. ursina*, among others, are small to large terrestrials with stiff, spine-edged, green leaves, and yellow to orange flowers on large lateral inflorescences, native to South America and especially Brazil; *Encholirium*, 27 species, all endemic to Brazil, including *E. bradeanum*, *E. horridum*, and others, not common in cultivation; the monotypic *Fascicularia bicolor*;

Fernseea, with only two species: *F. bocainensis* and *F. itatiaiae*; *Fosterella*, 31 species; *Greigia*, 35 species; *Navia*, 91 species; *Neoglaziovia burle-marxii*, *N. concolor*, and *N. variegata* are the known species; *Nidularium*, 45 species; *Ochagavia*, 4 species, including the terrestrial *O. litoralis* (= *O. carnea*, *O. lindleyana*); *Orthophytum*, 53 species, including *O. gerkenii*, *O. mucugense*, *O. saxicola*, and others; *Pitcairnia*, a large genus of 395 species, including *P. albiflos*, *P. andreana*, *P. angustifolia*, *P. aphelandriflora*, *P. atrorubens*, *P. bromeliifolia*, *P. bulbosa*, *P. carinata*, *P. corallina*, *P. flammea*, *P. integrifolia*, *P. maidifolia*, *P. punicea*, *P. straminea*, *P. tabuliformis*, *P. tuerckheimii*, *P. wendlandii*, and *P. xanthocalyx*, among many others, are morphologically diverse terrestrial or saxicolous, grass-like plants of shady-moist locations, sometimes with a tendency to climb, with long linear and entire or short and spiny leaves, and variously colored (except blue) tubular flowers on elongated racemes, native from West Indies to South America; *Portea*, 9 species, *P. alatisepala*, *P. petropolotiana* (=*Aechmea petropolitiana*); *Puya*, large genus of 220 species, including *P. aequatorialis*, *P. alpestris*, *P. berteroniana*, *P. chilensis*, *P. coerulea*, *P. ferruginea*, *P. humilis*, *P. raimondii*, *P. roezlii*, *P. spathacea*, *P. thomasiana*, and *P. venusta*, among others; they are large, clumping or arborescent, mostly monocarpic plants, sometimes requiring many years to flower (more than 100 years in *P. raimondii*), with a rosette of stiff, spiny, gray or blue-green leaves and with attractive flowers of various color combinations on terminal spikes or racemes, native primarily to the Andes of South America; *Quesnelia*, 22 species, *Q. arvensis*, *Q. edmundoi*, *Q. humilis*, and others; *Ronnbergia*, 11 species native to South and Central America, *R. campanulata*. *R. carvalhoi*, *R. hathewayi*, and others; *Wittrockia* 6 species, *W. cyathiformis*, *W. gigantea*, *W. stenuisepala*, and few others.

A collection of bromeliads at Fairchild Tropical Botanical Garden in Coral Gables, Florida

Selby Gardens in Sarasota Florida specializes in epiphytic plants, including bromeliads.

Aechmea fasciata

SILVER VASE
SYNONYMS *Aechmea hamata*, *A. leopoldii*, *A. rhodocyanea*, *Bilbergia fasciata*, *B. glazioviana*, *Hohenbergia fasciata*, *Hoplophytum fasciatum*, *Platyaechmea fasciata*, *Tillandsia bracteata*

ETYMOLOGY [From Greek *aichme*, peak or point, alluding to the pointed sepals] [From Latin, banded, in reference to the leaves]

GROWTH HABIT Evergreen epiphytic; round herbaceous perennial with leaves in a basal rosette; to 2 ft. ($\frac{2}{3}$ m) tall and equally as wide; medium-coarse texture; slow growth rate. **STEM/BARK**: stemless. **LEAVES**: simple, in a tubular rosette, linear, inner leaves rounded and mucronate at the tip, to 24 in. (60 cm) long and 3 in. (7 $\frac{1}{2}$ cm) wide, green and may be banded with silver cross or covered with silvery scales, stiff, spiny margins. **FLOWERS**: pale blue, sessile; sepals mucronate, long-lasting, persistent (3–5 months) pink bracts are most noticeable; borne in a spike-like, pyramidal inflorescence; blooms in March and April. **FRUIT**: berry. **NATIVE HABITAT**: on trees in mid-elevation forests of southern Brazil.

CULTURE Partial shade, in interiors under lower light intensities; tolerates short periods of exposure to full sun; requires a well-drained, well-aerated medium; slight salt tolerance; scale; rots if kept too wet. **PROPAGATION**: seed, separation of offsets, and probably tissue culture. **LANDSCAPE USES**: can be fastened to the branches of rough-barked trees; used as an interior or exterior potted plant or as a specimen or in groups. **HARDINESS ZONE**: 10A–13B.

COMMENTS *Aechmea fasciata* is noted in various publications as the most popular of all bromeliads. Many varieties (e.g., **var**. *purpurea* with red leaves and **var**. *variegata* with longitudinal stripes), and cultivars with different foliage variegation patterns are available, a few of the more common include *Aechmea fasciata* 'Checkers', 'Donna Marie' 'Frost', 'Ghost', 'Margarita', 'Mona', 'Morgana', 'Silver King', 'Starbrite', and 'White Head', among others. There are also a number of intergeneric hybrids in the trade, particularly with genus *Nidularium* (×**Nidumea**). *Aechmea*, with 268 species in 8 subgenera, as well as numerous hybrids and cultivars, because of their ease of culture, are among the more popular bromeliads. A few of the common species include: *A. allenii*, *A. amazonica*, *A. aquilega*, *A. aripensis*, *A. blanchetiana*, *A. blumenavii*, *A. bracteata*, *A. bromeliifolia*, *A. candida*, *A. caudata*, *A. chantinii*, *A. coelestis*, *A. distichantha*, *A. farinosa*, *A. fasciata* (urn plant), *A. fosteriana*, *A. fulgens* (coralberry), *A. gracilis*, *A. involucrata*, *A. lasseri*, *A. latifolia*, *A. lingulata*, *A. lueddemanniana*, *A. mariae-reginae*, *A. mertensii*, *A. mexicana*, *A. nallyi*, *A. nudicaulis*, *A. ornata*, *A. paniculigera*, *A. pectinata*, *A. penduliflora*, *A. pineliana*, *A. purpureorosea*, *A. ramosa*, *A. recurvata*, *A. rubens*, *A. serrata*, *A. setigera*, *A. tillandsioides*, *A. triangularis*, *A. triticina*, *A. veitchii*, *A. victoriana*, *A. weberbaueri*, and *A. zebrina*, among many others. These are small to large tubular or open rosette plants with green or variously colored, marked or spotted leaves with soft or spiny margins, and large, simple or branched, persistent terminal inflorescences and with brightly colored bracts. Native from central Mexico to Argentina.

Aechmea fasciata var. *fasciata* with green leaves (two center rows)

Aechmea fasciata

Aechmea blanchetiana

BLANCHET BROMELIAD
SYNONYMS *Aechmea laxiflora*, *A. remotiflora*, *Streptocalyx laxiflorus*, *Tillandsia blanchetiana*

ETYMOLOGY [From Greek *aichme*, point or peak, alluding to the pointed sepals] [Jacques Samuel Blanchet (1807–1875), Swedish botanist, collector of plants chiefly in Bahia, Brazil]

GROWTH HABIT Herbaceous epiphyte and terrestrial, to 3 ft. (1 m) tall and about 4 $\frac{1}{2}$ ft. (1 $\frac{1}{2}$ m) wide when mature. **STEM/BARK**: essentially stemless. **LEAVES**: orange to red, spirally arranged, coriaceous, lanceolate, broad at base; margins entire; sheathing; form a rosette that apparently hold significant amounts of water. **FLOWERS**: bisexual, red or yellow; on much branched long lasting spikes to 6 ft. (2 m) tall; bracts are red with yellow apices, petals yellow, and sepals purple. **FRUIT**: small purple berries, when mature. **NATIVE HABITAT**: epiphyte in the rainforests or in sand in shrubby back dunes, along the Atlantic coast of Brazil.

CULTURE Full sun provides most intense orange-red color – plants become yellowish-green in shady areas; in well-drained dryer soils. **PROPAGATION**: seed or division of side shoots. **LANDSCAPE Uses**: best when grown on raised mounds or with evergreen background trees and shrubs; often grown as a container plant for indoor or outdoor use. **HARDINESS ZONE**: 10A–13B.

COMMENTS Perhaps one of the most attractive of all bromeliads, this species is a favorite of nectar feeding birds. Available cultivars include *Aechmea blanchetiana* **'Lemon'** (yellow with red leaf tips), **'Red Raspberry'** (dark red leaves).

Aechmea blanchetiana (Photographed at Fairchild Tropical Botanical Garden)

Aechmea gamosepala

MATCH STICK PLANT; GAMOS BROMELIAD
SYNONYM *Ortgiesia gamosepala*

ETYMOLOGY [From Greek *aichme*, point or peak, alluding to the pointed sepals] [united sepals]

GROWTH HABIT Epiphytic clumping perennial; to about 25 in (60 cm) tall; fast growing; fine to medium texture. **STEM/BARK**: stemless, but produces lateral plantlets by stolons. **LEAVES**: lanceolate glossy green to 20 in (50 cm) long, forming a compact cluster of multiple rosettes with upright leaves; margins entire; apex nearly rounded but apiculate. **FLOWERS**: bisexual small pink and lavender-blue sessile flowers that resemble matchsticks with blue heads before opening, $\frac{1}{2}$–$\frac{2}{3}$ in (15–20 mm) long, on a many-flowered erect cylindrical spike inflorescence. **FRUIT**: blue berries. **NATIVE HABITAT**: in lower elevation forests of Southern Brazil, from Sao Paulo State to Rio Grande do Sul and Santa Catarina to northern Argentina.

CULTURE Tolerates direct sun but prefers partial shade to retain the natural leaf color; epiphytic but may be grown in containers, hanging baskets in well-drained but moist organic media, and on rocks. **PROPAGATION**: can be readily propagated by division. **LANDSCAPE USES**: on tree branches, rocks, hanging baskets, terrariums, and in borders and mixed plantings. **HARDINESS ZONE**: 8B–12B. Tolerates some frost but intolerant of long freezes.

COMMENTS An attractive relatively cold-tolerant species adept at growing on rocks. The few reported cultivars include '**Big Pinkie**' (larger plants without stolons and pinkish flowers), '**Lucky Stripes**', '**Mardi Gras**' (white strips on leaf centers), '**Ruby Red**' (leaves reddish-brown and wider), '**Big Matchstick**' (long inflorescences and stolons), '**Red Lips**' (red leaf margins and tips). *Aechmea gamosepala* **var**. *nivea* (inflorescence with dark red flowers as opposed to pink).

Aechmea gamosepala '**Lucky Stripes**' (Photographed at Selby Botanical Gardens, Sarasota, Florida)

Aechmea gamosepala growth habit (above) and stages of floral development (below, left to right)

Aechmea mariae-reginae

QUEEN AECHMEA

SYNONYMS *Aecmea gigas*, *A. lalindei*, *Pothuava mariae-reginae*

ETYMOLOGY [Greek *aichme*, "peak" or "point," alluding to the pointed sepals] [Latin "of Queen Mary"]

GROWTH HABIT Dioecious epiphytic herb, up to 7 $\frac{1}{2}$ ft. (2 $\frac{1}{2}$ m) tall; roots hard and wiry with which they attach to trees. **STEM/BARK**: stemless. **LEAVES**: alternate, lanceolate, coriaceous soft green with flecked darker green, in dense spiral forming a dense spreading rosette with trap pools of water that house a number of organisms including frogs, salamander, spiders, and various insects and debris. **FLOWERS**: unisexual, white with blue tip, about $\frac{1}{4}$ in. (8 mm) long, on densely crowded cylindrical spike to 6 in. (15 cm) long and with lanceolate-ovate bright pink 4 in.

(10 cm) long bracts on the peduncle. **FRUIT**: ovoid-spherical all along the inflorescence spike, turning red with white endosperm at maturity. **NATIVE HABITAT**: rainforest canopy of Nicaragua, Costa Rica, Columbia, and Venezuela.

CULTURE Prefers partial shade on tree branches, rocks, or well-drained moist organic media. **PROPAGATION**: Seed or division. **LANDSCAPE USES**: not a preferred choice for indoor use but forms an attractive addition to trees or special setups outdoors. It can, however, be grown in greenhouses and conservatories. **HARDINESS ZONE**: 10A–13B. Subject to freeze damage.

COMMENTS The dioecious nature of this species is notable since majority of bromeliads are hermaphroditic. A few cultivars have been reported, including *A. marieae-reginae* 'Jimmie Knight', 'Maygood Moir', 'Prince Albert', and the ×*Androlaechmea* 'Dean' (*Androlepis skinneri* × *Aechmea mariae-reginae*).

Aechmea mariae-reginae

Aechmea orlandiana

Aechmea bracteata   *Aechmea warasii*

Ananas comosus

PINEAPPLE

SYNONYMS *Ananas acetosae*, *A. ananas*, *A. argentata*, *A. aurata*, *A. bracteatus*, *A. coccineus*, *Ananassa ananas*, *Bromelia ananas*, *B. communis*, *Distianthus communis*, and many more...,

ETYMOLOGY [Guarani *nanas*, excellent fruit] [tufted, in reference to the fruit peduncle].

GROWTH HABIT Evergreen terrestrial herbaceous monocarpic perennial; to 4 ft. (1 $\frac{1}{3}$ m) tall; medium-coarse texture; medium-rapid growth rate. **STEM/BARK**: forms a short thick stem that ultimately continues growth that terminates in fruit. **LEAVES**: simple, spirally arranged into a basal rosette, to 3 ft. (1 m) long and 1 $\frac{1}{2}$ in (3 $\frac{1}{4}$ cm) wide, stiff, bright green (or colors in cultivars), channeled, with sharp spiny tips and margins. **FLOWERS**: bisexual, violet or reddish, to $\frac{1}{2}$ in. (1 $\frac{1}{2}$ cm) long, individually inconspicuous; in a dense head, borne on a stout scape that rises from the center of the plant. After pollination, it forms a fleshy multiple fruit (fused carpels of many flowers, coalesced berries) of inferior ovaries. **FRUIT**: the pineapple of commerce; fleshy, juicy multiple fruit, globose to cylindrical, to 1 ft. (25 cm) long and weighing 1–14 lb. (0.45–

6.35 kg). The fruit is topped with a small rosette of leaves that form a miniature plant. **NATIVE HABITAT**: origin uncertain, probably southern Brazil and Paraguay.

CULTURE In light intensities ranging from full sun down to interiors; well-drained soils with good aeration; slight salt tolerance; pineapple is more attractive as an ornamental if given some protection from full sun; scales, mealy-bugs, nematodes, and mites; root rots in poorly drained media. **PROPAGATION**: division of basal suckers, crowns of the fruit, and tissue culture; seeds are used only in experimental breeding programs. **LANDSCAPE USES**: a horticultural curiosity, the plant is grown as a specimen in outdoor rock gardens, in tubs, or as an interior plant. Except for spineless cultivars, the plant is not recommended for indoor use. **HARDINESS ZONE**: 10B–13B.

COMMENTS The most important commercial cultivars are the nearly spineless '**Smooth Cayenne**' in Hawaii and '**Red Spanish**' in the West Indies. A few cultivars are available with variegated foliage: '**Porteanus**' (leaves with yellow stripes), '**Variegatus**' (longitudinal stripes of green, white, and pink. However, commercially produced fruit include one of four cultivars: '**Queen**', '**Spanish**', '**Abacaxi**', and '**Maipure**', which are harvested for fresh market in South Africa and Australia, the Caribbean, South America, and Hawaii. Other *Ananas* species in cultivation include: *A. ananassoides*, *A. bracteatus* (red pineapple, listed as a synonym), *A. lucidus*, and *A. nanus*. These are terrestrial rosette plants with long, usually spiny, often brightly colored leaves, blue flowers, and multiple fruit terminating a simple inflorescence which itself bears a crown of small leaves. Native to tropical America.

Ananas comosus var. *comosus*

Ananas comosus 'Sugar Loaf'

Ananas comosus 'Variegatus'

Ananas comosus **'Variegatus,'** in a display

Ananas comosus **'Ivory Coast'**

Probably a variation of *Ananas comosus* 'Ivory Coast'

Ananas lucidus

Tissue cultured pineapples

Billbergia nutans

QUEEN'S TEARS, FRIENDSHIP PLANT
SYNONYMS *Billbergia linearifolia*, *B. minuta*

ETYMOLOGY [Gustaf Johan Billberg (1772–1844), a lawyer by profession and botanist, zoologist, and anatomist as a hobby, in Stockholm, Sweden] [nodding, pendent, in reference to inflorescences]

GROWTH HABIT Evergreen clumping epiphytic and terrestrial rosette, to 2 ft. (60 cm) tall and as wide; fine texture; rapid growth rate. **STEM/BARK**: stemless. **LEAVES**: basal, linear strap-shaped, about 12 in. (30 cm) long and 1 in. (2 $\frac{1}{2}$ cm) wide, somewhat coriaceous but not stiff, dark green, margins serrulate, apex acuminate. **FLOWERS**: bisexual, in clusters on pendant paniculate or racemose inflorescence; bracts pink, petals tubular-funnelform, green with blue margins; stamens distinct yellow, elongated and protrude the corolla. **FRUIT**: septicidal capsule. **NATIVE HABITAT**: southern Brazil to Uruguay, and south to Argentina.

CULTURE Direct morning sun in cooler areas or bright shade, in well-drained organic fertile soil; drought-tolerant; no serious insect or diseases reported. **PROPAGATION**: seed, division, tissue culture. **LANDSCAPE USES**: on tree crotches, in containers, in-ground on mass or in borders, rock and wall gardens, indoors, etc. **HARDINESS ZONE**: 8B–13B. Tolerates short freezing periods.

COMMENTS *Billbergia nutans* is among the most common bromeliads in cultivation due to its ease of culture. The vernacular name friendship plant is presumably in reference to its ease of **PROPAGATION** and sharing with friends. There are, however, 65 known species of *Billbergia*, of which several have been reported in cultivation: *B. fosteriana*, *B. horrida*, *B. iridifolia*, *B. lietzei*, *B. leptopoda*, *B. macrocalyx*, *B. macrolepis*, *B. meyeri*, *B. minarum*, *B. morelii*, *B. pyramidalis* (foolproof plant), *B. reichardtii*, *B. rosea*, *B. sanderiana*, *B. alfonsi-joannis*, *B. amoena*, *B. braziliensis*, *B. buchholtzii*, *B. decora*, *B. distachia*, *B. elegans*, *B. saundersii* (rainbow plant), *B. tweedieana*, *B. viridiflora*, *B. vittata*, and *B. zebrina*, and other species, as well varieties (e.g., *Billbergia nutans* var. *schimperiana*), interspecific and intergeneric hybrids, such as *Canistrum* × *Cryptbergia* (*Billbergia* × *Cryptanthus*), and cultivars, are epiphytic herbs, with comparatively few basal, usually marked or dotted grayish leaves, and showy tubular, variously colored flowers in terminal bracteate inflorescences. Native primarily to eastern Brazil, but occur from eastern North America to Argentina.

Billbergia nutans

Cryptanthus bivittatus

EARTH STAR
SYNONYMS *Acanthospora vittata*, *Billbergia bivittata*, *Acryptanthus atropurpureus*, *Nidularium bivittatum*, *Tillandsia bivittata*, *T. vittata*

ETYMOLOGY [From Greek *kryptos*, hidden and *anthos*, flower, hence hidden flower in reference to flowers that are obscured by the bracts] [with two stripes]

GROWTH HABIT Andromonoecious (plants with both male and bisexual flowers), evergreen; terrestrial or sometimes epiphytic (saxicolous – on rocks) herbaceous peren-

nial; from 2–10 in. (5–25 cm) tall and 6–10 in. (15–25 cm) spread; medium-fine texture; moderately slow growth rate. **STEM/BARK**: usually with short stems or none; rarely with leafy stems; may produce suckers. **LEAVES**: simple; crowded, linear, stiff, spreading; undulate with prickly margins, in flattened star-shaped rosettes; length varies with cultivar, from 2–18 in. (5–45 cm) long; generally, 1–2 in. (2 $\frac{1}{2}$ –5 cm) wide; colors vary from green to reddish to various striping or cross-banding patterns. **FLOWERS**: white or greenish-white; inconspicuous; male flowers are borne in small heads that nest in the center of the rosette and bisexual flowers in bract axils. **FRUIT**: small dry berries. **NATIVE HABITAT**: shady moist areas in dry forests of eastern Brazil.

CULTURE Does best in partial shade, in well-drained, high organic matter fertile soil; tolerant of some salt in the root zone; not tolerant of dune conditions (salt spray on the foliage); scale; root rot if overwatered. **PROPAGATION**: division of offsets and tissue culture. **LANDSCAPE USES**: small planters and dish gardens in interiors; may be used outdoors to decorate limbs of rough-barked trees or as a ground cover or rock garden plant. **HARDINESS ZONE**: 10B–13B

COMMENTS *Cryptanthus* includes about 60 species, several hybrids, and numerous cultivars. There are also cultivars of unknown but probably of interspecific or intergeneric hybrid origin between *Cryptanthus* and *Bilbergia* or *Vriesia*, such as '**Aloha**', '**Cascade**', '**It**', '**Koko**', '**Mirabilis**', '**Pink Starlight**', '**Ruby**', among others. The popular species include: *C. bivittatus*, and *C. zonatus* (zebra plant). Other species are also reported from cultivation, including: *Cryptanthus acaulis* (starfish plant), *C. bahianus*, *C. beuckeri*, *C. bromelioides*, *C. diversifolius*, *C. fosterianus*, *C. lacerdae* (silver-star), *C. marginatus*, *C. maritimus*, *C. pseudoscaposus*, and probably others. These are small terrestrial, sometimes stoloniferous plants with flat, basal, symmetrically arranged, variously colorfully mottled or striped leaves, and white tubular flowers in short terminal heads. The genus is endemic to Brazil.

Cryptanthus '**Ruby**' (left) and *C.* '**Elaine**' (right)

Cryptanthus bivittatus '**Pink Starlight**'

Cryptanthus bivittatus **'Ruby Star'** in Longwood Botanical Garden Conservatory; green plant on the left is *Anthurium superbum* (Araceae)

Cryptanthus bivittatus **'Minor'**

Cryptanthus bromelioides var. *tricolor*

Guzmania lingulata

GUZMANIA, SCARLET STAR

SYNONYMS *Caraguata lingulata*, *C. peacockii*, *C. splendens*, *Guzmania peacockii*, *Tillandsia clavata*, *T. lingulata*

ETYMOLOGY [Spanish pharmacist, naturalist, and plant collector Anastasio Guzman (17??–1807] [tongue-shaped]

GROWTH HABIT Epiphytic monocarpic herb, clustering rosette to 2 ft. (50 cm) tall and 12 in. (25 cm) wide; medium texture; slow growth rate. **STEM/BARK**: stemless, rarely with a short stem formed by leaf bases in older plants. **LEAVES**: linear-lanceolate, spirally arranged to form a basal rosette, stiff, entire, thin, leathery, 12–18 in. (25–45 cm) long, 1 in. (2 ½ cm) wide. **FLOWERS**: yellow or white, in vertical ranks in terminal spikes or panicles with distinctive red bracts; ovary superior; long floral bracts, bright orange-red. **FRUIT**: capsule. **NATIVE HABITAT**: tropical America, including Florida, Caribbean, and Central and South America, predominantly at higher elevations.

CULTURE Requires part shade; in very well-drained soil if in containers; no serious problems reported, although mites may be troublesome in dry interiors. **PROPAGATION**: seed, tissue culture, division of offsets after the mother plant dies. **LANDSCAPE USES**: ideal for house and conservatory plant, specimen plant, popular in interior and exterior landscapes on tree branches. **HARDINESS ZONE**: 10A–12B.

COMMENTS The genus *Guzmania* includes 200+ species and hybrids as well as many cultivars: **'Empire'** is one of the most popular in trade; '**Magnifica**' (selection from a hybrid between the typical species and var. *minor*) has red inflorescences and spreading bracts; '**Variegata**' has green and yellowish-white leaves; var. *splendens* has longitudinal maroon stripes; and var. *cardinalis* has green leaves and scarlet floral bracts. There are also a number of colorful cultivars of unknown origin such as '**Alerta**', '**Calypso**', '**Cherry**', '**Claret**', '**Fiesta**', '**Grand Prix**', '**Hilda**', '**Lemonade**', '**Lipstick**', '**Luna**', '**Marina**', '**Orangeade**', '**Passion**', '**Purple Knight**', '**Sunnytime**', '**Vella Red**', '**Wendy**', and '**Zamora**', among many others. Several species are also reported in cultivation, including: *Guzmania angustifolia*, *G. berteroniana*, *G. danielii*,

G. dissitiflora, *G. erythrolepis*, *G. fuerstenbergiana*, *G. gloriosa*, *G. lindenii*, *G. lingulata*, *G. monostachia*, *G. musaica*, *G. nicaraguensis*, *G. patula*, *G. sanguinea*, *G. vittata*, and *G. zahnii*, among others. These are mostly epiphytic rosette plants with entire, glabrous leaves that are often marked with brown or maroon longitudinal lines or striking horizontal crossbands, and white or yellow flowers with brightly colored bracts, on terminal racemes, native primarily to Andean rainforests of Colombia and Ecuador. Also, *G. lingulata* var. *cardinalis*, *G. lingulata* var. *concolor*, and *G. lingulata* var. *flammea* are recognized, as the intergeneric hybrid ×*Guzvriesia* (*Guzmania* × *Vriesia*).

Guzmania lingulata 'Minor Empire'

Guzmania
'Graaf Van Hoorn'

Guzmania 'Soledo'

Guzmania 'Limones'

Guzmania 'Rena Orange'

Guzmania 'Yellow Hilda'

Guzmania lingulata 'Gwendolyn'

Guzmania lingulata 'Yellow Lemon'

Guzmania wittmackii

Hechtia spp.

ETYMOLOGY [Julius Gottfried Conrad Hecht (1771–1837), a counselor to the King of Prussia]

GROWTH HABIT Dioecious terrestrial xeric monocarpic perennial herbs, from about 6 in. (15 cm) to 5 ft. (1 $\frac{2}{3}$ m) across; fine texture; slow growth rate (unless fertilized). **STEM/BARK**: stemless, but usually with tall flowering stalks. **LEAVES**: lanceolate, stiff coriaceous in a rosette, margins with numerous sharp spines, sometimes colorful. **FLOWERS**: unisexual but sometimes appear as bisexual

with either stamens or carpels none-functional; small white on branched inflorescences to 8 ft. (2 $\frac{2}{3}$ m) tall. **FRUIT**: ovoid or ellipsoid capsule; seeds brown, with wings. **NATIVE HABITAT**: on rocky arid slopes, from Texas to Central America, but all but four species are native to Mexico.

CULTURE Naturally grow and prefer full sun but can tolerate some bright shade, in well-drained sandy soil; other than avoidance of overwatering the plants are problem free. **PROPAGATION**: seed and division of side shoots (pups); should be handled with folded newspaper to avoid the painful spines. **LANDSCAPE USES**: in succulent gardens along with cacti and other xeric plants; may also be used in containers. Caution should be exercised in their landscape use location. **HARDINESS ZONE**: 10B–13B.

COMMENTS Although more than 50 known species of *Hectia* have been described, they are not commonly available in the trade for understandable reasons, in part because of their sharp painful spines and in part their requirement for

hot arid conditions. The genus is presented here to illustrate the diversity of the Bromeliaceae: *Hechtia argentea*, *H. desmetiana*, *H. glomerata*, *H. marnier-lapostollei*, *H. podantha*, *H. rosea*, *H. scariosa*, *H. schottii*, *H. texensis*, *H. tillandsioides*, and others, are robust xerophytic terrestrials with gray leaves becoming rosy-bronze in the sun, and mostly white, inconspicuous flowers on lateral inflorescences, native from Texas to Central and South America.

Hechtia Mexicana photographed in Tehuacana Valley, Mexico

Hechtia montana photographed in Comarapa, Bolivia

Hechtia mexicana in Tehuacana Valley, Mexico

Various terrestrial bromeliads, including *Hechtia* spp. in Chiapas, Mexico

Hohenbergia stellata

HOHENBERGIA
SYNONYMS *Aechmea longisepala*, *A. oligosphaera*,
Hohenbergia eryrthrostachya, *H. oligosphaera*,
Pironneava morreniana, *P. roseocoerulea*

ETYMOLOGY [Named for Prince Fredrich Paul Wilhelm
of Wurttemberg (1797–1860), German naturalist, explorer,
and patron of botany, named Hohenberg] [Starry or star-like,
in reference to the flowers]

GROWTH HABIT Comparatively large epiphyte or terrestrial rosette-forming herb, to 4 ft. (1 $\frac{1}{3}$ m) tall; medium
texture; relatively fast growth rate. **STEM/BARK**: essentially stemless but the long inflorescence stalks resemble
stems. **LEAVES**: broadly lanceolate, light green, 2–3 ft.
($\frac{2}{3}$–1 m) long; in a large rosette, margins finely serrate
(serrulate) and spiny; apex purplish-red. **FLOWERS**: blue,
in compact globular clusters with bright red, stiff, sharp,
spiny bracts on dazzling red spikes; to 3 ft. (1 m) long. **Fruit**:
berries. **NATIVE HABITAT**: cloud forests of eastern Brazil,
Martinique, Trinidad, and Venezuela.

CULTURE Direct sun or bright light and dry soils; no
major problems reported. **PROPAGATION**: seed or division of clumps. **LANDSCAPE USES**: almost exclusively in
outdoor landscapes and away from the general public; may
also be grown in large conservatories or greenhouses.
HARDINESS ZONE: 10B–13B.

COMMENTS *Hohenbergia*, with more than 60 species, is
one of the few bromeliads with only a few species in cultivation because they are mostly lacking color, as do species of
most other genera, and are quite spiny. Some taxonomists
argue for placing the genus in *Aechmea*. The few other listed
species in cultivation include *H. augusta* (with whitish
bracts), *H. blanchetii*, and *H. ridleyi*.

Hohenbergia stellata

Neoregelia spectabilis

PAINTED FINGERNAIL BROMELIAD
SYNONYMS *Aregelia spectabilis*, *Karatas spectabilis*,
Nidularium eximum, *N. spectabilis*, *Regelia spectabilis*

ETYMOLOGY [From Latin *neo*, new + Eduard August
von Regel (1815–1892), director of the Imperial Botanical
Garden at St. Petersburg, Russia] [Spectacular, showy].

GROWTH HABIT Evergreen epiphytic, round, clumping,
herbaceous perennial; 12–16 in. (30–40 cm) tall but variable
width depending on the number of rosettes; medium-coarse
texture; fast growth rate. **STEM/BARK**: stemless. **LEAVES**:
basal rosette; stiff, strap-shaped to 16 in. (40 cm) long and
1 $\frac{1}{2}$ in. (4 cm) wide; olive green above, with transverse
gray bands beneath, and a red acute apex; inner leaves
marked with maroon; margins nearly spineless. **FLOWERS**:
whitish-blue; tubular, to $\frac{1}{2}$ in. (1 $\frac{1}{4}$ cm) in diameter;
borne in a dense head that nests in the center of the leaf
bases. **FRUIT**: berry. **NATIVE HABITAT**: rainforests of
southern Brazil.

CULTURE Can be grown in light levels from full sun to
partial shade; more attractive if grown at bright lights or part
sun; in full sun or very high light the foliage will be bronze
colored; in well-drained, well-aerated soils; moderate salt
tolerance; scale; mites in interiors; rots in poorly drained
media; grows well on large rocks. **PROPAGATION**: separation of rosettes. **LANDSCAPE USES**: may be grown in
well-drained soil in landscapes but often used as indoor
plants and on tree branches, rocks, large planters, and hanging baskets. **HARDINESS ZONE**: 8B–12B. Surprisingly
cold-tolerant with some protection from trees but shows
some injury in prolonged freeze.

COMMENTS *Neoregelia spectabilis* and its many cultivars are among popular plants in warmer climates for indoor and outdoor use primarily because of ease of culture. Included in its many cultivars are *Neoregelia* '**Black Spectabilis**', '**Fiesta**', '**Inspiration**', '**Lady Racine**', '**Pinkie**', '**Pinstripe**', and '**Thunderball**'. In addition, there are some intergeneric hybrids such as ×*Neomea* '**Pink Cascade**', ×*Niduregelia* '**Sunset**', to mention a few. There are a significant number of listed cultivars that do not specify *N. spectabilis* as parent but most likely originated from it. Number of *Neoregelia* species exceed 100, of which many are in cultivation, including: *Neoregelia albiflora*, *N. ampullacea*, *N. bahiana*, *N. carolinae*, *N. compacta*, *N. concentrica*, *N. coriacea*, *N. farinosa*, *N. fosteriana*, *N. laevis*, *N. marmorata*, *N. melanodonta*, *N. mooreana*, *N. pauciflora*, *N. pineliana*, *N. princeps*, *N. sarmentosa*, *N. tigrina*, *N. tristis*, *N. wilsoniana*, and *N. zonata*, as well as other species, hybrids and cultivars. These are small to medium size, compact, epiphytic plants with variously marked, often brightly colored leaves, and simple or compound inflorescences which appear sessile. Native primarily to eastern Brazil but also in most of the Amazonian region.

Neoregelia carolinae **f.** *Tricolor* (blushing bromeliad), endemic to Brazil, but with a large number of forms, varieties, and cultivars. Photographed at Epcot, Orlando, Florida.

Neoregelia carolinae **f.** *tricolor* '**Striata**'   *Neoregelia carolinae* **f.** *tricolor* '**Perfecta**'

Neoregelia 'Royal Burgundy'

Neoregelia 'Aztec'

Neoregelia spectabilis
used to create an island

Neoregelia spectabilis in a large planter
at Harry P. Leu Botanical Gardens, Orlando, Florida

Neoregelia spectabilis growing on a rock, in partial sun (left,), and in full sun (right)

Neoregelia spectabilis small flowers are open only in the morning, typical flowers for the genus.

Neoregelia johannis

Neoregelia 'Fireball'

Neoregelia 'Takamura Princeps'

Neoregelia meyendorffii var. *tricolor*

Tillandsia guatemalensis

PINK QUILL
SYNONYMS *Allardtia cyanea, Tillandsia cyanea, T. uycensis*

ETYMOLOGY [Elias Tillandz (1640–1693), Swedish botanist and professor of medicine at the Academy of Turku, Finland] [From Guatemala].

GROWTH HABIT Epiphytic, monocarpic herb, suckering rosette; 12–18 in. (30–45 cm); fine texture; slow growth rate. **STEM/BARK**: stemless. **LEAVES**: linear, channeled, and recurved, with a reddish-brown line on the lower side, to 14 in. (35 cm) long and $\frac{1}{2}$ in. (1 $\frac{1}{4}$ cm) wide, forming a rosette. **FLOWERS**: violet-blue, arising from a long-lasting spike, 4–6 in. (10–15 cm) long and about 2 in. (5 cm) wide, with broad, flattened, clear pink overlapping bracts; flowers emerge singly or rarely only two at one time, to 2 in. (5 cm) across. **FRUIT**: capsule. **NATIVE HABITAT**: rainforests of Ecuador on tree tops.

CULTURE Requires part shade, in very well-drained media; no serious problem reported. **PROPAGATION**: seed, division, and tissue culture. **LANDSCAPE USES**: specimen plant, popular in containers for interior use or in exterior landscapes, ideal subject for conservatory and greenhouses. **HARDINESS ZONE**: 10A–13B.

COMMENTS With nearly 700 diverse species, *Tillandsia* is by far the largest genus in the family and perhaps the most common and popular bromeliad. Many of the smaller taxa are used in dish gardens. These require little care and remain alive and attractive for a long time. There are only a few cultivars of *Tillandsia cyanea* in the trade, primarily based on width and color (pink to magenta) of the bract, the best known of which are '**Paradise**', '**Pink Quill**', and '**Anita**'. The large number of species in cultivation include but are not limited to: *T. aeranthos, T. albida, T. anceps, T. andreana, T. araujei, T. argentea, T. baileyi, T. bandensis, T. bergeri, T. bryoides, T. bulbosa, T. capitata, T. caput-medusae, T. complanata, T. concolor, T. crispa, T. cyanea, T. deppeana, T. duratii, T. erecta, T. fasciculata* (wild pineapple), *T. filifolia, T. gardneri, T. geminiflora, T. grandis, T. imperialis, T. ionantha, T. jucunda, T. leucodepis, T. lindenii, T. lucida, T. maxima, T. monadelpha, T. multicaulis, T. paraensis, T. parryi, T. polystachia, T. pruinosa, T. punctata* (Mexican black-torch), *T. rhomboidea, T. schiedeana, T. streptophylla, T. stricta, T. tenuifolia, T. tricolor, T. vernicosa, T. violacea, T. viridiflora,* and *T. xiphioides*, and others. These are small to large, usually rosette, urn-shaped, or twisted plants with poorly developed or no roots, mostly with soft scale-covered, silvery-gray leaves, and tubular, usually blue flowers with brightly colored bracts in distichous spikes. Native from southern North America to southern Argentina.

Tillandsia guatemalensis

Tillandsia ionantha

Tillandsia funckiana   *Tillandsia fuchsia*

Tillandsia bergeri (photographed at Busch Gardens, Tampa, Florida)

Tillandsia dyeriana

Tillandsia usneoides

SPANISH MOSS

SYNONYMS *Dendropogon usneoides*, *Renealma usneoides*, *Stepsia usneoides*, *Tillandsia crinita*, *T. filiformis*, *T. trichoides*

ETYMOLOGY [Elias Tillandz (1640–1693), Swedish botanist and professor of medicine at the Academy of Turku, Finland] [like a lichen of the genus *Usnea*].

GROWTH HABIT Evergreen; epiphytic herbaceous perennial; with slender, hanging stems; often seen festooned in masses over trees, particularly oaks; fine texture; to 21 ft. (7 m) long; lacking any aerial roots. **STEM/BARK**: stems slender, branched, much elongated, and flexible and covered with silver-gray scales. **LEAVES**: simple; scattered at intervals along the stem; smooth, very narrowly linear, curved, to 2 in. (5 cm) long (rarely seen); silver-gray, peltate scales cover the leaves; entire margins, not readily recognizable. **FLOWERS**: pale green or yellowish; inconspicuous to $\frac{1}{2}$ in. ($1\frac{1}{4}$ cm) long; borne singly in the leaf axils; faintly fragrant, especially at night. **FRUIT**: capsule. **NATIVE HABITAT**: mostly on southern live oaks (*Quercus virginiana*) and to a lesser degree on other oak trees, ranges from eastern Virginia in U.S., south to Argentina.

CULTURE Tolerates full sun but grows best under partial shade; will tolerate low light intensities; the plant is rootless; the scales covering the stems and leaves absorb moisture. Moderate salt tolerance. **PROPAGATION**: in nature by tiny, wind-borne seeds and by bird- or wind-borne small pieces of the plant. **LANDSCAPE USES**: hanging from trees to create naturalistic settings, can also be used in hanging baskets and sometimes used as such in combination with other plants. Often used in conservatories on tree branches that also contains other epiphytic bromeliads and orchids. **HARDINESS ZONE**: 7B–13B.

COMMENTS *Tillandsia usneoides* is an epiphyte similar to other bromeliads and not a moss or parasitic plant. It uses tree branches on which it grows merely as support. Growing in massive quantities it may block sunlight and cause a reduction in the number of leaves on a tree. During hurricanes or periods of wet weather, the increased weight of large clumps of Spanish moss may contribute to breaking of tree limbs. It has been used as a packing material, upholstery stuffing, cattle feed, and in air conditioner and oil filters. It can be used as sandpaper when green. Spanish moss is one of the characteristic natural features of the Southern United States. It is included here to represent another example of bromeliad diversity.

Tillandsia usneoides (photograph below at Phipps Conservatory, Pittsburgh, Pennsylvania)

Vriesea spp. and Cultivars

VRIESEA
ETYMOLOGY [Dutch botanist Willem Hendrik de Vries (1806–1862), professor of botany at Amsterdam and Leiden]

GROWTH HABIT Smaller species predominantly epiphytic but larger taxa terrestrial, varying in size from 6 in. (15 cm) to 5 ft. (1 $\frac{1}{2}$ m) in height and spread; growth rates and textures vary with species and cultivars. **STEM/BARK**: essentially stemless but sometimes inflorescence stalks of the larger terrestrial species resemble stems. **LEAVES**: lanceolate to lanceolate-ovate, mostly soft and pliable; margins entire; light to dark green and variously blotched, spotted, or barred. **FLOWERS**: green, yellow, or white flowers with most species having brilliant variously colored bracts; on spike inflorescences. **FRUIT**: dry capsule with parachute-like seeds. **NATIVE HABITAT**: forests of Central and South America as well as the West Indies, but mostly in Brazil.

CULTURE Partial bright shade; the epiphytic species should be grown in very well-drained media with ability to retain moisture but not standing water, if in containers but preferably on tree branches where there is air movement. **PROPAGATION**: seeds, division of "pups," and tissue culture, with the latter perhaps most common. **LANDSCAPE USES**: most species and cultivars are suitable as container-grown plants particularly for indoors or on balconies and verandas; most epiphytic taxa are also suitable for trees or on "bromeliad trees," and of course they are ideal plants for greenhouses and conservatories. **HARDINESS ZONE**: 8B–13B. Some species tolerate colder temperatures than others but probably not withstand long freeze periods.

COMMENTS Despite the 360 listed species *Vriesea* has been used more in hybridization than any other bromeliad. The best-known species include *V. splendens* (with distinct banded foliage and bright red inflorescence), *V. fenestralis* and *V. zamorensis*. The photo index of Florida Council of Bromeliad Society illustrates the extent and diversity of the genus by including 198 species and 713 cultivars. Thus, inclusion of cultivar names in this discussion is not necessary. For an interesting historical account of hybridization of *Vriesea* and selection of cultivar see Samyn, (1995), and http://www.fcbs.org/articles/vriesea_hybrids.htm

Vriesea, is the second largest but the most hybridized and cultivated genus in the family. Some of the more common taxa include: *V. amazonica*, *V. barilletii*, *V. bituminosa*, *V. carinata* (lobster-claws), *V. corcovadensis*, *V. × elegans* (a complex hybrid), *V. ensiformis*, *V. erythrodactylon*, *V. fenestralis*, *V. flammea*, *V. fosteriana*, *V. gigantea*, *V. gladioliflora*, *V. glutinosa*, *V. heliconioides*, *V. heterostachys*, *V. hieroglyphica* (nicknamed "king of the bromeliads"), *V.*

imperialis, *V. incurvata*, *V. × kitteliana* (*V. barilletii × saundersii*), *V. mariae* (*V. bariletii × carinata*), *V. × morreniana* (*V. carinata × psittacina*), *V. platynema*, *V. psittacina*, *V. racinae*, *V. recurvata*, *V. regina*, *V. × retroflexa* (*V. psittacina × scalaris*), *V. ringens*, *V. rubra*, *V. saundersii*, *V. schwackeana*, *V. splendens*, *V. vagans*, and many others. These are medium size, mostly epiphytic plants with entire, soft or firm, variously green but often spotted, blotched or distinctly marked leaves, and yellow, green, or white flowers and brightly colored bracts on erect or pendent, usually long-lasting inflorescences. Native from Mexico to northern Argentina and the West Indies, but mostly in eastern Brazil and Costa Rica.

Vriesia poelmanii

Vriesia gigantea

Vriesea fosteriana

Vriesea hieroglyphica

ADDITIONAL TAXA

Alcantarea vinicolor, endemic to eastern Brazil, it is among the lerger bromeliad genera, named after Dom Pedro d'Alcântara, second emperor of Brazil.
Photographed at Selbey Botanical Garden, Sarasota, Florida.

Alcantarea imperialis, endemic to Brazil, it is the largest bromeliad with a diversity of leaf colors.
Photographed at New York Botanical Gardens

Nidularium innocentii

Orthophytum saxicola

Quesnelia testudo

Puya berteroniana (blue puya, terquise puya)
(Photographed at Huntington Botanic Gardens, California)

Dyckia fosteriana × frigida
(Photographed at Huntington Botanic Garden, California)

Deuterocohinia brevifolia (= *Abromeitiella brevifolia*), native to Bolivia and Argentina, dense
mat-forming succulent

CYPERACEAE

SEDGE FAMILY

113 GENERA AND ±6000 SPECIES

GEOGRAPHY Cosmopolitan, predominantly in poor wet soils of temperate zones.

GROWTH HABIT Hermaphroditic or monoecious rhizomatous, grass-like perennial herbs, mostly with solid triangular stems (culms). **LEAVES**: basal, 3-ranked, grass-like and sheathing around the stem. **FLOWERS**: florets small and inconspicuous, bisexual (plants hermaphroditic) or unisexual (plants monoecious); 3-merous, perianth absent or represented by bristles; ovary superior, situated in axil of a bract (glume), arranged in spikelets; stamens 1–3; ovary superior; stigmas 2–3; ovules one per flower. **FRUIT**: Single-seeded achene with bristles.

ECONOMIC USES Ornamentals; the source of papyrus paper, mat making, edible rhizomes, perfumes, and other uses such as hat and basket making and medicinal purposes.

CULTIVATED GENERA *Carex*, *Cyperus*, *Eleocharis*, *Rhynchospora*, and *Scripus*

COMMENTS Cyperaceae is a relatively large family with complicated taxonomic relationships (see detailed treatment of the family in APG IV, 2016), but with limited number of genera in cultivation. The genus *Cyperus* includes more than 700 species and may well be the only taxon in cultivation but with a vast majority being weedy. Several species and cultivars of *Carex* are commonly used in landscapes. A few examples include *C. acuta*, *C. alba* (cedar sedge grass), *C. aquatilis* (leafy tussock sedge), *C. aurea* (wet meadow sedge), *C. blanda* (eastern woodland sedge), *C. buxibaumii* (Club sedge), *C. cephalodea* (rough-clustered sedge), *C.*

eburnean (Cedar Sedge), *C. elata, C. hachijoensis, C. lacustris* (lake sedge), *C. morrowii* (Morrow's sedge and its cultivars), *C. muehlenbergii* (sand bracted sedge), *C. muskingumensis* (palm sedge and its cultivars), *C. oshimensis* (and its cultivars, including variegated 'Evergold'), *C. platyphylla* (silver sedge), *C. siderosticha* (and its cultivars), *C. typhina* (common sedge), among many others; *Eleocharis acicularis, E. baldwinii* (Baldwin's spicebush), and others are knowns as rushes; *Rhynchospora colorata* (=*Dichromena colorata*), *R. latifolia* (= *Dichromena latifolia*, white star grass), and *Scripus acutus* are used primarily in landscapes.

Cyperus alternifolius

UMBRELLA SEDGE, UMBRELLA PALM
SYNONYMS *Cyperus frondosus, C. onustus, C. rosemosus, Eucyprerus alternifolius*

ETYMOLOGY [Greek word meaning sedge] [Alternate leaves].

GROWTH HABIT Evergreen; herbaceous perennial shrub; rhizomatous, forms dense tufted clumps (caespitose); medium-fine texture. to 6 ft. (2 m) tall; rapid growth rate. **STEM/BARK**: slender, green, glabrous; triangular to rounded; finely grooved; terminate in a conspicuous umbrella-shaped flowering head. **Leaves**: basal; reduced to sheaths at base of plant; reddish-brown. **FLOWERS**: scapose inflorescence a compound spreading umbel of brown spikelets, with 12–20 conspicuously drooping, leaf-like bracts, 4–12 in. (10–30 cm) or longer and $\frac{1}{2} - \frac{3}{4}$ in. (12–18 mm) wide; bracts subtend by a shorter central cluster of rays bearing many minute flowers arranged in spikes to $\frac{1}{2}$ in. (12 mm) long; dull brown. **FRUIT**: inconspicuous black achenes. **NATIVE HABITAT**: Africa-Madagascar, Arabian Peninsula.

CULTURE Full sun to deep shade but best in partial shade, in varied fertile soils; grows well in standing water; tolerates some salt drift but is susceptible to mites in enclosed areas. **PROPAGATION**: division of clumps. **LANDSCAPE USES**: popular garden plant as accent or screen, in water, or indoors in containers. **HARDINESS ZONE**: 8B–11B.

COMMENTS *Cyperus alternifolius* is the most common species of the genus in cultivation and includes a few cultivars including the dwarf '**Gracilis**' (dwarf umbrella sedge), '**Nanus**', '**Giant**' as well as other noted cultivars '**Dick Albert**', '**Maja Dumat**', '**Shihchuan**'. Although less common several related species and cultivars are in the trade which include: *C. albostriatus* (dwarf umbrella sedge, including '**Variagatus**'), *C. flabelliformis, C. glaber, C. haspan, C. isocladus, C. laevigatus. C. longus, C. profiler, C. textilis*, and perhaps others. *Cyperus papyrus* ('**King Tut**', '**Little Tat**', '**Nanus**'), "wetland sedge papyrus" of Nile Delta, was used extensively by Egyptians to make paper, mats, ropes, blankets, etc., as well as medicine and perfume. Some of these materials, particularly the paper, is currently available for sale.

Cyperus alternifolius, growth habit and landscape use (above), in water (below left), and leaf close-up (below right)

Cyperus papyrus (papyrus, paper reed, Nile grass), native to Africa, an herbaceous perennial of swamp vegetation. Source of the papyrus paper, first type of paper used by Egyptians. Photographed at New York Botanical Garden

POACEAE (= GRAMINEAE)

GRASS FAMILY

ABOUT 650 GENERA AND 9000 SPECIES

GEOGRAPHY Truly cosmopolitan: from both poles to the equator, nearly all land masses on earth, rare in deserts.

GROWTH HABIT Hermaphroditic, monoecious, or rarely dioecious; from less than 1 in. (2 $\frac{1}{2}$ cm) to 18 ft. (6 m) or larger; annual and perennial herbs with fibrous roots, or sometimes woody and arborescent, with usually hollow stems and solid nodes, and with rhizomes and/or stolons, often forming vegetative shoots (= tillers). Stems may terminate in inflorescences (in which case they are called "culms"), or remain simple or branched, erect or prostrate. **LEAVES**: alternate, arising at the nodes and consisting of a blade and sheath; distichous, linear, ensiform, or rarely broad (lanceolate to oblong or ovate or elliptic), often crowded at the base forming open or closed overlapping sheath, usually with membranous or hairy ligules (located at the junction of the blade and the sheath); venation parallel, rarely with transverse veinlets (as in *Bambusa*). **FLOWERS**: called florets, may be bisexual or rarely unisexual, small and inconspicuous, in rarely solitary or more often variously arranged paniculate spikelets on a short axis (= rachilla), each subtended by a pair of small or relatively large bracts (= glumes); perianth of 2 or 3 scales called lodicules, surrounded by 2 outer scales, the outer called the lemma, the inner called the palea; stamens usually 3, sometimes 2 or 1, rarely 6 or more, hypogynous, filaments delicate, anthers bilocular, versatile, longitudinally dehiscent; carpel 1, unilocular, ovary superior, ovule with a single, one-seeded functional ovary. **FRUIT**: caryopsis or grain, rarely an achene or berry (in some bamboos).

ECONOMIC USES Perhaps the single most important food source for mankind and animals (e.g. wheat, rice, barley, corn, etc.), building and furniture materials from bamboos, and several species are used as ornamentals.

CULTIVATED GENERA Many, including all cereals, such as *Triticum* (wheat), *Hordeum* (barley), *Oryza* (rice), *Saccharum* (sugar cane), *Sorghum*, *Zea* (corn), etc. However, ornamental genera include *Agrostis*, *Arundinaria*, *Axonopus*, *Bambusa*, *Cortaderia*, *Cynodon*, *Dendrocalamus*, *Eragrostis*, *Eremochloa*, *Festuca*, *Lolium*, *Panicum*, *Pennisetum*, *Phragmites*, *Phyllostachys*, *Poa*, *Pseudosasa*, *Stenotaphrum*, and *Zoysia*, among many others. [See Christenhusz et al., 2017, for a complete list of Poaceae genera; for several beautiful water color illustrations of Poaceae see Otto Wilhelm Thomé (1840–1925), on line.]

COMMENTS A large and taxonomically complex family consisting of 780 genera and approximately 12,000 species, in nearly every conceivable habitat, from the driest deserts to the coldest regions, and throughout the tropics. Poaceae is the most economically important family in the plant kingdom, providing much of the staple food of the world's population (wheat, corn, rice, etc.) and contributing to man's visual pleasure (lawn grasses and other landscape plants). Many species are major weeds and invasive. The most recent worldwide phylogenetic classification of Poaceae (Soreng et al. 2017), recognizes 12 subfamilies, seven supertribes, and 52 tribes, five supersubtribes, and 90 subtribes. Understandably, no other plant group has been classified in as complex arrangement, and for obvious reasons, any discussion of the family as a whole is beyond the scope of this book. Complexity of Poaceae is further illustrated in specific terminology of its morphological features (see family description).

Ornamental grass collection at Oxford University Botanical Garden, England

Bamboos at Kanapaha Botanical Garden, Gainesville, Florida

Bambusa textilis

TEXTILE BAMBOO; WEAVER'S BAMBOO
SYNONYMS *Bambusa annulata*, *B. minutiligulata*, *B. varioaurita*

ETYMOLOGY [Latinized Malayan vernacular name] [Textile, referring to the use of the culms in furniture building]

GROWTH HABIT Evergreen pachymorph (= clumping) bamboo, 30–40 ft. (10–15 m) tall, but often smaller in culti-

vation; fine texture; slow growth to start but eventually fast growing and expands by producing additional culms. **STEM/BARK**: erect but gracefully arching (drooping) pale green culms (stems), culms 2 in. (5 cm) in diameter, thin-walled and without branches on the lower parts; long internodes, hollow; rhizomes erect; lateral branches at nodes, woody and usually one dominant; culm sheaths triangular, coriaceous, deciduous. **Leaves**: light green delicate fine-textured lanceolate leaves with the blade base connected to the sheath with a short petiole-like stalk, to about 4 in. (10 cm) long and 1 in. (2 $\frac{1}{2}$ cm) wide; margins scabrous, apex acuminate; less subject to damage by winds. **FLOWERS**: panicles of spikelets with 2 to many flowers, no specific details reported. **FRUIT**: typical seed-like caryopsis. **NATIVE HABITAT**: Guangxi and Guangdong Provinces of southeast China.

CULTURE Prefers full sun but tolerates partial shade, in moist sandy-loam soil; irrigate 2–3 times weekly until established; although a clumping species in time it requires room to expand but may be kept under control if planted near a hard surface such as paving or with use of plastic or aluminum sheathing; It is a low-maintenance bamboo once established. **PROPAGATION**: division of rhizome, nodal cuttings, air and trench layering; tissue culture has been mentioned. **LANDSCAPE USES**: excellent candidate for specimen or tall privacy or windbreak hedge planting; may also be used as potted plants in large containers. **HARDINESS ZONE**: 7–13A. Said to tolerate temperatures below freezing to −15 °C.

COMMENTS *Bambusa textilis* has pliable fibers, making it ideal for weaving baskets, ropes, and furniture. Silicic acid, which is collected from the internodes is used in Chinese medicine. *Bambusa textilis* **var**. *gracilis* (slender weaver's bamboo) has thinner culms and more pronounced nodding tops, and its 'Dwarf' cultivar as implied, is in fact shorter but apparently less tolerant of colder temperatures,

Bambusa textilis **var**. *glabra* is a shorter plant with thicker culms and withstands lower temperatures, *B*. *textilis* '**Kanapaha**' is a large cultivar with white markings on the culms, and *B*. *textilis* '**Scranton**' is vertical (not nodding) and has short lateral branches. The Bambusa Group, in addition to *Bambusa textilis* (textile bamboo), includes several other commonly cultivated species, including: *B*. *arundinacea* (giant thorny bamboo), *B*. *beecheyana* (beechey bamboo), *B*. *glaucescent* (hedge bamboo) and its many cultivars, *B*. *mutabilis*, *B*. *oldhamii* (giant timber bamboo), *B*. *tulda*, *B*. *tuldoides* (punting-pole bamboo). *B*. *ventricosa* (Buddha's belly bamboo), *B*. *vulgaris* (common bamboo), among others. These are tall, clump-forming perennial grasses with climbing or erect, sometimes spiny, hollow woody culms of great strength, short-petioled leaves, and panicles of spiklets with 2 to many bisexual or unisexual flowers. Native primarily to tropical and subtropical Asia but with few from North to South America. Widely used as ornamental and landscape plants, as timber for construction, cut into sections and used for milk churns or water pales, musical instruments, young shoots eaten as vegetable, as well as a number of other uses, including pulp for paper, fishing poles, etc. *Guadua* **spp**. are not reported from cultivation but *Guadua angustifolia* is noted as being the largest Neotropical bamboo, and said to yield prime timber in tropical South America.

Bambusa textilis (photographed at Kanapaha Botanical Garden, Gainesville, Florida).

Bambusa ventricosa

BUDDHA'S BELLY BAMBOO
SYNONYMS *Bambusa tuldoides* 'Ventricosa', *Leleba ventricosa*

ETYMOLOGY [Latinized Malayan vernacular name] [Having a swelling on one side; pot-bellied, in reference to swollen nodes of the culms]

GROWTH HABIT Evergreen pachymorph (clumping) bamboo with numerous tightly spaced culms, 24–30 ft. (8–10 m) tall when planted in ground. **STEM/BARK**: culms dimorphic, normal culms 24–30 ft. (8–10 m), and 1 $\frac{1}{4}$ –2 in. (3–5 cm) in diameter, flexible at base (can move side to side) and only slightly drooping at apex; internodes 12–14 in. (30–35 cm), slightly swollen; lower nodes with rings of gray-white silky hairs below and above sheath scar; 2 or 3 branches starting from 3rd or 4th node but several branches higher up; abnormal culms thinner and with shorter internodes and swollen at base; branches only on upper nodes; culm sheathers deciduous, glabrous, broadly arched or subtruncate. **LEAVES**: blade linear lanceolate to lanceolate, 4–7 in. (10–18 cm) long and less than 1 in. (2 cm) wide; densely pubescent below, glabrous above. **FLOWERS**: pseudospikelets solitary or several clustered at nodes; florets 6–8. Not seen in cultivation. **FRUIT**: not reported. **NATIVE HABITAT**: Guandong (southern China), Vietnam, and naturalized in Brazil and Malaysia.

CULTURE Full sun or partial shade in well-drained soil; said to do well in large containers that limit vertical growth and water-stressed for best-inflated nodes, some irrigation is required if the plant is to maintain its foliage; no major problems other than spider mites reported. **PROPAGATION**: division of rhizomes. **LANDSCAPE USES**: often used in bonsai, but widely cultivated as specimen in subtropical regions as a landscape plant; can be used as a screen for privacy. **HARDINESS ZONE**: 9B–13A. May be grown in zone 9A with some protection or in large containers that can be moved.

COMMENTS It is commonly stated that internodes swell when plants are grown in containers or in ground and water-stressed. However, plant shown in the accompanying photographs was a normal landscape plant and did not appear to be under water stress, but nodes of all its culms were swollen. *Bambusa ventricosa* '**Kimmei**' has yellow culms with narrow green stripes and foliage sometimes with narrow cream stripes as well. The culm sheaths of *B. ventricosa* and *B. tuldoides* are substantially different, hence each considered a distinct species. In The Plant List, *Bambusa ventricosa* is considered a synonym of *B. tuldoides*. There is, however, some similarity with *Bambusa vulgaris* '**Warmin**' that also has swollen internodes.

Bambusa ventricosa growth habit (above left), leaves close-up (above right), green-shaded stems (below left), and sun-exposed stems (below right)

Bambusa vulgaris

COMMON BAMBOO; GOLDEN BAMBOO

SYNONYMS *Arundarbor blancoi, A. fera, A. monogyna, A. striata, Arundo fera, Bambusa auriculata, B. blancoi, B. fera, B. mitis, B. monogyna, B. nguyenii, B. sieberi, B. striata, B. surinamensis, B. thouarsii, Gigantochloa auriculata, Leieba vulgaris, Nastus thouarsii, N. viviparous, Gigantochloa auriculata, Phylostachys striata*

ETYMOLOGY [Latinized Malayan vernacular name] [From Latin, meaning common]

GROWTH HABIT Pachymorph (clumping) with densely packed erect culms, 30–60 ft. (10–20 m) tall and 12–30 ft. (4–10 m) wide; rapid growth and fast-spreading. **STEM/BARK**: rhizomes extend up to 2 $\frac{1}{2}$ ft. (45 cm) before turning upward; culm internodes green, glabrous; culms bases with short aerial roots; branches many, dominant at the center; sheaths deciduous, straw colored, and covered with dark rigid hairs. **LEAVES**: green, medium-sized, glabrous on both surfaces, somewhat rough on the upper side. **FLOWERS**: not reported. **FRUIT**: fruit and fertile seed unknown. **NATIVE HABITAT**: uncertain origin, but probably from Java, Sumatra, or China but currently distributed worldwide and naturalized in many areas.

CULTURE Easy to grow in full sun or partial shade, does fine in well-drained but moist soils; needs relatively large space. **PROPAGATION**: can be easily propagated by division of rhizomes; stem cuttings, and tissue culture. **LANDSCAPE USES**: specimen in landscape, a hedge or privacy screen, and erosion control. **HARDINESS ZONE**: 8B–13A. Said to tolerate frost to 27 °F (−3 °C).

COMMENTS *Bambusa vulgaris* is one of the most commonly distributed and used bamboos in tropical and subtropical regions of the world. According to Meredith (2001), the plant was first cultivated in south Florida in the 1840s, and may have been the first foreign species introduced into America. It was also among the first plants introduced to Europe's hothouses. It is an attractive timber bamboo often used in construction, the pulp is used for paper making, and young shoots are edible. *Bambusa vulgaris* 'Vittata' has golden-yellow culms with green vertical striping, and the green culm leaf sheaths with yellow striping, it is among the most attractive of all bamboos, *B. vulgaris* 'Wamin' has short swollen lower internodes and shows some color variations and striping, *B. vulgaris* 'Striata' is considered synonym of 'Vittata' and appears to have nomenclatural priority.

Bambusa vulgaris 'Striata'

Cortaderia selloana

PAMPAS GRASS

SYNONYMS *Arundo dioica, A. kila, A. selloana, Cortaderia argentea, C. dioica, Gynerium argenteum, G. dioicum, G. purpureum, Moorea argentea*

ETYMOLOGY [Argentinian name] [Friedrich Sellow (1789–1831), German traveler and naturalist who made extensive collections in Brazil and Uruguay].

GROWTH HABIT Dioecious, perennial grass, large, clumping; reaches 6–9 ft. (2–3 m) in height with an equal spread; rapid growth. **LEAVES**: simple, 5–7 ft. ($1\frac{2}{3}$–$2\frac{1}{3}$ m) long, $\frac{1}{2}$ in. ($1\frac{1}{4}$ cm) wide at the base, tapering to a point at the drooping apex; margins are saw-toothed. **STEM/BARK**: dense, clumping, grass-like stem near ground. **FLOWERS**: silvery-white, tiny, in terminal panicles (plumes), 1–2 ft. ($\frac{1}{3}$–$\frac{2}{3}$ m) long and held 1–3 ft. ($\frac{1}{3}$–1 m) above the leaves. Plumes appear in August, persisting until January. **FRUIT**: caryopsis, small, usually non-viable in Florida. **NATIVE HABITAT**: South America.

CULTURE Grows and blooms best in full sun, but tolerates partial shade; in most soils except very wet ones and tolerates drought and salt spray; should be cut back to 18 in. (45 cm) high after blooms fade in late winter; relatively pest-free. **PROPAGATION**: seed, if available, or clump division. **LANDSCAPE USES**: makes an attractive, showy specimen, looks especially good in seaside landscapes; often used in dried arrangements; pampas grass plumes are grown commercially for this purpose in California, where it has become invasive; once planted, it is difficult to remove. **HARDINESS ZONE**: 5A–12B.

COMMENTS *Cortaderia selloana* 'Rosea' has pink plumes. Several cultivars selected for height or panicle color have been introduced, such as '**Andes Silver**', (tall, silvery white plumes), '**Bertini**' (compact, cold hardy), '**Gold Band**' (foliage with golden stripes, compact), '**Monvin**' (yellow striped leaves), '**Patagonia**' (silvery white plumes, tall), '**Pumila**' (compact, to 7 ft.), '**Pink Feather**' (pinkish large plumes), '**Silver Comet**' (foliage with white stripes), '**Sundale Silver**' (tall, silvery white plumes). This plant is invasive in some Mediterranean climate areas. It does not set seed in Florida and probably in other humid subtropical climates.

Cortaderia selloana

Cynodon dactylon

BERMUDA GRASS

ETYMOLOGY [Greek *kyon*, dog and *donti*, tooth, probably in reference to the narrow alternating long and short pointed leaves] [Greek *dactylos*, finger, in reference to the finger-like appearance of the inflorescence].

GROWTH HABIT Low-growing, perennial grass, dense, vigorous; fine-textured; rapid growth rate. **STEM/BARK**: spreads by stolons and rhizomes, internodes alternating long and short. **LEAVES**: simple, alternate at nodes, in two ranks, entire; linear to $1\frac{1}{2}$ in. (3 cm) long, sheathing at the base, folded in bud; green, often glaucous, turning brown in cold weather; ligule a fringe of hairs; leaves soft, somewhat pubescent on upper surface. **FLOWERS**: in terminal spikes 4–7, to 2 in. (5 cm) long, flattened, stalk to 16 in. (40 cm) tall. **FRUIT**: seed-like grain (caryopsis). **NATIVE HABITAT**: warm regions of both hemispheres, but apparently originated in the Middle East.

CULTURE Full sun, in fertile soils; highly salt and good drought tolerance if nematodes are controlled; poor shade tolerance, therefore not for shady sites; tolerates a lot of wear; recovers quickly when injured; high maintenance grass: must be mowed closer and more often than other grasses, requires frequent fertilization and spraying to control insect and disease problems, including mites, nematodes, sod webworms, army worms, scale, many fungal diseases, and mole crickets; cut to $\frac{1}{2}$ to $1\frac{1}{2}$ in. ($1\frac{1}{4}$–$3\frac{1}{4}$ cm) tall. **PROPAGATION**: seed of "improved" common types and cultivars. **LANDSCAPE USES**: fine-textured durable lawn grass, golf course greens, and athletic fields. **HARDINESS ZONE**: 7B–13B. Preferably in warmer areas.

COMMENTS Cultivar 'Ormond' and several cultivars of *C. dactylon* and *C. transvaalensis*, such as 'FLoraTeX', 'Tifdwarf', 'Tifgreen', 'Tifgreen II', 'Tiflawn', 'Tifway', and 'Tifway II', are used mostly in golf greens. It is considered invasive in some regions.

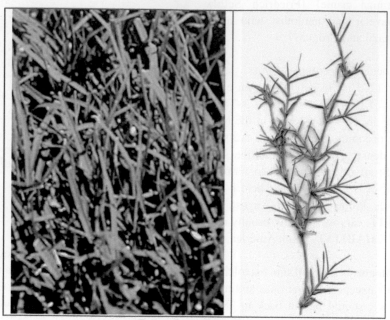

Cynodon dactylon

Eremochloa ophiuroides

CENTIPEDE GRASS
ETYMOLOGY [From Greek *eremia*, desert and *chloris*, green, probably in reference to its adaptability to sandy soils] [From Greek *ophis*, serpent, *oura*, tail, and *eidos*, probably in reference to the cylindrical inflorescences].

GROWTH HABIT Low-growing, perennial grass, somewhat coarse-textured, dense, vigorous; medium texture; slow growth rate. **STEM/BARK**: spreads by stolons, internodes short, more or less equidistant. **LEAVES**: simple, alternate at nodes, 2-ranked, entire, linear to 2 in. (5 cm) long; sheathing at the base, folded in bud; ligule a purple membrane; leaves soft, pubescent margins: green, turning brown in cold weather. **FLOWERS**: spikes solitary, cylindrical to $2\frac{1}{2}$ in. ($6\frac{1}{2}$ cm) long, terminal and axillary.

FRUIT: seed-like grain (caryopsis). **NATIVE HABITAT**: southern China.

CULTURE Full sun to partial shade, various soils; slightly salt-tolerant; fair to good shade tolerance; good drought tolerance if nematodes controlled; adapted to low fertility soils; iron chlorosis on soils with pH greater than 6.2; does not withstand heavy foot traffic; potential problems include nematodes (highly susceptible), mole crickets, sod webworms, scale insects called ground pearls (serious problem), brown patch and dollar spot (fungal diseases); cut to 2 in. (5 cm) tall. **PROPAGATION**: sod of cultivars, seeds. **LANDSCAPE USES**: lawn, low-maintenance grass. **HARDINESS ZONE**: 8B–13A. Warmer areas of zone 8.

COMMENTS A few cultivars: 'Oaklawn' and 'Centennial' are selected specifically for cold tolerance.

Eremochloa ophiuroides

Gigantochloa atroviolacea

TROPICAL BLACK BAMBOO; TIMOR BLACK BAMBOO
SYNONYMS *Gigantochloa atter* var. *nigra*, *Gigantochloa atter* f. *nigra*

ETYMOLOGY [From Greek *gigantas*, giant and *chloe*, grass] [Dark violet]

GROWTH HABIT Evergreen perennial pachymorph that resembles *Bambusa*, may bloom occasionally but it is not monocarpic; fast growing. **STEM/BARK**: culms initially green but become purplish black with narrow green stripes, to 50 ft. (17 m) tall and 4 in. (10 cm) in diameter; has a tendency to arch, giving the appearance of open growth; nodes whitish; sheaths deciduous, with dark brown hairs; aerial roots at nodes. **LEAVES**: dark green, blades lanceolate and pendulous with a short petiole-like connection to sheath, to about 10 in. (25 cm) long, glabrous, apex acute. **FLOWERS**: clustered spikelets at nodes with glumaceous (glume-like), subtending bracts, noted as readily hybridizing with other species. **FRUIT**: grain-like caryopsis. **NATIVE HABITAT**: India, Indonesia, and Sumatra and widely planted in many tropical areas.

CULTURE Full sun, in a drier area of well-drained limestone rich soil; no serious problems reported. **PROPAGATION**: division of rhizome or single node culm cuttings. **LANDSCAPE USES**: as landscape specimens or in containers. **HARDINESS ZONE**: 10A–13B.

COMMENTS The culm is chiefly used for handicraft and musical instruments. *Gigantochloa atroviolacea* 'Watupawan' has green culms with narrow black stripes and provides high quality culms for construction. This taxon should not be confused with the common black bamboo (*Phyllostachys nigra*) which is a leptomorph (= running, spreading) bamboo with extended rhizome, and is more cold tolerance. *Gigantochloa verticillata* yields valuable timber and its shoots said to be edible. However, genera related to *Gigantochloa* such as *Dendrocalamus stricta* (giant bamboo), and other species, are tall, clump-forming arborescent taxa with long, few to many-flowered panicles of spikelets in globose heads. Native to Southeast Asia and widely cultivated for paper pulp, used as timber, and yields tabashir (or tabasheer), a resin in the node. *Melocanna* spp. yield timber and are widely planted in Indo-Malaysia, *M. baccifera* has large, fleshy edible fruit; *Oxytenanthera* spp. and *Schizostachyum* spp. from Africa and Southeast Asia, respectively, also yield valuable timber.

Gigantochloa atroviolacea (black bamboo)

Hakonechloa macra 'Aureola'

GOLDEN HAKONE GRASS

ETYMOLOGY [Hakon, a region of Japan and Greek, *chloa*, a grass] [Large"] [From Latin *aureus*, golden]

GROWTH HABIT Rhizomatous deciduous perennial clumping grass in dense cascading mounds of 12–20 in. (30–60 cm) tall; fine texture; slow spreading. **STEM/ BARK**: spreads by rhizomes; thin wiry foliage stalks. **LEAVES**: arching linear-lanceolate, variegated golden (original species is bright green) stripes, to 10 in. (25 cm) long and about $\frac{1}{4}$ in. (1 cm) wide; margins entire or slightly involute, thin and chartaceous (= papery) texture, resembling those of bamboos; they arise from the rhizomes in thin wiry 1 to 2 ft. (30–60 cm) tall stalks. **FLOWERS**: yellow-green, nodding open panicles, from August to October; spikelets yellow-green. **FRUIT**: grain-like cary-opsis. **NATIVE HABITAT**: moist woodland and rocky cliffs in central Japan.

CULTURE Easy to maintain, prefers bright shady location for best color development but becomes yellow-green in deep shady locations particularly in warmer climates (apparently the original green form can tolerate direct sun and are more cold hardy); in moist but well-drained soil. **PROPAGATION**: division, but original green can be grown by seed. **LANDSCAPE USES**: mixed borders along paths and walkways, naturalized woodland gardens, in Japanese garden designs, and in containers. **HARDINESS ZONE**: 6A–10A. Does poorly in warmer climates.

COMMENTS Referred to as one of the most elegant grasses, in addition to the only known original green species in the genus, there are a few other cultivars, such as *H. macra* 'All Gold' (pale yellow to solid gold leaves), *H.* 'Albo-aurea' (leaves variegated white and yellow), *H.* 'Albostriata' (cream to white stripes), *H.* 'Albovariagata' (leaves variegated white), *H.* 'Nicolas' (dwarf, reaches to 6 in. (15 cm) tall and is orange to reddish), *H.* 'Naomi' (dwarf), and few others.

Hakonechloa macra 'Aureola'

H. macra 'All Gold'

Paspalum notatum

BAHIA GRASS
ETYMOLOGY [From Greek *paspalos*, name for millet] [spotted, marked].

GROWTH HABIT Tall-growing, perennial grass, moderately dense, vigorous; coarse texture; rapid growth rate. **STEM/BARK**: spreads by stout rhizomes, internodes short, compressed. **LEAVES**: simple, alternate at nodes, 2-ranked, entire, soft, somewhat papery, tough, slightly hairy, linear to 12 in. (30 cm) long, sheathing at base, rolled in bud; light green, browning only below 30 °F, ligule membranous. **FLOWERS**: spike-like raceme, predominantly 2 but sometimes 3 or rarely 5 per stalk, to 2–4 in. (5–10 cm), but the stalk to 20 in. (30 cm) tall, terminal. **FRUIT**: caryopsis (seed-like grain). **NATIVE HABITAT**: Paraguay, Argentina.

CULTURE Full sun to partial shade, various soils; slightly salt-tolerant; withstands drought better than most grasses due to its extensive root and thick rhizome system; relatively good shade tolerance but prefers sun; withstands heavy foot traffic; not well adapted to high pH soils; problems include sod webworms, army worms, mole crickets, brown patch, dollar spot, less susceptible to nematodes than most grasses, fewer insect and disease problems, tall unsightly seed heads; cut 2 $\frac{1}{2}$–4 in. (6 $\frac{1}{2}$–10 cm); **PROPAGATION**: seed (which is abundant and relatively inexpensive), sod. **LANDSCAPE USES**: lawn, road sides, forage, parks, athletic fields, minimum maintenance grass; extensively planted on disturbed sites and for erosion control. **HARDINESS ZONE**: 8A–13A.

COMMENTS Several cultivars including: 'Argentine', 'Paraguay', 'Paraguay Z-2' (short, narrow, hairy leaves), 'Pensacola', 'Tifton-9 Pensacola' (greater vigor in seedling stage), and 'Wilmington' (more cold hardy than others).

Paspalum notatum not mowed (left), mowed (middle), and close-up of rhizome.

Pennisetum alopecuroides

FOUNTAIN GRASS

SYNONYMS *Cenchrus purpurascens*, *Gymnotrix japonica* var. *viridescens*, *Panicum aloecuroides*, *Pennisetum alopecuroides* var. *erythrochaetum*, *P. chinense*, *P. compressum*, *P. dispiculatum*, *P. japonicum*, *P. purpurascens*

ETYMOLOGY [From Latin *pinna*, wing and *seta*, bristle, in reference to long feathery bristles of flowers] [*Alopecurus*-like, foxtail grass-like]

GROWTH HABIT Upright perennial clumping grass; $3\frac{1}{2}$ ft. ($1\frac{1}{3}$ m) tall and as wide; fine texture; rapid growth rate. **STEM/BARK**: culms stout, to 47 in. (10 m) tall. **LEAVES**: blades linear, flat or often involute, medium to deep green, to 31 in. (80 cm) tall and $\frac{1}{3}$ in. (1 cm) wide; apex long acuminate; ligule small; sheaths papery, keeled, imbricate at culms base; leaves turn golden yellow in fall. **FLOWERS**: inflorescence linear to 10 in. (25 cm) tall and $1\frac{1}{4}$ in ($3\frac{1}{2}$ cm) wide; peduncle short; each spikelet enclosed within an involucre; bristles greenish-purple, to $1\frac{1}{4}$ in. ($3\frac{1}{2}$ cm) long, slender, hispid, bottlebrush-like; flower in summer. **FRUIT**: caryopsis (seed-like grain). **NATIVE HABITAT**: grassy hillsides, roadside, field margins, in much of China, India, Indonesia, Japan, Southeast Asia, and Australia.

CULTURE Full sun for best flowering but tolerates part shade; in well-drained but moist to somewhat wet soils; cut foliage to ground in late winter to about 3 in. (8 cm); no serious insect or diseases reported. **PROPAGATION**: seed; division of cultivar clumps. **LANDSCAPE USE**: mass plantings, borders, foundations, containers, meadow landscapes, erosion control, and cut flowers. **HARDINESS ZONE**: 6B–10B. In zones 5B–6A with some protection.

COMMENTS *Pennisetum alopecuroides* is listed as a synonym of *Pennisetum polystachion*. No judgment is made as to nomenclatural status of these taxa, but the more prevalent name is used. Plants under both names are noted as invasive. It is also used for forage. A variety of cultivars ranging in height and flower and fall foliage colors are offered: '**Desert Plains**' (PP 20751– to 3 ft., brownish inflorescences), '**Fireworks**' (PP18504 – upright with variegated stripes of white, green, burgundy, and pink, and purple inflorescences), '**Hamlin**' (dwarf, compact, to 2 ft. tall), '**Little Bunny**' (more dwarf and compact, to 1 ft. tall and wide), '**Little Honey**' (upright clump of arching variegated green and white foliage and creamy-white inflorescence), '**Moudry**' (black fountain Grass, vigorous growth, inflorescence dark purple), '**National Arboretum**' (spreading clumps, dark-purple spikes), '**Pav300**' (= **Penn Stripe**™, PP21693– compact with gracefully arching white variegated foliage), '**Viridescens**' (densely tufted, arching leaves that turn golden-brown and purple-black spikes), among others. *Pennisetum* has recently been lumped into *Cenchrus*.

Pennisetum alopecuroides 'Red Head'

Pennisetum alopecuroides 'National Arboretum'
(Photo from U. S. National Arboretum on the WWW)

Pennisetum × *advena* 'Sky Rocket'

SKY ROCKET FOUNTAIN GRASS; VARIAGATED GREEN FOUNTAIN GRASS
ETYMOLOGY [Latin *pinna*, feather, and *seta*, bristle, alluding to the feathery bristles of some species] [Stranger, in reference to a species that is not native to or fully established in a new habitat]

GROWTH HABIT A clumping semievergreen (dormant in cooler climates) perennial grass that grows 24–36 in. (60–90 cm) tall; fine texture. **STEM/BARK**: rhizome. **LEAVES**: linear, green, arching, with white striped margins, 12–20 in (30–50 cm) long and $\frac{1}{2}$ in (1 $\frac{1}{3}$ cm) wide. **FLOWERS**: pink spike-like paniculate inflorescences comprised of interspersed spikelets and bristles, exceeding the height of the foliage, giving the appearance of smoky pink plumes that turn cream color as they age; flowers are sterile. **FRUIT**: Caryopsis (grain), apparently sterile. **NATIVE HABITAT**: parental taxa in savannas and woodlands of tropical and warm temperate regions.

CULTURE Full sun but tolerates some bright shade, in well-drained soil that should be irrigated periodically during summer months; should be topped in early spring in warmer climates. **PROPAGATION**: division. **LANDSCAPE USES**: best in mass plantings or mixed with other cultivars of the same hybrid origin such as '**Rubrum**', mixed herbaceous borders, and rock gardens, inflorescences are suitable for dried and fresh flower arrangements. **HARDINESS ZONE**: 9–13A. Recommended using as an annual in cooler climates.

COMMENTS *Pennisetum × advena* has been determined to be *P. setaceum × P. macrostachyum*. Its cultivars were first collectively referred to as the Celebration Series® by Proven Winners, now no longer marketed. These cultivars have since been bundled into their Graceful Grasses® collection, a conglomeration of different taxa, including sedges (*Cyperus* sp.), in a completely different family (Cyperaceae).

Included in the original series were sports from a plant of *P. × advena* 'Rubrum' (= *P. setaceum* 'Cupreum', with red/burgundy leaves): '**Sky Rocket**' (USPP 21497), 'Fireworks' (USPP 18504, variegated purple fountain grass) and '**Cherry Sparkler**' (USPP 22538). Most *Pennisetum* species are noxious weeds, but a few such as *P. setaceum* are used as ornamentals despite being invasive in many regions. Others such as *P. glaucum* (pearl millet) are important food crops or for animal feed. A few examples of other ornamental *Pennisetum* include *P. alopecuroides* (Chinese fountain grass, swamp foxtail and its several cold hardy cultivars), *P. glaucum* '**Purple Majesty**' (purple pearl millet), *P. macrostachyum* (giant fountain grass), *P. messiacum* (bunny tails), *P. orientale* (oriental fountain grass), and others.

Pennisetum × advena **'Sky Rocket'**

Pennisetum × advena **'Rubrum'**

Pennisetum × advena **'Cupreum'**

Stenotaphrum secundatum

ST. AUGUSTINE GRASS
ETYMOLOGY [From Greek *stenos*, narrow and *taphros*, a trench, in reference to the grooved racemose peduncles] [Latin for parts arranged on one side; one sided].

GROWTH HABIT Low growing lawn grass, very dense, vigorous; cut 1–2 in. (5 cm) tall; coarse texture; rapid growth rate. **STEM/BARK**: spreads by thick fleshy stolons, branching, internodes equidistant, leaf sheaths are flattened. **LEAVES**: simple, alternate or opposite at nodes, 2-ranked, linear but wider than most lawn grasses, green turning purple with cold weather; ligule a fringe of hairs; leaves soft, smooth, folded in the bud. **FLOWERS**: fleshy raceme, solitary, to 4 in. (10 cm) long, stalk to 3 in. (8 cm) tall, terminal. **FRUIT**: seed-like grain (caryopsis), sterile. **NATIVE HABITAT**: tropical and subtropical America.

CULTURE Medium to high maintenance grass; prefers full sun to moderate shade (tolerates shade better than most warm season grasses), various soils; highly salt-tolerant; little drought tolerance, therefore requires frequent irrigation; poor wear tolerance; chinch bugs may be a major problem, army worms, sod webworms, mole crickets, nematodes, several fungal diseases, thatch build-up, St. Augustine Decline (SAD virus). **PROPAGATION**: sod of cultivars; cultivars are essentially sterile, reproduced by plugs. **LANDSCAPE USES**: lawns – moderately high-maintenance grass. **HARDINESS ZONE**: 8A–13A.

COMMENTS '**Bitterblue**' (fine, dense texture and darker color), '**Captiva**' (dwarf and resistant to southern chinch bug), '**Floratam**' (resistant to nematodes and SADV), '**Raleigh**' (coarse texture, cold-hardy), '**Floratine**' (finer texture, denser grasses), '**Seville**' (dwarf, dark green, more shade-tolerant), '**Common**' and '**Roseland**' are pasture grasses. '**Floralawn**' (resistant to SADV, chinch bug, and webworm), '**Jade**' and '**Delmar**' (both with improved shade tolerance), '**FX-10**' (chinch bug resistant and deep roots), etc.

Stenotaphrum secundatum before establishment (left), stolon (middle), inflorescence (right)

Glossary

A

A- A Greek prefix meaning "without."

Ab- A Latin prefix meaning "away from."

Abaxial On the side away from the axis. For example, the lower side of a leaf is its abaxial side.

Acantha, Acantho A Greek word meaning "spine," "thorn," or "prickle."

Acanthocarpous Having fruit covered with thorns or prickles.

Acanthocephalous With a hooked beak.

Acanthocladous With thorny branches.

Acarpic, Acarpous Without fruit; not producing fruit.

Acaulescent Without a stem; or with a very short, scarcely evident stem, as in *Gerbera jamesonii* (Asteraceae).

Accessory Bud A bud additional to the normal axillary bud.

Accessory Organ Any organ additional to the normal number; sometimes specifically in reference to the calyx and/or corolla.

Accessory Fruit A fruit in which a major portion consists of tissue derived from other parts of the flower, as in apple, pear, strawberry, etc.

Accrescent Increasing in size with age, as calyx of some flowers after pollination.

Accumbent Lying against another body, as cotyledons against hypocotyl (syn. Reclinate).

Acephalous Without a head.

Acerate, Acerose, Acerous Needle-shaped.

Aceus, (-Aceous) A Latin suffix meaning 'resemblance', as in foliaceus (foliaceous), leaflike.

Achene A small, dry, one-chambered, one-seeded indehiscent fruit. For example, fruit of the Asteraceae.

Achenecetum An aggregation of achenes, as in Ranunculus (buttercups).

Achenodium A double achene, as the Cremocarp of Apiaceae (umbel family).

Achlamydeous Without a perianth, as in Salix spp. (willows).

Acicula (pl. Aciculae) The bristlelike continuation of the rachilla of grasses.

Acicular Needle-shaped (syn., Acerose).

Aciculate Superficially marked with fine lines, as if scratched.

Acorn Fruit of Quercus spp. (oaks), consisting of a nut embedded in a cup (Cupule).

Acra-, Acro- A Greek prefix meaning "apex," "apical."

Acranthous Having an inflorescence born at the tip of the main axis.

Acrid Sharp, bitter, or irritating acid taste.

Acrocaulous At the tip of the stem, as an inflorescence or a flower.

Acrocidal Capsule A capsule which dehisces through terminal slits or fissures.

Acrodromous A palmately veined leaf whose main veins are parallel but arching and terminate at or near the apex.

Acrolaminar Ethereal oil glands located near the leaf apex (cf. Basilaminar).

Acropetal Produced in succession upwards, towards the apex, as in an Acropetalous Inflorescence, where the oldest flowers are at the base and youngest at the apex. (opp. Basipetal).

Acropetiolar Etheral oil glands located near the apex of the petiole. (opp. Basipetiolar)

Acroramous Leaves terminal, near the tip of a branch. (opp. Basiramous)

Acroscopic Facing towards the apex.

Acrospire The first shoot of a germinating seed.

Acrostichoid Distributed in a solid mass. For example, in some ferns sporangia spread across the entire back of the frond, rather than organized in discrete sori.

Actinodromous A palmately veined leaf in which veins arise at the base and arch near the margin until they reach the apex.

Actinomorphic Flowers with radial symmetry (also see Regular). A line drawn through the center of such flower, in any direction, will divide the flower into two equal halves. For example in the Solanaceae (*Petunia*)

Aculeate Armed with a thorn or prickle, as the stem of a rose.

Acumen A sharp, tapering point.

Acuminate Narrowly tapering to a sharp point, as in a leaf apex.

Acute Tapering more broadly than acuminate to a sharp point.

Acyclic Arranged in a spiral, as alternate leaves on a stem.

Ad- A Latin prefix meaning "at" or "toward."

Adaptive Radiation Rapid diversification in response to specific environmental niche.

Adaxial Facing toward the primary axis. For example, the upper side of a leaf (syn. Ventral, opp. Dorsal or Abaxial).

Adelphous Having stamens united by their filaments.

Adherent Coming together of similar organs, touching but not fused, as in anthers (syn. Coherent, cf. Adnate).

Adnate, Adnation United, referring to the union of one organ with another, as ovary with receptacle, and stamen with corolla (cf. Adherent).

Adpressed, Appressed Pressed closely to an axis upwards with a narrow angle of divergence.

Adventitious Appearing in other than the usual place, as roots from the stem or stems from roots (suckers).

Aerating Roots Knee roots; horizontal or vertical aboveground roots.

Aerial Living above the surface of the ground, as in adventitious roots of some Araceae.

Aerocaulous Having above-ground stems.

Aerophyllous Having above-ground leaves.

Aerophyte See Epiphyte.

Aestival Summer flowering; with flowers appearing in summer.

Aestivation, Estivation The manner in which parts of a flower are folded in the bud before expansion.

Affinis, Affinitas A Latin word meaning "alliance," "closely related."

Affinity Close morphological similarity between two individuals.

Agamic, Agamous Neuter; without stamens and carpels, or sex organs abortive.

Agamandrous, Agamandrocephalous An inflorescence with neuter flowers inside or above, and staminate outside or below.

Agamagynous, Agamagynocephalous An inflorescence with neuter flowers inside or above, and pistillate outside or below.

Agamohermaphroditic, Agamohermaphrodicephalous An inflorescence with neuter flowers inside or above, and hermaphroditic outside or below.

Agamospermy The production of seeds without fertilization (syn. Apomictic).

Agglomerate Cluster; crowded; dense structures (syn. Aggregate, Conglomerate).

Aggregate See Agglomerate.

Aggregate Cup Fruit A fruit derived from an apocarpous (free carpel), perigynous flower; fruit composed of fruitlets.

Aggregate Free Fruit Fruit derived from an apocarpous (free carpel), hypogynous flower; fruit composed of fruitlets.

Aggregate Fruit Fruit formed by the coherence of the carpels that were distinct in the flower, as in raspberry or strawberry.

Agrostology The branch of systematic botany concerned with the study of grasses (Poaceae).

Aianthous Flowering continuously; ever-blooming.

Aigrette A feathery crown or tuft attached to a seed, as in some Asteraceae. This is an adaptation for wind dissemination.

Air Plants See Epiphytes.

Air Roots See Pneumatophores.

Ala (pl. Alae) One of the lateral petals of papilionaceous flower.

Alate Winged, as in stem of Liquidambar, ulmus, etc.; having a winglike extension, as a petiole.

Alba, Albus A Latin word meaning "white."

Albido The white tissue of the rind in citrus fruit.

Albuminose, Albuminous Starchy, as in seeds with endosperm.

Allagostemonous Having stamens alternately attached to petal and receptacle.

Allantoid Sausage-shaped.

Allautogamy Crossing and selfing in the same plant.

Allo- A Greek prefix meaning "different," "another."

Allochtonous Originating outside and transported into a given area; exogenous; introduced.

Allogamous Requiring a male and a female for reproduction; usually cross-fertilizing, although capable of self-fertilization.

Allogamy Cross-fertilization. (opp. Autogamy)

Allopatric, Allopatry Geographically separated populations of species without possibility of overlap (cf. Sympatric).

Allophylic Having flowers that can be pollinated indiscriminately by any animal visitor.

Allotropic Having flowers that are poorly adapted for insect pollination.

Alpha Taxonomy Descriptive taxonomy; That aspect of taxonomy concerned with description and naming of plants on the basis of morphological characteristics.

Alpine Refers to habitats or plants above the tree line in mountainous regions.

Alternate Arranged in a spiral manner, as when there is only one leaf at each node but alternating on the opposite side of the stem.

Alveolate Pitted, as the receptacle of many Asteraceae; honeycombed, with cavities separated by thin partitions, as pollen wall of Cycadales.

Alveola (pl. Alveolae), Alveolus Cavities or pits on a surface.

Ambiguous Of uncertain origin, or doubtful taxonomic position.

Ament A catkin, a type of pendulous, scaly spike, as in *Salix* (willows) and *Betula* (birches)

Amentiferous Bearing aments or catkins, as in Salicaceae and Betulaceae.

Amorphic, Amorphous Without a definite shape or form.

Amphanthium A dilated receptacle of an inflorescence, as that of *Dorstenia* spp. (Moraceae)

Amphi- A Greek prefix meaning "around," "on both sides," or "of both types."

Amphicarpous Possessing two kinds of fruit, differing morphologically or in time of ripening. Also said of plants with fruit above and below ground.

Amphiflorus Possessing flowers (and fruit) above and below ground.

Amphisarca An indehiscent, multilocular fruit with bony or dry outside and fleshy within, such as that of Cucurbitaceae.

Amphistomatic Leaves with stomata on both surfaces.

Amphitropous See Placentation.

Amplectant, Amplectans Clasping or winding tightly around an object, as with tendrils.

Amplexicaul Said of a sessile leaf whose base clasps the stem.

Ampliate Enlarged, as the ray florets of *Chrysanthemum* (Asteraceae).

Amygdaloid Almond-shaped.

Anacanthouse Without thorns.

Anadromous A type of venation in compound leaves of ferns, where the first set of veins in each pinna (leaflet) arises near the apex, as in *Asplenium, Aspedium,* etc.

Analog Similar features in taxa of different ancestry.

Analogous Structures Organs that are functionally similar but morphologically and phylogenetically different (cf. Homologous Structures).

Analytical key An orderly arrangement of contrasting or comparable statements about plants or plant structures, leading to identification.

Anandrous Without stamens; pistillate flower.

Anastomosing Joining by cross-connection to form a network, as in veins of leaves with reticulate (netted) venation.

Anatropous A bent or inverted ovule, with the helium and micropyle together at one end, and opposite the chalaza.

Ancestor An original taxon, a progenitor, from which others have evolved.

Ancipital, Ancipitus Two-edged; flattened and compressed.

Andragamocephalous, Andragamous An inflorescence with staminate flowers inside or above and neuter flowers outside or below.

Andro- A Greek prefix meaning "male" (cf. Gyno-).

Androdioecious A species in which some plants bear staminate flowers and some bear perfect flowers.

Androecium Collective term for the stamens of a flower.

Androgynecandrous Inflorescence with staminate flowers above and below pistillate, as in spikes of some *Carex* species.

Androgynophore An elongation of the receptacle, forming a stalk upon which pistil and stamens appear above the perianth.

Androgynecephalous, Androgynous Inflorescence bearing staminate flowers inside or above and pistillate flowers outside or below.

Androhermaphroditic, Androhermaphrodicephalous Inflorescence with staminate flowers inside or above and hermaphroditic outside or below.

Andromonoecious A plant with staminate and hermaphroditic flowers, but without pistillate flowers.

Androsporophyll See Microsporophyll.

Androstrobilus See Microstrobilus.

Anemophilous Wind-pollinated flower.

Anfractuosus, Anfractuose A Latin word meaning "wavy," or "twisted" (syn. Sinuous).

Angiosperm Any flowering plant; plants which produce seeds enclosed within a carpel.

Angustate Narrow.

Aniso- A Greek prefix meaning "unequal."

Anisocarpic, Anisocarpous With carpels of unequal size.

Anisocotylous, Anisocotyly With cotyledons of unequal size.

Anisolateral With unequal sides.

Anisomerous With unequal number of parts in different whorles of the flower.

Anisopetalous, Anisopetaly With petals of unequal size.

Anisophyllous With leaves of unequal size (cf. Heterophyllous).

Anisopterous With seeds having wings of unequal size.

Anisosepalous With sepals of unequal size

Anisosporous With male and female spores of unequal size

Anisostylous With styles of unequal length.

Annotinal, Annotinus Appearing yearly.

Annual A plant which completes its life cycle in one year. In horticulture, this term may refer to perennial plants that are used as bedding plants but are replaced seasonally (*Antirrhinum, Petunia, Impatiens,* etc.).

Annualar Ringlike; ring forming.

Annulus A ring of thick-walled cells around the sporangium of ferns.

Anodic The upward direction of a leaf-spiral.

Anomalicidal Capsule   A capsule which dehisces irregularly, at its weakest location.

Ante-   A Latin prefix meaning 'before'.

Anterior   Position of the flower when it is turned away from the axis and toward the bracts.

Anterior Grooves   Ridges or lines on the front (dorsal) side.

Anterior Lobes   Front lobes; the lobes away from the axis and toward the subtending bracts.

Anther   The pollen-bearing segment of the stamen.

Antheridium (pl. Antheridia)   In vascular plants, a multicellular organ responsible for production of male gametes (sperms).

Anthesis   Time when a flower begins to expand. (This term has been defined by various authors as the beginning of the opening of a flower, a fully expanded flower, when a flower is ready to be pollinated, or the period from flowering to fruit set).

Antho-, Anthos-   A Greek prefix meaning "flower."

Anthocarpous   An indehiscent fruit covered with persistent perianth or other floral parts, as in *Mirabilis* spp. (Nyctaginaceae).

Anthodium   See Capitulum.

Anthophyta   Flowering plants (syn. Angiospermae).

Anthotaxis   The arrangement of flowers on an axis.

Anthrophil, Anthropophily   A plant of disturbed habitats.

-Anthus   A Greek suffix meaning "flowered."

Anti-   A Latin prefix meaning "opposite."

Antipetalous   Opposite the petal, as in stamens situated in front of petals.

Antisepalous   Opposite the sepals.

Antitropus   Twisting in opposite direction, as with embryos which have the radicle pointing away from the hilum.

Antrorse   Pointing upwards and inward, usually in reference to stamens. (opp. Retrorse)

Apetalous   A flower lacking petals (cf. petaliferous, polypetalous).

Apex (pl. Apices)   Summit; the tip; the growing point of roots and shoots (apical meristem).

Aphyllopodic   In reference to plants without a basal rosette of leaves.

Aphyllous   leafless; with rudimentary leaves or none.

Apical   Belonging to or situated at the apex.

Apicula, Apiculus   A short, sharp point, but not stiff, as in a leaf ending.

Apiculate   Terminating in a sharp point.

Apocarpous, Apocarpy   Composed of distinct or separate carpels (cf. syncarpus).

Apochlamydeous   With perianth parts free, not united or fused.

Apodal   Without a foot-stalk, as in trichomes which do not have a basal stalk (cf. Sessile).

Apogamy   Reproduction without the fusion of gametes, by vegetative cells of the gametophyte (syn. Parthenogenesis).

Apogeotropic   Negatively geotropic, such as the specialized roots of cycads which grow upwards above the soil surface and with nodules that usually contain nitrogen-fixing algae.

Apomictic   Plants that produce seed without fertilization.

Apomixis   The ability of certain plants to produce seeds without fertilization.

Apomorphy   A novel evolutionary trait unique to a particular taxon and its descendants.

Apopetalous   Having many free petals (syn. Polypetalous).

Apophysis   The enlarged distal end of the scale of certain pine cones.

Aposepalous   Having many free sepals.

Apostemonous   Having many free stamens.

Appendicular Stamen   Stamens which have small appendages (connectives), as in Viola.

Applanate   Flattened.

Appressed   See Adpressed.

Aquatic   Plants that are adapted to growing in water.

Arachnoid   Entangled hairs having the appearance of cobweb.

Arbor   A Latin word meaning 'tree'.

Arborescent   Tree-like; attaining the size or character of a tree.

Arboretum (pl. Arboreta)   A place where trees, shrubs, and other primarily woody plants are grown for research, education, and introduction purposes.

Arbuscula   A dwarf tree or a shrub looking like a tree.

Arcuate   Like an arc; bent like a bow.

Arenaceous, Arenarius   Growing best on sandy soil.

Areole, Areola (pl. Areolae)   Small, irregular spaces formed by the veinlets of a leaf; small areas on the stem of cacti (Cactaceae) which bear the hair-like spines.

Areolate   With small pits; divided into many square spaces.

Argentate, Argentius   Silvery in appearance.

Arhizal, Arhizous   In reference to plants without true roots, as *Psilotum* (psilotaceae).

Aril   An outgrowth or covering of a seed originating from the helium or funiculus (cf. aril in Taxaceae and Podocarpaceae).

Arillate   Seed with an aril.

Arista   An awn; a bristle-like appendage on the glums of grasses.

Aristate   Awned; bearing a stiff, bristly appendage, as in Poaceae.

Armature   Bristles, barbs, hooks, prickles, spines, or thorns.

Armed   Any structure with stiff, sharp bristles, spines, or thorns.

Armor A covering of old leaf bases on cycads, palms, and some ferns.

Aromatic Having an aroma, as when some leaves are crushed; a group of organic compounds, as essential oils of plants.

Arrangement Disposition of organs or parts with respect to one another.

Articulate Jointed; easily separating at nodes or joints, as in spikelet of grasses.

Artificial classification A classification or key that does not consider natural relationships, and is based on few characteristics.

Arundinaceous Reed like, as in stem of some tall grasses.

Arvensis A Latin word meaning 'of arable land'.

Ascending, Ascendent Growing obliquely upward or curving upward as the plant grows. The term is applied to branches, hairs, and ovules.

Asepalous Without sepals.

Asexual Without sexual organs; without spores or seeds (cf. Vegetative).

Asperate, Asperous Having a rough surface (syn. Scabrous).

Asperulate Having minutely rough surface.

Assurgent Directed upward or forward (syn. Ascendant).

Astemonous Without stamens.

Astigmatic Without stigmas.

Astylocarpellous Without a style and a stipe.

Astylocarpepodic Without a style, but with a stipe.

Astylous Without a style.

Asymmetric Without symmetry; irregular in shape and outline (syn. Zygomorphic).

Atro- A Latin prefix meaning "black,", "dark"

Atropurpureus A Latin word meaning "dark purple."

Atrovirens A Latin word meaning "dark green."

Atropous With straight body so that funicular attachment is at one end and micropyle at the other (syn. Orthotropus).

Attenuate Narrowed; terminating in a long, slender point, as in the apex of a leaf.

Atypical Not typical; unusual; abnormal.

Aureus A Latin word meaning "golden."

Auricle An ear-shaped appendage, as in the top of the leaf sheath in grasses.

Auriculate With auricles or "ears" as in hastate or sagitate leaves.

Australis Southern.

Auto- A Greek prefix meaning "self."

Autogamy Self-pollination.

Autumnal Appearing in autumn.

Awl-shaped Leaf Narrow and gradually tapering to a sharp point, as in juvenile leaves of *Juniperus* (syn. Subulate).

Awn A stiff, bristle-like appendage. A beard, as on the tip of the glumes and lemmas of most grasses.

Axial Belonging to the main axis.

Axil The angle formed by a leaf with the stem to which it is attached.

Axile Attached to the central axis; central in position.

Axile Placentation See Placentation.

Axillary Situated in the axil of a leaf.

Axis The line running lengthwise through the center of an organ, as in a flower or a stem. The term is also applied to the stem itself or to the receptacle of a flower.

B

Bacca A berry; a fleshy fruit with seeds immersed in the pulp, as in blueberry (*Vaccinium*).

Baccacetum An aggregation of berries.

Baccate Berry-like; pulpy or fleshy as a berry.

Bacciferous Producing berries.

Baculate Pollen with pillar-like processes.

Baculum (pl. Bacula) The rod-shaped processes of baculate pollen.

Baculiform Rod-shaped.

Badious, Badius Latin word meaning "chestnut brown."

Balausta Many-seeded, many-loculed indehiscent fruit with a leathery pericarp, as in *Punica granatum* (pomogranate).

Ballochore A plant which is dispersed by explosive dehiscence of the fruit, as in many Acanthaceae, Euphorbiaceae, etc.

Banded Transverse stripes of one color crossing another.

Banner The large, broad, upper petal in the flower of Fabaceae (syn. Standard).

Barb A hooked or doubly hooked trichome.

Barbate Bearded, usually with a tuft of long, stiff hairs.

Barbed Having barbs.

Barbellate Minutely barbed, usually with short, stiff hairs.

Barbellulate, Barbulate Finely barbed.

Bark Protective suberized tissue on the stem of woody plants.

Barochore A plant dispersed by means of gravity, because of heavy seed.

Basal Situated at the base.

Basal Group A term used in phylogenetic studies in reference to early diverging ancestral group.

Basal Placentation See Placentation.

Base The end or the lower portion of a plant organ nearest to the point of attachment to another organ.

Basicaulous Near the point of attachment of a stem.

Basicidal Capsule A capsule that dehisces by basal slits or fissures, as in Aristolochia (Dutchman's pipe).

Basifixed Said of an anther which is attached by its base to the filament.

Basifugal See Acropetal.

Basipetiolar At the base of the petiole.

Basilaminar Oil glands at the base of the leaf blade.

Basipetal Development of leaves or flowers from the apex to the base of the axis. (opp. Acropetal)

Basipetiolar At the base of the petiole.

Basiramous Leaves at the lower part of the branch.

Basiscopic Facing basally, often in reference to stamens.

Beaked A prominent stiff projection on certain carpels, fruit, or seeds.

Beard An awn, as in the grasses. A cluster of hairs in some corollas, as in some species of *Iris*.

Bedding Plants A common term used in horticulture to designate annual or perennial flowering plants used in the garden for temporary color, usually replaced seasonally.

Bent Embryo Foliate embryo with expanded and usually thick cotyledons in an axile position bent upon the hypocotyl in a jacknife position.

Berry A fleshy indehiscent fruit, from one or more carpels and with one to many seeds. For example, *Persea* (avocado), *Lycopersicum* (tomato), etc. (syn. Bacca).

Bi- From the Latin word Bis, meaning "two," "twice," or "having two."

Bibacca A fused double berry, as in Lonicera (honeysuckle).

Bicalcarate With two spures or projections.

Bicarpellate A gynoecium which consists of two carpels.

Biconjugate Said of a compound leaf with two divisions, each having two leaflets.

Bicrenate A leaf with doubly crenate margin.

Bidentate A leaf margin with double teeth.

Biduous A plant lasting two days.

Biennial A plant that lives through two growing seasons, flowering and fruiting the second year, then dying. For example, *Dacus carrota* (carrot).

Biferous Appearing twice annually; as two crops in one season.

Bifid Two-cleft; forked into two limbs to near the middle. For example, stigmata of the Euphorbiaceae

Biflorus Two-flowered; flowering twice annually, as spring and fall.

Bifoliate A plant with two leaves, as Welwitchia mirabilis.

Bifoliolate A compound leaf with two leaflets.

Bifurcate Branched twice, as a Y-shaped trichome.

Bigeminate In two pairs; with two orders of leaflets (syn. Bijugate).

Bijugate See Bigeminate.

Bilabiate Two-lipped, referring especially to the corolla, as in Scrophulariaceae or Lamiaceae (Labiatae)..

Bilateral In reference to pollen with two vertical plains of symmetry. Also see Zygomorphic.

Bilocular Two-celled, as applied to an ovary or a fruit with two locules.

Bimestrial Lasting two months; flowering every two months.

Binate Occurring in pairs; in two pairs.

Binomial, Binomial Nomenclature, Binomial System According to the code of botanical nomenclature, which is fully accepted by horticultural taxonomists, the name of each plant species consists of two parts: A generic name and a specific epithet, both in Latin or Latinized form.

Biogeography The geographical distribution of living organisms (cf. Phytogeography).

Biological Classification Hierarchical system that reflects evolutionary history of a group.

Biosystematics The field of study concerned with variation and evolution of species or species complexes, involving among other studies, breeding systems, and cytogenetic.

Biovulate Having two ovules.

Bipartite Split into two, nearly to the base.

Bipalmate Palmately cleft or divided with lateral lobes cleft or divided (syn. Pedate).

Bipalmately Compound With two orders of leaflets, each palmately compound.

Bipinnate Twice-pinnate, referring to compound leaves.

Bipinnatifid Twice-pinnatifid. For example, *Nandina domestica* and certain ferns.

Bisected Completely divided into two parts.

Biseptate Having two partitions.

Biseriate In two rows or series; in two whorles or cycles, as calyx and corolla.

Biserrate Twice serrate; doubly serrate, as a leaf margin

Bisexual Having both stamens and carpels (a "perfect" flower).

Biternate Divided into three parts, each of which is divided into three.

Blade The expanded portion of a leaf, which is usually broad, flat, thin, and green.

Blasto- A prefix from the Greek word Blastos, meaning 'bud'.

Blastocarpous Germination of seeds while within the pericarp, as in *Rhizophora* (Rhizophoraceae) (cf. Vivipory).

Bloom The white waxy covering on many leaves and fruits; also synonymous with blossom or flower.

Blotched The color disposed in broad, irregular spots, on leaves or petals.

Bole The main trunk of a tree; a caudex.

Boll The fruit of the cotton (gossypium) plant.

Bolting The process of flower stalk formation in rosette plants.

Bonsai The exaggerated dwarfing of plants by means of root and shoot pruning; a plant so created.

Boraginaceous Hairs Uni- or multicellular, conical, calcified, and/or silicified bristles, often characteristic of Boraginaceae.

Bordered One color surrounded by a narrow margin of another, often characteristic of cultivars, as in *Euonymous* (Celastraceae).

Boreal Northern.

Boss A protuberance (syn. Umbo).

Bostryx A cymose inflorescence with flowers all born on one side of the axis; a helicoid cyme.

Botryose, Botryoid, Botrytic Resembling a bunch of grapes; racemose.

Botuliform Cylindrical, with rounded ends; sausage-shaped.

Brachy- A prefix from the Greek word Brachus, meaning 'short'.

Brachyblast A short shoot (syn. Spur).

Brachycladous With very short branches.

Bract A much-reduced leaf, usually subtending a flower or an inflorescence. For example, red bracts of *Euphorbia pulcherima* (poincettia), colored bracts of Bogainvillea (cf. Involucre; phyllaries).

Bracteal Pertaining to bracts, as bracteal nectary or glands.

Bracteate Furnished with bracts.

Bracteole A small bract.

Bractlet A small secondary bract born on the pedicel.

Bramble Any prickly, sprawling shrub or vine, as blackberry.

Brevi- A Latin prefix meaning "short."

Breviloba, Brevilobus With short lobes.

Bristle Stiff hairs.

Bristly Covered with stiff hairs; like a bristle.

Brochidodromus Leaf venation, consisting of a single primary vein, the secondary veins not terminating at the margin but joined together in a series of prominent upward arches or marginal loops on each side of the primary vein.

Bud A resting structure which may develop into an inflorescence, or stem.

Bud scales Specialized protective leaves which cover the buds during winter months to prevent desiccation and injury.

Bud Scale Scars Scars left on the branch by the abscission of the bud scales from the terminal buds of the previous year.

Bulb A short thick modified stem, the leaves of which are thickened and store reserved food, as in Amaryllidaceae (*Eucharis, Hyppeastrum*, etc.) and Alliaceae (*Allium, Tulbaghia*, etc.)

Bulbel A bulblet that arises from the mother bulb.

Bulbiferous A bulbous plant.

Bulblet A small bulb.

Bulbil Bulblets born singly or in clusters in the leaf axils (as in some Liliaceae, e.g., *Lilium*) or in the inflorescences (as in some Amaryllidaceae, e.g., *Crinum*).

Bulbous Having the characteristic of a bulb; producing bulb.

Bullate Blistered or puckered, as in surface of some leaves.

Bundle scars Scars left in a leaf scar by the severance of vascular bundles at leaf fall.

Bur A rough and spiny covering of a fruit; sometimes the fruit itself.

Bush A vernacular word referring to small shrubs or sub-shrubs. In plant geography, it denotes scrubby vegetation.

Buttress Tree trunk with a widening base, caused by the extension of the roots on the soil surface.

C

cDNA Complementary DNA synthesized from a messenger RNA (mRNA) template, used in cloning genes.

Caducous Falling off early, as the sepals or petals of certain flowers.

Caeruleus A Latin word meaning "pale blue."

Ceasious Having a grayish-blue waxy bloom.

Caesius A Latin word meaning "lavender."

Caespitose, Cespitose In tufts; in cushions.

Calcarate Spurred, as in flowers of *Impatiens* or *Delphinium*.

Calceiform, Calciolate Shaped like a shoe or slipper, as in flower of *Paphiopetalum* (Orchidaceae).

Callosity A callus; a thickened area on bark.

Callus A thickened, raised area, which is usually hard.

Calybium A hard unilocular, dry fruit derived from an inferior ovary, as in *Quercus* (Fagaceae).

Calyc- A Greek prefix meaning "a cup."

Calycanthemy An abnormal condition where the sepals become petaloid.

Calycicular, Calyciform Cuplike; calyx-like, as in short bracts of the outer involucre in some members of Asteraceae.

Calyciflorous A condition where stamens and petals are adnate to the calyx.

Calycle See Epicalyx.

Calycule, Calyculus A group of leaflike bracts below the calyx.

Calyptera A hood or a lid, as in the lid of Eucalyptus fruit.

Calyx (pl. Calyces) The outer set of the perianth segment of a flower; collective term for sepals. It may be distinct or connate, and in some plants, it may resemble the petals (cf. tepals).

Calyx lobe The free portion of a gamopetalous calyx.

Calyx tube The tube of a gamopetalous calyx. Sometimes used for the hypanthium or receptacle of perigynous and epigynous flowers.

Cambrian A geological period from 500 to 420 million years ago.

Camelinus A Latin word meaning "camel-colored," or "tawny."

Campanulate Bell-shaped, usually in reference to form of flowers, as in Campanulaceae.

Campestris A Latin word meaning 'growing in the field'.

Camptodromous A type of leaf venation in which secondary veins curve toward the margin but terminate before reaching it.

Campylodromous A type of leaf venation in which several primary veins originate from the same point, curve out, then converge to join again at the apex.

Campylotropous An ovule which is bent so as to bring the micropyle (apex) and funiculus (stalk) close together.

Canal A large air space in the stem, often containing resin or oil.

Canaliculate With channels or grooves, usually in reference to petioles or midribs.

Cancellate Being latticed, as in a network (syn. Clatrate).

Candidus A Latin word for "pure white."

Canescence, Canescent Bearing a hoary, grayish pubescence.

Canopy The cover formed by leaves and branches of plants.

Cantharophilous, Cantharophily Beetle pollination.

Canus A Latin word meaning "gray-white."

Cap The removable covering of a part. For example, operculum of a capsule (syn. Pileus).

Capillary Hair-like, thread-like (syn. Filiform).

Capillate Hair-shaped; resembling a hair.

Capitate Head-like; collected into a dense cluster, as the inflorescence of Asteraceae; Like a pinhead, as in gland-tipped hairs.

Capitulum A dense inflorescence comprised of an aggregation of sessile or subsessile flowers.

Capriolate, Capriolatus A Latin word meaning 'having tendrils'.

Capsular Belonging to or having the characteristics of a capsule.

Caprification The formation of an inflorescence composed of a number of flowers on a common receptacle, as in a *Ficus* or *Dorstinia* (Moraceae); artificial pollination of figs by using caprifig.

Caprifig A cultivar of fig which does not produce edible fig but houses wasps of the family Agagonidae, which pollinate the figs.

Capsule A dry dehiscent fruit (pod); with two or more carpels (compound ovary), usually with several to many seeds.

Character A heritable feature.

Character Polarity In cladistics analysis, primitive or derived state of characters used in study.

Character State Refers to presence or absence of a character, e.g., stipules (the character) present (the state).

Carboniferous A geological period from about 270 to 220 million years ago.

Cardinal, Cardinalis A Latin word meaning "red."

Carina A keel, as in the lower surface of glumes and lemmas of many grasses; the two fused lower petals of papilionaceous flowers.

Carinate Keeled; having a carina.

Carneus A Latin word meaning "flesh colored."

Carnivorous Plants Plants that are capable of catching and digesting insects and other small animals.

Carnose Fleshy in texture.

Carpel The ovuliferous organ of the flower; a simple pistil or one of the segments of a compound pistil. It is a modified seed-bearing leaf, consisting of three segments: The swollen basal portion (ovary), the elongated portion (style), and a receptive head (stigma) (cf. Gynoecium).

Carpellate Possessing carpel(s), in reference to pistilate (female) flowers.

Carpophore A thin wiry stalk that supports each half of the pendulous fruit (Mericarp) in Apiaceae.

Carpopodium Fruit stalk; short, thick, pistillate stalk.

Carpotaxus Arrangement of fruit on an axis.

Cartilaginous Hardened and tough, but capable of being bent.

Caruncle A white oily outgrowth at or near the hilium of certain seeds, as in some members of the Euphorbiaceae (*Ricinus, Jatropha*, etc.). Carunculate seeds are often dispersed by ants.

Caryopsis The one-seeded, indehiscent grain (fruit) of grasses.

Cassideous A Latin word meaning "helmet-shaped."

Castaneous A Latin word meaning 'chestnut-colored', 'dark brown'.

Casing See Tube

Catadromous Compound leaves in which the first leaflet arises toward the apex.

Cataphyll Small scale leaves of the rhizome in angiosperms; protective winter bud scales of trees and shrubs; scale leaves of cycads. They function in storage and protection (cf. Hypsophylls).

Catkin An ament; an elongate, pendulous, cluster (spike) of unisexual, apetalous, and often bracteate lowers, as in Salicaceae, Betulaceae, Fagaceae, etc.

Cauda A Latin word meaning "a tail" or "any tail-like appendage."

Caudate Bearing a tail-like appendage, as in leaves of some Araceae and Liliaceae (*Gloriosa* spp.).

Caudex The short thickened stem of some xerophytes. It may be above or below ground, function as water storage, and occurs in several angiosperm families. For example, Agavaceae (*Nolina recurvata*), Apocynaceae (*Pachypodium, Adenium*), Passifloraceae (*Adenia*),etc. The word is sometimes used in reference to trunk of some cycads.

Caul- A Latin prefix meaning "stem."

Caulescent With a leafy stem (cf. Acaulescent).

Cauliflory Bearing flowers on old stems, as in some Fabaceae (*Cercis canadensis*).

Cauline Growing on a stem; belonging to a stem or branch.

-Caulis A Latin suffix meaning 'stemmed'.

Caulocarpic A plant which survives after flowering to produce more flowers in subsequent years.

Caulous With branches more or less evenly spaced along the trunk.

Centionate Spotted with different colors, resembling a patch work.

Centralium The central longitudinal cavity found in some palm seeds.

Centrifugal In inflorescences, blooming from the inside outward or from top to bottom (syn. Basipetal).

Centripetal In inflorescences, blooming from the outside inward or toward the axis (syn. Acropetal).

Centroramous At the center of the branch.

Ceraceous, Ceriferous Waxy; wax bearing (syn. Ceriferous).

Cereal Any grass whose seeds (fruits) serve as food.

Cernuous Drooping.

Cespitose or Caespitose Matted; forming a dense cushion; forming dense clumps, as in ground covers or turf.

Chaff Small, more or less dry, membranaceus bract; especially the small bracts at the base of disk flowers of some Asteraceae. Flower parts of the cereal grains removed during milling.

Chalaza The basal part of an ovule where the stalk of the funiculus enters. The end of a seed opposite the micropyle.

Chalazagamy The entry of pollen tube through the chalazal end of an ovule (cf. Porogamy).

Chamaephyte A perennial herb or low shrub, with buds less than one-half meter above ground.

Chambered Pith In stems, where a solid core is replaced by a partitioned pith.

Channeled With one or more longitudinal grooves. Hollowed like a gutter, as in the petiole of some monocotyledons.

Character Taxonomically, any well-defined feature which distinguishes one taxon from another; genetically, any feature which is transmitted from parents to offspring, or may result from gene- environment interaction.

Chartaceous Papery in texture.

Chasmantheric Pollination Normal mode of pollination, where pollen is transferred from a dehisced anther and germinates on the stigma (cf. Cleistantheric Pollination).

Chasmogamous Said of a flower which opens before pollination (cf. Cleistogamous).

Chasmogamy The opening of the perianth at flowering time.

Chasmophyte A plant that grows on rocks, rooted in debris in the crevices.

Cheirostemonous With five stamens united at the base.

Chemotaxonomy The use of chemical evidence to establish or ascertain taxonomic relationships in plants.

Chimera A plant composed of two genetically distinct tissues adjacent to one another. For example, by spontaneous mutation, as in branches with variegated leaves in otherwise normally green plants (e.g., *Sansevieria trifasciata* 'Laurentii'), or may be artificially induced by grafting (Periclinal chimera) or colchicine treatment.

Chiropterophily, Cheiropterophily Pollination by bats.

Chlamydeous Having a perianth.

Choripetalous, Choriopetalous Having distinct and separate petals; polypetalous.

Chorisepalous, Choriosepalous Having distinct and separate sepals; polysepalous.

Chromosome The rod-like structures of definite number formed during cell division, carrying the genes and containing DNA.

Chylocauly Having a succulent stem.

Chilophylly Having succulent leaves.

Ciliate Fringed with hairs along the margins.

Cilia Minute short trichomes.

Ciliatulate Having widely dispersed cilia.

Ciliolate Fringed with very short fine hairs.

Cincinnal, Cincinnate Curled; rolled around.

Cincinnus A cymose inflorescence having a lateral axis arising on alternate sides of the main axis, with short, tightly arranged internodes; a helicoid cyme.

Cinerious, Cinerus A Latin word meaning "Ash-gray color."

Cingulate Edged all around.

Cinnabarine A Greek word meaning "bright orange-red."

Cinnamomeous A Latin word meaning "light yellowish-brown."

Circinate Said of a leaf that is coiled or rolled from the tip toward the base with the lower surface outermost and the tip near the center. For example, circinate vernation of ferns.

Circum- A Latin prefix meaning "all around."

Circumcinct Having a band around the middle

Circumalate Winged circumferentially.

Circumferential At or near the circumference; around a circular structure.

Circumscissile Dehiscing or separating by a circular zone, as the valve of a capsule coming off as a lid.

Cirrhous, Cirrhus A coiled apex; a curl, tendril-like.

Citriform Lemon-shaped.

Clade From the Greek word "klados", meaning branch. Studies used to denote a monophyletic taxon where all its branches (e.g., species of a genus) have a common ancestor.

Cladocarpus Having the fruit at the end of a lateral branch.

Cladistic Genetic relationships which reflect or imply recent common ancestry.

Cladistic Analysis Numerical analysis of morphological data, used as a taxonomic tool to prove or disprove common ancestry.

Cladode, Cladophyll A branch modified in the form of a leaf; a flattened photosynthetic stem as in certain species of *Asparagus* (Asparagaceae) and *Epiphyllum* (Cactaceae).

Cladodromous A type of leaf venation where there is but a single primary vein, and secondary veins freely branch toward but do not terminate at the margin.

Cladogenesis Creation of new clade; speciation.

Cladogenous Producing flowers at the end of a branch.

Cladogram A branching diagram that shows the cladistics relationship between species; a diagram that shows common ancestry of related groups.

Cladoptosis, Cladoptosic The shedding of branches, as in some deciduous conifers, for example, *Larix*, *Metaseqouia*, etc.

Clambering Sprawling across objects, but not climbing, as Wisteria spp. and other vines.

Clasping A sessile leaf which partly or wholly surrounds the stem, as in most monocots.

Class A group of plants ranking above an order and below a division.

Classification The orderly (systematic) arrangement of plants into groups, or their assignment to a hierarchy, based on common characteristics. For example, a group of species makes up a genus, and a group of genera constitutes a family, etc.

Clathrate, Clathroid Like a net; latticed.

Clavate, Claviform Club-shaped; gradually thickened upwards.

Claviculate Having tendrils or hooks.

Claw The narrow stalk (base) of some petals, resembling a petiole, as in some members of Brassicaceae (mustard family).

Cleft Divided to the middle, as in palmately lobed leaves in which the sinus is only about halfway to the midrib.

Cleistantheric Pollination Pollen not transfered from a normally dehiscent anther by a pollen agent, pollen germinates within the anther, and the pollen tube subsequently penetrates the anther wall and ovary wall into the ovule and the embryo sac (cf. Chasmentheric Pollination).

Cleistogamous Self-pollination or self-fertilization in the bud stage, or before anthesis, as in *Viola* (Violaceae).

Cleistogene A plant which produces cleistogamous flower.

Climber, Climbing A plant which grows upwards and uses other plants or objects as support. Various specialized organs may be used for this purpose, such as roots, claws, or growing tip of the plant, etc.

Clinanthium The compound receptacle of the Asteraceae flower.

Cline A gradual and more or less continuous change in phenotypic expression of a character over the geographical range of a species. In horticulture, clinal variations are often used as cultivars.

Clinium See Clinanthium.

Clone A group of genetically identical individuals resulting from vegetative propagation (cuttings, grafting, tissue culture, or division). In horticulture, this is considered one of the categories of cultivar.

Closed Venation Venation of a leaf characterized by anastomosing veins.

Clouded Not clear, as some plant latex which is not milky but appears opaque.

Cluster, Clustered A general term used to designate a closely crowded inflorescence of small flowers.

Clypioli Shield-shaped basses from which leaves arise.

Coalescence The union of like parts or organs.

Coccineus A Latin word meaning "scarlet."

Coccoid Coconut-shaped; rounded.

Cochilea A tightly coiled leguminous fruit.

Cochleate Coiled, like a snail shell, in reference to position of an embryo.

Coelospermous Hollow-seeded; having boat-shaped seeds; seedlike carpels of Apiaceae with incurved ventral face.

Coeno- A Greek prefix meaning "living together."

Coenocarpium Multiple fruit derived from ovaries, flower parts, and receptacle as in Annanas (pineapple).

Coetaneous Flowering while the leaves are expanding (syn. Synantherous).

Coherent Union (but not actual fusion) of the same kind of organs, as in petals or sepals (cf. Connate).

Coleoptile The sheath surrounding the epicotyl (embryonic stem) of grasses.

Coleorhiza The sheath surrounding the radicle (embryonic root) of grasses

Collar The transition zone between the primary stem and root (syn. Collet).

Collateral Placed side by side, as some carpels and ovules, or as in the glume of some grasses, or as the xylem and phloem in certain vascular bundles.

Collet See Collar.

Convergence Independent development of similar characteristic in unrelated taxa (cf. Homology).

Copate A pollen grain bearing colpi.

Colporate Said of pollen grains with compound apertures

Colpus (pl. Colpi) An oblong furrow or aperture in the exine of pollen grains, running from one pole to another and having a length/width ratio greater than two. Gymnosperms, monocotyledons, and primitive dicotyledons (magnoliales) have monocolpate pollen, whereas most of dicotyledons have two or more colpi.

Columella The central axis of a fruit, as in Apiaceae.

Column The part of the flower of an orchid which is formed by the fusion of the style and the filaments, and which supports the anthers and the stigma (syn. Gynostemium); the basal twisted portion of an awn in grasses; the fused staminal tube in Malvaceae; the structure that supports the hood of a pitcher plant leaf.

Columnar Trees which have an erect main trunk but missing or self-pruning lateral branches, as *Pinus* (pines) and arborescent single-trunked palms.

Coma The trichomes attached to the testa of some seeds; the leafy crown of such plants as palms and cycads.

Communis A Latin word meaning "social" or "general."

Comose Having tufted hairs.

Companulate Bell-shaped, as in flowers of Companulaceae.

Complanate Flattened; compressed.

Complete Flower A flower with all parts present: sepals, petals, stamens, and carpels.

Complete Leaf Leaf with blade, petiole, and stipules.

Compound Composed of a number of similar united parts, as carpels in a syncarpus gynoecium; divided into a number of similar parts or divisions, as leaflets of a compound leaf.

Compound Carpel An ovary produced by connation of two or more carpels (syncarpus gynoecium).

Compound Corymb A branched corymb.

Compound Cyme A branched cyme.

Compound Fruit See Aggregate Fruit.

Compound Inflorescence An inflorescence with secondary branches (co-florescences), as in some Euphorbiaceae (e. g. *Jatropha*).

Compound Leaf A leaf with two or more leaflets; in some cases, the lateral leaflet may have been lost (e.g., *Citrus*) and only the terminal leaflet remains. Ternately Compound, when the leaflets are in 3's. Palmately Compound, when all leaflets arise from a common point at the end of the petiole, as in *Schefflera*. Pinnately Compound, when leaflets are arranged along a rachis, as in species of Anacardiaceae (e.g., *Pistacia*). Odd-Pinnate, when the total number of leaflets is an odd number and a single leaflet terminates the leaf (e.g., *Carya, Juglans*; Juglandaceae). Even-Pinnate, when the total number of leaflets is an even number and there is no terminal leaflet (e.g., *Koelreuteria*, goldenrain tree).

Compound Umbel A raceme consisting of a large number of heads rising close together at the end of a main branch, as in *Fatsia* (Araliaceae).

Compressed Flattened (syn. Complanate).

Con- A Latin prefix meaning "with"; "together."

Concinnus A Latin word meaning "neat."

Concolor Of uniform color.

Concrescent Two organs that are separate when immature but become fused when mature, as in fused stamens and carpels.

Condensed An inflorescence with nearly or completely sessile flowers crowded together.

Conduplicate A leaf or leaflet in which the blade is folded lengthwise along the midrib, as in leaflets of some palms (cf. Induplicate and Reduplicate).

Cone (see Strobilus).

Conferted Crowded, close together.

Configuration Pattern usually resulting from internal structural form. The term is not applicable to venation of leaves.

Confluent Blending, running into one another.

Congeneric Belonging to the same genus.

Congested Crowded (cf. Lax).

Conglomerate Clustered.

Conglutinate Stuck together in a sticky mass.

Conical Cone-shaped.

Conifer A general term referring to the Coniferales, one of the four groups of gymnosperms, as *Juniperus, Cupressus, Taxodium, Taxus, Podocarpus*, etc.

Coniferous Cone-bearing, as in conifers.

Conjugate Joined; occurring in pairs.

Connate Union or fusion of similar structures, as in petals, sepals, or leaves.

Connate-Perfoliate When the bases of two opposite, sessile leaves appear to have fused around the stem, as in the juvenile form of *Eucalyptus* spp., or terminal leaves of *Lonicera sempervivum*.

Connective An extension of the anther, connecting the two cells of the anther.

Connivent Touching but not fused (syn. Adherent, Coherent).

Conopodium A cone-shaped floral receptacle.

Conspecific Within or belonging to the same species.

Constricted Narrowed suddenly.

Conocarpium An aggregate fruit on a conical receptacle, as in strawberry.

Contiguous Touching but not fused, regardless of whether parts are similar or dissimilar.

Continuous Variation Variation between individuals of a taxon, so that differences intergrade into each other.

Contorted Twisted around a central axis, often in reference to sepals or petals in bud which are twisted so that they overlap on one side (aestivated) (syn. Convolute).

Contortuplicate Twisted and folded, in reference to cotyledons.

Contractile Roots Roots of certain plants that contract and pull the plant deeper into the soil. For example, in bulbous plants (Amarillydaceae), ferns (*Botrychium*), cycads (*Zamia*), etc.

Convolute   Said of a leaf or floral envelope in which the blade or petals and/or sepals are rolled lengthwise from side to side like a scroll, often occurring in the bud.

Coralloid   Coral-like.

Coralloid Roots   The dichotomously branched, nodulated roots of cycads which occur on or near the soil surface (Apogeotropic) and are associated with blue-green algae (*Nostoc, Anabeana*), which are nitrogen fixers.

Cordate (Cordiform)   Heart-shaped, in reference to the base of a leaf.

Coriacious   Thick and leathery, often in reference to leaf texture.

Corm   A short, erect, thick, solid, subterranean stem which is enclosed within dry scale-like leaves, and functions in food storage (e. g. *Gladiolus*).

Cormel   A corm arising vegetatively from a mother corm.

Cormophyte   A plant with a stem, root, and leaf.

Corneous   Having the texture of horn.

Corniculate   Bearing or terminating in a horn-like protuberance or process.

Corolla   A collective term referring to petals of a flower; the inner perianth segments. If the petals are separate, the corolla is said to be polypetalous, but if the corolla is fused (connate) then it is gamopetalous or sympetalous.

Corolla Tube   A tube-like structure resulting from fusion of the petals along their edges, as in tubular flowers of some Amaryllidaceae and Rubiaceae.

Corolliform   When the calyx resembles a corolla, as in flowers of *Hydrangea* , *Mirabilis*, etc.

Corona   A collar-like or tubular appendage of the corolla and the stamens, as in *Narcissus* (not to be confused with staminal cup of such genera as *Eucharis*). The outgrowth of the staminal part of the flowers in milkweeds (Apocynaceae-Asclepiadaceae).

Coronate   Having a corona.

Coroniform   Having the shape of a corona; crown-shaped.

Corrugate   Irregularly folded or wrinkled.

Cortex   Parenchymatous tissue in the stem or root between the vascular bundles and the epidermis.

Corymb   A flat-topped racemose inflorescence, the main axis of which is elongated, but the pedicels of the older flowers longer than those of the younger flowers (the outer flowers open first).

Corymbose   Arranged in corymbs.

Cosmopolitan   An organism which is worldwide in distribution.

Costa (pl. Costae)   The midvein of a single leaf in angiosperms or the pinna or pinnule of ferns, or the rachis of a pinnately compound leaf.

Costal   along the main veins of a leaf, in reference to oil or resin glands or trichomes.

Costapalmate   A petiole which extends into the palmately compound leaf of certain palms, as in *Sabel palmetto*.

Costate   Coarsely ribbed.

Cotyledon   Seed leaf; embryonic leaves which often appear upon germination of seeds.

Cotylespermous   With food reserve in cotyledons, derived from zygote.

Cotyliform   Disk-shape.

Cotyloid   Concave or cup-shaped.

Crampon   A hook or adventitious root which acts as support.

Craspidodromous Venation   A type of leaf venation characterized by a single primary vein and with secondary veins terminating at the margin.

Crassus   A Latin word meaning "thick."

Crateriform   Saucer- or cup-shaped.

Creeper   A plant with trailing shoots which root along the length of the stem as it grows. Such plants are often used as ground covers in the landscape.

Cremocarp   A fruit splitting into two or more one-seeded portions, specifically in reference to Apiaceae (a schizocarp).

Crenate   Shallowly round-toothed; scalloped.

Crenulate   Minutely or finely crenate.

Crest   A ridge or an outgrowth on a structure, as in the hornlike projection from the hood of the corona in milkweed (*Asclepias*) flowers. In some plants such as cacti or cactoid euphorbs, the term refers to a grotesque form of flattened stems with irregular ridges (Crested, Cristate).

Cretaceous   A geological period from 140 to 70 million years ago. According to some authorities, this is the period in which flowering plants arose.

Cribiform   Like a sieve.

Crinite, Crinitus   A Latin word meaning "having soft hair."

Crispate   Curled, as in leaf margins or trichomes; an extreme form of undulate.

Cristate, Cristatus   A Latin word meaning "crested."

Cross   The interbreeding of two individuals; the product of such interbreeding.

Cross-Pollination   The transfer of pollen from the anther of one plant to the stigma of another.

Crown   The persistent base of a herbaceous perennial; the top of a tree; a corona; the junction between a stem and a root in a seed plant; the part of a rhizome with a large bud suitable for propagation.

Crown Group   A phylogenetic term that refers collectively to living and extinct species and their common ancestor (cf. Stem Group).

Crownshaft   A green pillarlike extention of the trunk in certain palms, formed by the overlapping petiole bases of new leaves, as in *Roystonea regia*.

Crozier   The curled end of the developing frond of a fern.

Cruciate   Cross-shaped, as in the arrangement of petals in flowers of Brassicaceae (Cruciferae).

Crucifer   Plants of the family Brassicaceae (Cruciferae).

Cruciform   See Cruciate.

Crustaceous   Hard and brittle.

Cryptanthous   With hidden flowers; the stamens remaining enclosed in the flower (syn. Cleistogamous).

Crypto-   A Latin prefix meaning 'hidden'.

Cryptocotylar Seedling   Having cotyledons remaining inside the seed coat; cotyledons usually remaining below ground (cf. Phanerocotylar).

Cryptogam   Plants that reproduce by spore, such as ferns, mosses, and algae (cf. Phanerogam).

Cryptophyte   A plant which forms dormant buds below the soil surface. For example, bulbs, corms, rhizomes, etc.

Ctenoid   Comb-like.

Cucullate, Cucullus   Hooded. Having a hoodlike structure, as in seeds with a caruncle.

Culm   The stem of grasses or bamboos.

Cultiform   See Cultivar.

Cultigen   Plants or group of plants originating in cultivation and known only from cultivation. For example, maize and cabbage (cf. indigen).

Cultivar   Cultivated variety. Plants selected from the wild or individuals cultivated because of a particularly desirable feature(s). This has been done for food, fiber, and landscape plants. For example, *Ilex vomitoria* "Nana."

Cultriform   Shaped like a knife.

Cuneate   Wedge-shaped; triangular, with the narrow end at the point of attachment, as the bases of leaves or petals.

Cuneiform   See Cuneate.

Cup   A hollow floral receptacle.

Cupreous   A Latin word meaning "copper-colored."

Cupulate   Cup-shaped, as the involucre of an acorn.

Cupuliform   See Cupulate.

Cupule   Cuplike structure at the base of some fruits, formed by the fusion of involucral bracts at their bases, as in some palms and oaks.

Curvative   With leaf lamina folded transversely into an arc (syn. arcuate).

Cusp   A sharp point as on teeth.

Cuspidate   Having a rigid, sharp point.

Cuticle   A layer of waxy or fatty material on the outer walls of epidermal cells, as in leaf surfaces.

Cutting   A method of vegetative propagation, using pieces of leaves, stems, or roots.

Cyaneus   A Latin word meaning 'Bright blue'.

Cyathiform   Cup-shaped.

Cyathium   A reduced inflorescence resembling a single flower, as in the subfamily Euphorbioidae, family Euphorbiaceae (*Euphorbia*, *Pedilanthus*, etc.). A cyathium includes one carpelate and several staminate flowers, as well as one or more large nectariferous gland. Not all members of the Euphorbiaceae possess a cyathium.

Cyclic   Having floral parts in whorls.

Cycly   The absolute number of whorls or parts, such as that of floral parts or leaves.

Cylindric   Any elongated structure with a circular cross-section.

Cymba   A woody boat-shaped spathe that encloses the inflorescences of many palms (cf. Manubrium).

Cymbiform   Boat-shaped.

Cyme   A broad, more or less flat-topped determinate inflorescence, with the central flowers opening first.

Cymule   A few flowered small cyme.

Cynarrhodion   A fleshy, hollow fruit, enclosing achenes, as in Rosa.

Cypsela   An achene derived from a unilocular, inferior ovary, as in the indehiscent fruits of Asteraceae.

Cytodeme   A group of individuals of a taxon differing from others cytologically, usually in chromosome number.

Cytogenetics   The combined study of chromosome cytology and genetic methods of breeding analysis.

Cytogeography   The study of the relationship between chromosome morphology (karyology) and geographical distribution.

Cytotaxonomy   The use of chromosome numbers and morphology as an aid to determining taxonomic and evolutionary relationships.

Cytotype   A plant or group of plants distinguished from others of the same species by one or more cytological feature.

D

Dactyloid   Fingerlike.

Dasycarpous   Having thick fruit.

Dasyphyllous   Having thick leaves.

Dealbate   Whitened, usually by a covering of hairs.

Deca   A Greek prefix meaning 'ten'.

Decandrous   Having ten stamens.

Decem (Deci)   A latin prefix meaning ten.

Deciduous   The falling of leaves at the end of a growing season by trees and shrubs, usually in autumn. The term may also refer to early shedding of petals, sepals, or stipules.

Decipiens   A Latin word meaning "deceiving."

Declinate   Bent downward or forward.

Decompound   More than once compound.

Decumbent   Reclining or lying on the ground, but with branch tips ascending and without formation of adventitious roots.

Decurrent   Said of trees without a central leader and many lateral branches (cf. Deliquescent, Excurrent). A leaf base which extends down the stem, resembling two wings, as in many Asteraceae.

Decussate Opposite leaves alternating at right angles with those above and below. A four-ranked leaf arrangement (cf. Distichous).

Definite When parts are always the same number in a given species, as in stamens.

Deflexed Bent sharply outward and downward.

Defoliation Loss of leaves, resulting from natural shedding, insects and diseases, or herbicides.

Dehiscence The method or process of opening of a fruit (seed pod) or anther; Loculicidal when the split opens into a cavity or locule, Septicidal when opening at the point of union of septum (partition) to the side, Circumscissile when the top valve comes off as a lid, and Poricidal when opening by means of pores.

Deliquescent Trees having many lateral branches but lacking a central leader, as in maple (*Acer*), elm (*Ulmus*), etc. (cf. Decurrent, Excurrent).

Deltoid Triangular.

Dendritic Much branched; branching like a tree as that of trichomes in species of Bromeliaceae or Brassicaceae.

Dendrogram A graphic method of illustrating taxonomic relationships based on similarity of characteristics.

Dendroid Tree-like.

Dendron A Greek word meaning tree.

Dentate A leaf margin with teeth.

Denticidal Capsule A capsule that dehisces apically, leaving a ring of teeth.

Denticulate Of a leaf margin having small teeth.

Denuded Naked, as a tree having lost nearly all its leaves.

Depauperate Dwarfed, stunted, or poorly developed.

Dependent Hanging down.

Depilation The natural loss of pubescence as the plant matures.

Deplanate Flattened.

Depressed Pressed downward close to the axis; more or less flattened endwise or from above.

Derived An organism descended from an ancestral form; modified version of a primitive characteristic.

Descending Growing or hanging downward, as in the branches of certain trees.

Desert A region of scant rainfall and with poor or more or less xerophytic vegetation.

Determinate Growth Growth of limited duration, Characteristic of leaves and inflorescences (cf. Indeterminate Growth).

Determinate Inflorescence An inflorescence in which the terminal flower develops first, thereby arresting further elongation of the axis (cf. Indeterminate Inflorescence).

Devonian A geological period 310 to 270 million years ago.

Dewy Covered with waxy platelets, appearing dewy.

Dextrorse Twining or twisting to the right, as in the stem of certain vines.

Di- A Greek prefix meaning two.

Diadelphous Stamens which have filaments grouped into two separate bundles. In legumes, there is one bundle with nine stamens and a solitary free stamen.

Diadromous Venation A fan-shaped venation pattern, as in Ginkgo biloba.

Diagnosis A brief description of a taxon (often in Latin), with special reference to the characteristics that distinguish it from related, morphologically similar taxa.

Diagnostic Characteristics Clearly defined characteristics which separate one plant from another.

Diageotropic Said of leaves or roots which grow horizontally.

Dialycarpic Having a fruit composed of distinct carpels.

Dialypetalous Polypetalous, the corolla being composed of many distinct petals.

Diandrous With two stamens per flower.

Diaphanous Translucent.

Diaspore Any unit of dessimination, such as seeds, spores, stem pieces, etc.

Dicarpellate Having two carpels per flower.

Dichasium A determinate cymose inflorescence with a central flower situated between two laterals develops first (usually female if plant is monoecious).

Dichlamydeous Having the perianth composed of distinct calyx and corolla.

Dichogamous, Dichogamy In a perfect flower, stamens and stigma maturing at different times so that self - pollination cannot occur.

Dichotomous, Dichotomy Forked branching, produced by division of the apical meristem into two branches and may be repeated several times. This feature is primarily characteristic of such primitive plants as Psilotum, but also occurs in advanced plants such as *Adenium* (Apocynaceae).

Dichotomous Key An identification tool in which each division is divided into two contrasting statements.

Diclesium An achene or nut enclosed within a free but persistent calyx, as in Mirabilis (Nyctaginaceae).

Diclinous Having unisexual flowers either on the same (Monoecious) plant or on different (Dioecious) plants.

Dicot An abbreviated term for dicotyledons.

Dicotyledon, Dicotyledonous One of the two divisions of angiosperms, which is characterized by two cotyledons, netted (reticulate) leaves, well-organized vascular bundles, a tap root, and flower parts in fours, fives, or rarely in twos (or multiples thereof) (cf. Monocotyledons).

Dicyclic A series of organs arranged in two whorles, as in calyx and corolla.

Didymous Occurring in pairs, as a fruit composed of two similar parts, or two stamens of equal size.

Didynamous With four stamens in two pairs of two different lengths, as in Lamiaceae.

Diffuse Loosely branching or spreading; open growth habit.

Digitate Diverging from a central point as in the fingers of a hand. For example, leaves of *Shefflera*.

Dilated Expanded or flattened.

Dimerous Whorl with two members.

Dimidiate Divided into two unequal halves; lop-sided.

Dimorphic Having two distinct forms.

Dioecious Having pistillate and staminate flowers on different plants, as in *Ilex* spp., Cycadales, etc.

Diplecolobal With cotyledons folded two or more times.

Diplostemonous Having the stamens in two whorls, the outer whorl alternate with the petals, and those of the inner whorl opposite the petals.

Diplotegium A pyxis derived from an inferior ovary.

Dipterous Two-winged.

Disarticulate To separate at a joint.

Disc A fleshy outgrowth from the receptacle of a flower beneath carpels or stamens; the receptacle in the head of Asteraceae (in this case the word is spelled Disk).

Disciform Disc-shaped.

Discoid Having only disk flowers, as in some Asteraceae, where the capitulum has no ray florets.

Discolor Not the same color throughout.

Discontinuous Distribution Occurrence of the same or related species in widely separated geographical areas (cf. Continuous Distribution).

Discrete Separate, not coalesced.

Disjunct Distribution The occurrence of related taxa in widely separated geographical areas.

Disk Flower The tubular flowers in the center of the heads of most Asteraceae flowers.

Dispersal Dissemination and spread of seeds or spores.

Dissected Divided into many slender segments.

Disseminule Any part of the plant that serves for dissemination and establishment.

Dissepiment See Septum.

Distal Opposite from the point of attachment, toward the apex; farthest from the point of attachment (cf. Proximal).

Distichous In two rows, on opposite sides of the stem; leaves that are arranged in two vertical rows, as in two-ranked leaves (cf. Decussate).

Distinct Separate, not united with parts in the same series, as in petals of a polypetalous flower (cf. Free).

Distribution Geographical area inhabited by a given taxonomic unit (a taxon).

Distylous, Distyly Said of a flower of a species which possesses one of the two style types: a long style ("pin" flowers), or a short style ("thrum" flowers). This arrangement promotes outcrossing.

Diurnal Opening only during daylight hours.

Divaricate Spreading far apart; forked.

Divergence A Latin word meaning separating.

Diversity Heterogeneity among individuals of a species.

Divided Cut or separated nearly to the base or to the midrib, as in a leaf.

Division The largest unit in the hierarchy of plants.

Dolabriform, Dolabrate Axe-shaped.

Doliform Barrel-shaped.

Domestication To bring plants into cultivation for a specific purpose.

Dorsal Back, referring to the back or outer surface of an organ, as the lower side of a leaf (syn. Abaxial; opp. Ventral).

Dorsifixed Attached medially to the back or dorsal surface, as in anthers of Lilium.

Dorsilaminar On the back (dorsal) side of the blade.

Dorsiventral Flattened and provided with a definite dorsal and ventral surface, such as a leaf

Dotted The color disposed in small round spots.

Double Flower A flower with more than the usual number of petals.

Double Cross A cross between two F1 hybrids.

Double-Serrate With course serrations bearing minute teeth in their margins.

Downy Covered with a fine coat of short soft hairs.

Drepanium A monochasial inflorescence in which all of the branches arise on one side of the main axis.

Drupaceous Of the nature or texture of a drupe, but not necessarily with the structure of one.

Drupe A fleshy one-seeded indehiscent fruit with a single seed enclosed within a stony endocarp, as in *Prunus* spp. (peach, cherry, etc.) (syn. Stone Fruit).

Drupecetum An aggregation of druplets.

Druplet A single fruit of an aggregate drupe, as in raspberry or blackberry.

Dry Fruit Any fruit which at maturity has dry ovary walls.

Dulcis Latin word meaning sweet.

Duration Length of time that a plant or any of its component parts exist.

Dwarf Very small, as selected cultivars of fruit trees or landscape plants.

Dyad Pollen grains occurring in clusters of two rather than four (tetrad).

E

E- or Ex- A Latin prefix meaning without; missing parts, as in estipulate or exstipulate (without stipules).

Ebracteate Without bracts.

Eccentric, Excentric One-sided; out of the center (syn. Abaxial).

Echinate Having a covering of spines or prickles.

Echma (pl. Echmata) The hardened, hook-shaped funicle in most Acanthaceae, which supports the seed (cf. Retinaculum).

Ecostate Leaves which lack a costa or midrib.

Ecotype An ecological variant of a species which is adapted to a particular environment. Although infrequently used in horticulture, this term is most useful in plant selection.

Ecto- A Greek prefix meaning "outside."

Ectocarp See Exocarp.

Ectosexine The outer part of the sexine of a pollen grain.

Edaphic Any soil factor that influences plant growth.

Edged Having a margin of a different color than the rest, as in the leaves of many cultivars (e.g., *Euonymus*).

Edulis A Latin word meaning "edible."

Ektexine The outer layer (surface) of a pollen grain, which is frequently sculptured.

Elaminate Leaf without a blade.

Electrophoresis A chemical method often used in solving complex taxonomic problems, particularly in species complexes. Electrically charged molecules migrate differentially along an electric field. Similarities and differences in the banding patterns are indicative of taxonomic relationships. The method is used to separate DNA molecules.

Ellipsoid Shaped like an ellipse, oblong with regularly rounded ends (cf. Elliptic).

Elliptic (Elliptical) Like an ellipse, broadest at the middle, tapering broadly and evenly toward each end.

Elongate Stretched out, lengthened.

Emarginate With a shallow notch at the apex.

Embryo A young plant developing from an egg cell (zygote) within a seed.

Embryo Sac The female gametophyte (Megaspore) of angiosperms.

Embryostega, Embryotega A disk-like callus on the seed coat of species in Commelinaceae.

Emergence, Emergent Outgrowth from epidermal or subepidermal cells, as in prickles.

Emersifolius With emergent leaves.

Enation An epidermal outgrowth, such as prickles on rose stems.

Endarch The direction of differentiation of xylem toward the center of the axis, as in the stem of seed plants.

Endemic Native and geographically restricted to a given area (cf. Indigenous).

Endo- A Greek prefix meaning "within".

Endocarp The inner layer of a fruit wall, usually woody, as in the stoney part of a drupe.

Endogamy Inbreeding between closely related individuals, as in Line or Line Hybrids which result in homozygocity.

Endogenous Growing from the inside, as in growth of the stem in thickness or initiation of lateral roots (cf. Exogenous).

Endosperm The tissue with stored food (albumen) or oil in seeds of angiosperms. In gymnosperm seeds, similar material is referred to as "female gametophyte tissue."

Ensate, Ensiform Sword-shape.

Entire Without indentation or lobes on the margins, as in leaves of Magnolia.

Entomophilous, Entomophily Pollination by insects (cf. Anemophilous).

Eocene A geological period 60–45 million years ago.

Eophyll The first few transitional leaves in seedlings before the growth of adult leaves.

Epappose Without a Pappus.

Epetiolate Without a petiole (cf. Sessile).

Epetiolulate Without a petiololule (cf. sessile).

Ephemeral A plant which completes its life cycle in a very short time.

Epi- A Greek prefix meaning "upon" or "above."

Epibiotic An endemic species which is remnant of a lost flora.

Epicalyx A series of bracts below and alternating with sepals but resembling a true calyx.

Epicarp The outer layer of the pericarp.

Epicormic Shoots A branch which develops from a dormant bud on the trunk of a tree, usually as a result of injury to the apical portion of that tree.

Epicotyl That portion of an embryo above the cotyledonary node (syn. Plumule).

Epidermis The layer of tissue which covers the surface of plant organs, such as leaves, stem, etc.

Epigeal Seed germination in which the cotyledons are above ground (cf. Hypogeal).

Epigynous, Epigyny Flowers with an inferior ovary are said to be epigynous because all floral parts arise on top of the ovary or gynoecium, as in the Onagraceae (e.g., *Fuchsia*).

Epihyperigyny The condition in which the sepals, petals, and stamens are attached to the floral tube or hypanthium surrounding the ovary; a combination of perigyny and partly inferior ovary.

Epihypogyny The condition in which the sepals, petals, and stamens are attached about halfway from the base of the ovary to the partly adnate hypanthium tube; half inferior insertion of parts.

Epimatium The ovuliferous scale of conifers.

Epiperigyny The condition in which the sepals, petals, and stamens are attached to the floral or hypanthium cup above the ovary with the lower part of the hypanthium completely adnate to the ovary.

Epipetalous Born on or attached to the petals, often in reference to the stamens.

Epipetiolar Born on or attached to the petiole, often in reference to branches that arise from the petiole.

Epipetric Growing on rocks.

Epiphyll (Epiphyllous)   Growing upon or arising from a leaf.

Epiphyte, Epiphytic   Plants that grow on branches of trees (or on other objects) but are not parasitic, as in most orchids, ferns, bromeliads, etc.

Epirhizous   Growing on roots, as certain parasites. The term has also been used to mean "with roots upon another plant."

Episepalous   Born on or attached to the sepals, usually in reference to the stamens.

Epistomatic   Referring to leaves in which stomata occur only on the adaxial (upper) surface.

Epithet, Specific epithet   The specific name applied to a species binomial. For example, the word "virginiana" is the specific epithet for *Magnolia virginiana*.

Epitropous Ovule   A pendulous or hanging ovule.

Equisetoid Hairs   Trichomes that have the appearance of horsetails (Equisetum).

Equilateral   With sides equal in shape and size.

Equinoctial   Having flowers which open and close at a regular hour each day.

Equitant   Two ranked leaves with overlapping bases, as in Iris.

Eramous   Having unbranched stems.

Erect   Upright. straight up from the ground.

Erose   A leaf margin which appears eroded and jagged, but not toothed or fringed.

Espalier   Plants that are trained to grow in a geometric design against walls or trellises.

Estipitate   Without a stipe, as in stemless glands on leaves.

Estipulate   Without stipules.

Etaerio   An aggregate fruit composed of achenes or druplets, as in Ranunculus, Rubus, etc.

Ethnobotany   Study of plants in relation to their utilization by various cultures.

Ethological Isolation   Reproductive isolation due to pollinator specificity.

Eucamptodronous   With a single primary vein, the secondary veins gradually curved upwards.

Euphilic   A flowering plant which is especially adapted for pollination by specific vectors.

Eusporangiate   In reference to a group of ferns, having the sporangium developed from a large amount of leaf tissue as opposed to one or few cells (cf. Leptosporangiate).

Evanescent   Disappearing early; lasting only a short time, usually in reference to parts rather than whole plants. (cf. Ephemeral)

Even-Pinnate   see compound leaves.

Evergreen   A plant which remains green during the dormant season. Although this term is properly applied to plants rather than leaves, the reference is made to the green leaves that remain on the plant for more than one growing season (cf. Deciduous).

Evergrowing   Said of an evergreen plant that grows continuously, such as most conifers.

Evolution   The gradual change in the characteristics of successive generations of a species, ultimately giving rise to a form or characteristic different from that of the common ancestor.

Evolutionary Tree   A phylogenetic diagram that depicts the evolutionary lines of descent.

Ex-   A Latin prefix meaning "out," "lack of," or "without."

Exalbuminous   Without endosperm.

Exarch   Said of a xylem strand which has the protoxylem farthest away from the center of the axis.

Excentric   See Eccentric.

Excurrent   Extending beyond the apex, as in a vein which runs out beyond the lamina to form a mucrinate apex; a tree with a central leader and self-pruning or relatively short lateral branches. For example, most conifers or *Liquidambar styraciflua*.

Exfoliate   The falling away in flakes, layers, or scales, usually in reference to tree bark.

Exindusiate   Without an endusium, as in the fern Polypodiaceae.

Exine   The outermost layer of pollen grain or spore wall.

Exo-   A Greek prefix meaning 'without' or 'outside of'.

Exocarp   The outermost layer of a fruit (syn. Epicarp).

Exogamy   Out-breeding; cross-fertilization, usually leading to heterozygocity.

Exogenous   Produced on the outside; developed from a superficial tissue.

Exospore   The outer spore wall layer (syn. Exine).

Exotic   Not native, introduced from a distant area.

Explanate   Flattened; spread out flat.

Expressed Sequence Tag (EST)   A short sub-sequence of a cDNA sequence; randomly selected clones from cDNA libraries.

Exserted   Extending beyond; protruding, as when the stamens project beyond the corolla.

Exstipellate   Without stiples.

Exstipulate   Without stipules.

Exsiccatus   A dried herbarium specimen.

Extant   Living; currently existing (cf. Extinct).

Extinct   No longer living.

Extra-   A Latin prefix meaning 'beyond'.

Extrafloral Nectary   A nectariferous gland found on parts other than the flower, usually on leaves, petioles, or bracts.

Extrorse   Facing outward, as in anthers that open away from the center of the flower.

Eye   A vegetative bud, as in potato.

F

F1   The first generation of a cross between two individuals.

F2 The second generation of a cross resulting from self-fertilization of the F1.

Facultative Descriptive of an individual with the ability to survive under a different environmental condition. For example, a facultative annual is one which can survive for a longer period of time when moved into a greenhouse.

Falcate Sickle-shaped.

Falcate-Secund Leaves falcate and all turned toward one side of the stem.

Falls The drooping portion of the perianth of an *Iris* flower (cf. Standard).

False Indusium A covering formed by the inrolling of the leaf margin of some ferns, as in Adiantum.

Family A taxonomic category between an order and a genus. A group of related genera constitute a family. Familial names of plants terminate with the suffix -aceae. Eight family names have been conserved and terminate with the suffix -ea, although their use is optional (e.g., Palmae = Arecaceae).

Farinose, Farinaceous Having a mealy surface; of a starchy character, such as endosperm.

Fasciation An unnatural widening and flattening of the stem, as in some cacti or cactoid *Euphorbia*.

Fascicle, Fasciculate A tuft of leaves or flowers arising from the same location.

Fastigiate With erect branches which are more or less appressed to form an exaggerated form of Excurrent growth habit. For example, Populus nigra "Italica" (lombardy poplar) or *Cephalotaxus harringtonia* "Fastigiata."

Faucal, Fauces The throat of gamopetalous corolla.

Faviolate Honeycombed, as in the receptacle of several species of Asteraceae (e.g., sunflower) (cf. Alveolate)

Felted Closely covered with fine intertwined hairs.

Fenestrate Perforated; having openings; with translucent areas.

-Fer, -Ferus A Latin suffix meaning "bearing," as in Florifer, Floriferus (bearing flowers).

Feral Wild; not cultivated; indigenous.

Ferruginous Rust-colored.

Fertilization The fusion of gametes, as egg and sperm.

Fetid, Foetidius Having an unpleasant odor.

Fibrous Having fibers, as in the leaves or trunk of some palms.

Fibrillos See Fibrous.

Fibrous Root A mass of fine adventitious roots, of more or less equal thickness. These are primarily characteristic of monocotyledons.

-Fid, -Fidus A Latin suffix meaning "deeply cut" or "cleft."

Filament The stalk of a stamen which supports the anther.

Filamentous With fibrous or threadlike appendages (syn. Filiferous).

Filantherous Stamen A typical stamen with distinct anther and filament.

Filicoid Fern-like

Filiferous With threadlike appendages, often used in reference to leaf margins (syn. Filamentous).

Filiform Threadlike.

Fimbriate Fringed along the margins.

Fimbriolate Minutely fimbriate.

Fissured Split or cracked, as in some tree barks

Fistular, Fistulose Hollow and cylindrical, as in the stem and leaves of an onion.

Flabellate, Flabelliform Fan-shaped, as in the leaves of Ginkgo biloba and Phoenix canariensis.

Flaccid Lax, weak or limp.

Flagelliform Whip-shaped.

Flavescent Yellowish; having yellow spots on a normal green surface.

Flavous Latin word meaning "yellow."

Flexuous A zig-zag stem which changes direction at the nodes.

Floccose With tufts of soft silky hair; cottony.

Flora The plants of a particular area; a descriptive treatment of plants of a particular area which includes a key for identification.

Floral Axis The receptacle.

Floral Diagram A composite diagram that shows position and number of flower parts in relation to other parts.

Floral Envelope The perianth; the calyx and the corolla.

Floral Formula A summary of the characteristics of a flower, including number, arrangement, etc. of parts and their relationship to one another.

Floral Nectary Nectariferous glands which occur in one or more segments of a flower.

Floral Tube A tubular flower formed by the fusion of sepals, petals, and stamens, usually characteristic of perigynous or epigynous flowers.

Florescence Refers to various parts of the inflorescence, as in main-florescence (the primary branch), co-florescence (the lateral branch), etc. It has also been defined as the period of flowering.

Floret A small flower, as in the spiklet of grasses or the ray and disk florets of Asteraceae.

Floricane The flowering and fruiting stem, especially of Rubus (brambles).

Floribundus A Latin word meaning "producing many flowers."

Floridus A Latin word meaning "showy."

Floriculture The growing of flowers under controlled conditions, including cut flowers and potted flowering plants.

Floriferous Producing a profusion of flowers.

Flotophyllous With floating leaves.

Flower The reproductive structure of angiosperms, typically made up of a Calyx (the sepals), Corolla (the petals), Androecium (the stamens), and Gynoecium (the carpels). The axis upon which these organs are attached is the Receptacle.

Foliacious, Foliose Having characteristics of a leaf; bearing leaves.

Foliage A collective term for leaves of a plant; leafy plants specifically grown for indoor use.

Foliar Pertaining to leaves or leaf-like parts.

Foliate Having leaves.

Foliolate Having leaflets, as in compound leaves.

Follicetum An aggregate of follicles, the product of multiple carpels.

Follicle A dry fruit formed from a single carpel, usually dehiscing along the ventral suture and often with many seeds.

Fornicate Provided with scale-like appendages in the corolla tube.

Forked Separating into two distinct and more or less equal divisions (syn. Dichotomous).

Formicophily Pollination by ants.

Fovea The seat of the pollinium in orchids.

Foveolate Pitted, having small depressions.

Free Distinct, not united, as in floral organs.

Free Central Placentation See Placentation.

Fringe Margins with hairlike appendages (syn. Fimbriate).

Frond The leaf of a fern. Although sometimes used to designate leaves of palms and cycads, the term should be reserved for ferns.

Fructiferous, Fructification Producing or bearing fruit.

Fruit The ripened ovary of angiosperms.

Fruticose Shrubby; shrub-like.

Fucous Grayish-brown; dull-brown.

Fugacious Falling off early, early deciduous.

Fulcra Any appendage of leaves, such as tendril, prickles, stipules, etc.

Fulvous Tawny, dull yellowish-brown.

Fungiform Having the shape of a mushroom.

Funiculus The stalk of an ovule in angiosperms, by which it is attached to the placenta.

Funnelform Funnel-shaped, gradually widening upward.

Furcate Forked.

Furfurate, Furfuraceous Covered with scales (syn. Scurfy).

Furrowed With longitudinal channels or grooves (syn. Sulcate).

Fusiform Spindle-shaped; elongated and tapering toward each end.

G

Galbulus (Galbalus) A strobilus with fleshy cone-scale, as that of Cupressaceae.

Galea A hooded or helmet-shaped sepal or petal, as in Aconitum.

Galeate Shaped like a hood or a helmet; having a galea.

Gall An abnormal growth caused by an insect or a disease organism.

Gamo A Latin prefix meaning "united."

Gamopetalous Having united petals; referring to a Sympetalous corolla.

Gamophyllous With leaves united by their edges.

Gamosepalous Having united sepals; referring to synsepalous calyx.

Geitonogamy Pollination between flowers of the same individual plant (cf. Xenogamy).

Geminate Paired, in pairs, as in a pair of branches arising from the same node (e.g., *Parkinsonia aculeata*, Jerusalem thorn).

Gemma A young bud, either of flower or leaf, often used for buds on the root crown, bulbs, corms, or rhizomes.

Genetic Marker A short DNA sequence that is used for identification or locating genes on a chromosome.

Geniculate Bent, like a knee.

Genotype The genetic makeup of an individual (cf. Phenotype).

Genotypic Variation Genetic variation among individuals of a species population.

Genus (pl. Genera) A taxonomic rank between Family and Species. Related species are grouped into a genus and related genera into a family.

Geocarpy, Geocarpous The ripening of fruit underground. The young fruits are pushed into the soil by a post-fertilization curvature of the stalk, as in *Arachis hypogaea*, (peanut).

Geoflory, Geoflorous The flowering occurring underground or on soil surface, as in *Aspidastra elatior*.

Geographic Isolation When populations of a given species are isolated from each other by some distance. This may or may not result in reproductive isolation. For example, populations of *Liquidambar styraciflua* occur in S.E. United States as well as S.E. Mexico.

Geonasty Curving toward the ground.

Geophyte A plant which has subterranean buds on a bulb, a rhizome, or a caudex.

Germinal Pores, Germinal Furrows Thin areas on pollen grains through which the pollen tube is produced. The existence of pores or furrows is of considerable taxonomic significance.

Gibbous Swollen at the base (cf. Caudex).

Glabrate, Glabrescent Becoming Glabrous at maturity.

Glabrous Without pubescence, not hairy.

Gladiate Shaped like a sword blade.

Gland A uni- or multicellular secretory structure often found on various plant organs.

Glandular, Glanduliferous Having glands.

Glandular Hair   A trichome with a uni- or multicellular gland.

Glandular-Pubescent   Having glands and hairs intermixed.

Glaucescent   Slightly glaucous.

Glaucous   Covered with a whitish or bluish "waxy bloom" that rubs off.

Globose, Globular   Spherical or rounded.

Glochid   Barbed bristles, often occurring in tufts, as in the cacti (Cactaceae).

Glochidiate   Having glochids.

Glomerate   In dense or compact clusters.

Glomerule   A compact cluster of more or less sessile flowers.

Glumaceous   Bearing glumes; of the character of glumes; thin, brown, and papery in texture.

Glume   Rigid, chaff-like, or scale-like bracts, referring especially to the two empty bracts at the base of the spiklet in grasses.

Glutinous   Sticky or mucilaginous; with a waxy exudate.

Glycophyte   A moderately salt-tolerant plant (cf. Halophyte).

Graft, Grafting   The union of a small piece of meristematic tissue (usually a bud) or actively growing stem (the scion) with another plant (the Stock). Successful grafts usually indicate a close taxonomic relationship.

Graft Hybrid   See Chimaera.

Grain   The fruit or seed of Poaceae (grasses) (syn. Caryopsis).

Granular   Covered with fine mealy granules.

Growth Form   Habit or shape of plants in the landscape, such as trees, shrubs, etc. (cf. Life Form).

Gymno-   A Greek prefix meaning "naked."

Gymnospermae   A class of vascular plants the seeds of which are not enclosed within a carpel (fruit) but are born on sporophylls, a group of which are organized into a cone. These include Cycadales, Coniferales, Gingkoales, and Gnetales (when and if considered as a single order).

Gynagamocephalous   See Gynagamous.

Gynagamous   Inflorescence with pistillate flowers inside or above and neuter flowers outside or below. (syn. Gynagamocephalous).

Gynandrium   Said of the fused stamens and carpels (stigma and style), as in Orchidaceae (syn. Column; Gynostemium).

Gynandrous   Having stamens and carpels united into a column, as in Orchidaceae

Gynecandrous   With both staminate and pistillate flowers in the same inflorescence, as in Cyperaceae.

Gynehermaphrodicephalous   See Gynehermaphroditic.

Gynehermaphroditic   Inflorescence with pistillate flowers inside or above, hermaphroditic outside or below (syn. Gynehermphrodicephalous).

Gynobase   An elongation of the receptacle of a flower forming a short stock upon which nutlets are located, as in Boraginaceae.

Gynobasic Style   Said of a style which arises near the base of the carpels or ovary-lobes, as in Lamiaceae and Boraginaceae.

Gynodioecy, Gynodioecious   A species in which some plants bear perfect flowers, while others bear only pistillate flowers.

Gynoecium   A carpel or an aggregation of carpels, whether free or united.

Gynomonoecy, Gynomonoecious   A species in which individual plants bear perfect and pistillate flowers.

Gynophore   A stalk or stipe bearing an ovary or fruit.

Gynostemium   A compound structure in orchid flowers formed by the adnation of stamens and carpels (syn. Column).

Gypsophil   A plant which lives on chalky or limestone soil.

H

Habit   The general appearance of a plant in terms of its characteristic form, as in erect, prostrate, climbing, etc.

Haft   A leaf stalk with a strip of photosynthetic tissue running along each side of it to form a wing; the stalk of a spathulate leaf; the claw of a petal.

Half-Inferior   In reference to the position of the ovary, when floral organs are attached around ovary with hypanthium adnate to the lower half of ovary

Half- Terete   Said of calyx, corolla, and perianth, when flat on one side, terete on the other; semicircular in cross-section.

Halophyte   A plant which grows in saline or alkaline soil.

Hamate   A narrow leaf or a trichome hooked or bent at the tip.

Hamulate, Hamulous   Having hooked or bent processes.

Haplo   A Latin prefix meaning "single" or "simple."

Hepaxanthic   Perennial plants which grow for several years, but die after flowering once (syn. Monocarpic).

Haplomorphic   Flowers with many parts, spirally arranged in a semihemispheric or hemispheric form, as in *Magnolia*, *Ranunculus*, etc.

Haplostemonous.   With a single whorl of stamens.

Hastate   Shaped like an arrowhead, but with the basal lobes pointed or narrow and at right angles.

Hastiform   Triangular with two basal lobes; the condition of being hastate.

Hastula   Terminal part of the petiole on upper surface of leaf blade in palmately lobed leaves of palms.

Haustorial   Absorbing roots, within host of some parasitic plants.

Head The inflorescence of Asteraceae; a compact inflorescence (syn. Capitulum).

Helicoid Coiled like a spring; curved.

Helicoid Cyme A sympodial inflorescence in which all branches develop on the same side of the main axis, although not in the same plane.

Heliophyte A plant which grows in full sun.

Helophyte A bog plant or freshwater marshes.

Hemi- A Greek prefix meaning "half" or "partial."

Hemianatropous Orientation of the ovule when it is half-inverted so that funiculus is attached near the middle with micropyle terminal and at right angle (syn. Hemitropus).

Hemicarp A half-carpel (syn. Mericarp).

Hemicryptophyte A plant which develops its buds just above or just below ground surface so that buds are protected.

Hemiepiphytic A plant that is epiphytic early but eventually roots in soil and becomes terrestrial.

Hemitropous See Hemianatropus.

Herb An annual, biennial, or perennial plant without woody parts above the ground.

Herbaceous A plant which is soft and green and has little or no woody tissue.

Herbals Botanical writings of the 15th, 16th, and 17th century in which specific plants were recommended for medicinal purposes.

Herbarium A collection of preserved, or dried and pressed plant specimens kept for identification and taxonomic studies.

Herkogamy The condition of a flower when stamens and stigmas are so placed as to make self-pollination impossible.

Hermaphrodagamous Inflorescence with hermaphroditic flowers inside or above and neuter outside or below (syn. Hermaphrodagamocephalous).

Hermaphrodagamocephalous See Hermaphrodagamous.

Hermaphrodrandrous Inflorescence with hermaphroditic flowers inside or above and staminate outside or below (syn. Hermaphrodandrocephalous).

Hermaphrodandrocephalous See Hermaphrodrandrous.

Hermaphrodigynous Inflorescence with hermaphroditic flowers inside or above and pistillate flower outside or below (syn. Hermaphrodigynecephalous).

Hermaphrodigynecephalous See Hermaphrodygynous.

Hermaphrodite Bisexual; having both the androecium and gynoecium present in the same flower.

Hesperidium The fruit of the orange and other citrus plants.

Heterandrous Having stamens of different lengths (cf. Homandrous).

Heteranthous Having distinct kinds of stamens in different flowers of the same species.

Hetero- A Greek prefix meaning "different.".

Heteroblastic, Heteroblasty Having distinctly different juvenile and adult foliage, as in *Eucalyptus*, *Hedera*, *Juniperus*, etc.

Heterocarpous Having morphologically different types of fruit.

Heterocephalous Flower heads with flowers of different sexual conditions

Heterocladous With stems of different sizes and/or shapes.

Heterocymous Cymose inflorescence with flowers of different sexual condition.

Heterogamous Having male, female, and/or hermaphroditic flowers in the same inflorescence (cf. Polygamous, Anisogamous).

Heterogeneous Lack of uniformity in a population of plants or other organisms (cf. Homogeneous).

Heteromerous Having a different number of parts in different whorles, in reference to flowers.

Heteromorphic Existing in more than one form; having more than one kind of flower on the same plant (cf. Polymorphic, Monomorphic).

Heteropetalous With petals of different sizes and/or shapes.

Heterophylly, Heterophyllous With leaves of different forms or sizes.

Heterophytous Plants which bear leaves and flowers on separate stems; plants which appear different when in different habitats.

Heteropolar Pollen A pollen grain with two polar faces, one with an aperture, the other without.

Heterosepalous With sepals of different sizes and/or shapes.

Heterosexual Inflorescences or flowers with different sexual conditions in the same plant.

Heterospicuous Spikes with flowers of different sexual conditions.

Heterospory, Heterosporous A plant which produces two different kinds of spores (cf. Homospory).

Heterostichous With unequal rows.

Heterostyly, Heterostylous A species with flowers that have styles of different length.

Heterotropous Position of ovule not fixed in ovary.

Hexa- A Greek prefix meaning "six."

Hexandrous A flower with six stamens.

Hibernal Appearing in the winter (syn. Hiemal).

Hiemal See Hibernal.

Hilum The scar on a seed, marking the point where the seed broke from the stalk.

Hip Fruit of the genus Rosa.

Hippocrepiform Horseshoe-shaped.

Hirsute With stiff hairs.

Hirsutullous Minutely hirsute (syn. Hirtellous).

Hirtellous Minutely hirsute (syn. Hirsurullous).

Hispid With bristly hair.

Hispidulous   Minutely hispid.

Hoary   Covered with grayish-white pubescence.

Holotype   A single specimen designated as the type of a species, upon which that species is named.

Holosericeous   Covered with fine silky pubescence.

Homandrous, Homoeandrous   Having stamens the same size and shape (cf. Heterandrous).

Homanthous   Having more than one state in each individual flower, any one segment may be different in some respect.

Homo-, Homio- (Homeo), Homolo-   A Greek prefix meaning "alike," "similar."

Homocarpous   Having fruit of only one kind.

Homocephalous   When the anthers fertilize the stigma of another flower within the same inflorescence; flowers sexually uniform (syn. Homogamy, Homocymous, Homospicous).

Homochlamydeous   With a perianth of tepals, undifferentiated into calyx and corolla, as in Amaryllidaceae.

Homocymous   See Homocephalous.

Homodromous   Having leaves arranged in spirals, all running in the same direction; having all leaves turned in the same direction.

Homogamous   Bearing one kind of flower; simultaneous ripening of pollen (syn. Homocephalous).

Homogeneous   Uniform, alike (cf. Heterogeneous).

Homologous, Homology   Similar organs of phylogenetically common origin, but different structure and/or function.

Homomorphic   Having the same form; monomorphic (cf. Heteromorphic).

Homonym   In nomenclature, a name rejected because it duplicates a name previously and validly published for the same taxon based on a different type.

Homophytous   Plants are all the same, but there may be different states within individual plants.

Homosexual   Inflorescence or flowers sexually uniform; having the same sex in all flowers of the same individual.

Homospicous   Spikes with flowers sexually uniform.

Homospory, Homosporous   A plant which produces only one kind of spore (cf. Heterospory).

Homostylous   Having styles of all the same length.

Hood   The lid which hangs over the pitcher of the pitcher plant trap; the concave segment of the corona in milkweed (Asclepiadaceae) flower.

Horn   A projection which is part of the corona of milkweed flower (Asclepiadaceae).

Horticulture (Hortus = Garden)   Literally culture of gardens, in reference to vegetable, fruit, and ornamental crops.

Humistrate   See Procumbent.

Hyaline   Thin and translucent, as in some leaf margins.

Hybrid   The offspring of a cross between two taxa, at the generic, specific, or lower rank.

Hydathode   Specialized vein openings through which water is lost under certain atmospheric conditions.

Hydrochore   A plant disseminated by water.

Hydrophily   Pollination by water.

Hydrophyte   A vascular plant that lives in water or a very moist habitat.

Hylophyte, Hylophylous   A plant that lives in forest or woodland.

Hymenopterophily   Pollination by bees.

Hypanepigyny   The condition in which the sepals, petals, and stamens are attached to the elongate floral tube or hypanthium above the inferior ovary, as in *Oenothera* (Onagraceae).

Hypanthium   The tube of the receptacle upon which the calyx, corolla, and stamens are born; "calyx tube."

Hypanthodium   An inflorescence with flowers on the wall of a concave capitulum, as in *Ficus* (Moraceae).

Hypercrateriform   See Salverform.

Hyphodromous   A type of leaf venation with a single primary vein, as in Cycadaceae (Cycas).

Hypo-   A Latin prefix meaning "under,", "below."

Hypocotyl   The part of an embryo below the cotyledons, including the cotyledonary node; the embryonic stem.

Hypocotylespermous   Said of seeds which have food reserve stored in hypocotyl (syn. Macropodial).

Hypogeal Germination   Seed germination in which cotyledons remain below the soil surface.

Hypogeous   Growing below the soil surface.

Hypogyny, Hypogenous   Born below the gynoecium or the ovary, in reference to calyx, corolla, and stamens. Flowers with this arrangement have a superior ovary.

Hyponym   In nomenclature, a name rejected when not based on a type specimen, or a type species.

Hypophyllous   With small leaves, as bracts, scales, or catophylls.

Hypostomatic   A leaf in which stomata occur only on the abaxial (lower) surface.

Hypotropous   Position of the ovule where the ovule is erect above the micropyle and toward the ventral bundle (syn. Ventral).

Hypsophyll   Leaves born on the upper part of the branch below the flowers, as in floral bracts (cf. Catophylls).

Hysteranthous   Plants in which leaves appear after the flowers, as in *Cornus florida*.

I

Ianthinus   A Greek word meaning "bluish purple," "violet."

Icones Plantarum   Botanical figures; pictorial representation of plants.

Igneus A Latin word meaning "flame-colored" (a combination of red and yellow).

Idiogram A diagram illustrating the morphology of chromosome complements of a cell, including size, position of centromeres, satellites, and secondary constrictions.

Illegitimate In botanical nomenclature, a name improperly used.

Im- A Latin prefix meaning "not," "without."

Imbricate Overlapping like tiles of a roof, referring to sepals and petals in the bud (cf. Valvate).

Imparipinnate An odd-pinnately compound leaf with a terminal leaflet.

Imperfect Flower A flower which lacks stamens or carpels, without regard to condition of the perianth; a unisexual flower (cf. Perfect Flower; bisexual flower).

Imperforate Having no opening or perforation.

Implicate Interwoven; entangled.

In- A Latin prefix meaning "without."

Inarticulate Not jointed; continuous.

Inaparturate Pollen grains which lack germinal pores.

Inbred Line Homozygous plants produced by repeated selfing (inbreeding) or backcrossing between closely related individuals, as in most bedding plants which are uniform in habit and flower color. It is designated "In" by the code of nomenclature.

Incanuous A Latin word meaning "quite gray" or "hoary."

Incarnate, Incarnatus A Latin word meaning "flesh-colored" (syn. Carneus).

Incised Margins which are deeply cut, irregular, or have jagged teeth.

Inclinate With lamina folded or curved downwards near the apex.

Inclined Bent down.

Included Not protruding beyond surrounding members, as in stamens not projecting beyond the corolla (cf. Exserted).

Incomplete Flower A flower which is lacking one or more of the four regular sets of parts; absence of sepals, petals, stamens, or carpels.

Inconspicuous Not easily seen, because of small size or lack of color.

Incrassate Thickened.

Incumbent Leaning on or upon; with anthers leaning forward and against the filament; with cotyledons leaning backwards and against the radicle.

Incurved Curved inwards.

Indefinite Inconstant with numbers; applied also to the continuous (Indeterminate) growth of a racemose inflorescence.

Indehiscent Not splitting open; remaining closed, as a drupe or an achene.

Indeterminate Growth Of indefinite growth, as a racemose inflorescence whose terminal flower opens last so that there is no restriction to continued growth (cf. Determinate Growth).

Indigen, Indigenous Native to a region, not introduced.

Indumentum Pubescence or other coverings on plant surfaces.

Induplicate Folded inwards, as in leaflets of some palms where a V-shaped groove is formed by the margins (cf. Reduplicate).

Indurate Hardened.

Indusium (pl. Indusia) The cover growing over the sporangia of some ferns.

Inermis, Inermous Unarmed, thornless.

Inequilateral With unequal margins; uneven; asymmetrical.

Inferior Below, usually referring to the position of the ovary in an epigynous flower.

Inflated Puffed up; swollen; bladdery.

Inflexed Said of a leaf in which the upper part of the blade is bent over the lower part; bent inwards (cf. Reclinate).

Inflorescence A flower cluster, such as a panicle, spike, raceme, etc. An inflorescence may consist of only one (solitary) or many flowers.

Infossus Sunken, as in veins of some leaves.

Infra- A Latin prefix meaning "below."

Infrafoliar On the stem below the leaves, usually in reference to inflorescence, as in Arecaceae (palms).

Infrapetiolar Bud Axillary bud surrounded by the base of the petiole (syn. Subpetiolar).

Infraspecific Any taxon below the rank of species, such as subspecies, variety, etc.

Infundibular Funnel-shaped.

Ingroup Used in cladistics analysis in reference to taxa that are hypothetically closely related to each other (cf. Outgroup).

Innocuous Unarmed, spineless.

Insectivorous In reference to insect trapping plants, such as Dionaea, Sarracenia, etc.

Inserted Growing on or attached to another organ, as in stamens growing on the corolla.

Insignis Notable, significant.

In Situ In the normal or natural position

Intectate Pollen Pollen grains that lack a tectum.

Integument The jacket or the outer covering of an ovule.

Inter- A Latin prefix meaning "between."

Intercalary Said of growth, which is not apical but occurs between the apex and the base, as in leaves of grasses.

Intercostal The space between veinlets of a leaf.

Interfoliar Between or among the leaves.

Intergeneric Hybrid A hybrid between species of two genera of the same family. For example, ×*Fatshedera* which is a hybrid between *Fatsia* and *Hedera* in Araliaceae

Intermittent Discontinuous; irregular, usually in reference to periodicity of growth.

Internode   Part of the stem lying between two successive nodes.

Interpetiolar   Between the petioles.

Interpetiolar Stipules   Connate stipules which have coalesced from two opposite leaves.

Interrupted   Broken or irregular in arrangement; not evenly spaced, as in leaves or inflorescences.

Interspecific   The relationship between two populations or two species.

Interspecific Hybrid   A hybrid between two or related species of the same genus. For example, *Osmanthus* × *fortunei*, which is a hybrid of *O. fragrance* and *O. heterophyllus*.

Interstitial Growth   The kind of growth which does not involve an organized meristematic region, but occurs all over, as in some fruits.

Intine   The inner wall of pollen grains.

Intraspecific   The relationship between members of the same population or the same species.

Introduced   A plant(s) not native to an area but brought from another region.

Introrse   Facing inward toward the axis, as an anther which dehisces toward the stigma.

Invaginated   Enclosed within a sheath.

Inverted   Turned top side down; turned over.

Involucel   A small involucre, as bracts that subtend the secondary umbels in the Apiaceae (Umbelliferae).

Involucral   Belonging to an involucre.

Involucrate   Bearing an involucre.

Involucre   Cluster of bracts subtending an inflorescence, as in the heads of Asteraceae or umbels of Apiaceae.

Involute   A leaf in which the margins are rolled upwards toward the midrib (cf. Revolute).

Irregular   Differing in size and shape; asymmetrical, referring to flowers which are not divisable into halves by an indefinite number of longitudinal planes (cf. Zygomorphic).

Iso-   A Latin prefix meaning "equal."

Isobilateral   Having two equal sides, as in leaves which have palisade parenchyma on both sides of the blade.

Isocotylous   Having equally developed cotyledons.

Isodiametric   Equal on all sides.

Isodynamous   Any equally developed structure(s).

Isogenous   Said of individuals that belong to the same genotype.

Isomerous Flower   Flowers which have the same number of parts in each whorl.

Isomorphic   With all parts similar so that they are indistinguishable.

Isopetalous   With petals of the same size and shape.

Isophyllous   With leaves of the same size and shape.

Isopolar Pollen   A pollen grain in which the equitorial plane divides the grain into identical halves so that proximal and distal faces are similar.

Isosepalous   With sepals of the same size and shape.

Isostichous   With rows of equal length and in the same direction.

Isotype   A herbarium specimen of the type collection; a duplicate of the holotype.

Ithyphyllus   Straight and stiff-leaved.

Ixious   A Greek word meaning "sticky," "viscous.".

J

Jaculiferous   Having dartlike spines.

Joint   A node, as in a grass culm.

Jordanism   An excessive propagation of varieties which are relatively constant under cultivation, considered as microspecies.

Jugate   Having a pair of opposite leaves; a pair of leaves originating from the same point.

Jugum   The ridges on the fruit of Apiaceae (umbel family).

Juniperinus   Bluish-brown, as in berries of Juniperus.

Juvenile Form   A young plant with features different from the adult form of the same plant, as in *Eucalyptus*, *Hedera*, etc. which differ in their leaf characteristics.

Juxtaposition   The relative position in which organs are placed.

K

Karyology   Nuclear cytology; study of chromosomal characteristics.

Karyotype   Characteristics of chromosomal complex as defined by the size, shape, and the number of mitotic chromosomes.

Katablast   A shoot from an underground stock.

Keel   A prominent dorsal rib or ridge, as in some carpels or in glumes of grasses, the lower petals of the flowers of legumes (Fabaceae).

Keeled   Ridged like the bottom of a boat (syn. Carinate).

Kernel   An entire seed of a grass, as in corn.

Key   See Dichotomous Key.

Kingdom   The highest taxonomic rank in living organisms, including plants.

Knee   An above-ground outgrowth of roots in certain plants when grown in wet habitats, as in Bald Cypress (*Taxodium distichum*) (cf. Pneumatophore).

Krummholz   The creeping, often stunted growth habit of plants in high altitudes, caused by cold and wind.

L

Labellum Lip; the modified, often enlarged lowermost petal of an orchid flower.

Labiate Lipped; when the corolla forms an upper and a lower lip, as in flowers of Lamiaceae.

Labium (pl. Labia) A lip.

Laccate Having a shining surface.

Lacerate Cut irregularly; appearing torn or cut, as in certain leaves.

Laciniate Cut into narrow deep lobes.

Lactiferous Containing latex.

Lacuna (pl. Lacunae) A cavity; a depression on the leaf surface.

Lacunose Having a pitted surface.

Laevigate Smooth as if polished.

Laevis Smooth; not rough.

Lageniform Flask-shaped, gourd-shaped.

Lagopus Densely covered with long hairs.

Lamina Leaf blade.

Laminar Thin and flattened, as in a leaf blade.

Lanate, Lanatus Covered with long and loosely tangled hairs; woolly.

Lanceolate Lance-shaped, broadest below the middle and tapering gradually to the apex.

Lanose See Lanate.

Lanuginose, Lanuginous Woolly, cottony; the same as Lanate but with shorter hair.

Lanulose with very short-woolly trichomes.

Lapideus Stoney, as the seed of stone fruits.

Lapidose Growing among stones.

Lasio- A Latin prefix meaning "woolly."

Lasianthus Woolly flowered.

Lasiocarpus Woolly fruited.

Lateral Arising from or attached to the side of an axis, as a lateral bud (syn. Axillary).

Laterifolius On the side of the leaf, near the base.

Laterinervis, Laterinervius Straight-veined; parallel-veined, as in grasses.

Laterospermous On the side of the seed.

Latex A viscous fluid contained in special structures (Laticifer) of certain groups of plants. For example, in the Euphorbiaceae, the latex may be milky (*Euphorbia*) or may be red, yellow, and cloudy or clear (*Jatropha*).

Lati- A Latin prefix meaning "broad."

Laticifer A "pluming" system in various tissues or organs of certain plants, containing latex. Used as a taxonomic characteristic; morphologically, laticifers may be Articulated (with walls separating the cells which are attached end to end), Nonarticulated (without walls), or they may be Idioblastic (solitary, variously shaped cells).

Laticiferous Having laticifers; latex producing.

Latiflorous Having broad flowers.

Latifolius Broad-leaved.

Latrorse Dehiscence Dehiscing longitudinally and laterally, in reference to anthers.

Lax Arranged loosely (cf. Congested).

Layering A method of propagation whereby a small section of the stem is buried shallowly below the soil surface or variously treated and covered (air-layering) with moist materials to promote formation of adventitious roots.

Leader The main stem or trunk of trees.

Leaf The usually green-colored, expanded, flattened portion of plants. It consists of a Lamina or Blade, the Petiole, and when present, the Stipules.

Leaf Axil The angle between the leaf petiole and the stem from which it arose.

Leaf Bud See Vegetative Bud.

Leaflet A segment of a compound leaf.

Leaf Scar A scar on the stem left after the leaf falls.

Leaf Sheath The expanded base of the leaf which covers the stem in most monocots and some dicots.

Lectotype A herbarium specimen selected to represent the type when the original specimen is missing or there is none designated as such.

Legume The pod of members of the Fabaceae (Leguminosae), a pod dehiscent on two sides; any member of the Fabaceae.

Lemma The outer (lower) bract of the floret of grasses.

Lenticels Porous spots in the periderm of woody plants, giving the appearance of a rough surface on the stem or other plant parts.

Lenticular Lens-shaped.

Lentiginose, Lentiginous Minutely dotted, as if freckled; with exfoliating scaly incrustations (syn. Scurfy).

Lepidote Covered with small scurfy scales.

Lepto- A Greek prefix meaning "slender"; "narrow."

Leptoma A thin region of the pollen grain from which the tube usually emerges.

Leptomorph A spreading rhizome, with slender stem and long internodes, as in *Convallaria* spp. (cf. Pachymorph).

Leptophyllous Slender-leaved; narrow-leaved.

Leptosporangiate Ferns The majority of ferns in which sporangia arise from single cells of the epidermis, as in Polypodiaceae, Osmondaceae, etc.

Leuco- A Greek prefix meaning "white."

Leucanthous White-flowered.

Leucaphyllum White-leaved.

Levigate Having a smooth polished surface.

Liana Woody vines found in tropical forests.

Ligneous Woody.

Ligulate Strap-shaped or tongue-shaped, as in a petal or a leaf; having a ligule.

Ligule The strap-shaped part of the corolla in Asteraceae; the annular collar-like projection at the junction of the leaf blade and the sheath in grasses.

Limb   The expanded flat part of a gamopetalous corolla; a large tree branch.

Line   In horticulture, homozygous plants with uniform characteristic(s), resulting from repeated selfing, as in most bedding plants; a measure of length, 1/12th of an inch.

Lineage   Ancestry, line of descent from an ancestor.

Linear   Long and narrow with nearly parallel sides.

Lineate   Marked with lines.

Lineolate   Marked with fine lines.

Linguiformis, Lingulate   Tongue-shaped.

Lip   See Labellum.

Lobate   Having lobes.

Lobe   Any segment of an organ, particularly when rounded, as in leaves or perianth.

Lobed   Margin cut less than halfway to the center, incurved or angular segments.

Location   The approximate position of an organ in relation to other organs.

Lobulate   Having small lobes.

Locular, Loculate   Having a locule (chamber), as in unilocular, bilocular, trilocular, etc.

Locule   Compartment or cell of an ovary or an anther.

Loculicidal   See Dehiscence.

Lodicule   A rudimentary organ at the base of the ovary of grasses, believed to be remnants of the ancestral perianth part.

Loment   A leguminous fruit (pod) that is constricted between the seeds, as in Parkinsonia aculeata.

Long Shoot   The normal branches of woody plants, with considerable distance between the nodes (cf. Short Shoot; spur).

Lorate, Loriform   Strap-shaped, with flexuous, wavy margins, as in certain leaves.

Lunate   Half-moon shaped (syn. Selenoid).

Lurid   Dingy yellowish-brown.

Luteus   Yellow color.

Lutescent   Yellowish.

Lyrate   Lyre-shaped; a leaf which is pinnately lobed and has a terminal lobe much larger than the lateral lobes.

M

Macro-   A prefix meaning "long" or "large" (syn. Mega-; opp. Micro-).

Macrophyll   Large leaves (syn. Megaphyll).

Macrocephalous   Big-headed; dicotyledonous embryos with consolidated cotyledons.

Macropodal (Macropodial), Macropodous   Said of an embryo with enlarged hypocotyl, in which reserved food is stored; a leaf with a long petiole.

Macroscopic   Large enough to be visible with the naked eye.

Macrosporangium   See Megasporangium.

Macrospore   See Megaspore.

Maculatus   Spotted, blotched.

Malacoid   Mucilaginous; with soft or fleshy leaves.

Malacophily   Pollination by snails or slugs.

Malacophyllous   Xerophytic plants with fleshy leaves (syn. Leaf Succulent).

Malpighiaceous Hair   In reference to the branched unicellular trichomes of the taxa in Malpighiaceae, which are attached by their middle.

Majus   Greater; larger.

Mammilar,   Mammiliform,   Mammilate, Mammose   Having a rounded outgrowth or apex; breast-shaped.

Manicate   With a very thick interwoven pubescence.

Manubrium   The long, thin, more or less cylindrical base of palm spathes.

Marbled   Marked with irregular streaks of color.

Marcescent   Withered but remaining attached to the plant, in reference to flowers or leaves.

Margin   The edge or boundary line of a body.

Marginal   Placed upon or attached to the margin.

Marginate   Having a distinct margin or border, often differing in color from the rest of the member.

Marmorate, Marmoratus   Marked like marble, with veins of color.

Massula   A group of cohering pollen grain (syn. Pllinium), as in Orchidaceae; a clump of microspores, as in Heterosporous. Aquatic ferns that produce smaller male and larger female spores, e.g., Azolla (cf. Homosporous).

Matutinal   Having flowers that open early in the morning.

Mealy   Covered with a scurfy powder (syn. Farinaceous).

Medial   Central; middle.

Medulate   Having pith.

Mega-   A Greek prefix meaning "very large."

Megaphyll, Megaphyllous   Having very large leaves, as in ferns, cycads, and most angiosperms (cf. Microphyll, Microphyllous).

Megasporangium   A sporangium in which megaspores (ovules) are produced, as in ovary.

Megaspore   Ovules of angiosperms and gymnosperms; the larger of the two spores (female) in heterosporous ferns.

Megasporophyll   The leaf-like structure of cycads which bear the megasporangia or seeds; a carpel in angiosperms.

Megastrobilus   The female strobilus (cone) of gymnosperms.

Melittophily   Pollination by bees.

Membranaceous, Membranous   Thin, dry, and flexible.

Meniscoidal   Thin and concave-convex, like a watch glass.

Mentum   A chin or an extension of the base of the column in the flower of some orchids.

Mericarp   One of the halves of the fruit of umbellifers (Apiaceae).

Merosity The absolute number of parts within a whorle, as the number of parts in a flower.

-Merous A Latin suffix which indicates the number of flower parts. For example, 3-merous (in monocots), 4-merous or 5-merous (in dicots).

Mesic A term which denotes moist habitats (cf. Xeric).

Mesocarp The middle layer of the Pericarp in fruit, as in *Prunus* (peach, apricot, etc.) (cf. Endocarp, Exocarp).

Mesomorphic Having the structure of a mesophyte.

Mesophyte A plant that occurs in moist (but not wet) habitats.

Mesozoic A geological period 65 to 230 million years ago.

Metaphyll An adult leaf.

Micro- A prefix meaning "small"

Micromelittophily Pollination by small bees (cf. Melittophily).

Microphyll, Microphyllous Small leaves with a single central vein, as in *Psilotum* (cf. Megaphyll).

Micropyle The canal into the nucellus; a minute pore through which water enters the seed prior to germination.

Microsporangium A sporangium in which Microspores (pollen) are produced, as in the anther locule and its walls.

Microspore The pollen grain in angiosperms or the smaller of the two spores (male) in pteridophytes (cf. Megaspore).

Microsporophyll The leaf-like structure of gymnosperm cones or the anther in angiosperms, which bear the pollen (cf. Megasporophyll).

Microstrobilus Male cone of gymnosperms (cf. Megastrobilus).

Midrib The largest vein of a leaf, longitudinally running through the blade (syn. Midvein).

Midvein See Midrib.

Milliaris A minute glandular spot on a leaf.

Miocene A geological period (subdivision of tertiary) from 35 to 15 million years ago.

Mixed Buds Buds that produce leaves as well as flowers.

Mollis A Latin word meaning "soft" or "pliant" "pubescent."

Monad A pollen grain that occurs singly rather than in tetrads (units of four).

Monadelphous Having stamens united by their filaments into one body.

Moniliform Elongate roots with regularly arranged swollen areas.

Mono- A Latin prefix meaning "one."

Monoaxial Having a single axis.

Monocarpic A perennial plant which may grow for several years but dies once it flowers, as *Agave Americana*, *Corypha umbraculifera* (tailpot palm), most Bromeliaceae, etc.

Monocarpous, Monocarpellate Having a single carpel.

Monocephalic, Monocephalous Having only one head, as in scapose inflorescence of *Gerbera*.

Monochasium A cymose inflorescence reduced to a single flower.

Monoclamydeous Having the perianth reduced to a single series, as in *Hydrangea*.

Monoclinous Hermaphroditic or perfect flower, having both stamens and carpels present.

Monocolpate A pollen grain with a single colpus (furrow or groove), as in gymnosperms, monocotyledons, and some woody Ranales.

Monocotyledon One of the subdivisions of the angiosperms, which is characterized by one cotyledon, parallel venation, scattered vascular bundles, a fibrous root system, and flower parts in threes or multiples thereof (cf. Dicotyledon).

Monoculture Uniform planting of a single species, such as turf, orchards, citrus groves, etc.

Monocyclic Having each of the floral parts in one whorl.

Monodesmic Said of a petiole with a single vascular strand.

Monoecious Having the androecium (stamens) and the gynoecium (carpels) in separate flowers, but on the same plant; a plant with male and female unisexual flowers.

Monograph A comprehensive taxonomic treatment of a taxon (subspecies, varieties, and/or cultivars of a species, a genus, or a family) (cf. Revision).

Monogynous Having one carpel or gynoecium.

Monolete Said of a pollen which is bean-shaped and has a single scar line.

Monomerous A whorle consisting of one series.

Monomorphic All of the same shape and size.

Monophyletic Descended from a single ancestral line; all members of a group descendants of a common ancestor (cf. Polyphyletic).

Monophyllous With a single leaf.

Monopodial Branching with a main axis and reduced or missing laterals, as in pines, palms, etc.; orchids that grow as upright stem (cf. sympodial).

Monostichous Forming one row, line, or series.

Monotypic A genus which consists of only one species (cf. Monogeneric).

Moriform Shaped like a mulberry fruit.

Morphology The study of form and structure of plants.

Mucilaginous Slimy, sticky when wet; pertaining to or containing mucilage.

Mucro A short sharp point formed by the continuation of the midrib.

Mucronate Said of a leaf terminating in a sharp point.

Mucronulate Shortly mucronate.

Multi- A Latin prefix meaning "many" or "several."

Multicarpelate A flower with compound ovary of many carpels.

Multicellular Composed of many cells, as in a trichome or a gland.

Multicipital A rootstock or a caudex from which several branches arise.

Multicostal Many-ribbed, the ribs running from base to the apex.

Multifid A leaf with many narrow lobes, as in Jatropha multifida.

Multiflorus With many flowers.

Multifoliate A leaf with many leaflets.

Multilocular With many chambers, in reference to ovary.

Multiperennial Monocarpic but living several to many years before flowering (syn. Pliestesial).

Multiple Epidermis In certain leaves or roots, there exist more than a single epidermal layer (syn. Velamen).

Multiple Fruit A fruit formed from several flowers into a single structure, as in mulberry (Morus).

Multiseriate Having many rows of cells, as in trichomes (cf. Uniseriate).

Multistriate With many lines, as in wax extrusions on some leaf surfaces.

Muricate A rough surface caused by short sharp points.

Muriculate Minutely muricate.

Muriform Having bricklike markings or reticulations, as in some seed coat or leaf surfaces.

Mutabilis Changeable, either in form or color.

Mutation The sudden alteration of one or more characteristics caused by a change in the genetic makeup of a plant, often as a result of external, as opposed to natural, factors.

Muticous Pointless; blunt; awnless.

Myophily (Myiophily) Pollination by flies, gnats, or mosquitoes.

Myrmecophily Pollination by ants.

Myrmecosymbiosis A symbiotic association between plants and ants, as in Mermecodia caudex which is inhabited by ants.

N

Naked Lacking a perianth; not enclosed within a pericarp, as in seeds of gymnosperms.

Nana, Nanus A Latin word meaning "dwarf".

Nanophyllous With leaves to 225 sq. mm. in size.

Napiform Turnip-shaped.

Native Said of plants that are indigenous to an area, such as Zamia floridana (Florida coontie), Eschscholtzia californica (California poppy), etc. (cf. Naturalized, Indigenous).

Naturalized A plant which was originally introduced from another region but now grows wild, such as Casuarina in Florida or Eucalyptus in California, both native to Australia (cf. Indigenous).

Natural Selection An evolutionary process by which certain individuals of a population are favored by a given environment and are thus more adaptable to that environment; "survival of the fittest."

Natural Classification A system of classification that utilizes natural (Phylogenetic) relationships.

Navicular, Naviculate Boat-shaped, as the glume of grasses or pollen grain of gymnosperms.

Neck Narrowed portion of hypanthium, between the base and the flared limb.

Necrocoleopterophily Pollination by carrion beetles.

Nectar A glandular secretion, containing sugars and amino acids, produced by insect-pollinated flowers or by extra-floral nectaries.

Nectar Guides Markings of flowers which are often invisible to the human eye, but function as orientation cues to insects.

Nectariferous Nectar producing; having a nectary.

Nectary Glands of flowers (Floral) or other parts of the plant (Extrafloral) which secrete nectar.

Needle The long and narrow leaves of pines and some species of Hakea (Proteaceae) (syn. Acicular).

Neo- A Greek prefix meaning "new."

Neoteny An evolutionary theory which assumes the juvenile characteristics of a taxon to resemble that taxon's ancestral form (cf. Paedomorphosis).

Neotropics The American tropics.

Neotype A specimen selected as the nomenclatural type when the original type of the taxon is known to have been destroyed.

Nervation See Venation.

Netted See Reticulate.

Nexine The inner layer of the Exine in pollen. (syn. Endexine).

Nidose Having an unpleasant odor.

Niger A Latin word meaning "black."

Nigrescent, Nigresense, Nigricans Blackish, or becoming black.

Nitid, Nitidous Smooth and clear, lustrous.

Nivalis A Latin word meaning "growing near snow."

Niveous Snow-white.

Nocturnal Said of flowers that open at night and close during the day, e.g., Mirabilis jalapa (the four o'clock flower).

Nodal Nectary With nectary glands at the nodes.

Node A joint, as in a stem where buds and leaves occur.

Nodiform See Nodulose.

Nodose Bearing knot-like swellings.

Nodules Enlargements or swellings on roots of nitrogen-fixing plants, as in legumes, Alnus, Ceonothus, etc.

Nodulose Having nodules.

Nomen A Latin word meaning "name".

Nomenclature The naming of plants and other organisms.

Nomen Ambiguum An ambiguous name; a name with different meanings.

Nomen Confusum The name of a taxon that is based on more than one element.

Nomen Conservandum A name conserved by the decision of International Congress for Botanical Nomenclature (ICBN).

Nomen Conservandum Propositum A name proposed for conservation.

Nomen Dubium A name of uncertain origin, no plant known to which such name may be applied.

Nomen Nudum A name applied to a plant without proper Latin diagnosis or description.

Nonarticulated Without walls separating a long unit, a single cell, as in Nonarticulated Laticifer (cf. Articulated).

Notate Said of a surface with spots or lines.

Notorhizal See Incumbent.

Nototribic, Nototribous Flowers whose stamens and styles turn so as to strike their visitors on the back.

Novirame A flowering or fruiting shoot arising from a Primocane, as in blackberries.

Nuclear Beak A conical protrusion of the nucellus that forms a cavity or chamber in which pollen grains may collect, as in cycads.

Nucellus The Megasporangium; the tissue of the ovule which contains the embryo-sac and is surrounded by the integuments.

Nuciferous Bearing a nut.

Numerical Taxonomy A taxonomic technique which employs statistical methods by assigning numerical values of 1 and 0 (presence or absence) to characteristics of taxonomic units, and thus avoids a priori weighing of evidence in making decisions with regard to relationships between taxa.

Nut A dry, indehiscent, one-seeded fruit, derived from a single or a compound ovary.

Nutant, Nutans A Latin word meaning "nodding" or "hanging down."

Nutation The spiral growth of a plant organ, as in the shoot or tendrils of vines.

Nutlet A small nut.

O

Ob- A Latin prefix meaning "reversed" or "inverted."

Obclavate Club-shaped, but widest at the base.

Obcompressed Flattened at right angles to the primary axis; flattened dorsiventrally, as the achenes of some Asteraceae that are flattened at right angles to the receptacle.

Obconic, Obconical Inverse of Conical, but attached at the narrow point.

Obcordate, Obcordiform Inversely Cordate; wide at the apex but narrow at the base.

Obdeltate, Obdeltoid Inversely triangular.

Obdiplostemonous Having two rows of stamens, those of the outer whorl opposite the petals and the inner opposite the sepals.

Oblanceolate Inverse of Lanceolate; wide at the apex, tapering toward the base.

Oblate Globose, but noticeably wider than long.

Oblique With the two sides unequal, especially at the base of leaf blades, as in some species of Betulaceae.

Oblong With nearly parallel sides, and two to four times longer than broad.

Obovate Inversely Ovate, the apical half broader than the basal.

Obovoid Egg-shaped, and attached at the narrow end.

Obscure Not visible to the naked eye; not distinct.

Obsolete, Obsolescent Vestigial; rudimentary; not evident, such as stipules of certain species.

Obtuse Blunt; rounded, such as a leaf apex.

Occidentalis A Latin word meaning "western."

Ocellus, Ocellated An enlarged discolored spot on a leaf; a large spot on a leaf with another spot of a different color within it.

Ochraceous, Ochreous A Latin word meaning "Yellowish-brown."

Ochroleucous A Latin word meaning "Yellowish-white."

Ochrea, Ocrea A nodal sheath around the stem, formed by united stipules or united leaf bases, as in Polygonaceae.

Ocreate Having stipular sheathing around the stem.

Ocreolate Having a small stipular sheathing around the stem, usually on reference to bract bases.

Oculus An eye (bud) of a tuber, as in potato.

Odd-Pinnate See Compound Leaf.

Officinalis A Latin word meaning "medicinal."

-Oid A suffix meaning "like," "similar," "resembling," or "imitating."

Oleiferous Bearing oil, as in exocarp of an orange fruit.

Oleraceous Edible.

Oligo- A Greek prefix meaning "few" or "small."

Oligomerous Having fewer than the usual number of parts.

Oligophylic A flowering plant which is pollinated by very few related insect species.

Oligotaxy The decrease in the number of whorles in a flower.

Olivaceous Olive-colored; grayish-green with a touch of orange.

Ontogeny, Ontogenetic The developmental history of an organism or parts of it.

Oo- A Latin prefix meaning 'egg'.

Opaque Not transparent; dull and not shining.

Open Dichotomous Venation Dichotomous venation with free vein endings, as in Stangeria eriopus.

Open Venation A type of leaf venation where primary veins terminate blindly in the mesophyll.

Operculate Possessing a lid; opening by means of a lid.

Operculum A lid, as in capsule-type fruit of certain angiosperms.

Opposite Occurrence of organs on facing sides of an axis, as in opposite leaves which arise from the same node.

-Opsis A Greek suffix meaning "like."

Orbicular, Orbiculate Nearly circular in outline.

Order The taxonomic unit below class which includes one or more closely related families. The suffix for an ordinal name is -ales.

Orientale A Latin word meaning "eastern."

Orientation Relative position, as applied to organs

Ornamental Horticulture The science and the art of growing plants for environmental, psychological, and esthetic purposes.

Ornothophily, Ornithophilous Pollination by birds.

Ortho- A Greek prefix meaning "straight" or "upright."

Orthocladous Having long upright branches.

Orthostichy Arrangement of plant organs in vertical rows, as in phylotaxy of leaves (cf. Parastichy).

Orthotropus Ovule An erect, straight ovule, with micropyle at the apex and hilium at the base.

Osseous Boney.

Outgroup In phylogenetic studies, a reference group (e.g., a genus) that serves to resolve evolutionary relationships of the group being studied.

Oval Broadly elliptical, two times longer than wide.

Ovary Part of the carpel or gynoecium containing the ovule.

Ovate Egg-shaped, much broader below the middle.

Ovoid A solid oval organ that is attached by the broader end.

Ovulate Possessing an ovary.

Ovule The megasporangium; the egg which upon fertilization develops into a seed.

Ovuliferous Bearing ovules.

Ovuliferous Scale The scale-like structure in the cones of gymnosperms which bear the ovule and later the seed; the Megasporophyll.

P

Pachy- A Greek prefix meaning "thick."

Pachycarpous Having a thick pericarp.

Pachycauly Having a short, thick, often succulent stem, as in most species of Cactaceae; *Pachypodium*.

Pachyphyllous Having a thick leaf.

Pachymorph A short, thick, and fleshy multibranched rhizome (cf. Leptomorph).

Paedomorphosis Said of a plant that possesses some mature and some juvenile characteristics (syn. Neoteny).

Palate The projection of the lower lip in sympetalous corolla which nearly closes the throat of the flower, as in the Scrophulariaceae.

Palea (=Palet) The upper or inner of the two bracts of the floret in grasses, often partly enclosed by the lemma.

Paleaceous Chaffy; furnished with a palea; chafflike in texture.

Paleobotany The study of fossil plants.

Palaeocene A geological period (a subdivision of tertiary) which lasted approximately from 70 to 60 million years ago.

Paleomorphis Resembling a palea.

Paleozoic A geological period which lasted from 600 to 260 million years ago.

Palinactinodromous Having primary veins with one or more subsidiary radiation above the primary one (syn. Actinodromous).

Palmate Palm-like; applied to venation of a simple leaf when the major veins radiate from a common point at the base of the blade (Palmately veined) (syn. Digitate).

Palmately Compound See Compound Leaf.

Palmatifid Said of a leaf-blade cut halfway down, so that a number of lobes are formed.

Palmatisect Said of a leaf-blade cut nearly to the base, so that a number of distinct lobes are formed.

Palynology The study of pollen grains and spores.

Pandurrate, Panduriform Fiddle-shaped; obovate but with distinct shallow lobes near the base, as in certain leaves.

Panicle A compound racemose inflorescence, as in oat.

Paniculate Of the character of a panicle; bearing panicles.

Pannose, Panniform Looking like felt; having a matted layer of trichomes.

Pantoaperturate Pollen grains with apertures on their entire surface.

Papilionaceous Butterfly-shaped, referring to the corolla of subfamily Paplionoideae of Fabaceae.

Papilla (pl. Papillae) A small nipple-shaped protuberance or trichome.

Papillate, Papillose Bearing papillae.

Pappose Having a pappus.

Pappus The modified bristly or scale-like corolla of Asteraceae, which persists on the fruit and aids in dispersal.

Para- A Greek prefix meaning "beside."

Parallelodromous Having parallel venation; primary veins running side-by-side, as in most monocotyledons.

Paraphyly, Paraphyletic In phylogenetic studies, refers to a separate group originating from a common ancestor as another.

Paraphyses The rays of the corolla in *Passiflora*; hair or hair-like structures in the sorus of ferns.

Paratype   One of the group of specimens from which the original description was prepared (cf. Holotype, Isotype).

Parietal   Born on the sides or walls of a locule, as ovules in the ovary (Parietal Placentation).

Parted   Margin cut nearly to the base or the midrib, as in some leaves.

Parsimony   In phylogenetic studies, the simplest clado-gram with the least number of character states is best when it can explain evolutionary relationship of the taxa.

Parthenocarpy   Formation or enlargement of fruit without fertilization, as in seedless grapes, naval oranges, or the commercial banana.

Patellate, Patelliform   Saucer or disk-shaped.

Patens, Patent   Spreading, as in leaves which spread out widely from a branch.

Pectinate   Comb like.

Pedate   A palmately lobed or divided leaf with lateral lobes divided or cleft.

Pedicel   The stalk of a flower or a fruit (cf. Peduncle).

Pedicellate   Said of a flower or a fruit which is born on a pedicel.

Peduncle   The stalk of a solitary flower or an inflorescence.

Pellucid   Clear, nearly transparent in transmitted light.

Peliose   A Greek word meaning "black."

Peltate, Peltiform   Shield-shape, as leaves which have the petiole attached at or near the center of the blade (peltate leaf).

Pendent, Pendulous   Drooping, hanging downward, as in branches of *Ilex vomitoria* "Pendula."

Penicillate   Ending in a tuft of hairs or branchlets.

Penninerved   Pinnately veined.

Penta-   A Greek prefix meaning "five."

Pentacyclic   A flower with five whorles of members (cf. Pentamerous)

Pentamerous   Having five members in a whorl; 5-merous.

Pentandrous   A flower with five stamens.

Pepo   The fruit of Cucurbitaceae; a berry with a hard rind derived from an inferior ovary.

Perennate, Perennation   Survival from season to season, usually self-renewing by lateral shoots which arise from the base of the plant.

Perennial   A plant which lives for more than two years and flowers more than once (Polycarpic), or a plant which lives for several years, but dies after the first flowering (Monocarpic).

Perfect Flower   Having both androecium (stamens) and gynoecium (carpels) in the same flower; bisexual.

Perfoliate   Said of a sessile leaf which surrounds a stem completely so that the stem appears to pass through it.

Pergamentaceous, Pergameneous, Pergamenous   Paper or parchment-like in appearance or texture.

Peri-   A Greek prefix meaning "around."

Perianth   A collective term for the floral envelope, including calys or corolla, but usually both.

Pericarp   The wall of the ovary (ripened fruit), which consists of three distinct layers: exocarp, mesocarp, and endocarp.

Pericarpous   Fruit which has a pericarp.

Pericladium   The sheathing base of a leaf when it surrounds the supporting branch.

Periclinal   Curved in the same direction as the surface or circumference.

Periclinal Chimera   Graft hybrid, a mixture of cells of different gynotypes, e.g., thornless blackberry.

Perigynium   The inflated bracts that enclose the gynoecium or carpellate flower in Carex.

Perigynous, Perigyny   A condition where floral parts are born around the gynoecium, as when the calyx, corolla, and stamens arise from the edge of a cup-shaped hypanthium (e.g., Rosaceae) (cf. pigynous, Hypogenous).

Perine   The outermost sculpturing of fern spores (syn. Perispore).

Periodicity   Rhythmic activity, as seasonal growth or flowering.

Peripheral   Surrounding; on the outer surface or edge.

Perisperm   A nutritive tissue derived from a diploid nucellus, present outside the embryo sac in some seeds, as in coffee, spinach, etc. (cf. Endosperm).

Perispore   The wrinkled outer covering of some fern spores (e.g., Polypodiaceae).

Persistent   Remaining attached, not falling off.

Personate   A two-lipped (Bilabiate) corolla the upper lip of which is arched and the lower lip protrudes and nearly closes the throat of the corolla tube.

Pertusate   Perforated, pierced by slits.

Perulate   Scale-bearing, as in most buds.

Petal   The usually colored inner perianth (Corolla) segments.

Petalantherous Stamen   With a terminal anther and a distinct petaloid filament, as in Saxifraga.

Petaliferous Nectary   Nectaries located at the base of the petal.

Petaloid   Petal-like, resembling a petal.

Petalostemanous   A condition where staminal filaments are fused to the corolla but the anthers are free.

Petiolar   Referring to a petiole.

Petiolate   Having a petiole (cf. Sessile).

Petiole   The leaf stalk. The stem with which a leaf attaches to the stem.

Petiolulate   Having a petiolule.

Petiolule   The leaflet stalk in a compound leaf.

Petrorhizous   With roots growing on rocks.

Phalaenophyly   Pollination by moths.

Phanerantherous   When the anthers protrude beyond the perianth.

Phanerocotylar   A type of seed germination where coty-ledons fully emerge from the seed coat and appear above ground (cf. Cryptocotylar).

Phanerogam   A seed plant (Spermatophyte) (cf. Cryptogam).

Phanerophyte   Woody plants or herbaceous evergreen perennials that grow taller than 0.5 m and whose shoots do not die back periodically (cf. Chamaephyte).

Phenetic   A classification system based on total similarity of all characteristics, including morphological, anatomical, chemical, etc.

Phenogram   A graph illustrating phenotypic relationships.

Phenology   The study of periodical phenomena in plants, as the time of flowering in relation to climate.

Phenotype   A kind or type of plant produced as a result of interaction between genotype and the environment, such as growth habit or shape.

Phenotypic   Characteristics which are caused or produced by environmental factors.

Phoeniceus   A Latin word meaning "scarlet."

Phylad   An evolutionary line; a natural group with common ancestry.

Phyllary   One of the bracts of the involucre, as in Asteraceae.

Phyllo-, Phyll-, -Phyll   A prefix or suffix meaning "leaf."

Phylloclade   A more or less flattened stem which functions as a leaf, as in Epiphyllum (Cactaceae).

Phyllode, Phyllodial, Phyllodium   A flattened leaf-like petiole, as in some Acacia species.

Phyllody   Transformation of parts of a flower into leaves.

Phyllopodic   Having well-developed leaves at the base of the plant.

Phyllotaxy   The arrangement of leaves on the stem, such as alternate, opposite, or whorled.

Phylogeny, Phylogenetic   The evolutionary relationship of a group of organisms.

Phylum (pl. Phyla)   A major division of the plant or animal kingdom, whose members are presumed to have common ancestry.

Physiognomy   The characteristic appearance of a plant community by which it can be identified from a distant, such as a forest, grassland, etc.

Phyto-, -Phyte   A Latin prefix or suffix meaning "plant."

Phytogeography   The study of the distribution of plants.

Pilaris   Composed of small hairs (syn. Pilose).

Pileate, Pileus   Having the form of a cap.

Piliferous   Having fine hairs; ending in a delicate hair-like point.

Pilose   With soft slender hairs.

Pilus (pl. Pili)   A Latin word meaning "hair."

Pin Flower   In heterostylous plants, the flower having the longer style (cf. Thrum Flower).

Pinna (pl. Pinnae)   The primary division of a compound leaf; a leaflet. The term is primarily used for fern fronds (cf. Pinnule).

Pinnate   Feather-like; applied to venation of a simple leaf when the principle veins are extended from the midrib toward the margin (Pinnately Veined), or arrangement of leaflets of a compound leaf when they are on opposite side of a rachis (Pinnately Compound).

Pinnatifid   Said of a leaf blade which is cut halfway toward the midrib, into a number of pinnately arranged lobes.

Pinnatisect   Pinnatifid, but with the cuts reaching nearly to the midrib.

Pinninerved   Pinnately veined.

Pinnule   One of the segments when the pinna (leaflet) is itself divided. The term is most often used for fern fronds.

Piriform   See Pyriform.

Pisiform   Pea-shaped.

Pistil   The ovuliferous or seed-bearing organ of a flower; carpel; gynoecium (collective term), consisting of ovary, style, and stigma.

Pistillate   Bearing the pistil or pistils only; carpelate; a unisexual (female) flower.

Pistillode   An abortive or vestigial pistil in some staminate (male) flowers.

Pitcher Plants   Plants of the genus Sarracenia which trap insects in pitcher-like, inflated appendages at their leaf apex.

Pith   The soft spongy central tissue in the stem of most angiosperms.

Placenta   The tissue within an ovary to which the ovules are attached.

Placentation   The arrangement of ovules within an ovary: Axile Placentation, the margins of the compound ovary fold inward fusing together at the center, forming a single placenta onto which the ovules are attached; Basal Placentation, ovules are few or reduced to one and attached at the base of the ovary; Free Central Placentation, the ovules are born on a central column which arises from the base of the ovary; Parietal Placentation, the carpels are fused only by their margins so that the placenta appears as ridges onto which the ovules are attached.

Plantlet   A small plant, such as those formed by propagation of Begonia leaves.

Platy-   A Greek prefix meaning "broad".

Platyform   Flattened.

Platycanthous   With flattened spines.

Pleiandrous   Having a large number of stamens.

Pleio-, Pleo-   A Greek prefix meaning "several."

Pleiochasium   A cymose inflorescence in which each branch bears two or more lateral branches.

Pleiomerous   Having a large number of parts or organs.

Pleiotaxy   An increase in the number of whorls of a flower.

Pleistocene A geological period lasting approximately from one million to 10,000 years ago, during which the four major ice ages occurred.

Pleomorphic, Pleomorphous, Pleomorphy Mutability of shape; having more than one form in the life cycle.

Plesiomorphic, Plesiomorphous, Plesiomorphy Nearly of the same form; shared ancestral or primitive character (cf. Apomorphy, Sympesiomorphy, Synapomorphy).

Pleura-, Pleuro- A Greek prefix meaning "side" or "rib.".

Pleurogamy Pollen tube entrance through the side of the ovule.

Pleurogenous Born in a lateral position.

Pleurorhizal Said of the embryo of a flowering plant, when the radical is placed against the edge of the cotyledons.

Pleurotropous Said of a horizontal ovule.

Plexus A network, often of anastomosing veins.

Plicate Said of a leaf in which the blade is folded back and forth along the main veins like pleats in an accordion.

Pliocene A geologic period, subdivision of tertiary which lasted from about 15 to 1 million years ago.

-Ploid A suffix used to designate a particular multiple of the chromosome set, as in diploid, tetraploid, etc.

Ploidy The number of chromosomes in a cell, as in polyploidy.

Plumose Covered with fine hairs, like a feather.

Plumule The terminal bud of an embryo in seed plants.

Pluri- A Latin prefix meaning "many."

Plurilocular Having many locules, as in some fruits.

Pneumatophore A specialized root which grows vertically upwards, often occurring in wet habitats. For example, *Taxodium distichum* (swamp cyprus) (syn. Knee).

Pod A dry fruit formed from a single carpel, containing one to many seeds, and opening along two sutures. The term, however, is most often used in reference to any dry dehiscent fruit.

Podo- A Greek prefix meaning "stalk," "stipe."

Podogyne Carpel with a basal stalk (syn. Carpopodium).

Poikilohydry Under dry conditions, some fern species roll into brown balls but become green under moist conditions. In this state, they may be uprooted and replanted (*cf.* resurrection plants).

Pollarding A method of pruning where branches are severely cut in order to promote growth of a thick mass of laterals, thus creating an umbrella-shaped plant.

Pollen The grains or male spores of angiosperms and gymnosperms (syn. Microspore).

Pollination The transfer of pollen from anthers to a stigma in angiosperms or from male to female cones in gymnosperms.

Pollinium (pl. Pollinia) A mass of pollen held together by a sticky substance, and transported as a unit by pollinating agents, as in Asclepiadaceae and Orchidaceae.

Poly- A Greek prefix meaning "many."

Polyad Pollen grains which occur in groups of more than four (cf. tetrad).

Polyandrous, Polyandry Having a large and indefinite number of stamens.

Polyanthous Having many flowers.

Polycarpous, Polycarpellate Having many separate carpels.

Polycephalous Bearing many heads or capitula.

Polychromatic Having many colors on the same organ.

Polycormic Said of a woody plant with several strong trunks.

Polycyclic When the member of a series, such as calyx or corolla, are in several circles.

Polydelphous With several groups of stamens connate by their filaments.

Polygamodioecious Plants that are functionally dioecious but have a few flowers of the opposite sex or a few bisexual flowers.

Polygamomonoecious Plants monoecious, but with some perfect flowers.

Polygamous Bearing unisexual and bisexual flowers on the same plant, as in some species of *Acer* (Aceraceae).

Polyheterophytous Having different states in several (more than three) different individual or sets of plants, only one set present in each set.

Polymerous With numerous members of each series or cycle.

Polymorphic, Polymorphism The occurrence of several forms in the same species.

Polypetalous, Polypetaly Having separate petals.

Polyphore A receptacle bearing many carpels, as in strawberry or rose.

Polyphyletic A group of species classified together but having different lines of descent or evolutionary history.

Polyploid, Polyploidy An individual with more than the diploid number of chromosomes.

Polysepalous, Polysepaly Having separate sepals.

Polysomic A normal diploid chromosome set which has one or more extra chromosome, as in 2n + 1 (trisomic) or 2n + 2 (tetrasomic).

Polystichous Arranged in several rows.

Pome A fruit, like an apple or pear, in which most of the edible part is the enlarged axis of the flower , rather than the ovary.

Porate Referring to a pollen grain which has one or more pores.

Pore A small hole.

Poricidal See Dehiscence.

Porogamy The entry of the pollen tube through the micropyle.

Porrect Extending forward, usually in reference to teeth (syn. Salient, Projected).

Posterior   The upper side; the side toward the axis (cf. Adaxial).

Pouch   A sac-like structure, as the spur in many orchids.

Praecox   A Latin word meaning "appearing early."

Praefloration, prefloration   See Aestivation.

Praefoliation, Prefoliation   See Vernation.

Pratensis   A Latin word meaning "of meadows."

Precocious   Appearing early, as when a very young plant flowers.

Prickle   A small and hard spine appearing on the bark or epidermis as an enation.

Primary axis   The main stem.

Primary Root   The main root developed from the radicle.

Primary shoot   The main stem developed from the plumule.

Primitive   Most similar to the ancestral condition; not highly evolved.

Primocane   The first year's shoot or cane of a woody biennial stem which normally does not flower.

Proanthesis   Flowering in advance of the normal period, as in plants that flower in winter when unusually warm.

Process   A general term for an extension or projection.

Procumbent   Trailing or lying flat on the ground, but not rooting (cf. Decumbent).

Prodromus   Botanical works which are intended as preliminary and are to be followed by a more complete treatises.

Projected   See Porrect.

Prolate   Somewhat globular, but flattened equatorially.

Proleptic Shoots   Abnormal late season shoots that develop from the lateral buds immediately below the terminal.

Propagation   The reproduction of plants by sexual (seed) or asexual (vegetative) means.

Propagule   Any structure which becomes detached from the mother plant and grows into a new plant.

Prophyll, Prophyllum   A small scale-like appendage at the base of an individual flower (syn. Bracteole).

Prop Roots   The adventitious roots that are formed on the trunk of certain plants, as in *Ficus*, *Pandanus*, etc. (syn. Stilt Roots, Aerial Roots).

Prostrate   A general term for plants whose branches lie flat on the ground (cf. Procumbent, Decumbent).

Protandrous, Protandry   Said of a flower in which the stamens mature before the pistil.

Protantherous   With leaves appearing before flowers.

Protected Bud   Shoot or flower primordia surrounded by scales.

Protective Stipules   Sheathing stipules which protect a bud or a flower.

Proterogynous, Proterogyny   Said of a flower in which the pistils mature before the stamens.

Prothallus   Gametophyte of lower vascular plants.

Proto-   A prefix meaning "first," "primitive," or "primordial,"

Proximal   Situated toward the point of attachment (opp. Distal).

Pruinose   Dusted with coarse granular waxy material.

Prune, Pruning   The selective removal of branches or stems of plants to improve their health, vigor, and appearance.

Pruniform   Shaped like a plum.

Pseud-, Pseudo-   A Latin prefix meaning "false", "untrue", "atypical".

Pseudanthium   An inflorescence which resembles a flower because the individual flowers are reduced to single stamens or carpels, as in the genus *Euphorbia* (Euphorbiaceae) and its related genera.

Pseudobulb   The thickened or bulb-like basal portion of some orchids.

Pseudocarp   A fruit type in which an aggregation of achenes is enclosed in a fleshy receptacle, as in *Fragaria* (strawberry).

Pseudocleistogamy   Failure of flowers to open because of ecological rather than inherent factors.

Pseudodrupe   A two to four loculed nut surrounded by a fleshy involucre, as in Juglance (walnuts).

Pseudogamy   The condition in which seed development is stimulated by the presence of the pollen, although fertilization does not actually take place (cf. Apomyctic).

Pseudomonomerous   Whorle seemingly with one member, but usually as a result of fusion of two or more parts.

Pseudomonopodial   A plant which appears to be monopodial when young but becomes sympodial once mature, as in most tree seedlings.

Pseudoviviparous, Pseudovivipary   Plants in which lateral buds or parts of the inflorescence develop into vegetative propagules, as in some species of *Lilium* (Liliaceae).

Psilate   A pollen grain which appears smooth, without visible external features, as in some cycads, Betula, etc.

Psilophytes   Plants which live in grasslands.

Psychophily   Butterfly pollination.

Pterate   Having wings.

Ptero-   A Greek prefix meaning "wing."

Pterocauly   With winged stems.

Pterodophyta   An obsolete botanical term indicating a division of vascular plants more primitive than the gymnosperms. In vernacular usage, it refers to the ferns in general.

Ptyxis   The rolling or folding of leaves or floral parts in the bud (syn. Vernation).

Puberulent   With very fine down-like hairs.

Pubescence, Pubescent   Covered with fine hairs.

Pulverulent   Covered with fine powdery wax.

Pulvinate   Shaped like a cushion.

Pulvinule   The small pulvinus at the base of a petiolule.

Pulvinus   A swelling at the base of a petiole, which is responsible for movement of leaves in the presence of a stimulus, as in *Mimosa pudica* (sensitive plant).

Pumilus   A Latin word meaning 'small' or 'low'.

Punctate   Dotted or marked with small dots.

Pungens   A Latin word meaning 'ending in a sharp point'.

Pungent   Ending in a rigid and sharp point, as in leaves of Ilex opaca (American holly).

Puniceus   A Latin word meaning 'light red'

Pure line   The descendant of a homozygous plant obtained by repeated self-fertilization; inbred homozygous individuals, as in most annual bedding plants which have uniform growth habit and flower color.

Purpureus   A Latin word meaning "purple."

Pustule, Pustular, Pustulate, Pustulose   With pimples or blisters.

Pygmaeus   A Latin word meaning "dwarf."

Pyramidal   Shaped like a pyramid.

Pyrene   The nutlet or seed of a drupe; a seed surrounded by a bony endocarp, as in a cherry or peach pit.

Pyriform   Shaped like a pear.

Pyrophyte   Plants that live in savannahs or open woodlands and are protected from forest fire by their thick bark.

Pyxis   A capsule whose top comes off as a lid or opens circumscissilely.

Q

Quadri   A Latin word meaning "four."

Quadrifid   Divided into four lobes or parts.

Quilled   Florets which are normally, ligulate but have become tubular.

Quinate   Growing together in fives, as in leaflets from the same point.

Quincuncial   Having five structures, two of which are exterior, two interior, and the fifth with one margin covering the interior structure and the other margin covered by that of one of exterior structures.

Quinque   A Latin word meaning "five."

Quinquefoliate   Having five leaflets.

R

Race   A genetically and often geographically distinct division of species, varying by a minor characteristic such as flower color.

Raceme   A simple, elongated, indeterminate inflorescence with pedicellate (stalked) flowers.

Racemose   Having flowers in a raceme or raceme-like inflorescence.

Racemule   A small raceme.

Rachilla   A small raceme; in grasses and sedges the spiklet that bears the florets.

Rachis   The axis of a compound leaf, above the petiole to which the leaflets are attached; the main axis of an inflorescence.

Radial symmetry   When a flower can be dissected into two similar halves in any direction, it is radially symmetrica. (syn. Actinomorphy).

Radiate   Spreading from the center, as in ray florets of Asteraceae flowers.

Radical   Arising from the root crown, as in basal leaves of Amaryllidaceae.

Radicant, Radicance   Latin word meaning "rooting."

Radicle   The embryonic root in seeds.

Ramentaceous   Covered with ramenta, brown chaffy scales.

Ramentum (pl. Ramenta)   The small brown scale on the stem of ferns.

Ramet   An individual of a clone.

Ramification   Branching, as in leaf veins; sometimes to branching in general.

Ramose   Having branches.

Ramous   Branching; having many branches.

Ramular, Ramulose   With many branches.

Ramulus   A very small branch.

Random   Said of branches that arise from buds without relation to leaves; any event which may occur by chance without discrimination, as in pollination.

Rank   A vertical row, as in 2-ranked or 4-ranked leaves; referring to levels of hierarchy such as species, genus, etc.

Raphe   An elongated mass of tissue which lies on the side of an anatropus ovule and appears as a ridge on the seed.

Raphide   Needle-shaped crystals of calcium oxalate occurring singly or in bundles in leaves or stems of certain plants, as in Diffienbachia (Araceae). Direct contact may be irritating to the skin.

Ray Floret   One of the zygomorphic flowers radiating from the margin of the capitulum in Asteraceae.. (cf. Disk Floret).

Receptacle, Receptaculum   The enlarged part of the floral axis that bears the floral organs; the hypanthium of epigynous and perigynous flowers; the common axis of the flowers in the inflorescence of Asteraceae (syn. Torus).

Reclinate, Reclining   Turned or bent downward, as in a branch.

Reclined   Descending at 106–135 degree angle of divergence; said of a plant pressed against another.

Recurved   Bent or curved backward, as in revolute leaf margins.

Reduced   Simplified in structure or diminished in number as compared to the ancestral form.

Reflexed   Turned backward or downward abruptly.

Regular   See Actinomorphic.

Relatedness   Relationship of clades by common ancestry.

Relict Remnants of plants from an earlier geological period, usually restricted in geographical distribution, such as cycads.

Reniform Kidney-shaped.

Repand Having a slightly wavy margin; weakly sinuate.

Repens, Repent, Reptans Prostrate plants whose stems produce adventitious roots at the nodes when in contact with moist soil.

Replicate Folded back along the middle, as in the leaf or leaflet of some legumes.

Replum The partition between the two locules of the cruciferous fruits.

Reproduction The process of creating new individuals by sexual or asexual means.

Reproductive Isolation The prevention of inbreeding by spatial or mechanical means.

Resinous, Resiniferous Producing or containing resin, as in sticky buds of Aesculus spp.

Resupinate Upside down; twisted 180 degrees, as in flowers of Orchidaceae.

Reticulate Having a network, as in venation of some leaves. (syn. Netted); in phylogenetic trees, joining together of clades.

Reticulodromous With a single primary vein, the secondary veins not terminating at the margin but branching profusely near it, yielding a dense reticulum.

Reticulum A network.

Retinaculum The persistent, hook-like funiculus, as in fruit of Acanthaceae; the structure to which pollinium is attached, as in Asclepiadaceae and Orchidaceae.

Retrorse Pointing backward or in opposite direction than normal, as in stamens which face out rather than toward the center of the flower.

Retroserrate Having marginal teeth bent backwards.

Retuse Slightly notched at a usually rounded apex.

Revolute A leaf in which margins are rolled backward and usually downward (cf. Involute).

Rhipidium A fan-shaped cymose inflorescence.

Rhizanthous Appearing to flower from roots, as in *Aspidistra* (Asparagaceae), *Maxillaria* (Orchidaceae),

Rhizocarpic, Rhizocarpous Producing subterranean as well as aerial flowers and fruit, as in *Annona rhizantha* (Annonaceae). The term is also used to denote perennial herbs.

Rhizocorm See Corm.

Rhizoid A hairlike organ which functions as a root in gametophyte of lower vascular plants.

Rhizomatous Producing or bearing rhizome.

Rhizome An elongated subterranean, usually horizontal stem of rootlike appearance.

Rhizophore A leafless stem of Selaginella, from which roots arise.

Rhizotaxis The position or arrangement of lateral roots.

Rhodo- A Greek prefix meaning 'rose-red'.

Rhombate, Rhombeus, Rhombic, Rhomboid, Rhomboidal Diamond-shaped and attached by one of the acute angles.

Rib The primary or any of the larger veins of a leaf.

Rigens A Latin word meaning "rigid."

Rimose Having the surface marked by a network of intersecting cracks.

Rimulose Having small cracks.

Rind The outer layer of the bark of a tree; the outer layers of a fruit.

Ringent Said of a corolla which has two distinct, gaping lips.

Riparial, Riparian Living on the banks of rivers and streams.

Rivalis A Latin word meaning "growing by rivers."

Rivulose Marked with lines, giving the appearance of rivers on a map.

Robust Large; healthy.

Rogue In horticulture, a term used for plants that do not come true from seed.

Root The descending axis of vascular plants, growing in the opposite direction from the stem, and function in absorption of water and nutrients.

Root Nodule The small swelling on roots which are caused by nitrogen-fixing bacteria, as in legumes, cycads, etc.

Rootlet A very small root; the branch of a root.

Rootstock Subterranean stem; rhizome; in horticulture, the seedling or rooted cuttings onto which the desired cultivar is grafted.

Roridus Dewy; covered with particles which resemble dew drops.

Roridulate Covered with waxy platelets; appearing dewy.

Rosette A cluster of closely crowded leaves radiating from a very short stem near the surface of the ground, as in *Lactuca* (lettuce), most Crassulaceae, etc.

Rostellate Beaked.

Rostellum A beaklike outgrowth from the column in flowers of Orchidaceae.

Rostrate, Rostrum Any beaklike projection.

Rosulate Referring to the rosette condition.

Rotate Wheel-shaped, referring to a sympetalous corolla with a short tube, as in that of Solanaceae.

Rotund Round or nearly circular in outline.

Rough Scabrous; covered with stiff hairs.

Rubens, Ruber Latin word meaning "red."

Rubescent Turning pink or red.

Rubigenos, Rubiginous Rust-colored.

Ruderal, Ruderalis Said of plants that grow in disturbed habitats or waste places.

Rudimentary Imperfectly developed and nonfunctional, such as staminodia (syn. Vestigial).

Rufescent Becoming reddish-brown.

Ruffled   Having a strongly wavy margin.

Rufidulus   Somewhat red.

Rufous   A Latin word meaning "reddish-brown."

Rugose   Having a wrinkled surface.

Rugulose   Somewhat wrinkled.

Ruminate   Appearing chewed; having dark and light spots of color or mottled.

Runcinate   Coarsely serrate or sharply incised or cleft, with the teeth pointing toward the base of the leaf.

Runner   A prostrate shoot with roots at the end, giving rise to new plants, as in *Fragaria* (strawberry), *Chlorophytum* (spider plant), etc. (syn. Stolon).

Rupestral   Plants that grow on walls or rocks.

Ruptured Capsule   See Anomalicidal Capsule.

Rushlike   Resembling rushes (Juncus, Typha, etc.).

S

Sabulose, Sabuline   Growing in sandy places.

Saccate   Pouchlike, saclike.

Saccus (pl. Sacci)   The wing-like extension of the exine in pollen grains of conifers.

Sagittate   Arrow-shaped, with the basal lobes directed downward.

Salient   Projecting forward. In common usage, the word is used to indicate conspicuous features.

Salverform   Trumpet-shaped; said of a corolla which is long and tubular, and the upper part spreads horizontally, as in *Phlox*. (syn. Hypocrateriform).

Samara   An indehiscent, dry, one- or two-seeded, winged fruit, as in *Acer* (maples), *Fraxinus* (ashes).

Samaracetum   A aggregation of samaras, as in *Liriodendron*.

Sanguineous   A Latin word meaning "blood red" or "crimson."

Sapidus, Sapid   A Latin word meaning "with a pleasant taste."

Sapling   A young tree.

Sapromyophily, (Sapromyiophily)   Pollination by carrion or dung flies.

Saprophyte, Saprophytic   A plant living on and, deriving its food from decaying organic matter.

Sarcocauly   Said of plants with fleshy stem, as in Cactaceae, Euphorbiaceae, etc.

Sarcotesta   The outer fleshy seed coat, such as that of cycads (cf. Sclerotesta).

Sarcous   Having a fleshy seed coat.

Sarmentose   Producing long runners or stolons.

Savanna (Savannah)   Grassland with scattered trees or patches of forest.

Saxicole, Saxicolous, Saxitilis   Growing on rocks or stones.

Scaberulous,   Scabrellate,   Scabridulous   Slightly scabrous.

Scabrid, Scabrous   Rough; rough pubescent; having a surface covered with wart-like projections (syn. Scurfy).

Scale   Any small, usually dry, appressed leaf or bract, as in scale leaves of pines; in common usage, it refers to peltate hairs.

Scalloped   Said of margins with rounded teeth.

Scaly Bulb   Bulbs which have several segments (leaves) arising from a common base, such as that of *Lilium longiflorum* (Easter lily) (syn. Non-Tunicate).

Scandens, Scandent   Climbing.

Scape   A leafless flowering stalk (peduncle) of an acaulescent plant, such as that of Amaryllidaceae, Asteraceae, etc.

Scaphoid   Boat-shaped.

Scapigerous, Scapose   A plant which has one or more scapes.

Scarious   A thin, nongreen, dry, membranaceous structure.

Scarred   Said of a stem which is marked by separation of the leaves, or on a seed by its detachment (a Cicatrix).

Scarlet   See Coccineus.

Scariose, Scarious   Thin, dry, and membranous, not green.

Scatter diagram   A graphic method of showing the relationship between two characteristics. The distance between points (or groups of points) reflects degree of similarity or dissimilarity.

Scattered   Without apparent order.

Schizo-   A Greek prefix meaning "split.".

Schizocarp   A dry fruit formed from a syncarpous ovary that splits at maturity into its constituent one- seeded carpels (mericarps).

Schizopetalous   Having divided or split petals.

Scion   The portion of a stem that is grafted onto a rootstock.

Sciophyll   A shade leaf (usually larger, thinner, and less green).

Sciophyllous   Having leaves which can tolerate shade.

Sciophyte   Shade-loving plant.

Sciuroid   Curved and bushy, like a squirrel's tail.

Sclero-   A Greek prefix meaning "hard" or "stony."

Scleranthium   An achene enclosed within a hardened (indurated) portion of the calyx tube, as in *Mirabilis* (four o'clock flower).

Sclerocauly   The possession of dry hard stem, as in *Ephedra* (Mormon tea).

Sclerous   Having a hard texture.

Sclerophyll, Sclerophyllous   A hard, leathery leaf, charateristic of dry habitats.

Sclerophyllous Vegetation   Woody plants characteristic of dry habitats, such as that of Califonia Chaparral.

Sclerotesta   The stony layer of the seed coat, as in cycads (cf. Sarcotesta).

Scobinate   A surface which is roughened.

Scrumbler   A plant with long, weak shoots which grow over other plants.

Scorpoid, Scorpoidal   Resembling a scorpion; coiled (circinate) like the tail of a scorpion.

Scorpoid Cyme   When the main axis of an inflorescence is coiled so that the flowers occur alternately on opposite sides, as in Boraginaceae.

Scrobiculate   Dotted with small depressions.

Scrotiform   Pouch-shaped.

Scurfy   Covered with small branlike scales.

Scutate, Scutiform   Shield-shaped.

Scutellate   Saucer-shaped.

Scutellum   The cotyledon of a grass embryo, which is in contact with the endosperm and acts as an absorbing organ.

Scutum   The dilated apex of the style in Asclepiadaceae (milkweeds).

Seasonally   Occurring during a seasonal cycle; each season.

Seasonally deciduous   Falling after one growing season.

Secondary Root   Lateral roots with root caps and hairs, derived from the pericycle.

Section   A division of a genus, containing one or many related species.

Seculate   Sickle-shaped (syn. Falcate).

Secund   One-sided; in reference to an inflorescence which has flowers born only on one side.

Seed   The ripened ovule of seed plants, which develops after fertilization, containing the embryo and endosperm or female gametophyte tissue.

Seed Coat   The outer protective covering of seeds, consisting of one or more layers, such as Sarcotesta and Sclerotesta (syn. Testa).

Seed Leaf   See Cotyledon.

Seedling   The young plant developing from a germinating seed.

Seed Plant   A plant that produces seeds, as in gymnosperms and angiosperms.

Segetalis   Growing in fields of crops.

Segment   A portion of the lamina of leaf which is deeply lobed but not divided into leaflets; one of a series, as one petal of a polypetalous flower.

Sejugous   Having six pairs of leaflets, as some pinnately compound leaves.

Selection   In horticulture, the deliberate (non-random) process of favoring a particular genotype for a specific purpose, such as food, fiber, or ornamental; in evolutionary biology, natural preference of feature(s) that favors environmental adaptation.

Semataxis   Arrangement of semaphylls (petals, sepals, or tepals) for the specific purpose of advertisement, to attract pollinators

Semi-   A Latin prefix meaning "half"; "somewhat"; "more or less."

Semiamplexicaul   Said of a leaf base which only partially clasps the stem.

Semianatropous   See Amphitropous.

Semicalyciform   Half cup-shaped.

Semicarpous   With adjacent ovaries partly fused, so as stigmas and styles are free.

Semiconnate   Applied to such structures as the half-united filaments of certain plants.

Semicraspedodromous   With a single primary vein, secondary veins branching just within the margin, one branch terminating at the margin, the other forming a loop and joining the ones immediately above it.

Semidigynus   When two carpels cohere only near the base.

Semidouble   When the inner stamens are normal but the outer ones petaloid, half changed to a double flower.

Semifloscular, Semiflosculous   Having the corolla split and turned to one side, as in ligulate florets of Asteraceae.

Semiglobose   Half-globose; hemispherical.

Semilunate   Shaped like a half-moon; crescent-shaped.

Semiterete   Half-cylindric.

Semperflorent, Semperflorous   With flowers appearing throughout the year, as *Begonia semperflorens* (syn. Aianthous).

Sempervirent, Sempervirens   Evergreen.

Sepal   A segment of the calyx, usually green and foliaceous.

Sepaliferous Nectary   Floral nectaries that occur at the base of the sepals.

Sepaloid   Resembling a sepal.

Septal Nectary   Floral nectaries which occur at the junction of the septa in the ovary.

Septate   Divided by a partition.

Septemfid   Deeply divided into seven parts.

Septempartite   divided into seven lobes.

Septenate   Having parts in sevens.

Septicidal   Dehiscence along the septa, rather than directly into the locules.

Septifragal   Said of a capsule whose valves break away from the septa at maturity, as in *Ipomoea* (morning glory) (syn. Valvular).

Septum   Partitions or segments of a compound ovary.

Seriate   In rows or series.

Sericate, Sericeous   With silky hairs.

Serology   A chemical method of analyzing relationships of plants.

Serotaxonomy   Analysis of plant secretions to study taxonomic relationship. Method is also used in virology.

Serotinal, Serotinous, Serotinus   Appearing late in the year.

Serrate   With sharp marginal teeth which point forward.

Serrulate   Serrate, but with very small teeth.

Sesamoid   Granular.

Sessile   Without a stalk, as in a leaf without a petiole.

Seta (pl. Setae)   A bristle; a slender straight prickle.

Setaceous, Setiferous Having bristles.

Setiform bristle-shape.

Setose Covered with bristles.

Setulose Covered with minute bristles.

Sexine The outer layer of the exine of pollen grain (syn. Ektexine).

Sexual Propagation Reproduction of plants by seed (cf. Vegetative).

Shape An outline of specific form which is plain or two-dimensional; a specific three-dimensional figure.

Sheath The tubular leaf base which forms a casing around the stem.

Sheathing Enclosing; covering; surrounding.

Shining A clear and polished surface; lucid.

Shoot Aerial portions of the plant to which leaves are attached.

Short shoot A small shoot with short internodes and numerous leaves, as in *Pinus* (syn. Spur).

Shrub A woody plant with multiple trunks, not exceeding 5 m in height (cf. Subshrub, Tree).

Siccus A Latin word meaning "dry"; "juiceless.". The word applied by Linnaeus to herbarium specimens (*Hortus Siccus*).

Sigmoid Curved like the letter S, in reference to leaves that are so shaped.

Silicle A short, dry, dehiscent fruit formed from a superior ovary of two united carpels, united by a septum between the two loculi, as in many Brassicaceae.

Silique The same as Silicle, but long.

Silky See Sericeous.

Silurian A geologic period from 340 to 310 million years ago.

Simple Not branched or divided into segments, as in gynoecia, inflorescences, leaves, or venation.

Sinistrorse Turned to the left; a counterclockwise helical growth pattern characteristic of some twining stems.

Sinuate, Sinous With abrupt wavy margin (cf. Repand).

Sinus The cleft or indentation between the lobes of leaf blade.

Sister Groups In cladistics, represents closest relatives indicated in a phylogenetic tree.

Smooth See Glabrous.

Soboliferous Bearing suckers; producing shoots from the ground; clump-forming.

Solitary Single or alone, as when the flowers are born one per axil.

Sordid, Sordidus Dirty-colored; dull white.

Sorediate Having small patches on the surface.

Sorosis A fleshy fruit formed from coalesced ovaries, as in Annanas (pineapple).

Sorus (pl. Sori) A group of sporangia in ferns.

Spadiceous (Spadiceus) Bearing a spadix; like a spadix.

Spadix A succulent axis supporting an inflorescence, which is subtended by a Spathe, as in Araceae.

Spathaceous Bearing a spathe; like a spathe.

Spathe The leaflike, often showy bract which encloses or subtends a spadix, as in Araceae (cf. Cymba in palms).

Spathe Valves The herbaceous or scarious bract(s) which subtends or encloses a flower or an inflorescence in the bud, as in some monocotyledons.

Spathulate, Spatulate Spoon-shaped; broad and rounded in the middle and tapering gradually to a narrow base.

Specialized Having special adaptation to a particular habitat or bearing modified organs as a result of divergence from the ancestral form; evolutionarily advanced.

Speciation The evolutionary process of species formation.

Species Interbreeding group of individuals that are morphologically similar but not necessarily identical. The word is abbreviated as sp. when singular, and spp. when plural.

Species Nova New species, often noted in publications

Specific Of or pertaining to a species.

Specific Epithet Species name; the second part of binomials.

Specimen An individual plant representing a population of a species; a dried or preserved sample of a plant, such as a herbarium sheet.

Speciosus A Latin word meaning 'showy'; 'handsome'.

Spectabilis A Latin word meaning 'remarkable'.

Spermatophyte A seed plant; any angiosperm or gymnosperm.

Sphenoid, Sphenoidal wedge-shaped

Spherical Relating to a sphere; with multidimentional radial symmetry (syn. Orbicular).

Spheroid Perfectly rounded, with a 1:1 ratio.

Sphingophily Pollination by hawkmoths and nocturnal lepidoptera.

Spicate Having the form of a spike; in spikes.

Spicula, Spiculum A small spike.

Spiculate, Spiculose Having the surface covered with small spikes.

Spike An elongated, usually unbranched, inflorescence bearing sessile flowers.

Spikelet A small spike; the unit of inflorescence of grasses (Poaceae) and sedges (Cyperaceae).

Spine A hard, sharp-pointed, modified tip of a branch, leaf, or stipule (cf. Thorn, Prickle).

Spinescent Having spines; terminating in a spine.

Spinose Spiny; spinelike.

Spinule A small spine.

Spinulose With small spines over the surface.

Spiral Arranged in a coiled series around an axis, as in leaves around a stem.

Spiricle A delicate, coiled, hygroscopic filament found on the suface of certain seeds and achenes which uncoils when moistened, as in *Ruellia* (Acanthaceae).

Spirolobal, Spirolobous With the cotyledons spirally rolled up.

Splendens   A Latin word meaning "glittering" or "shining."

Spongy   Tissue whose cells are separated by large intercellular spaces; sponglike, as in many seed coats.

Sporangium (pl. Sporangia)   The saclike structure which contains the spores; a spore case.

Spore   A haploid uni- or multicellular reproductive unit of ferns and lower plants.

Sporocarp   A hard, nutlike structure containing the sporangia in heterosporous ferns.

Sporoderm   The wall of a spore; in seed plants, the wall of pollen grain which consists of the exine (outer wall) and the intine (inner wall).

Sporagenous   A tissue capable of producing reproductive organs.

Sporophyll   A modified leaf that bears sporangia, as in stamens and carpels of angiosperm flowers, mega- or microsporophyll of gymnosperm cones, or fronds of ferns.

Sporophyte   The foliaceous vegetative plants in ferns and seed plants.

Sport   A plant or portion of a plant which arises by spontaneous mutation and differs from the mother plant by one or more significant features, such as leaf coloration, growth habit, etc. Sports are important sources of new plants in horticulture and are propagated vegetatively.

Spotted   The color disposed in small spots, usually on a ground of different color.

Spreading   Growing outward or horizontally, as in some ground cover plants.

Sprig   A small shoot or twig; in horticulture, it refers to small pieces of lawn grass rhizomes used for turf planting (sprigging).

Sprout   The newly emerging shoots of plants; the beginning of growth in bulbs, corms, and other modified stems. The term is sometimes used incorrectly for seed germination.

Spur   The tubular projection of corolla in certain plants, such as *Impatiens* (Balsaminaceae); also see Short Shoot.

Squama   A scale or scalelike structure.

Squamellate   Appearing like a small scale; with small scales.

Squamose, Squamaceous, Squamate   Scaly; covered with scales.

Squamula   A small scale.

Squamulose   Covered with small scales.

Squarrose   Spreading in all directions and bent backwards, as in phyllaries of some Asteraceae

Squarrulose   Barely or minutely squarrose.

Stalk   The stem or main supporting axis of an organ.

Stamen   The pollen bearing organ of a flower, consisting of an anther and a filament (cf. Androecium).

Staminal Disc   A fleshy cushion of tissue found at the base of an ovary, formed by the coalesced staminodia or nectaries.

Staminal Nectary   Said of nectaries that occur at the base of the filaments.

Staminate   Having stamens but no carpels; a male flower.

Staminode, Staminodia, Staminodium   An abortive or sterile stamen, in some species may be petallike and showy, as in Cannaceae (*Canna*) and Mesembranthemaceae.

Standard   The broad upper petal of Papilionaceous (Fabaceae) corolla (syn. Banner); the narrow erect or ascending portion of the *Iris* (Iridaceae) perianth (syn. Falls). The term is used in horticulture in reference to perfectly formed rosette flowers of such plants as *Chrysanthemum, Gardenia, Camellia,* etc.

Stele   The central primary vascular cylinder of the stem and associated tissue.

Stellate   Star-shaped, in reference to trichomes which have radiating branches.

Stellulate   Resembling a small star; minutely stellate.

Stem   The above or below ground (bulb, corm, rhizome, and tuber) axis of a plant bearing leaves with buds in their axils.

Stem Group   A phylogenetic, term that refers to extinct members of a group (cf. Crown Group).

Steno-   A Latin prefix meaning "narrow."

Stenocarpous   With narrow friut.

Stenopetalous   With narrow petals.

Stenophyllous   With narrow leaves.

Stereomorphic   In reference to flowers which appear three-dimensional and actinomorphic in surface view, such as that of *Narcissus, Eucharis,* etc. which have a corona or staminal cup.

Sterigma, Sterigmata   Any persistent prolongation or projection along the stem, to which leaves are attached. Leaves may be sessile, as in *Picea* or petiolate, as in *Tsuga* (syn. Peg).

Sterile   Not productive; not capable of producing seeds or spores.

Stenotribic   Flowers with stamens (and/or stigmas) arranged in such a way as to dust pollen on the underside of the visiting insects, so that it can be deposited on the stigmatic surface of another flower (cf. Nototribic).

Stigma   That part of the style which is modified for the reception and germination of the pollen.

Stigmatic   Belonging to, or of the nature of the stigma.

Stilt Roots   See Prop Roots.

Stimulose   Bearing stinging hairs.

Stinging Hair   The stiff hairs of certain plants at the tip of which irritating compounds are stored so that upon contact they are released causing a stinging sensation. For example, *Cnidoscolus, Tragia, Dalechampia,* etc. (Euphorbiaceae), and most members of the Urticaceae.

Stipate   Crowded.

Stipe   The stalk (petiole) of a fern frond.

Stipel The stipulelike structure at the base of the petiolule of a leaflet.

Stipellate Having stipels.

Stipitate Having a stalk, as in some glandular trichomes.

Stipulate Having stipules.

Stipulose Having conspicuous stipules.

Stock See Root Stock.

Stolon A slender modified stem growing along surface of the ground and rooting at the nodes, as in *Saxifraga sarmentosa* (Saxifragaceae), *Chlorophytum* spp. (Asparagaceae), *Fragaria* spp. (Rosaceae), etc.

Stoloniferous Producing or bearing stolons.

Stoma (pl. Stomata) Minute pore on the epidermis of leaves, stems, etc. through which gaseous exchange takes place.

Stomium The region in the wall of the sporangium of ferns whose rupture results in release of the spores, or the slit or pore of the anther through which pollen is discharged.

Stone The hard endocarp of a drupe.

Stone Fruit The drupe or druplet of such fruits as peach, plum, etc.

Stool A plant which produces several stems together, or from which off-sets may be taken.

Strain A natural or artificial taxonomic group, uniform in some particular feature.

Straminous Straw-colored.

Strand Plant A dune plant which is exposed to seawater but never submerged.

Strap The ligule of the ray floret in Asteraceae.

Strap-shaped Long and narrow, as in leaves of Amaryllidaceae or ligules of the Asteraceae flowers.

Striate Marked with parallel, longitudinal lines, furrows, ridges, or streaks of color.

Strict Straight and upright; rigid and stiff.

Strigillose, Strigulose Slightly strigose.

Strigose With rigid, straight hairs, all pointing in the same direction.

Striolate Finely striate.

Striped Having longitudinal lines of color.

Strobilate, Strobiliform Of the nature of a cone.

Strobile, Strobilus (pl. Strobili) A cone, made up of sporophylls which are tightly arranged around an axis.

Strombiform, Strombuliform Said of a spirally twisted, snail-shaped fruit.

Strombus A spirally coiled pod, as in Medicago (Fabaceae).

Strophiole, Strophiola An outgrowth which occurs on seeds of certain plants, as in *Euonymous*, *Acacia*, and most members of the Euphorbiaceae (syn. Caruncle).

Strophular, Strophulate Having a small opening through which water enters the seed before germination can occur.

Strumose Having cushion-like swelling (syn. Bullate).

Stylar Relating to the style.

Style The contracted upper part of a carpel or gynoecium that supports the stigma, often considerably elongated, sometimes lacking.

Stylocarpellous The normal carpel, with a style but without a stipe (stalk).

Stylocarpepodic A carpel with a style and stipe (stalk).

Stylopodium, Stylodious Furnished with a style-like stigma, as in grasses and Asteraceae.

Stylopodic Possessing a stylopodium.

Stylopodium The swelling at the base of the style, as in Apiaceae (umbel family).

Suaveolens, Suaveolent A Latin word meaning "fragrant."

Sub- A Latin prefix meaning "under," "below," "almost."

Subapical Position of any structure when it is on one side near the apex, as in ovary or inflorescence.

Subbasal Position of any structure when it is near the base.

Subacrocaulous With branches located at or near the tip of the stem.

Subbasicaulous With branches located at or near the base of the stem.

Subbasifixed When anther is attached near its base to the apex of the filament.

Subentire Said of a leaf which is only slightly indented.

Suberect Upright below, nodding at the top.

Suberose Appearing as if somewhat gnawed and eroded.

Suberous Corky in texture.

Subfamily A taxonomic category below the rank of family.

Subgenus A taxonomic category below the rank of genus.

Subglabrate, Subglabrous Nearly glabrous.

Subglobose Almost round or spherical; nearly globose.

Subherbaceous Herbaceous, but becoming woody as the plant becomes older, such as most Pelargonium spp.

Subinferior Somewhat inferior; partially inferior, as half inferior ovary.

Submarginal Close to the margin.

Submerged, Submersed Growing under water.

Submersibilis Capable of existing under water.

Submersicaulous With submersed stem.

Subpetiolar Said of a bud which is located under the petiole and often covered by it, as in *Platanus* (Platanaceae); or below the petiole, as in *Metasequoia glyptosteroboides* (Taxodiaceae).

Subshrub A woody plant with several branches from the ground and not exceeding 1 m in height.

Subspecies A taxonomic category below the rank of species, usually based on geographical disjunction.

Subtend Occurring immediately below or close to, as a bract below a flower.

Subterminal See Subapical.

Subterranean Growing below the soil surface.

Subtropical Nearly tropical, but somewhat seasonal, as in central and southern Florida. In such areas, freezing temperatures are rare.

Subula   A delicate, sharp-pointed projection of an organ.

Subulate   Awl-shaped, tapering from base to apex.

Succineus   A Latin word meaning 'amber-colored'.

Succulent   Juicy; fleshy, soft, and thickened; generally refers to plants which store water in their leaves, stem, or root and are most often characteristic of dry habitats, as in Cactaceae, Euphorbiaceae, Apocynaceae, Asclepiadaceae, etc.

Sucker   A shoot that originates from the older part of an existing shoot below ground.

Suffrutescent, Suffruticose   Perennial plants with persistent woody base but herbaceous flowering shoots.

Sulcate   Marked by distinct longitudinal parallel grooves or furrows.

Sulcus (pl. Sulci)   A small longitudinal groove or channel, as on the surface of some pollen grains.

Sulphureous   A Latin word meaning "Pale" or "clear yellow".

Summer Annual   In horticulture, plants that are grown as bedding plants only for duration of the summer, although they may be perennials, such as *Impatiens*, *Petunia*, etc.; true annuals which germinate in spring and grow and flower during the summer months (cf. Winter Annual).

Super-   A Latin prefix meaning "above."

Superficial   On or near the surface; shallow.

Superior Ovary   See Hypogenous.

Superposed   One placed above another, as in superposed bud, where a lateral bud is positioned above an axillary bud.

Supervolute   Leaf blade with one edge tightly enrolled while the other loosely enrolled, so as to cover the first; loosely convolute.

Supine, Supinus   A Latin word meaning 'lying face upward'; prostrate but with branches upward, as in some species of *Juniperus*.

Suppression   Failure to develop or vestigial presence of organs which were present in the ancestral forms.

Supra-   A Latin prefix meaning 'above'.

Suprafoliar   On the stem above the leaves, in reference to position of the cotyledons, as in Arecaceae.

Suprarhizous   On top of the root

Surcarpous   With fruit born on or near the soil surface.

Surculose   Bearing suckers.

Surculus   A sucker.

Surcurrent   Having winged expansions from the base of the leaf prolonged up the stem.

Surficial   Spread over the surface of the ground.

Sursum   Directed upward and forward, as in spines or teeth of leaves.

Suspended   Hanging directly downward.

Suspensor   A structure which develops with the embryo and pushes it into the endosperm or the female gametophyte tissue, as in cycads.

Suture   The line of dehiscence of dry fruits; the line of junction or cleavage of two united organs.

Syconium   The fleshy fruit of Ficus (fig), which consists of achenes embedded in an enlarged receptacle.

Syleptic Shoot   Abnormal shoots which develop from lateral buds before they have reached maturity.

Sylvestral, Sylvestris   Growing in woods or in the forest.

Sym-   A modified form of the prefix Syn-.

Symbiosis   The living of two or more organisms together, in close association, for their mutual benefit.

Symmetrical, Symmetry   In reference to flowers which are regular in shape, size, and number of parts (syn. Actinomorphic).

Sympatric, Sympatry   Said of populations of two taxa which grow in the same geographical locality (cf. Allopatric).

Sympetalous, Sympetaly   Partially or completely united petals (syn. Gamopetalous).

Symphysis   Growing together; coalescence; fusion or connation of parts.

Sympleisiomorphy   In cladistics, an ancestral characteristic shared by two or more taxa.

Sympodial, Sympodium   Growth pattern of a plant in which the central leader produces successive superposed branches, so that in time a main axis does not appear to exist (cf. Monopodial).

Sympodial Inflorescence   A determinate inflorescence which simulates a sympodial growth pattern thus appears indeterminate (cf. Scorpioid Cyme).

Syn-   A Greek prefix meaning "adhesion" or "growing together"

Synandrous, Synandry   Having the anthers united, but not necessarily fused.

Synangium (pl. Synangia)   A compound structure formed by the union of sporangia, as in some ferns.

Synantherous   Simultaneous appearance of leaves and flowers.

Synanthesis   Simultaneous maturation of pistils and stamens.

Synapomorphy   In cladistics, shared derived characteristic that indicates common ancestry.

Syncarp   A multiple fleshy fruit.

Syncarpous, Syncarpy   Having united carpels (cf. Apocarpous).

Syncladous   Said of branchlets when they grow in tufts from the same point.

Syncotyly   The state of cohesion of cotyledons by only one margin.

Synema (pl. Synemata)   The column formed by monodelphous stamens, as in *Hibiscus* (Malvaceae).

Syngenious, Syngenesious   Flowers with coherent anthers, as that of Asteraceae

Synonym An untenable taxonomic name, rejected in favor of another based on the International Rules of Botanical Nomenclature.

Synovarious With ovaries of adjacent carpels completely fused, styles and stigmata separate.

Synsepalous, Synsepaly Having united sepals (cf. Gamosepalous).

Syntropous The position of the radicle when it is pointing toward the hilum.

Synstylovarious With ovaries and styles of adjacent carpels completely fused, stigmas separate.

Syntype One of several specimens from which the species was described, but actual type specimen was not designated.

Systematics The scientific study of evolutionary relationships of organisms. This term has been used interchangeably with Taxonomy and variously defined by different authors.

T

Taiga Coniferous forest belt of the northern hemisphere.

Tailed Bearing a long slender appendage.

Tapering Gradually becoming narrower toward one end.

Taproot A permanent, more or less thickened, often fleshy, primary root.

Tassel The staminate inflorescence of corn.

Tautonym An illegitimate binomial in which the specific epithet is a repetition of the generic name.

Tawny Dull yellowish-brown (syn. Fulvous).

Taxon (pl. Taxa) Any taxonomic unit irrespective of rank, such as family, genus, species, etc.

Taxonomy The principles and procedures of classification, including nomenclature (cf. Systematics, Phylogeny).

Tectum The outermost layer of some pollen grains.

Tendril A slender, coiling, branched or unbranched, modified leaf or stem, used by many vines for attachment, such as claws (as in *Macfadyena unguis-cati*), or simply twining (as in *Clytostoma callistegioides* or *Vitis vinifera*).

Tenellis, Tenellus A Latin word meaning "very slender" or "dainty."

Tentacle One of the hairs on the leaves of insectivorous plants. The term is more common in zoology.

Tentacular Having tentacles

Tepal A perianth segment, not differentiated into calyx and corolla, as in most members of Amaryllidaceae.

Terete Cylindrical; cylindrical-conical, tapering at one end.

Tergeminate With three orders of leaflets, each bifoliolate; with geminate leaflets ternately compound.

Termifolius A Latin word meaning "thin-leaved."

Terminal, Terminus Situated at the tip, such as the apical bud.

Ternate Arranged in threes, such as a compound leaf with three leaflets.

Terrestrial Living in soil (cf. Epiphyte).

Tertiary A geologic period from 70 to 1 million years ago; third rank, such as branches of secondary roots.

Tessellate Having a surface marked with a checkered pattern, as in arrangement of epidermal cells.

Testa The outer coat of a seed (cf. Sclerotesta).

Testaceous Brick-red.

Testicular Shaped like the pseudobulbs of orchids.

Tetra- A Greek prefix meaning "four."

Tetracyclic With four whorles.

Tetrad A group of four; the four-celled pollen mother cell.

Tetradynamous Having two long and two short stamens.

Tetragonal With four angles.

Tetrahedral Having the form of a pyramid.

Tetralocular Having four locules.

Tetramerous Having parts in fours, or multiples of four, as in flowers of Onagraceae.

Tetrandrous Having four stamens.

Tetraploid Having twice the usual number of chromosomes.

Tetrasomic When one of the chromosomes is represented four times, in an otherwise diploid plant.

Tetrastichous Arranged in four rows.

Texture Consistency or feel (to the touch) of an organ. In horticulture, this term is used to designate visual appearance of plants in the landscape: fine, plants with many small leaves; medium, plants with relatively large leaves; coarse, plants with few but very large leaves.

Thalmus The receptacle of a flower.

Theca A pollen sac; the locule of an anther.

Thermophylic, Thermophyllous Producing leaves in the summer.

Therophyte An annual plant.

Thorn A hard sharp-pointed modified shoot (cf. Spine, Prickle

Throat The opening of a gamopetalous corolla.

Thrum Flower The flower with the shorter style, in heterostylous plants (cf. Pin Flower

Thyrse, Thyrsus A compact inflorescence in which lateral branches are determinate, but the main branch indeterminate, as in *Syringa*, *Ligustrum* (Oleaceae), etc.

Tiller A grass shoot growing from the base of the plant.

Tomentose Covered with dense, short, fine hairs.

Tomentulose Somewhat tomentose.

Tomentum A covering of short, wooly hairs.

Topiary The pruning of dense trees or shrubs to make sculptured forms, such as animal figures.

Topotype A specimen collected at the original type locality of a species.

Tormentose   Cottony, with flattened, matted hairs

Torose   Elongated and cylindrical with constrictions at more or less regular intervals.

Tortuose, Tortuous   Irregularly twisted.

Torus (pl. Tori)   The receptacle of a flower

Tracheophyta   Plants with vascular tissue, such as ferns and seed plants.

Trachycarpous   Rough-skinned fruit.

Trachyspermous   Rough-surfaced seed.

Trailing   Prostrate, but not rooting

Translator   The structure which connects the pollinia in Asclepiadaceae.

Translucent   Semitransparent, as in the leaf margin of Ligustrum lucidum.

Transverse   Broader than long; perpendicular to the long axis.

Trap   The specialized leaves, flowers, or trichomes which entrap visiting insects. For example, *Nepenthes* spp. (pitcher plants), *Dionaea* spp. (venus fly trap), *Aristolochia* spp. (Dutchman's pipe), etc.

Tree   A woody perennial plant with a single trunk and a height which exceeds 5 m (15 feet). In horticulture, trees are divided into Small (5–10 m), Medium (10–20 m), and Large (greater than 20 m). A different set of heights is also used to designate tree sizes: 15, 30, and 45 feet, respectively. Number of trunks is irrelevant in horticulture and subject to manipulation by pruning; terms used in reference to phylogenetic diagrams (evolutionary trees).

Tri-   A Latin or Greek prefix meaning "three" or "triple."

Triadelphous   Having stamens in three tiers (cf. Monodelphous and Diadelphous).

Triangular   With three sides and three angles.

Triandrous   Having three stamens.

Triassic   A geologic period from 190 to 170 million years ago.

Tribe   A subdivision of a family consisting of several genera.

Tricamarus   Having a fruit with three locules (syn. Trilocular).

Tricarpellate   Having three carpels.

Trichasium   A cymose inflorescence with three branches.

Tricho-   A Greek prefix meaning "hair."

Trichocarpous   Having hairy fruit.

Trichome   A hair.

Trichotomous   Having three equal or nearly equal branches.

Tricolpate   Said of pollen grains with three furrows or groves, as in pollen of some dicotyledons.

Tricyclic   In three whorles.

Tridynamous   Having six stamens in two groups of three each.

Trifid   Divided about halfway down into three parts.

Triflorous   With three flowers.

Trifoliate   Having three leaves.

Trifoliolate   A compound leaf with three leaflets.

Trifurcate   Having three forks or branches.

Trigonal, Trigonous   Having three angles.

Triheteranthous   Having sets of flowers with three different states of sex expression, only one state in each set.

Triheterophytous   Having sets of plants with three different states of sex expression, only one present in each set

Trihybrid   A trigenic hybrid.

Trilete   Basically tetrahedral, but often appearing round or triangular, with three scar lines forming a Y.

Trilobate   With three lobes.

Trilocular   A fruit consisting of three locules.

Trimerous   Having parts in threes or multiples thereof.

Trimonoecious   Said of a plant which has staminate, pistillate, and hermaphroditic flowers.

Trimorphic   Existing in three forms.

Trioecious   Having hermaphroditic, staminate, and pistillate flowers on separate plants of the same species.

Tripalmate   Said of compound leaves with three orders of leaflets, each palmately compound.

Tripartite   Divided into three parts nearly to the base.

Tripinnate   A pinnately compound leaf with three pinnately divided leaflets which are pinnately divided.

Triploid   Having three complete sets of chromosomes per nucleus.

Tripterous   Having three wings, as in certain seeds or fruits.

Triquetrous   Having three angles.

Trisomic   An otherwise diploid individual with a chromosome that is represented three times.

Tristichous   In three rows.

Tristis   A Latin word meaning "dull colored."

Tristyly, Tristylous   A flower with styles of three different lengths and anthers in two different lengths.

Triternate   Leaflets of a ternate leaf which are themselves twice ternate (3 X 3).

Trullate   Leaves in which the widest axis is below the middle and which have straight margins; trowel- shaped.

Trulloid   Having a trullate shape.

Truncate   An apex or base which ends bluntly, as if cut off abruptly.

Trunk   The primary axis of a tree.

Tryma (pl. Trymata)   A nut with two or four locules and surrounded by a dehiscent involucre, as in *Juglans* or *Carya* (Juglandaceae).

Tube   The cylindrical part of the perianth.

Tuber   A much thickened, usually short, subterranean, modified stem, as in Solanum tuberosum (potato).

Tubercle   A rounded protuberance.

Tubercule   A small swelling (Nodule) on the root of leguminous plants.

Tuberculate   Having tubercles.

Tuberous Having tubers or resembling one.

Tuberous Root Fleshy roots resembling stem tubers, but lacking buds or other stem characteristics, as in *Dahlia*.

Tubular Cylindrical and hollow.

Tufted Having many short branches; occurring in clumps; clustered (syn. Cespitose).

Tumescent Somewhat tumid.

Tumid Swollen; inflated.

Tunic The membranous outer skin of some bulbs or corms.

Tunicate, Tunicated Having a layer or layers of membranous outer skin.

Turbinate Inversely conical; top-shaped.

Turgid Swollen or inflated, filled with water.

Turion A young shoot or sucker arising from underground, such as the vegetative bud of a rhizome in spring.

Tussock A tuft of grass or grass-like plant.

Twig A young woody stem; current season's shoot.

Twining Climbing by means of a spirally coiling stem.

Two-lipped See Bilabiate.

Type, Type Specimen A typical example from which new species are described, usually referring to the Holotype (cf. Isotype, Syntype, Lectotype, Neotype, and Paratype).

Typology The study of types, no longer an accepted practice since it ignored the variation within a species population.

U

Ubiquitous Widespread; occurring everywhere.

Ulginose, Ulginous Growing in wet places.

Umbel An umbrella-shaped inflorescence, in which the pedicels radiate from a common point at the summit of the peduncle, as in Apiaceae.

Umbellate Of the form of an umbel.

Umbellet A secondary umbel; one of a set of smaller umbels of a compound umbel.

Umbelliferous Bearing umbels.

Umbelliform Umbel-shaped; resembling an umbel.

Umbilicate Peltate and with a depression in the center.

Umbo A Latin word meaning 'a convex elevation'; a conical projection arising from any surface.

Umbonate With a central umbo.

Umbonulate With a small umbo.

Umbraculate, Umbraculiferous Resembling an expanded umbrella.

Umbraculiform Having the general form of a parasol, as the stigmas of *Sarracenia*.

Umbrinus A Latin word meaning "umber-colored."

Umberosus A Latin word meaning "of shady places."

Unarmed Lacking spines, thorns, prickles, or any other type of armature.

Uncate, Uncinate Hooked; bent at the tip in the form of a hook.

Unctuous Oily or greasy to the touch.

Undershrub See Subshrub.

Understory Small trees and shrubs under the main tree canopy of a forest.

Undulate Wavy at the margin.

Ungulate, Unguicular, Unguiculate Furnished with a claw.

Uni- A Latin prefix meaning "one"; "single."

Unicarpellate, Unicarpellous Said of a fruit with a single carpel.

Unicellular Having or being composed of a single cell, as in some trichomes.

Unifoliolate Said of a compound leaf which has been reduced to a single terminal leaflet.

Unilateral One-sided, as in a raceme with all flowers born on one side.

Unilocular Having a single locule, as in an ovary.

Uniseriate Occurring in a single row or series, as in some multicellular trichomes which are made up of a single row of cells.

Unisexual Said of a flower which has either stamens or pistils, but not both, although male and female flowers may occur in the same inflorescence.

Unigeneric See Monogeneric.

Unit Cup Fruit A fruit derived from a flower with inferior ovary and united carpels. The fruit wall consists of an ovary wall (pericarp) and accessory parts.

Unitegmic Said of an ovule with a single integument.

Unit Free Fruit A fruit derived from a flower with superior ovary and united carpels. Fruit wall consists of ovary wall (pericarp) only.

Urceolate Urn-shaped, as in flowers of certain Ericaceae.

Urent With irritating, long trichomes.

Urticating Hairs Said of stinging hairs of Urticaceae (nettles), Loasaceae, and such genera as *Cnidoscolus* (Euphorbiaceae).

Urticle An inflated, bladderlike envelope; a one-seeded, dry, indehiscent fruit, as in some Amaranthaceae.

Utricular Inflated; bladderlike, as in leaves of Utricularia spp.

Utriform Bladderlike.

Uvarious A Latin word meaning "like a bunch of grapes."

V

Vagina A sheathing leaf base, as in grasses.

Vaginate Sheathed.

Valid A name which is designated in accord with the International Rules of Botanical Nomenclature; an acceptable name.

Vallecular   Referring to the grooves between the ridges, as in fruits of the Apiaceae (umbel family).

Valvate   Dehiscing by valves or equal sections; in aestivation, when the segments of the perianth are so placed that they touch, but do not overlap (cf. Imbricate).

Valve   A portion of the fruit wall of a capsule or legume separated in dehiscence; the tissue which covers the pore in anthers which open by pores.

Valvular   Opening by means of valves.

Variant   Individual(s) which differs in some respect from the norm.

Variation   The occurrence of differences in individuals of a population.

Varicose   Abnormally and irregularly enlarged or swollen; irregularly dialated or inflated.

Variegated   Said of leaves or flowers which lack uniform pigmentation, appearing blotched, yellowish, or white.

Variety   A subdivision of species, differing in certain genetically fixed characteristic(s) from the rest of the population (cf, Subspecies, Cultivar).

Vascular   Pertaining to the presence of vessels or conducting tissue (xylem and phloem).

Vascular Bundle Scar   The scars left by vascular bundles of detached leaves on the stem.

Vascular Plant   A plant having a vascular system, as in Pteridophyta and Spermatophyta (cf. Trachiophyta).

Vasiform   Having the shape of an elongated funnel.

Vegetation   All the plants in a given area or region.

Vegetative   Pertaining to all but reproductive organs, such as leaves, stems, roots, etc.

Vegetative Propagation   Reproduction of plants by means of stem, leaves, roots, or any meristematic tissue (tissue culture). Plants reproduced in such a manner are genetically uniform and identical to the mother plant.

Vein   Vascular strand of leaves.

Veinlet   A small vein.

Velamen   The multiple epidermis of the aerial roots of epiphytic Orchidaceae and Araceae, functioning in water absorption.

Velum   A veil; a membranous structure involved in the closure of the trap in *Urticularia* (bladder pod).

Velutinate, Velutinous   Having a velvety surface; covered with fine, dense, straight, long, and soft trichomes.

Venation   The arrangement of veins in a leaf blade.

Ventral   The upper surface of a leaf; the inner surface of an organ (syn. Abaxial; opp. Dorsal).

Ventricose   Swollen in the middle; with a bulge on one side, near the middle.

Ventristipular   On the ventral side of stipule.

Venulose   Having many veinlets.

Vermicular, Vermiform   Worm-shaped.

Vernal, Vernalis   Appearing in or belonging to the spring season.

Vernalization   The process of dormancy breaks in buds, requiring low temperature.

Vernation   The arrangement or mode of folding of leaves in the bud (syn. Ptyxis; cf. Aestivation).

Vernicose   Shiny, appearing varnished.

Verruca (pl. Verucae)   A granular or wartlike outgrowth on pollen grains.

Verrucate   Warty.

Verrucose   With minute warts or blunt projections.

Verruculose   Slightly warty.

Versatile   Attached by the middle so as to swing freely, as an anther on the filament.

Versicolor   Variegated; variously colored; changing color with age.

Verticil   A whorl; a whorled arrangement of similar parts, as in flowers.

Verticillaster   Whorled dichasia at the nodes of an elongated rachis.

Verticillate   Arranged in a whorled cluster.

Vesicle   A small sac.

Vesicular   Bearing a small sac, or resembling one.

Vesiculose   Swollen like a bladder.

Vespertine   Night bloomer.

Vessel   Water and mineral conducting, tubelike elements of the xylem in angiosperms.

Vestibulum   A cavity inside the pollen pore, caused by separation of the exine layers.

Vestigial   Nonfunctional rudimentary organ.

Vestiture, Vesture   Any covering of the surface, especially trichomes.

Vexillate   Having a standard or vexillum.

Vexillum (pl. Vexilla)   The broad upper petal of a papilionaceous corolla (syn. Standard).

Viable   Capable of survival and development or germination.

Viability   A measure of seed germination capability.

Vicariants   Closely related taxa isolated geographically from one another by a natural barrier or event.

Villose, Villous   Covered with long shaggy trichomes.

Villusulous   Minutely villose.

Villus (pl. Villi)   A long, soft, fine trichome.

Vine   A plant of weak structure, often with climbing, clambering, or twining stem.

Vinous   Of the color of red wine.

Virens   A Latin word meaning "green."

Virescent   An abnormal green coloration accompanied by abnormal development of sepals or petals in virus-infected plants.

Virgate   Long, slender, and stiff, with many branches.

Virgatus   A Latin word meaning "twiggy."

Virose   Poisonous.

Viridiscent   Becoming green.

Viscid, Viscous   Sticky (syn. Glutinous).

Visidulous Slightly viscid.

Viscin Interconnecting threads between small groups of pollen grains of certain orchids.

Vitreous A Latin word meaning "transparent" (syn. Hyaline).

Vitta A resin or oil canal in the pericarp of certain fruit, as in Apiaceae (umbel family).

Vittate Having longitudinal ridges or stripes.

Viviparous, Vivipary Germination of seeds while still attached to the plant, as in some Amaryllidaceae and *Avecinia* or other mangrove species.

Volatile Unstable, vaporizing readily, as in volatile oil of some plants.

Voluble, Volubilis A Latin word meaning "twining."

Volute, Volutus A Latin word meaning "rolling."

Volution A spiral; a turn.

Voucher Specimen A preserved sample of plant(s) used in taxonomic studies (cf. Type Specimen).

Vulgaris A Latin word meaning 'common'.

W

Water Excreting Gland Specialized vein openings through which water is lost, particularly at night (syn. Hydathodes)

Weed A plant growing where it is not wanted.

Whorl A circle or ring of organs inserted around an axis, as floral segments or leaves on a stem.

Wind Pollination See Anemophily.

Wing An outgrowth from the side of an organ, such as seed and fruit; the lateral petals of legumes and orchids.

Winged Bark An outgrowth on the stem of certain species, as in *Liquidambar styraciflua*, *Ulmus parvifolia*, etc.

Winter Annual An annual which usually germinates in the fall, flowers in the winter, sets fruit in early spring, and dies. In horticulture, winter annuals used as bedding plants are not necessarily true annuals, such as *Viola* spp.

Woody In reference to stems or branches which have secondary growth. The term is also used in a vernacular sense for very hard fruits.

Woolly See Tomentose.

X

X The basic or haploid number of chromosomes.

Xanth-, Xantho- A Greek prefix meaning "yellow.".

Xanthophyll The yellow pigment of various organs.

Xenogamy Cross-pollination between flowers of separate individuals of the same species.

Xeric Dry habitat; plants of dry habitats (cf. Mesic and Aquatic).

Xeromorphic Having the characteristics of plants adapted to xeric habitats, such as thick cuticle, succulent habit, etc.

Xerophyte A plant of dry habitat, such as a desert.

Xylem The vascular tissue responsible for water conduction and production of wood for mechanical strength.

Xylotomy The comparative study of anatomical features of the xylem, usually for establishment of taxonomic relationships.

Z

Zonoapeturate A pollen grain in which apertures are located on the equatorial zone.

Zonocolpate XXX

Zonocolporate XXX

Zonoporate XXX

Zonopororate XXX

Zonocaulous With branches intermittently spaced along the main stem.

Zoochore, Zoochorous A plant which is disseminated by animals.

Zygomorphic Said of a flower having bilateral symmetry; divisible in half by one longitudinal plane only, as in flowers of *Antirrhinum* (Scrophulariaceae), Orchidaceae, etc. (syn. Irregular; opp. Actinomorphic).

References

Acevedo-Rodriguez, P. and M. T. Strong (eds.). 2005. Monocotyledons and Gymnosperms of Puerto Rico and the Virgin Islands. Smithsonian Institution, Contributions from the United States National Herbarium 52: 1–415.

Adams, C. D. 1973. Flowering Plants of Jamaica. University of the West Indies, Mona, Jamaica.

Adams, R. P. 1986. Geographic variation in *Juniperus silicicola* and *J. virginiana* of the Southeastern United States: Multivariance analyses of morphology and terpenoides. Taxon 35: 61–75.

Addison, S. 1992. A Passion for Daylilies: The Flowers and the People. Henry Holt and Company. New York.

Alcock, R. H. 1876. Botanical Names for English Readers. L. Reave Press and Co., London. (Reprinted by Grand River Books, Detroit, 1971).

Allen, O. N. and E. K. Allen. 1981. The Leguminosae: A Source Book of Characteristics, Uses, and Nodulation. The University of Wisconsin Press.

Alrich, P., W. Higgins, B. Hansen, R. L. Dressler, T. Sheehan, and J. Atwood. 2008. The Marie Selby Botanical Gardens Illustrated Dictionary of Orchid Genera. Comstock Publishing Associates (Cornell University Press), New York.

Anderson E. 2001. The Cactus Family. Timber Press, Portland, Oregon.

Anonymous. 1971. Plants of the Philippines. A Project of the Science Education Center, University of the Philippines Press.

Anonymous. 1997. Rock Garden Plants: A Photographic Guide to More Than 450 Rock Garden Plants by Type, Size, Season of Interest, and Color. Eyewitness Garden Handbooks, DK Publishing Inc., New York.

Andrews, S., A. Leslie, and C. Alexander (eds.). 2000. Taxonomy of Cultivated Plants. Royal Botanic Gardens, Kew. London.

APG IV 2016–2021. An update of the Angiosperm Phylogeny Group classification for the orders and families of flowering plants: APG IV. *Botanical Journal of the Linnean Society* 181: 1–20. [for complete treatment of the APG IV, it is advisable to consult the Missouri Botanical Garden site]

Armitage, A. M. 1989. Herbaceous Perennial Plants: A Treatise on their Identification, Culture, and Garden Attributes. Varsity Press, Inc., Athens, Georgia.

Armitage, A. M. 2009. The Color Encyclopedia of Garden Plants: Annuals. Perennials. Timber Press, Portland, Oregon, London.

Arnold, H. L. M.D. 1968, Poisonous Plants of Hawaii. Charles E. Tuttle Company, Rutland, Vermont.

Asakawa, B. and S. Asakawa. 2000. California Gardener's Guide. Cool Spring Press, Nashville, Tennessee.

Atkinson, R. E. 1970. The Complete Book of Groundcovers: Lawns you don't have to mow. David McKay Company, Inc., New York.

Austin, D. 2007. The English Roses: Classic Favorites & new Selections. Firefly Books, New York.

Ayensue, E. S. and R. A. DeFilipps. 1978. Endangered and Threatened Plants of the United States. Smithsonian Institution and the World Wildlife Fund, Inc., Washington, DC.

Ayensue, E. S. V. H. Heywood, G. L. Lucas, and R. A. Defilipps. 1984. Our Green and Living World: The wisdom to Save it. Smithsonian Inst. Press.

Ayers, D. M. 1977. Bioscientific Terminology. Words from Latin and Greek Stems. The University of Arizona Press. Tucson, AZ.

Azani, N. et al. (too many authors). 2017. The "Legume Phylogeny Working Group - LPWG", A new subfamily classification of the Leguminosae based on a taxonomically comprehensive phylogeny. Taxon 66: 44–77.

Bailey, L. H. 1933. How Plants Get Their Names. Macmillan Company, New York. (Dover Edition, 1963).

Bailey, L. H. 1978. The Cultivated Conifers of North America: Comprising the Pine Family and the Taxads. Allanheld, Osmun/ Universe Books, New York.

Bailey, L. H. 1949. Manual of Cultivated Plants. Macmillan Publishers Co. Inc., New York.

Bailey, L. H. and E. Z. Bailey. (Revised and expanded by The Staff of the liberty Hyde Bailey Hortorium, Cornell university). 1976. Hortus Third: A Concise Dictionary of Plants Cultivated in the United States and Canada. Macmillan Publishing Company, New York.

Baranov, A. 1971. Basic Latin for Plant Taxonomists. Strauss and Cramer, Leutershausen, Germany.

Barash, C. W. 1997. Vines and Climbers. Crescent Books, New Jersey.

Barwick, M. 2004. Tropical & Subtropical Trees: An Encyclopedia. Timber Press, Portland, Oregon.

Bayer, M. B. 1982. The New *Haworthia* Handbook: A Revised Guide to the Literature of the Genus, with Discussion of the Species, Identification Keys and Color Illustrations. The National Botanical Gardens of South Africa, Cape Town.

Beacham, W, F. V. Castronova, and S. Sessine. (eds.). 2001. Beacham's guide to the Endangered Species of North America. Vol. 6: Dicots, Monocots, Glossary, Organization, Index. Gale Group, Detroit, New York.

Beales, P. 1988. Twentieth-Century Roses: An Illustrated Encyclopedia and Growers Manual of Classic Roses from the Twentieth Century. Harper & Row, Publishers, New York.

Beales, P, T. Cairns, W. Duncan, G. Fagan, W. Grant, K. Grapes, P. Harkness, K. Hughes, J. Mattock, and D. Ruston. 1998. Botanica's Roses. Random Hose Australia Pty Ltd., NSW, Australia.

Beales, P. 1983. Classic Roses: An Illustrated Encyclopedia and Growers Manual of Old Roses, Shrub Roses and Climbers. Henry Holt and Company, New York.

Bean, W. J. 1992. Trees and Shrubs Hardy in the British Isles. 8th edition. John Murray Publishers, Ltd., London.

Bechtel, H., P. Cribb, and E. Launert. 1981. The Manual of Cultivated Orchid Species. The MIT Press. Cambridge, Massachusetts.

Becker, R. 1984. The Identification of Hawaiian Tree Ferns of the Genus *Cibotium*. American Fern Journal 74: 97–100.

Bell, C. R. and B. J. Taylor. 1982. Florida Wild flowers and Roadside Plants. Laurel Hill Press, Chapel Hill.

Bell, M. 2003. The Gardener's Guide to Growing Temperate Bamboos. Timber Press, Portland, Oregon.

Benson, L. 1969. The Cacti of Arizona. The University of Arizona Press. Tucson, Arizona.

Benthall, A. P. 1933. The Trees of Calcutta and its Neighborhood. Thacker Spink & Co., Ltd. London.

Berg, R. G. van den. 1999. Cultivar-group classification. In: A. C. Andrews, A. C. Leslie, and C. Alexander. Taxonomy of Cultivated Plants: Third International Symposium, pp. 135–143.

Berliocchi, L. 1996. The Orchid in Lore and Legend. Translated by L. Rosenberg and A. Weston, edited by M. Griffiths. Timber Press. Portland, Oregon.

Berry, F. and W. J. Kress. 1991. *Heliconia*: An Identification Guide. Smithsonian Institution Press, Washington, DC.

Bessey, C. E. 1915. Phylogenetic taxonomy of flowering plants. Annals of the Miss. Bot. Garden 2:109–164.

Bird, R. 1991. Lilies: An Illustrated Identifier and Guide to Cultivation. Chartwell Books, Secaucus, New Jersey.

Black, R. J. and E. F. Gilman. 2004. Landscape Plants for the Gulf and South Atlantic Coasts: Selection and Establishment. University Press of Florida.

Blackburn, B. 1952. Trees and Shrubs in Eastern North America: Keys to the Wild and Cultivated Woody Plants Growing in the Temperate Regions Exclusive of Conifers. Oxford University Press, New York.

Bond, S. 1992. Hostas: Foliage Plants in Garden Design. Sterling Publishing Company, New York.

Bornman, C. B. 1978. Paradox Eines Verdorten Paradleses (Paradox of Parched paradise), *Welwitschia*. C. Strulk Publishers, Cape Town, Johannesburg.

Bowers, J. E. 1993. Shrubs and Trees of the Southwest Deserts. Southwest Parks and Monuments Association, Tucson, Arizona.

Bramwell, D. and Z. L. Bramwell. 1990. Flores Silvestres de las Islas Canarias. Puerto Cristo, Alcorcon, Madrid.

Brandenburg, W. A. (1986). "Classification of cultivated plants". Acta Horticulturae. 182: 109–115.

Brawner, F. 2003. Geraniums: The Complete Encyclopedia. Schiffer Publishing, Atglen, PA.

Bremer, K. 1994. Asteraceae: Cladistic & Classification. Timber Press, Portland, Oregon.

Bremmer, B. 2009. A review of molecular phylogenetic studies of Rubiaceae. Ann. Mis. Bot. Gard. 96: 4–26.

Brickell, C. (Ed.). 2000. Royal Horticultural Society A-Z Encyclopedia of Garden Plants (2nd ed.). DK Publishing, Inc., New York.

Brickell, C. 1995. Garden Plants: The Expert Guide to Over 500 Plants for All Seasons. Published in Association with the Royal Horticultural Society. Pavilion Books Limited, London.

Brickell, C. Alexander, J. C. David, W.L.A. Hetterscheid, A. C. Leslie, V. Malecot, Xiaobai Jin (eds.). 2016. International Code of Nomenclature for Cultivated Plants: Incorporating the Rules and Recommendations for Naming Plants in Cultivation. 8th Ed. Regnum Vegetabile Vol. 151. International Association for Plant Taxonomy, University of Vienna. Published by ISHS.

Brickell, C. and J. D. Zuk (Eds.). 2004. The American Horticultural Society A-Z encyclopedia of Garden Plants. DK Publishing, Inc., New York.

Brickell, C., T. Cole, and H. M. Catthey. 2002. The American Horticultural Society Encyclopedia of Plants and Flowers (American Horticultural Society Practical Guides). DK Publishing, Inc. New York.

Brilmayer, B. 1960. All About Begonias. Doubleday & Company, Inc. Garden city, New York.

Britton, N. L. and H. A. Brown. 1970 (1896–1898). An Illustrated Flora of the Northern United States and Canada. 3 vol. Dover Edition, Dover Publications, Inc. New York.

Britton, N. L. and J. N. Rose. 1937. The Cactaceae: Description and Illustrations of Plants of the Cactus Family. 2nd Edition. Carnegie Institution of Washington. Publ. No. 248.

Broschat, T. K. and A. W. Meerow. 2000. Ornamental Palm Horticulture. University Presses of Florida, Gainesville.

Brown, C. L. and L. C. Kirkman. 1990. Trees of Georgia and Adjacent States. Timber Press, Portland, Oregon.

Brown, D. 2000. Aroids: Plants of the Arum Family. Timber Press. Portland, Oregon.

Brown, E. 1986. Landscaping with Perennials. Timber Press, Portland, Oregon.

Brown, P. M. 2004. Wild Orchids of the Southeastern United States, North of the Peninsular Florida. University Press of Florida, Gainesville, Florida.

Brown, R. W. 1956. Composition of Scientific Words. Reese Press. Cambridge, Baltimore.

Bryan, J. and M. Griffiths (eds.). 1995. The New Royal Horticultural Society Dictionary: Manual of Bulbs. Timber Press, Portland, Oregon.

Bryan, J. E. 1989. Bulbs. Vol. I & II. Timber Press, Portland, Oregon.

Bryan, J. E. (ed.). 1992. Bulbs. Hearst Books, New York.

Bryant, K. (Consul. ed.). 2003. The Encyclopedia of Garden Flowers. Thunder Bay Press, San Diego, California.

Burras, J. K. and M. Griffitths (eds.). 1994. The New Royal Horticultural Society Dictionary: Manual of Climbers and Wall Plants. Timber Press, Portland, Oregon.

Burrell, C. C. (ed.). 1994. Ferns: Wild Things Make a Comeback in the Garden. Brooklyn Botanic Garden

Burrell, C. C. 1999. Perennial Combinations: Stunning Combinations that Make Your Garden Look Fantastic Right from the Start. Rodale Press, Inc., Emmaus, Pennsylvania.

Bush-Brown, L, and J. Bush-Brown. 1996. America's Garden Book. Macmillan, New York.

Caballero Ruano, M. (ed.). 1999. Ornamental Palms and other Monocots from the Tropics. Proc. Sec. Inter. Symp. On Ornamental Palms and other Monocots from the Tropics. Tenerife, Spain, 3–6 February 1997. Acta Hort. 1–348.

Caillet, M. and J. K. Mertzweiller (eds.). 1988. The Louisiana *Iris*: The History and Culture of Five Native American Species and Their Hybrids. A Publication of the Society for Louisiana Irises.

Calderon, C. E. and T. R. Soderstrom. 1980. The Genera of Bombusoideae (Poaceae) of the American Continent: Keys and Comments. Smithsonian Contr. Bot. 44: 1–23.

Callaway, D. 1994. The World of Magnolias. Timber Press, Portland, Oregon.

Cannon, J. and M. Cannon. 1994. Dye Plants and Dyeing. Published in Association with The Royal Botanic Gardens, Kew.

Cathey, H. M. and L. Bellamy. 1998. Heat Zone Gardening: how to Choose Plants that Thrive in Your Region's Warmest Weather. American Horticultural Society. Time-Life Books.

Chacōne, J., A. Sousa, C. M. Baeza, and S. S. Renner. 2012. Ribosomal DNA and gene-wide phylogeny reveal patterns of chromosomal evolution in *Alstroemeria* (Alstroemeriaceae). Amer. J. Bot. 99: 1501–1512.

Chandlee, D. K. and L. Lauren. 1988. A guide to *Artocarpus* fruit. The Archives of the Rare Fruit Council of Australia, pp. 1–7. http://rfcarchives.org.au/Next/Fruits/Jakfruit/ArtocarpusGuide.htm

Charles, G. 2007. Cacti and Succulents: An Illustrated Guide to the Plants and their Cultivation. Cordwood Press, London.

Chase, M. W. 2004. Monocot relationships: an overview. Amer. J. Bot. 91: 1645–1655.

Chase, M. W., K. M. Cameron, R. L. Barrett, and J. V. Freudenstein. 2003. "DNA data and Orchidaceae systematics: a new phylogenetic classification". pages 69–89. In: Kingsley W. Dixon, Shelagh P. Kell, Russell L. Barrett, and Phillip J. Cribb (editors). 2003. *Orchid Conservation*. Natural History Publications, Kota Kinabalu, Sabah, Malaysia.

Chase, M. S., K. M. Cameron, J. V. Freudenstein, A. M. Pridgeon, G. A. Salazar, C. van den Berg, and A. Schuiteman. 2015. "An updated classification of Orchidaceae". *Botanical Journal of the Linnean Society* 177: 151–174.

Chase, M. W., M.J.M. Christenhusz, M. F. Fay, J. W. Byng, W. S. Judd, D. E. Soltis, D. J. Mabberley, A. N. Sennikov, P. S. Soltis, and P. F. Stevens. 2016. An update to the Angiosperm Phylogeny Group classification for the orders and families of flowering plants. Bot. J. Linn. Soc. 181: 1–20.

Cheifetz, A., C. Double, L. Bernard, and D. Imwold (Eds.). 1999. Britanica's Trees and Shrubs. Laurel Glen Publishing, San Diego, California.

Chin, W. V. 2005. Ferns of the Tropics. Revised Edition. Times Editions, Marshal Cavandish International (Asia) Private Limited.

Christenhusz, M. J. M., J. L. Reveal, A. Farjon, M. F. Gardner, R. R. Mill, and M. W. Chase. 2011. A new classification and linear sequence of extant gymnosperms. Phytotaxa 19: 55–70.

Christenhusz, M. J. M. and M. W. Chase. 2014. Trends and Concepts in ffern classification. Annals Bot. 113: 571–594.

Christenhusz, M. J. M., M. F. Fay, and M. W. Chase, 2017. Plants of the World: An Illustrated Encyclopedia of Vascular Plants. Kew Publishing (Royal Botanic Gardens, Kew), Richmond & Chicago University Press, Chicago.

Christopher, T. (ed.). 1993. A-Z of Deciduous Trees and Shrubs. The Reader's Digest Association, Inc., Pleasantville, New York.

Church, G. 2007. Complete Hydrangeas. Firefly Books, Ontario, Canada.

Clarke, C. B. 1977. Edible and Useful Plants of California. University of California Press, Berkeley.

Clark, J. R. 1988. Fuchsias. The Globe Pequot Press, Chester, Connecticut.

Clausen, R. R. and N. H. Ekstrom. 1989. Perennials for American Gardens. Random House, New York.

Clausen, R. T. 1975. Sedum of North America North of the Mexican Plateau. Cornell University Press, Ithaca and London.

Clay, H. F. and J. C. Hubbard. 1977. The Hawaii's Garden: Tropical Exotics. University of Hawaii Press, Honolulu.

Clewell, A. F. 1985. Guide to Vascular Plants of the Florida Panhandle. University Press of Florida, Florida State University Press, Tallahassee.

Cline, C. 1998. The Rose. A Friedman/Fairfax Publishing Group, Inc., New York.

Coats, P. 1973. Roses. Octopus Books Limited, London.

Cobb, B. 1963. A Field Guide to the Ferns and Related Families of Northeastern and Central North America with A Section on Species Also Found in the British Isles and Western Europe. Houghton Mifflin Company, Boston.

Coombes, A. J. 1985. Dictionary of Plant Names. Timber Press, Portland, Oregon.

Coombs, A. J. 2001. The Illustrated Encyclopedia of Trees and Shrubs. Salamander Books, Limited. London.

Coombs, R. E. 2003. Violets: The History and Cultivation of Scented Violets. B. T. Basford, London.

Courtright, G. 1982. Trees and Shrubs for Temperate Climates. 3rd. Edition. Timber Press, Oregon.

Courtright, G. 1988. Tropicals. Timber Press, Portland, Oregon.

Cox, P. A. 1985. The Smaller Rhododendrons. Timber Press, Portland, Oregon.

Cox, P. A. 1990. The Larger *Rhododendron* Species. Timber Press, Portland, Oregon.

Craighead, F. C. Sr. 1971. The Trees of South Florida: Vol. 1. The Natural Environment and their Succession. University of Miami Press, Coral Gables, Miami, Florida.

Craigmyle, M. 1999. The Illustrated Encyclopedia of Perennials: A Unique Guide to More Than 1500 Plants with Comprehensive Descriptions and Planting Information. Salamander Books, Limited, London.

Craigmyle, M. 2003. Perennials: The Comprehensive Guide to over 2700 plants. Colin Gower Enterprises, Kent, United Kingdom. [Highly recommended reference]

Crandall, C. and B. Crandall. 1995. Flowering, Fruiting, and Foliage Vines: A Gardener's Guide. Sterling Publishing Co, Inc., New York.

Cribb, P. J., J. Greatwood, and P. F. Hunt. 1994. Handbook on Orchid Nomenclature and Registration. 4th Ed. Published by International Orchid Commission, London.

Cronquist, A. 1980a. An Integrated system of Classification of Flowering Plants. New York, Columbia University Press

Cronquist, A. 1980b. Vascular Flora of Southeastern United States. Vol. 1, Asteraceae. The University of North Carolina Press, Chapel Hill.

Cronquist, A. 1988. Evolution and Classification of Flowering Plants. 2nd ed., New York Botanical Garden.

Cullen, J., S. G. Knees, and H. S. Cubey. 2011. The European Garden Flora Flowering Plants: A Manual for the Identification of Plants Cultivated in Europe, Both Out-of-Doors and Under Glass. 2nd Ed. Cambridge University Press, New York.

Cusack, V. 1999. Bamboo World: The Growing and Use of Clumping Bamboos. Kangaroo Press, Sydney, Australia.

D'Amato, P. 1998. The Savage Garden: Cultivating Carnivorous Plants. TCN Speed Press, Berkeley, California.

Damp, P. 1987. Dahlias. The Globe Pequot Press, Chester, Connecticut.

Daniel, T. 2019. New and reconsidered Mexican Acanthaceae XIII. *Justicia*. Proc. Calif. Acad. Sci. Series 4, 66: 61–85.

D'Arcy, W. G. 1986. Solanaceae. Columbia University Press.

Davidian, H. H. 1989. The *Rhdodendron* species. Vol. I–III. Timber Press, Portland, Oregon.

Davies, D. 1992. Alliums: The Ornamental Onions. Timber Press, Portland, Oregon.

Dehgan, B. 1998. Landscape Plants for Subtropical Climates. University Press of Florida, Gainesville.

Dehgan, B. 1999. Propagation and Culture of Cycads: A Practical Approach. Pp. 123–131. In: *The 2nd Symp. On Ornamental Palms and Other Monocots from the Tropics.* Ed. M. Caballero Ruano. Acta Hort. 486, ISHS.

Dehgan, B. 2012. *Jatropha* (Euphorbiaceae). Flora Neotropica Monograph 110. Published for the Organization for Flora Neotropica by The New York Botanical Garden Press.

Dehgan, B. 2014. Public Garden Management: A Global Perspective. Vol. I & II. Xlibris, Bloomington, Indiana.

Dehgan, B. and C. K. K. H. Yuen. 1983. Morphology of the seed in relation to dispersal, evolution, and propagation in the genus *Cycas*. *Bot. Gaz.* 144: 412–418.

Dehgan, B. and F. Almira. 1993. Horticultural practices and conservation of cycads. In: *The Biology, Structure, and Systematics of Cycadales.* Pp. 332–328. In. Proc. Cycad 90, 2nd Intern. Conf. Cycad Biology. D. W. Stevenson and D. J. Norstog, eds.

Dehgan, B. and C. R. Johnson. 1983. Improved seed germination of *Zamia floridana* (sesu lato) with H2SO4 and GA3. Scientia Hort. 357–361

Dehgan, B. and G. L. Webster. 1979. Morphology and Infrageneric Relationships of Genus *Jatropha* (Euphorbiaceae). University of California Press, Berkeley, Los Angeles, London.

Dehgan, B. and N.B. Dehgan. 1988. Comparative pollen morphology and taxonomic affinities in Cycadales. Am. J. Bot. 75:1501–1516.

Dickerson, B. 1999. The Old Rose Adventurer: The Once-Blooming European Roses and More. Timber Press, Portland, Oregon.

Dirr, M. A. 1997. Dirr's Hardy Trees and Shrubs: An Illustrated Encyclopedia. Timber Press. Inc., Portland, Oregon.

Dirr, M. A. 2002. Dirr's Trees and Shrubs for Warm Climates: An Illustrated Encyclopedia. Timber Press, Portland, Oregon.

Dirr, M. A. 2004. Hydrangeas for American Gardens. Timber Press, Portland, Oregon.

Dirr, M. A. 2007. Viburnums: Flowering Shrubs for Every Season. Timber Press, Portland, Oregon.

Dirr, M. A. 2011. Dirr's Encyclopedia of Trees and Shrubs. Timber Press, Portland, Oregon.

Doerflinger, F. 1973. Complete Book of Bulb Gardening. Stackpole Books, David and Charles, Limited, Devon, England.

Dormon, C. 1959. Flowers Native to the Deep South. J. Horace McFarland Company, Mount Pleasant Press, Harrisburg, Pennsylvania.

Dransfield, J., N. W. Uhl, C. B. Osmussen-Lange, W. Baker, M. M. Harley, and C. E. Lewis. 2008. Genera Palmarum: The Evolution and Classification of Palms. Kew Publishing in Association with the International Palm Society and L. H. Bailey Hortorum, Cornell University.

Dressler, R. L. 1993. Phylogeny and Classification of the Orchid Family. Timber Press. Portland, Oregon.

Druse, K. 1993. Water Gardening. Burpee American Gardening Series. Prentice Hall Gardening, New York.

Duncan, W. H. and M. B. Duncan. 1988. Trees of the Southeastern United States. The University of Georgia Press, Athens, Georgia.

Dunk, G. 1982. Ferns for the Home and Garden. Angus & Robertson Publishers, NSW, Australia.

Dunn, T. and W. Reeves. 2003a. Jackson and Perkin's Beautiful Roses: Southern Edition. Coll Springs Press, Nashville, Tennessee.

Dunn, T. and W. Reeves. 2003b. Jackson and Prerkin's Beautiful Roses: Northwestern Edition Coll Springs Press, Nashville, Tennessee.

Earle, W. H. 1980. Cacti of the Southwest. Printed by Ironwood Lithographers, Inc., Scottsdale, Arizona.

Eggenberger, R. and M. H. Eggenberger. 1988. The Handbook on *Plumeria* Culture. Tropical Plant Specialists.

Eggli, U. 2001. Illustrated Handbook of Succulent Plants: Vol. I. Monocotyledons. Springer Verlag. New York.

Eggli, U. (ed.). 2002. Illustrated Handbook of Succulent Plants: Vol. IV. Dicotyledons, Springer Verlag. New York.

Eggli, U. (ed.). 2003. Illustrated Handbook of Succulent Plants: VI. Crassulaceae, Springer Verlag. New York.

Eggli, U. and L. E. Newton. 2004. Etymological Dictionary of Succulent Plant Names. Springer Verlag.

Eliovson, S. 1983. Proteas for Pleasure. 6th Edition. Macmillan South Africa Publishers (Pty) Ltd., Johannesburg.

Ellias, T. S. 1980. The Complete Trees of North America: Field Guide and Natural History. Outdoor Life/Nature Books, Van Niostrand Reinhold Company, New York.

Ellis, B. 2001. Bulbs: How to Select and Grow More Than 400 Summer-Hardy and Tender Bulbs. Houghton Mifflin Company, Boston, New York.

Ellison, D. 2002. An Illustrated Reference to Garden Plants of the World. New Holland Publishers, London.

Ellison, D. and A. Ellison. 2001. Betrock's Cultivated Palms of the World. Betrock Information Systems, Inc., Cooper City, Florida.

Elmore, F. H. 1976. Shrubs and Trees of the Southwest Uplands. Southwest Parks and Monuments Association, Tucson, Arizona.

Ettekoven, van, K. 2004. The future of the taxonomy of cultivated plants. Acta Hortic. 634: 211–216.

Everett, T. H. 1969. Living Trees of the World. Doubleday & Company, New York.

Everett. T. H. 1981. The New York Botanical Garden Illustrated Encyclopedia of Horticulture. Vol. 1–10. Garland Publishing, Inc. New York & London.

Ewart, R. 1987. *Fuchsia* Lexicon. 2nd ed. Cassel Publishers, Limited, London.

Fassett, N. C. 2006. A Manual of Aquatic Plants. University of Wisconsin Press, Madison, Wisconsin.

Feanley-Shittingstall, J. 1999. Peonies. Harry N. Abrams, Inc. Publishers, New York.

Feathers, D. L. (Ed.). 1978. The *Camellia*: Its History, Culture, Genetics, and a Look into Its Future Development. The Camellia Society, Printed by R. L. Bryan Company, Columbia, South Carolina.

Feininger, A. 1968. Trees. The Viking Press, New York.

Fiala, J. L. Jr. 1994. Flowering Crabapples: The Genus *Malus*. Timber Press, Portland, Oregon.

Fischer E, B. Schäferhoff, and K. Müller. 2013. The phylogeny of Linderniaceae - the new genus *Linderniella* and new combinations within *Bonnaya, Craterostigma, Lindernia, Micranthemum, Torenia* and *Vandellia. Willdenowia. Willdenowia* 43: 209–238.

Fisk, J. 1991. Clematis the Queen of Climbers. Cassell Publishers Limited, London.

Flint, H. L. 1983. Landscape Plants for Eastern North America, Exclusive of Florida and the immediate Gulf Coast. John Wiley & Sons., New York.

Flora of North America Editorial Committee. 1993–2016. Flora of North America North of Mexico. Oxford University Press, New York, Oxford.

Focke, A. and U. Meve (eds.). 2002. Illustrated Handbook of Succulent Plants: Ascelpiadaceae. Springer Verlag, New York

Foote, L. E. and S. B. Jones, Jr. 1989. Native Shrubs and Woody Vines of the Southeast: Landscaping Uses and Identification. Timber Press, Portland, Oregon.

Foster, G. F. 1984. Ferns to Know and Grow. Timber Press, Portland, Oregon.

Free, M. (Revised by C. M. Fitch). 1979. All About African Violets: The complete Guide to Success with Saintpaulias. Doubleday & Company, Inc. Garden City, New York.

Gaill, D. and J. White. 2002. Louisiana Gardener's Guide. Cool Springs Press, Nashville, Tennessee.

Galle, F. C. 1987. Azaleas. Revised and Enlarged Edition. Timber Press, Portland Oregon.

Galle, F. C. 1997. Hollies: The Genus *Ilex*. Published in Association with the Holly Society of America, Inc. Timber Press, Portland, Oregon.

Garcia, N., A. W. Meerow, S. Arroyo-Leuenberger, R. S. Oliveira, J. H. Dutilh, P. S. Soltis, and W. S. Judd. 2019. Generic classification of Amaryllidaceae tribe Hippeastreae. Taxon 68: 1–18.

Garrett, H. 1996. Plants for Texas. University of Texas Press, Austin.

Garcetti, G. and L. Livingston. 2018. *Protea*: The Magic and Mystery. Balcony Press, Pasadena, California.

Gault, S. M. and P. M. Synge. 1971. The Dictionary of Roses in Color. Madison Square Press, Grosset & Dunlap Publishers, New York.

Gelderen, D. M. van, and J. R. P. van Hoey Smith. Conifers: The Illustrated Encyclopedia. 2 vols. Timber Press, Portland, Oregon.

Gelderen, D. M. van. and J. R. P. van Hoey Smith. 1992. *Rhododendron* Portraits. Translated from Dutch language by N. Handgraaf and T. Handgraaf. Timber Press, Inc., Portland, Oregon.

Genders, R. 1965. The Rose: A Complete Handbook. The Dobbs-Merrill Company, Inc., Indianapolis.

Gentry, H. S. 1972. The Agave Family in Sonora. Agricultural Research Service, USDA, Agriculture Handbook No. 399.

Gentry, H. S. 2004. Agaves of Continental North America. University of Arizona Press, Tucson, Arizona.

George, A. S. 1996. The *Banksia* Book. 3rd. Ed. Kangaroo Press, in Association with The Society for Growing Australian Plants-NSW Ltd.

Gerbing, G. G. 1945. Camellias. Published by G. G. Gerbing, Fernandina, Florida.

Gibbons, M. 2003. A Pocket Guide to Palms. Chartwell Books, Inc. Edison, New Jersey.

Giddy, C. 1974. *Cycads of South Africa*. Purnell and Sons (S.A.) (Pty) Ltd., Cape Town, South Africa.

Gillespie, E. and K. A. Kron. 2010. Molecular phylogenetic relationships and revised classification of the subfamily Ericoideae (Ericaceae). Mol. Phylog. Evol. 56: 343–354.

Gilman, E. F. 1997. Trees for Urban and Suburban Landscapes. Delmar Publishers, Albany, New York.

Gledhill, D. 1985. The Names of Plants. Cambridge University Press, New York, N. Y.

Godfrey, R. K. 1988. Trees, Shrubs, and Woody Vines of Northern Florida and Adjacent Georgia and Alabama. The University of Georgia Press, Athens and London.

Godfrey, R. K. and J. W. Wooten. 1981. Aquatic and Wetland Plants of the Southeastern United States: Dicotyledons. The University of Georgia Press, Athens.

Goldberg, A. 2003. Character Variation in Angiosperm Families. Contributions from the United States Smithsonian Institution, National Herbarium 47: 1–185.

Goldenberg, J. 1992. Enjoying Roses. Ortho Books, San Ramon, California

Goode, D. 1989. Cycads of Africa. An Operating Division of Struik Group (Pty) Ltd., Cape Town, South Africa.

Goudey, C. J. 1985. Maidenhair Ferns in Cultivation. Lothian Publishing Company, Melbourne, Australia.

Goulding, E. 1995. Fuchsias: The Complete Guide. Timber Press, Portland, Oregon.

Graf, A. B. 1986. Exotica, Series 4: Pictorial Cyclopedia of Exotic Plants from Tropical and Near-Tropic Regions, A Treasury of Indoor Ornamentals. 12th Edition. Scribner Book Company. New York.

Graf, A. B. 1992a. Tropica: Color Cyclopedia of Exotic Plants and Trees from the Tropics and Subtropics, for Warm Region Horticulture in Cool Climate: The Sheltered Indoors. 4th Edition. Roehrs & Company, Exton, Pennsylvania.

Graf, A. B. 1992b. Hortica: Color Cyclopedia of Garden Flora and Indoor Plants. 1st Edition. Roehrs & Company, Exton, Pennsylvania.

Grant, J. A. and C. L. Grant. 1990. Trees and Shrubs for Pacific Northwest Gardens. 2nd Edition (revised by M. E. Black, B. O. Mulligan, J. A. Witt, and J. G. Witt). Timber Press, Portland, Oregon.

Gray-Wilson and V. Matthews. 1997. Gardening with Climbers. Timber Press, Portland, Oregon.

Green, D. 2003. Perennials All Season: Planning and Planting an Ever-Blooming Garden. Contemporary Books, Chicago, New York.

Green, W. F. and H. L. Blomquist. 1953. Flowers of the South: Native and Exotic. The University of North Carolina Press, Chapel Hill.

Grenfell, D. 1990. *Hosta*: The flowering Foliage Plant. Timber Press, Portland, Oregon.

Grenfell, D. and M. Shadrack. 2004. The Color Encyclopedia of Hostas. Timber Press, Portland, Oregon.

Grey-Wilson, C. 1988. The Genus *Cyclamen*. The Royal Botanical Garden, Kew in Association with Christopher Helm and Timber Press.

Grey-Wilson, C. 1993. Poppies: The Poppy Family in the Wild and In Cultivation. Timber Press, Portland, Oregon.

Grieve, M. 1992. A Modern Herbal. Dorset Press, New York.

Griffiths, M. 1994. The New Royal Horticultural Society Dictionary: Index of Garden Plants. Macmillan Press, London.

Griffiths, T. 1986. The Book of Classic Old Roses. Michael Joseph, London.

Grifiths, M. and J. Stewart. 1995. Manual of Orchids. Derived from the New Royal Horticultural Society Dictionary. Timber Press. Portland, Oregon.

Grimm, W. C. 1967. Familiar Trees of America. Harper & Row, Publishers, New York.

Groom, D. 2002. Texas Gardening Guide. Cool Springs Press, Nashville, Tennessee.

Grose, S. O. and R. G. Olmstead. 2007. Taxonomic revision in the polyphyletic genus *Tabebuia* s. l. (Bignoniaceae). Systematic Botany 32: 660–670.

Grounds, R. 1974. Ferns. Pelham Books, Ltd., London.

Guennel, G. K. 1995. Guide to Colorado Wildflowers: Plains and Foothills. Westcliffe Publishers, Inc. Englewood, Colorado.

Hadak, J. 1980. Trees for Every Purpose. McGraw-Hill Book Company, New York.

Haehle, R. G. and J. Brookwell. 1999. Native Florida Plants. Gulf Publishing Company, Houston, Texas.

Hansen R. and F. Stahl. 1993. Perennials and their Garden Habitats. 4th ed., translated by R. Ward. Timber Press, Portland, Oregon.

Harkens, J. 1978. Roses. J. M. Dent & Sons Ltd., London. [This book includes a list of the older literature on roses].

Harris, S. 2007. The Magnificent Flora Graeca: How the Mediterranean came to the English Garden. Bodleian Library, University of Oxford.

Haslam, S. M. 1978. River Plants. Cambridge University Press, Cambridge, London.

Hawkes, A. D. 1961. Orchids: Their Botany and Culture. Harper & Brothers, Publishers, New York.

Hawkes, J. G., R. N. Lester, and A. D. Skelding (eds.). 1979. The Biology and Taxonomy of Solanaceae. pp. 3–47. Linnean Society Symposium Series No. 7. Published for the Linnean Society of London by Academic Press.

Hay, R. and P. M. Synge: 1975. The Color Dictionary of Garden Plants, in Collaboration with the Royal Horticultural Society. Bloomsbury Books, London.

Haynes, J. 2013. Etymological Compendium of Cycad Names. http://www.cycad.org/documents/etymological_compendium.pdf

Haynes, R. R. and C. B. Hellquist. 2000. Alismataceae Ventenat. Water-plantain or arrowhead family. Flora of North America, Vol. 22.

Hedrick, U. P. (ed.). 1919 (Dover ed., 1972). Sturtevant's Edible Plants of the World. Dover Publications, Inc. New York.

Heidrun, E., K. Hartman, U. Egli, and H. E. K. Hartman (eds.). 2001. Illustrated Handbook of Succulent Plants: Aizoaceae F-Z. Springer, Berlin Heidelberg.

Heiser, C. B. Jr. 1969. Systematics and the origin of cultivated Plants. Taxon 18: 36–45.

Henderson, A., G. Galeano, and R. Bernal. 1995. Field Guide to the Palms of the Americas. Princeton University Press. Princeton, New Jersey.

Herwig, R. 1985. 2850 House and Garden Plants. Crescent Books, New York.

Hill, L. and N. Hill. 1991. Daylilies: The Perfect Perennial. Story Communications, Inc. Penal, Vermont.

Hillard, O. M. and B. L. Burtt. 1991. *Dierama*: The Hairbells of Africa. Acorn Books, CC, Johannesburg, London.

Hoffman, M.H.A. 2004. Cultivar classification of *Taxus* L. (Taxaceae). Acta Hortic. 634: 91–96.

Hodgson, L. 2002. Annuals for Every Purpose: Choose the Right Plants for Your Conditions, Your Garden, and Your taste. St. Martin Press, New York.

Hogan, S. (Chief Consultant). 2003. Flora: A Gardener's Encyclopedia. Volume I. A-K, Volume II. L-Z. Timber Press. Portland, Oregon.

Holliman, J. (ed.). 2002. Botanica's Orchids: Over 1200 Species (Botanica's Gardening Series). Paurel Glen Publishing, San Diego, California.

Holtum, R. E. and I. Enoch. 1997. Gardening in the Tropics. Timber Press, Portland, Oregon.

Hora, B. [Consulting Ed.], 1980. The Oxford Encyclopedia of Trees of the World. Crescent Books, New York.

Hoshizaki, B. J. and R. C. Moran. 2001. Fern Growers Manual: Revised and Expanded Edition. Timber Press, Portland, Oregon.

Hosie, R. C. 1979. Native Trees of Canada. 8th edition. Fitzhenry & Whieside Ltd., Don Mills, Ontario, Canada.

Hough, J. N. 1953. Scientific Terminology. Reinhart and Company Inc. Publishers, New York.

Hui-Lin Li. 1974. Plant taxonomy and the origin of cultivated plants. Taxon 23: 715–724.

Humphries, C. J., J. R. Press, and D. A. Sutton. 2000. Guide to Trees of Britain and Europe. Hamlyn, Octopus Publishing Group, London.

Hung Ta, C. and B. Bartholomew. 1984. Camellias. Timber Press, Portland, Oregon.

Hutchens, A. R. 1991. Indian Herbology of North America. Shambhala, Boston & London.

Huxley, A. (ed.). 1992. The New Royal Horticultural Society Dictionary of Gardening. 4 vol. Macmillan Press Ltd. London, Stockton Press, USA.

Huxley, C. R. 1978. The ant plants *Myrmecodia* and *Hydnophytum* (Rubiaceae) and the relationship between their morphology and occupants, physiology and ecology. New Phytologist 80: 231–268.

Huxley, J. M. 2009, A revision of the ant-plant genus *Hydnophytum* (Rubiaceae). National Botanic Gardens Glasnevin, Dublin, Ireland

Hyam, R. and R. Pankhurst. 1995. Plants and Their Names: A Concise Dictionary. Oxford University Press.

Innes, C. and C. Glass. 1991. Cacti: over 1200 Species Illustrated and Identified. Portland House, New York.

Innes, C. and C. Glass. 2001. Illustrated Encyclopedia of Cacti. Portland House, New York.

Iredell, J. 1994. Growing Bougainvilleas. Cassel Publishers, Limited, London.

Jacobson, A. L. 1996. North American Landscape Trees. Ten Speed Press, Berkeley, California.

Jacobson, H. 1974. A Handbook of Succulent Plants: Descriptions, Synonyms, and Cultural Details for Succulents other Than Cactaceae. Vol. 1–3. Blandford Press, London.

Jacobson, H. 1977. Lexicon of Succulent Plants. Offizin Anderson Nexö. Leipzig, Germany.

Jacquemyn, H., P. Endels, R. Brys, M. Hermy. and S. R. J. Woodell. 2009. Biological Flora of the British Isles: *Primula vulgaris* Huds. (*P. acaulis* (L.) Hill). Journal of Ecology 97: 812–833.

Jamieson, B. G. M. and J. F. Reynolds. 1967. Tropical Plant Types. Pergamum Press, Oxford.

Jefferson-Brown, M. 1991. *Narcissus*. B. T. Batsford, Ltd., London.

Jefferson-Brown, M. 1999. Ramblers, Scramblers, and Twiners: High Performance Climbing Plants and Wall Shrubs. David and Charles, UK.

Jeffrey, C. 1977. Biological Nomenclature. Second Edition. Crane, Russak and Company, Inc., New York, N.Y.

Jekyll, G. 1916. Annuals and Biennials: The Best Annual and Biennial Plants and their Uses in the Garden. Offices of Country Life, George Newnes, Covent Garden, London.

Jelitto, L. and W. Schacht. 1990. Hardy Herbaceous Perennials. 3rd Edition. Translated by M. E. Epp. Volume I. A-K; Volume II. L-Z. Timber Press, Portland, Oregon.

Jingwei, Z. (ed.). 1982. Alpine Plants of China. Science Press, Beijing, China; Gordon and Breach, Science Publishers, Inc. New York.

Jobson, R.W., J. Playford, K. M. Cameron, and V. A. Albert. (2003). "Molecular Phylogenetics of Lentibulariaceae Inferred from Plastid rps16 Intron and trnL-F DNA Sequences: Implications for Character Evolution and Biogeography". *Systematic Botany*. 28:157–171.

Johnson, H. 1990. Hugh Johnson's Encyclopedia of Trees. Portland House, New York.

Jones, C. 1989. The Complete Guide to Bedding Plants for Amateurs and Experts. Voyageur Press, Stillwater, Minnesota.

Jones, D. L. 1987a. Encyclopedia of Ferns: An Introduction to Ferns, their Structure, Biology, Economic Importance, Cultivation and Propagation. Timber Press, Portland, Oregon.

Jones, D. L. 1993. Cycads of the World: Ancient Plants for Today's Landscapes. Smithsonian Institution Press, Washington, DC.

Jones, D. L. 1998. Palms Throughout the World. Smithsonian Institution Press, Washington, DC.

Jones, D. L. 1987b. Encyclopedia of Ferns: An Introduction to Ferns, their Structure, Biology, Economic Importance, Cultivation, and Propagation. Timber Press, Portland, Oregon.

Jones, D. L. and B. Gray. 1988. Climbing Plants in Australia. Reed Books Pty Ltd. NSW.

Joyce, D. 1998. The Perfect Plant for Every site, Habitat, and Garden Style. Stewart, Tabori, and Chang. New York.

Judd, W. S. and S. R. Manchester. (1998). Circumscription of Malvaceae (Malvales) as determined by a preliminary cladistic analysis employing morphological, palynological, and chemical characters. Brittonia 49:384–405.

Judd, W. S., S. C. S. Campbell, E. A. Kellogg, and P. F. Stevens. 2015. Plant Systematics: A Phylogenetic Approach. 4th Ed. Sinauer Associates, Inc. Publishers, Sunderland, Massachusetts.

Judzhewicz, E. J., L. G. Clark, X. Londono, and M. J. Stern. American Bamboos. Smithsonian Institution, Washington, DC.

Kapitany, A. and L. Schutz (2005). *Echeveria* Cultivars. Schultz Publishing.

Keenan, P. E. 1998. Wild Orchids Across North America: A Botanical Travelogue. Timber Press, Portland, Oregon.

Kennedy, P. 2013. http://www.pacsoa.org.au/wiki/Lepidozamia_peroffskyana

Kenyon, J. and J. Walker. Vireyas: A Practical Gardening Guide. Timber Press, Portland, Oregon.

Kepler, A. K. and J. R. Mau. 1988. Proteas in Hawaii. Mutual Publishing, Honolulu, Hawaii.

Kester, D. E. 1983. The clone in horticulture. HortScience 18: 831–837.

Key, H. 2000. 1001. Pelargoniums. B. T. Batsford, London.

Kiaer, E. 1966. The Concise Handbook of Roses. Translated by G. Vevers. E. P. Dutton & Co., Inc., New York.

King, G. 2003. Trees: Natural Wonders of North America. Courage Books, an Imprint of the Running Press, Philadelphia.

Kingsbury, J. M. 1964. Poisonous Plants of the United States. Prentice-Hall, Inc. Englewood Cliffs, New Jersey.

Kirkbride, J. H. Jr., C. R. Gunn, and A. L. Weitzman. 2003. Fruits and Seeds of Genera in the Subfamily Faboidea (Fabaceae). 2 vol. United States Department of Agriculture, Tech. Bull.1890.

Kneller, M. 1995. The Book of Rhododendrons. Timber Press, Portland, Oregon

Köhlein, F. 1987. *Iris*. Timber Press, Portland, Oregon.

Kohlein, F. and P. Menzel. 1994. Color Encyclopedia of Garden Plants and Habitats. Timber Press, Portland, Oregon.

Kondo, K. 1976. A historical review of taxonomic complexes of cultivated taxa of *Camellia*. Amer. Camellia Yearbook, vol. 16–17: 102–115.

Kral, R. 1983. A Report on Some Rare, Threatened, or Endangered Forest-Related Vascular Plant of the South. Vol. 1: Isoetaceae through Euphorbiaceae; Vol. 2: Aquifoliaceae through Asteraceae and Glosssary. USDA Forest Service, Technical Publication R8-TP. Atlanta, Georgia.

Kress, W. J. 1984. Systematic of Central American *Heliconia* (Heliconiaceae) with pendent inflorescences. J. Arnold Arb. 65: 429–532.

Kron, K.A., W.S. Judd,; P. F. Stevens, D.M. Crayn, A.A. Anderberg, P. A. Gadek, C. J. Quinn, and J.L. Luteyn. 2002. Phylogenetic Classification of Ericaceae: Molecular and Morphological Evidence. *The Botanical Review*. 68 (3): 335–423.

Krüssman, G. 1985. *Manual of Cultivated Conifers*. Translated by M. E. Epp. Timber Press. Portland, Oregon.

Krüssmann, G. 1984. Manual of Cultivated Broad-Leaved Trees and Shrubs. 3 vols. Translated by M. E. Epp. Timber Press, Beaverton, Oregon.

Lakela, O. and R. W. Long. 1976. Ferns of Florida: An Illustrated Manual and Identification Guide. Banyan Books, Miami, Florida.

Lampe, K. F. and M A. McCann. 1985. AMA Handbook of Poisonous and Injurious Plants. American Medical Association, Chicago, Illinois.

Large, M. F. and J. E. Braggins. 2009. Tree Ferns. Timber Press, Portland, Oregon.

Lassak, E. V. and T. McCarty. 2001. Australian Medicinal Plants. Reed New Holland Publishers, Sydney.

Laubenfels, D. J. De. 1985. A Taxonomic Revision of the Genus *Podocarpus*. Blumea 30: 251–278.

Lecoufle, M. 1993. Carnivorous Plants: Care and Cultivation. Cosell Publishers, Ltd., London.

Lee, F. P. 1965. The Azalea Book (2nd ed.). D. Van Nostrand Company, Inc. Princeton, New Jersey. [includes extensive list of older cultivars]

Lellinger, D. B. 1985. A Field Manual of the Ferns and Fern Allies of the United States & Canada. Smithsonian Institution Press, Washington, DC.

Leue, M. 1987. *Epiphyllum*: The Splendor of Leaf Cacti. Marga Leue, Publisher. (Originally in German, with stunning photographs).

Lewington, A. 1990. Plants for People. Oxford University Press, New York.

Lindsay, T. S. 1923. Plant Names. The Sheldon Press, London. (Reprinted by Gate Research Company, Book Tower, U. K., 1976)

Lindstrom, A. J. (ed.). 2004. The Biology, Structure, and Systematics of the Cycadales. *Proceedings of the Sixth International Conference on Cycad Biology*. Nong Nooch Tropical Botanical Garden, Thailand.

Line, L., A. Sutton, and M. Sutton. 1981. The Audubon Society Book of Trees. Harry N. Abrams, Inc. Publishers, New York.

Little, E. L. Jr. 1979. Checklist of United States Trees (Native and Naturalized). Agriculture Handbook No. 541. Forest Service United States Dept. of Agric., Washington, DC.

Little, E. L. Jr. 1968. Southwestern Trees: A Guide to the Native Species of New Mexico and Arizona. U. S. Dept. Agric., Agric. Handb. No.9. Washington, DC.

Little, R. J. and C. E. Jones. 1980. A Dictionary of Botany. Van Nostrand Reinhold Company, New York.

Llamas, K. A. 2003. Tropical Flowering Plants: A Guide to Identification and Cultivation. Timber Press, Portland, Oregon.

Lloyd, F. E. 1976 (1942). The Carnivorous Plants. Dover Publications Inc., New York.

Long, R. W. 1970. The Genera of Acanthaceae in the Southeastern United States. J. Arnold Arb. 51: 257–309.

Lodé, J. 2021. Taxonomy of the Cactaceae: A New Classification of Cacti Mainly Based on Molecular Data: Description of the Species. Vol. 3–4. Publisher unknown – probably self-published. Also see Taxonomy_of_the_Cactaceae_Index_of_Synonyms_&_errata.pdf

Lord, E. E. 1970. Shrubs and Trees for Australian Gardens. Lothian Publishing Co., Pty., Ltd., Melbourne and Sydney, Australia.

Luteyn, J. L. and M. E. O'Brien (eds.). 1980. Contributions Toward a Classification of *Rhododendron*. Proc. Inter. Rhododendron Conference, The New York Botanical Garden May 15–17, 1978. The New York botanical Garden.

Mabberley, D. J. 1997. The Plant-Book: A Portable Dictionary of the Vascular Plants. 2nd ed. Cambridge University Press., UK.

Mabberley, D. J. 2008. Mabberley's Plant Book: A Portable Dictionary of Plants, Their Classification and Uses. Cambridge University Press., UK.

Mackenzie, S. D. 1997. Perennial Ground Covers. Timber Press, Portland, Oregon.

Macoboy, S. 1981. The Color Dictionary of Camellias. Lansdowne Press, Sydney, New York.

Macoboy, S. 1986. What Flower is That? Portland House, New York.

Macoboy, S. 1989. What Shrub is That? Portland House, New York

Macoboy, S. 1991. What Tree is That? Crescent Books, New York.

Macoboy, S. and R. Mann. 1998. The Illustrated Encyclopedia of Camellias. Timber Press, Portland, Oregon.

Manning, J. 2007. The Genus *Colchicum* L. redefined to include *Androcymbium* Willd. based on molecular evidence. *Taxon* 56: 872–882

Marie-Victorin, Frere, and P. Leon. 1942. Itineraires Botaniques Dans L'ile de Cuba. Institute Botanique de L'universite de Montreal, Canada.

Martin, M. J., P. R. Chapman, and H. A. Auger. 1971. Cacti and their Cultivation. Winchester Press, New York.

Martin, R. 2002. The Plant Finder's Guide to Garden Ferns. Timber Press, Portland, Oregon.

Mastalerz, J. W. and E. J. Holocomb (eds.). 1982. Geraniums: A Manual of the Culture of Geraniums as Greenhouse Crop, Third Edition. Pennsylvania Flower Growers. University Park, PA.

Mathew, B. 1973. Dwarf Bulbs. Published in Association with Royal Horticultural Society. Arco Publishing Company, Inc., New York.

Mathew, B. and P. Swindells. 1994. The Complete Book of Bulbs, Corms, Tubers, and Rhizomes: A Step-by-Step Guide to Nature's Easiest and Most Rewarding Plants. Reader's Digest Association, Inc., Pleasantville, New York.

Mathew, L. J. 2016. *Protea*: A Guide to Cultivated species and varieties. Latitude 20, University of Hawaii Press.

Mattock, J., S. McCann, F. Mitchell, and P. Wood. 1994. The Complete Book of Roses. Ward Lock Limited, London.

McCann, S. 1985. Miniature Roses for Home and Garden. Prentice Hall Press, New York.

McCann, S. 1991. Miniature Roses: Their Care and Cultivation. Cassell Publishers Limited, London.

McClaren, B. 2004. The Encyclopedia of Dahlias. Timber Press, Inc. Portland, Oregon.

McClintock, E. and A. T. Leiser. 1979. An Annotated Checklist of Woody Ornamental Plants of California, Oregon, and Washington. Agricultural Sciences Publications, University of California, Berkeley, California.

McDonald, E. 1991. Roses and Rose Gardens. Pp. 1–34, In: P. Hobhouse and E. McDonald (eds.), Gardens of the World: the art and Practice of Gardening. McMillan Publishing Company, New York.

McDonald, E. 1995. The 400 Best Garden Plants: A practical Encyclopedia of Annuals, Perennials, Bulbs, Trees, and Shrubs. Random House, New York.

McDonold, E. 1998. The Rose Garden: Tea Roses. Smithmark Publishers, New York.

McNeill, J. 2004. Nomenclature of cultivated plants: A historical botanical statement. Acta Hortic. 634: 29–36.

McNeill, J. 2008. The Taxonomy of cultivated Plants. Acta Hortic. 799, 21–28.

McRae, E. A. 1998. Lilies: A Guide for Growers and Collectors. Timber Press, Portland, Oregon.

McSwain, M. J. 1997. Florida Gardening by the Sea. University Press of Florida.

Meehan, T. 1853. The American Handbook of Ornamental Trees. Lippincott, Grambo, and Co., Philadelphia.

Meerow, A. W. 1985. Notes on Florida *Zephyranthes*. Herbertia 1985: 87–94.

Meerow, A. W. 1987. A monograph of Eucrosia (Amaryllidaceae). Syst. Bot. 12: 460–492.

Meerow, A. W. 1989. Systematics of the Amazon lilies. *Eucharis* and *Caliphuraria* (Amaryllidaceae)

Meerow, A. W. 2006. Betrock's Guide to Landscape Palms. Betrock Information Systems, Inc., Cooper City, Florida.

Meerow, A. W. 2012. Ornamental Geophytes Taxonomy and Phylogeny, In: Kamenetsky, R. and H. Okubo (eds.). Ornamental Geophytes: From Basic Science to Sustainable Production. Taylor & Francis Inc. Bosa Roca, U.S.

Meerow, A. W. and B. Dehgan. 1984. Re-establishment and lectotypification of *Eucharis amazonica* Linden ex Planchon (Amaryllidaceae). Taxon 33: 416–422.

Meerow, A. W., E. M. Gardner, and K. Nakamura. 2020. Phylogenomics of the Andean tetraploid clade of American Amaryllidaceae (subfamily Amaryllidoideae): Unlocking a polyploid generic radiation abetted by continental geodynamics. Frontiers Plant Sci. 11: 1–26.

Menninger, E. A. 1962. Flowering Trees of the World for Tropics and Warm Climates. Hearthside Press Incorporated Publishers, New York.

Menninger, E. A. 1970. Flowering Vines of the World: An Encyclopedia of Climbing Plants. Hearthside Press Incorporated, Publishers, New York.

Metcalf, L. 2006. Herbs: A Guide to Species, Hybrids, and Allied Genera. Timber Press, Portland, Oregon.

Meredith, T. J. 2001a. Bamboo for Gardens. Timber Press, Portland, Oregon.

Mickel, J. T. 1994. Ferns for American Gardens: The Definitive Guide to Selecting and Growing More Than 500 Kinds of Hardy Ferns. Macmillan Publishing Company, New York.

Miep, N. (ed.). 1985. Fuchsias: The Complete Handbook. Cassel Publishers, Limited, London.

Miller, D. 1996. Pelargoniums: A Gardener's Guide to the Species and their Cultivars and Hybrids. Timber Press, Inc. Portland, Oregon.

Millspaugh, C. 1974. American Medicinal Plants: An Illustrated and Descriptive Guide to Plants Indigenous to and Naturalized in the United States Which are used in Medicine. Dover Publications, Inc. New York.

Mineo, B. 1999. Rock Garden Plants: A Color Encyclopedia. Timber Press, Portland, Oregon.

Möller, M. and Q. C. B. Cronk. 1999. New approaches to the systematics of *Sainpaulia* and *Streptocarpus*. In: S. Andrews. A.C. Leslie, and C. Alexander (Eds.). Taxonomy of Cultivated Plants: Third International Symposium, pp. 253–264. Royal Botanic Garden Kew.

Moonlight, P.W, W.H. Ardi, L.A. Padilla, K.F. Chung, D. Fuller, D. Girmansyah, R. Hollands, A. Jara-Muñoz, R. Kiew, W.C. Leong, Y. Liu, A. Mahardika, L.D.K. Marasinghe, M. O'Connor, C. I. Peng, A. J. Pérez, T. Phutthai, M. Pullan, S. Rajbhandary, C. Reynel, R. R. Rubite, J. Sang, D. Scherberich, Y.M. Shui, M.C. Tebbitt, D. C. Thomas, H. P. Wilson, N.H. Zaini, M. Hughes. 2018, "Dividing and conquering the fastest-growing genus: Towards a natural sectional classification of the mega-diverse genus Begonia (Begoniaceae)". Taxon. 67: 267–323.

Moran, R. C. 2004. A Natural History of Ferns. Timber Press, Portland, Oregon.

Meredith, T. J. 2001b. Bamboo for Gardens. Timber Press, Portland, Oregon.

More, D. and J. White. 2002. The Illustrated Encyclopedia of Trees, Timber Press. Portland, Oregon.

Mori, S. A.; C. H.Tsou, C. C. Wu, B. Cronholm, A. A. Anderberg, 2007. "Evolution of Lecythidaceae with an emphasis on the circumscription of neotropical genera: Information from combined *ndhF* and *trnL-F* sequence data". *American Journal of Botany*. 94: 289–301.

Morton, J. 1974. 500 Plants of South Florida. E. A. Semann Publishing, Inc. Miami, Florida.

Morton, J. F. 1990. Trees, Shrubs, and Plants for Florida Landscaping: Native and Exotic. Florida Dept. of Agriculture.

Muller, K. K., R. E. Broder, and W. Beittel. 1974. Trees of Santa Barbara. Santa Barbara Botanic Garden, California.

Munson, R. W. Jr. 1989. *Hemerocallis*: The Daylily. Timber Press, Portland, Oregon.

Nash, N. and I. La Croix (Consul.). 2005. Flora's Orchids. Timber Press, Portland, Oregon.

Nehrling, A. and I. Nehrling. 1980. Peonies, Outdoors and In. Hearthside Press, Inc., Publishers, New York.

Nelson, G. 1994. The Trees of Florida: A Reference and Guide. Pineapple Press, Inc., Sarasota, Florida.

Nelson, S. 2000. Rhododendrons in the Landscape. Timber Press, Portland, Oregon.

Nicolson B. E. and A. R. Clapham. 1979. The Oxford Book of Trees. Oxford University Press, London.

Noblick, L. R. 2004. *Syagrus cearensis*, A twin-stemmed new palm from Brazil. Palms 48: 70–76.

Nybakken, O. E. 1960. Greek and Latin in Scientific Terminology. The Iowa State University Press, Ames, Iowa.

Nyffeler, R. and U. Eggli. 2010, "A farewell to dated ideas and concepts: molecular phylogenetics and a revised suprageneric classification of the family Cactaceae", Schumannia, 6: 109–149.

Ochsman, J. 2004. Current problems in nomenclature and taxonomy of cultivated plans. Acta Hortic. 634: 53–61.

Okita, Y. and J. L. Hoolenberg. 1981. The Miniature Palms of Japan. Wetherhill, New York.

Olde, P. and N. Marriott. 1995. The *Grevillea* Book. 3 vols. Timber Press, Portland, Oregon.

Olmstead, R. G., C. W. de Pamphillis, A. D. Wolfe, N. D. Young, W. J. Elisons, and P. A. Reeves. 2001. Disintegration of the Scrophulariaceae. Am. J. Bot. 88: 348–361.

Olmstead, R.G. and L. Bohs, 2007. "A Summary of molecular systematic research in Solanaceae: 1982–2006". *Acta Horticulturae*. 745:11

Olmstead, R. G.; M. L. Zjhra, L. G. Lohmann, S.O. Grose, and A. J. Eckert. 2009. A molecular phylogeny and classification of Bignoniaceae. *Amer. J. Bot*. 96: 1731–1743.

Osti, G. L. 2004. The Book of Mediterranean Peonies. Umberto Allemandi & C. Toreno.

Ouden, P. D. and B. K. Boom. 1965. Manual of Cultivated Conifers Hardy in the Cold and Warm Temperate Zones. The Hague/Martinus Nijhoff, Netherlands.

Page, S. and M. Olds (eds.). 2001. The Plant Book: The World of Plants in a Single Volume. Random House Australia Pty Ltd.

Palgrave, K. C. 1977. Trees of Southern Africa. C. Struik Publishers, Cape Town, Johannesburg.

Panczk, T. D. 1996. Are taxonomists an endangered species? Amer. Nurseryman 37 (5): 12–13.

Parer, R. 2016. The Plant Lovers Guide to Hardy Geraniums. Timber Press, Inc., Portland, Oregon.

Parham, J. W. 1972. Plants of the Fiji Islands. Published by the Authority of the Governments Printer, Suva.

Parker, H. 2000. Perennials. Dorling Kindersley Publishing, Inc., London.

Patterson, A. 1991. Tulips and the Spring Bulbs. Pp. 73–94, In: P. Hobhouse and E. McDonald (eds.), Gardens of the World: the art and Practice of Gardening. McMillan Publishing Company, New York.

Pelczar, R. and W. E. Barrick. 2004. American Horticultural Society, Southeast: Smart Garden Regional Guide. DK Publishing, Inc., New York.

Phillips, R. 1978a. Trees of North America and Europe. Random House, New York.

Phillips, R. 1978b. A Photographic Guide to More than 500 Trees of North America and Europe. Random House, New York.

Phillips, R. and M. Rix. 2002. Perennials: The Definitive Reference with over 2,500 Photographs. Firefly Books, Buffalo, New York.

Phillips, R. and M. Rix. 2002a. The Botanical Garden. Volume I. Trees and Shrubs; Volume II. Perennials and Annuals. Firefly Books, Ltd. Ontario, Canada.

Phillips, R. and M. Rix. 2002b. Annuals and Biennials: The Definitive Reference with over 1000 Photographs. Firefly Books, Buffalo, New York.

Pickersgill, B. and D. A. Karamura. 1999. Issues and options in the classification of cultivated bananas, with particular reference to the East African Highland bananas. In: S. Andrews, A.C. Leslie, and C. Alexander (Eds.). Taxonomy of Cultivated Plants: Third International Symposium, pp. 169–157. Royal Botanic Garden, Kew.

Pierot, S. and M. Mullaly 1974. The Ivy Book: The Growing and Care of Ivy and Ivy topiary. MacMillan Publishing Company, Stuttgart, Germany.

Pietropaulo, J. and P. Pietropaulo. 1986. Carnivorous Plants of the World. Timber Press, Portland, Oregon.

Pizzeti, I. and H. Cocker. 1968. Flowers: A Guide for Your Garden. Vol. 1–2. Harry N. Abrams, Inc, Publishers, New York.

Polunin, O. 1976. Trees and Bushes of Britain and Europe. Oxford University Press, London.

Poor, J. M. (ed.). 1984. Plants that Merit Attention: Volume I- Trees. The Garden Club of America. Timber Press, Portland, Oregon.

Poor, J. M. and N. P. Brewster. 1996. Plants that Merit attention: Volume II – Shrubs. The Garden Club of America. Timber Press, Portland, Oregon.

Potter, D., T. Ericksson, R. C. Evans, S. Oh, J.E.E. Smedmark, D. R. Morgan, M. Kerr, K. R. Robertson, M. Arsenault, T. A. Dickson, and C. S. Campbell. 2007. Phylogeny and Classification of Rosaceae. Pl. Syst. Evol. 266: 5–43.

Pratt, M. B. 1878. Shade and Ornamental Trees of California. California State Board of Forestry, Sacramento.

Preissel, U. and H-G. Preissel. 2002. *Brugmansia* and *Datura*: Angels Trumpets and Thorn Apples. Firefly Books, Ontario, Canada.

Preston, J. R., Jr. 1976. North American Trees (Exclusive of Mexico and Tropical United States). The Iowa State University Press, Ames, Iowa.

Preston-Mafham, K. 1994. Cacti and Succulents in Habitat. Cassel Villiers House. London.

Preston-Mafham, K. 2007. 500 Cacti: Species and Varieties in Cultivation. Duncan Baird Publishers. New York.

Preston-Mafham, R. and K. Preston-Mafham. 1991. Cacti: The Illustrated Dictionary. Castel Publishers Limited. London.

Price, M. The Iris Book. 1966. D. Van Nostrand Company, Inc. Princeton, New Jersey.

Pridgeon, A. (Ed.). 1995. The Illustrated Encyclopedia of Orchids. Timber Press. Portland, Oregon.

Putz, F. E. and H. A. Mooney (Eds.). 1991. The Biology of Vines. Cambridge University Press, Cambridge, New York.

Radford, A. E., W. C. Dickson, J. R. Massey, and C. R. Bell. 1974. Vascular Plant Systematics. Harper & Row, Publishers, New York.

Ranker, T. A. and C. H. Haufler (eds.). 2008. Biology and Evolution of Ferns and Lycophytes. Cambridge University Press, New York.

Rauch, F. D. and P. R. Weissich. 2000. Plants for Tropical Landscapes: A Gardeners' Guide. University of Hawai'i Press, Honolulu.

Rauh, W. 1979. The Wonderful World of Succulents: Cultivation and Description of Selected Succulent Plants other than Cacti. 2nd Edition, Revised. Translated by H. L. Kendall. Smithsonian Institution Press, Washington D.C.

Rebelo, T. 1995. Proteas: A Field Guide to Proteas of Southern Africa. Fernwood Press, Vlaeberg, Cape Town.

Rehder, A. 1960. Manual of Cultivated Trees and Shrubs Hardy in North America. 2nd Ed. The Macmillan Company, New York.

Reinikka, M. A. 1995. A History of Orchids. Timber Press, Portland, Oregon.

Reynolds, T. (ed.). 2004. Aloes: The Genus *Aloe* (Medicinal and Aromatic Plants – Industrial Profiles). CRC Press. Boca Raton, Florida.

Rice, G. 1995. Hardy Perennials. Timber Press, Portland, Oregon.

Rice, G. 2002. Hellebores. RHS Wesley Handbook, Cassel Illustrated, Octopus publishing Group, London.

Richards, J. 1993. *Primula*. Timber Press, Portland, Oregon.

Rickard, M. 2000. The Plant Finders Guide to Garden Ferns. Timber Press, Portland, Oregon.

Riffle, R. L. 1998. The Tropical Look: An Encyclopedia of Dramatic Landscape Plants. Timber Press, Portland, Oregon.

Riffle, R. L. and P. Craft. 2003. An Encyclopedia of Cultivated Palms. Timber Press, Portland, Oregon.

Riha, J. and R. Subik. 1981. The Illustrated Encyclopedia of Cacti & Other Succulents. Chartwell Books, Inc. Secaucus, New Jersey.

Rintz, R. E. 1980. The Peninsular Malayan Species of *Dischidia* (Asclepiadaceae). Blumea: 26: 81–126.

Robey, M. J. 1987. African Violets: Gifts from Nature. Cornell Books, New York, London.

Robinson, M. A. 2000. Auriculas for Everyone: How to Grow and Show Perfect Plants. Guild of Master Craftsman Publications, Ltd. East Sussex, UK.

Robinson, P. 1997. The American Horticultural Society Complete Guide to Water Gardening. A DK Publishing Book, New York.

Robinson, W. 1933. The English Flower Garden. 15th ed. Reprint. Sagapress Inc./Timber Press, Portland, Oregon.

Rodd, T. (Consul.). 2001. The Plant Book: The World of Plants in a Single Volume. James Mills-Hicks, Australia.

Rogers, A. 1995. Peonies. Timber Press, Portland, Oregon.

Rogers, D. J. and C. Rogers. 1991. Woody Ornamentals for Deep South Gardens. University of West Florida Press, Pensacola.

Rogers, G. K. 1985. The Genera of Phytolaccaceae in Southeastern United States. J. Arnold Arb.: 66: 1–37.

Rogers, G. K. 1986. The Genera of Loganiaceae in the Southeastern United States. J. Arnold Arb. 67: 143–185.

Rogers, G. K. 1982. The Casuranaceae in the Southeastern United States. J. Arnold Arb. 63. 357–373.

Rogers, J. E. 1905. The Tree Book: A Popular Guide to Knowledge of the Trees of North America and to their Uses and Cultivation. Doubleday, Page & Company, New York.

Rolfe, J. 1992. Gardening with Camellias: A New Zealand Guide. Godwit.

Romanowski, N. 2002. Gardening with Carnivores: *Sarracenia* Pitcher Plants in Cultivation & Wild. University Press of Florida, Gainesville.

Rosart, S. (ed.). 2000. Pansies. MetroBooks, Michael Freedman Publishing Group, Inc., New York.

Rose, Q. P. 1990a. Climbers and Wall Plants: Including *Clematis*, Roses, and Wisterias. Blandford, An Imprint of Cosell, London.

Rose, Q. P. 1990b. Ivies. Blandford, An Imprint of Cassel, London.

Rosenfeld, R. 2001. Gardening Encyclopedia. Chancellor Press, Octopus Publishing Group, London.

Rowley, G. D. 1987. Caudiciform & Pachycaul Succulents: Pachycauls, Bottle-, Barrel, and Elephant-Trees and Their Kin: A Collector's Miscellany. Strawberry Press. Mill Valley, California.

Roxburgh, W. and N. Wallich. 1975. Flora Indica or Description of Indian Plants. Oriole Editions, New York.

Rushforth, K. 1999. Trees of Britain and Europe: Collins Wildlife Trust Guide Trees. Harper Collins Publishers, London.

Russell, T., C. Cutler, and M. Walters. 2007. Trees of the World: An Illustrated Encyclopedia and Identifier. Anness Publishing Ltd., London.

Sajeva, M. and M. Costanzo. 1994. Succulents: An Illustrated Dictionary. Timber Press. Portland, Oregon.

Salmon, J. T. 1963. New Zealand Flowers in Color. A. H. & A. W. Reed, Wellington: Auckland.

Samyn, G. 1995. *Vriesea* hybrids of today and yesteryear. A paper presented at the Australian Bromeliad conference held in Adelaide 1995. (online)

Sanders, R. W. 2006. Taxonomy of Lantana sect. Lantana (Verbenaceae): I. correct application of *Lantana camara* and associated names. SIDA 22: 381–421.

Sargent, C. S. 1922. Manual of the Trees of North America (Exclusive of Mexico). 2nd edition. 2 vols. Houghton Mifflin Company, New York.

Schery, R. W, 1972. Plants for Man. 2nd ed. Prentice-Hall, Inc. Englewood Cliffs, New Jersey.

Schmid, W. G. 1991. The Genus *Hosta* Giboshi Zoku. Timber Press, Portland, Oregon.

Schneck, M. 1992. Cacti: An Illustrated Guide to Over 150 Representative Species. Crescent Books, New York.

Schnell, D. E. 1976. Carnivorous Plants of the United States and Canada. John F. Blair, Publisher, North Carolina.

Schwarts, R. 1974. Carnivorous Plants. Prager Publishers, New York, Washington.

Science Education Center, University of the Philippines. 1971. Plants of the Philippines. University of the Philippines Press.

Sheehan, T. J. and M. R. Sheehan. 1995. An Illustrated Survey of Orchid Genera. Timber Press, Portland, Oregon.

Sheehan, T. J. and R. J. Black. 2007. Orchids to Know and Grow. University Press of Florida, Gainesville.

Shoup, G. M. 2000. Roses in the Southern Garden. The Antique Rose Emporium, Inc.

Silberhorn, G. M. 1982. Common Plants of the Mid-Atlantic Coast: A field Guide. The Johns Hopkins University Press, Baltimore and London.

Simões, A. R.; Staples, G. 2017. Dissolution of Convolvulaceae tribe Merremieae and a new classification of the constituent genera. *Botanical Journal of the Linnean Society*. 183: 561–586.

Simpson, B. B. and M. C. Ogorzaly. 1986. Economic Botany: Plants in Our World. McGraw-Hill, Inc. New York.

Simpson, D. P. 1979, Cassell's Latin Dictionary. Macmillan Publishing Co., Inc. New York.

Simpson, M. G. 2010. Plant Systematics. 2nd. Ed. Academic Press, Elsevier, Amsterdam, London, New York.

Slaba, R. 1992. The Illustrated Guide to Cacti. Sterling Publishing Co., Inc. New York.

Slade, N. and G. Lane. 2018. Dahlias: Beautiful Varieties for Home & Garden. Gibbs Smith, Publisher, Layton, Utah.

Slocum, P. D., P. Robinson, and F. Perry. 1996. Water Gardening: Water Lilies and Lotuses. Timber Press, Portland, Oregon.

Small, J. K. (1933) 1972. Manual of Southeastern Flora, parts one and two. Hafner Publishing Company, New York.

Smith A. W. and W. T. Stearn. 1963. A Gardener's Dictionary of Plant Names. St. Martin's Press, New York.

Smith, A. R., K. M. Pryer, E. Schuettpelz, P. Korall, H. Schneider, and P. G. Wolf. 2006. A classification for extant ferns. Taxon 55: 705–731.

Smith, N, S. A. Mori, A. Henderson, D. Wm. Stevenson, and Scott V. Heald. 2004. Flowering Plants of the Neotropics. Princeton University Press, Princeton and Oxford.

Soltis, P. A. and D. E. Soltis. 2009. The role of hybridization in plant speciation. Ann. Rev. Plant Biol. 60: 561–588.

Soltis, D., P. Soltis, P. Endress, M. Chase, S. Manchester, W. Judd, L. Majure, and E, Mavrodiev. 2018. Phylogeny and Evolution of the Angiosperms. The University of Chicago Press, Chicago and London.

Spooner, D. and R. G. van der Berg. 2003. Plant Nomenclature and Taxonomy: An Horticultural and Agronomic Perspective. Hort. Rev. Vol. 28. John Wiley & Sons, Inc. Hoboken, New Jersey.

Soreng, R.J., P.M, Peterson, K. Romaschenko, G. Davidse, J.K. Teisher, L.G. Clark, P, Barberá, L.J. Gillespie, and F.O. Zuloaga. 2017. A worldwide phylogenetic classification of the Poaceae (Gramineae) II: An update and a comparison of two 2015 classifications. J. Syst. Evol. 55(4): 259–290.

Sprunger, S. 2004. Orchideentafeln aus Curtis's Botanical Magazine. Eugen Ulmer GmbH & Co. Stuttgart, Germany. [This a compilation of orchid paintings from Curtis' Botanical Magazine, from 1787 to 1985]

Staff of the Bailey Hortorium, Cornell University. 1976, Hortus Third: A Concise Dictionary of Plants Cultivated in the United States and Canada. Macmillan Publishing Company, New York.

Staples, G. W. and D, R. Herbst. 2005. A Tropical Garden Flora: Plants Cultivated in the Hawaiian Islands and Other Tropical Places. Bishop Museum Press, Honolulu, Hawaii.

Starosta, P. and Y, Crouzet. 1998. Bamboos. Benedikt Taschen Verlag, GmbH, Hohenzollernring, Röln.

Stearn, W. T. 1992a. Botanical Latin: History, Grammar, Syntac, Terminology, and Vocabulary. 4th ed. Timber Press, Portland, Oregon.

Stearn, W. T. 1992b. Stearn's Dictionary of Plant Names for Gardeners: A Handbook on the Origin and Meaning of Botanical Names of Some Cultivated Plants. Cassel Publishing Limited, London.

Stephenson, R. 1994. *Sedum*: Cultivated Stonecrops. Timber Press. Portland, Oregon.

Steinmann, V. W. 2003. The Submersion of *Pedilanthus* into *Euphorbia* (Euphorbiaceae). Acta Botanica Mexicana 65: 45–50.

Stepnell, K. and T. James. 1986. Australia's Native Flowers. Chiold and Associates, An All-Australian Publisher, NSW, Australia.

Stevens, P. F. 2001–2021. Angiosperm Phylogeny Website, Version 14, July 2017 [and more or less continuously updated since] www.mobot.org/MOBOT/Research/APweb/

Stevenson, D. W. (ed.). 1990. *The Biology, Structure, and Systematic of Cycadales*. Proc. Symp. Cycad 87, Beaulieu-sur-Mer, France, April 17–22, 1087. Memoirs of the New York Bot. Gard. Vol. 57.

Struwe L, VA Albert, 2002. Gentianaceae: Systematics and Natural History. Cambridge University Press.

Stuart, M. [Ed.] 1982. VNR Color Dictionary of Herbs and Herbalism. Van Nostrand Reinhold Company, New York.

Styles, B. T. (ed). 1986. Inter specific Classification of Wild and Cultivated Plants. The Systematics Association Special Volume. Oxford University Press.

Stuppy, W., P. C. Van Welzen, P. Klinratana, and M. C. T. Posa. 1999. Revision of the Genera *Aleurites, Reutealis* and *Vernicia* (Euphorbiaceae). Blumea 44: 73–98.

Swahn, J. O. 1991. The Lore of Spices: their History and Uses Around the World. Crescent Books, Randomhouse Company, New York.

Swerdlow, J. L. 2000. Natures Medicine: Plants that Heal. National Geographic Society.

Swindells, P. 2003. The Water Garden Encyclopedia: The Ultimate Guide to Designing, Constructing, Planting, and Maintaining Garden Ponds and Water Features. Firefly Books (US) Inc., New York.

Tan, K. 1980. What are taxonomists and why are people saying these terrible things about us. Amer. Orchid Soc. Bull. 49: 120–125.

Taylor, J. 1987. Kew Gardening Guides: Climbing Plants. Timber Press, Portland, Oregon.

Taylor, J. 1995. Drought Tolerant Plants: Waterwise Gardening for Every Climate. Prentice Hall, New York.

Taylor, J. 1998. Special Plants. Quadrille Publishing Limited, London.

Taylor, N. 1965. The Guide to Garden Shrubs and Trees (including woody vines): Their Identity and Culture. Houghton Mifflin Company, Boston.

Tenenbaum, F. (Ed.). 2003. Taylor's Encyclopedia of Garden Plants: The Most Authoritative Guide to the Best Flowers, Trees, and Shrubs for North American Gardens. Houghton Mifflin Company, Boston, New York.

Tenenbaum, F. (Ed.). 2004. Taylor's Master Guide to Gardening. Houghton Mifflin Company, Boston, New York.

The Plant List. 2010. Version 1. Published on the Internet; http://www.theplantlisit.org

The Woman's Club of Havana. 1958. Flowering Plants from Cuban Gardens. (Plantas Floridas de los Jardines Cubanos). Criterion Books, New York.

Thomas, G. S. 1970. Plants for Ground-Cover. Published in Association with the Royal Horticultural Society. Sagapress, Inc./Timber Press, Portland, Oregon.

Thomas, G. S. 1989. The Rock Garden and Its Plants: From Alpine-to-Alpine House. Saga press, Inc./Timber Press, Inc. Portland, Oregon.

Thomas, G. S. 1990. Perennial Garden Plants or the Modern Florilegium. Saga Press, Inc./Timber Press, Portland Oregon.

Thompson, M. L. and E. J. Thompson. 1981. Begonias: The Complete Reference Guide. Times Books, New York.

Tiner, R. W. 1993. Coastal Wetland Plants of the Southeastern United States. The University of Massachusetts Press, Amherst.

Tomlinson, P. B. 1980. The Biology of Trees Native to Tropical Florida. Harvard University Printing Office, Allston, Massachusetts.

Tourjé, E. C. 1958. *Camelia* Culture. The Macmillan Company, New York.

Towe, L. C. 2004. American Azaleas. Timber Press, Portland. Cambridge.

Trehane, P. 1990. Index Hortensis. Vol. I: Perennials. Quarterjack Publishing, Wimborne, East Dorset, UK. [this is a valuable reference work that comprehensively lists cultivars of various perennial plants to date. Regretfully the list was never updated or other volumes published].

Trehane, P. 2004. 50 years of the International Code of Nomenclature for cultivated plans: future prospects for the code. Acta Hort. 634: 17–27.

Tripp, K. and A. Coombes. 1998. The complete Book of Shrubs. The Readers Digest Association, Inc. Pleasantville, New York.

Tripp, K. E. and J. C. Roulston. 1995. The Year in Trees: Superb Woody Plants for Four-Season Gardens. Timber Press, Portland, Oregon.

Tryon, R. 1986. The Biogeography of Species with Special Reference to Ferns. Botanical Review 52: 117–156.

Turland, N. J., Wiersema, J. H., Barrie, F. R., Greuter, W., Hawksworth, D. L., Herendeen, P. S., Knapp, S., Kusber, W.-H., Li, D.-Z., Marhold, K., May, T. W., McNeill, J., Monro, A. M., Prado, J., Price, M. J. & Smith, G. F. (eds.) 2018. *International Code of Nomenclature for algae, fungi, and plants (Shenzhen Code) adopted by the Nineteenth International Botanical Congress. Shenzhen, China, July 2017.* Regnum Vegetabile 159. Glashütten: Koeltz Botanical Books. https://doi.org/10.12705/Code.2018.

Uhl. N. W. and J. Dransfield. 1987. Genera Palmarum: A Classification of Palms Based on the Work of Harold E. Moore, Jr. The L. H. Bailey Hortorum and the International Palm Society. Allen Press, Lawrence, Kansas.

Usher, G. 1974. A Dictionary of Plants Used by Man. Constable, London.

Valder, P. 1995. Wisterias: A Comprehensive Guide. Timber Press, Portland, Oregon.

Valder, P. 1999. Garden Plants of China. The Orion Publishing Group, London.

Valleau, J. 2004. Plant name changes: good science, angry growers, and confused gardeners. Acta Hortic. 634: 63–66.

Van der Kooij, P.A.C.E. 2004. Use and misuse of trade marks in European nursery industry. Acta Hortic. 634: 37–43.

Van Gelderen, D. M., P. C. de Jong, and H. J. Oterdoom. 1994. Maples of the World. Timber Press, Portland, Oregon.

Van Jaarsveld, E. J. 1994. Gasterias of South Africa: A New Revision of a Major Succulent Group. Fernwood Press, Vlaeberg, South Africa.

Van Gelderen, D. M. and J. R. P. van Hoey Smith. 1996. Conifers: The Illustrated Encyclopedia. Volume I. A-K; Volume II. L-Z. Published in cooperation with The Royal Boskoop Horticultural Society by Timber Press. Portland, Oregon.

Vermeulen, N. 1998. The Complete Rose Encyclopedia. Gramercy Books, New York.

Vertrees, J. D. 1987. Japanese Maples. 2nd Ed. Timber Press, Portland, Oregon.

Viard, M. 1996. Flowers of the World. Longmeadow Press, Ann Arbor, Minnesota.

Vinnersten, A. and J. Manning, 2007. A new classification of Colchicaceae. *Taxon* 56: 163–169

Waite, S. (ed.). 1998. Orchid population biology: conservation and challenges. Bot. J. Linn. Soc. 126: 1–190.

Walker, J. and G. Hanley. 1992. The Subtropical Garden. Timber Press, Portland, Oregon.

Walters, T. and R. Osborn (eds.). 2004. Cycad Classification: Concepts and Recommendations. CABI, Oxfordhire, UK.

Warren, W. 1997. Tropical Plants for Home and Garden. Thomas and Hudson, Ltd., London.

Wasowski, S. and A. Wasaowski. 1991. Native Landscaping: Texas Region by Region Plants. Gulf Publishing Company, Houston, Texas.

Wasowski, S. and A. Wasaowski. 1994. Gardening with Native Plants of the South. Taylor Publishing Company, Dallas, Texas.

Wasshausen, D. C. and J. R. I. Wood. 2004. Acanthaceae of Bolivia. Contributions from the United States National Herbarium 49: 1–152. Smithsonian Institution.

Wasson, E. (Chief Consul.). 2001. Trees and Shrubs: Illustrated A-Z of over 8500 Plants. Global Book Publishing, Pty., Ltd., NSW Australia.

Watkins, J. V., T. J. Sheehan, and R. J. Black. 2005. Florida Landscape Plants: native and Exotic. 2nd. Ed. University Presses of Florida. Gainesville, Florida.

Weber, C. and W. J. Dress. 1968. Notes on the nomenclature of some cultivated begonias (Begoniaceae). Baileya 16: 113–136.

Webber, S. (Ed.). 1988. Daylily Encyclopedia. Webber Gardens, Damascus, Maryland.

Webster, G. L. 1956–1958. A Monographic Study of the West Indian Species of *Phylllanthus.* Vol XXXVI, J. Arnold Arb.

Webster, G. L. 1994. Systematics of the Euphorbiaceae. Ann. Miss. Bot. Gard. 81(1–2): 1–144 & 145 401.

Wee, Y. C. 2003. Tropical Trees and Shrubs: A Selection for Urban Plantings. Sun Tree Publishing Limited, London.

Wen, J. 2011. Systematic and Biogeography of *Aralia* L. (Araliaceae): Revision of Section *Aralia, Humiles, Nanae,* and *Sciadodendron.* Contribution from the United States National Herbarium 57: 1–172.

West, E. and L. E. Arnold. 1956. The Native Trees of Florida. University of Florida Press, Gainesville, Florida.

Wharton, M. E. and R. W. Barbour. 1973. Trees and Shrubs of Kentucky. The University Press of Kentucky, Lexington.

Wettstein, R. von. 1895. Solanaceae. In: Die Naturlichen Planzenfamilien. Engler, A. and K. Prantl (eds.).

Wherry, E. T. 1961. The Fern Guide: Northeastern and Midland United States and Adjacent Canada. Doubleday Nature Guides, Doubleday & Company, Inc., Garden City, New York.

Whistler, W. A. 2000. Tropical Ornamentals: A Guide. Timber Press, Portland, Oregon.

Whitelock, L. M. 2002. *The Cycads.* Timber Press, Portland, Oregon.

Whiteside, K. 1991. Classic Bulbs: Hidden Treasures for Modern Garden. Villard Books, New York.

Whitmore, T. C. 1980. A Monograph of *Agathis.* Pl. Syst. Ecol. 135: 41–69.

Whittingstall, J. F. 1999. Peonies. Harry N. Abrams, Inc., Publishers, New York.

Widrlechner, M. P., C. Daly, M. Keller, and K. Kaplan. 2012. Horticultural Applications of a Newly Revised USDA Plant Hardiness Zone Map. Hort Technology 22: 6–19.

Williams, W. 1981. Methods of production of new varieties. Phil. Trans. R. Soc. Lond. B 292: 421–430.

Wills, M. M. and H. S. Irwin. 1961. Roadside Flowers of Texas. University of Texas Press, Austin.

Wikipedia. 2001–2020. The Free Online Encyclopedia. Articles by various authors.

Wikipedia, Cultivated plant taxonomy – Wikipedia

Withner, C. A. 1980. How to write type or print an orchid name properly. Amer. Orchid Soc. Bull. 49: 384–387.

Woods, R. F. 1986. Pictorial Library of Landscape Plants. Vol. 2, Southern Hardiness Zones 6–10. Merchants Publishing Company, Kalamazoo, Michigan.

Wunderlin, R. 1998. Guide to Vascular Plants of Florida. University Press of Florida, Gainesville.

Yao-Wu Y., D. J. Mabberley, D.A. Steane and R. G. Olmstead. 2010. Further disintegration and redefinition of *Clerodendrum* (Lamiaceae): Implications for the understanding of the evolution of an intriguing breeding strategy. Taxon 59: 125–133.

Yeo, P. E. 1992. Hardy Geraniums. Timber Press, Inc., Portland, Oregon.

Yoshihiro O. and J. L. Hollinberg. 1981. The Miniature Palms of Japan: Cultivating Kannonchiku and Shurochiku. Weatherhill Publishing, Inc. Fairfield, CT.

Young, M. A., P. Schorr, and R. Baer. 2007. Modern Roses 12. American Rose Society.

Zhang, D., M. A. Dirr, and R. A. Price. 1999. Classification of cultivated *Cephalotaxus* species based on *rcbL* sequences. In: S. Andrews, A. C. Leslie, and C. Alexander (Eds.), Taxonomy of Cultivated Plants: Third International Symposium, pp. 265–275. Royal Botanic Garden Kew.

Zomlefer, W.B. 1994. Guide to Flowering Plant Families. The University of North Carolina Press, Chapel Hill & London.

References for Glossary

Henderson, I. F. and W. D. Henderson. 1920. A Dictionary of Scientific Terms. Seventh Edition, by J. H. Kenneth. D. Van Nostrand Company, Inc., Princeton, New Jersey.

Jackson, B.D. 1939. A Glossary of Botanic Terms. Fourth Edition. Duckworth, London.

Lawrence, G. H. M. 1951. Taxonomy of Vascular Plants. The Macmillan Company, New York.

Lincoln, R. J., G. A. Boxshall, and P. F. Clark. 1982. A Dictionary of Ecology, Evolution and Systematics. Cambridge University Press, Cambridge, London.

Little, R.J. and C.E. Jones. 1980. A Dictionary of Botany. Van Nostrand Reinhold Company, New York.

Radford, A. E., W. C. Dickison, J. R. Massey, and C. R. Bell. 1974. Vascular Plant Systematics. Harper and Row Publishers, New York.

Stearn, W. T. 1983. Botanical Latin. Third Edition. David and Charles, Inc. North Pomfret, Vermont.

Usher, G. 1966. A Dictionary of Botany. Constable, London.

Punt, W, S. Blackmore, S. Nilsson, and A. Le Thomas. 1993. Glossary of pollen and spore terminology. http://www.pollen.mtu.edu/glos-gtx/glos-p1.htm [NOTE: Much of the pollen and spore terminology has not been included in this glossary].

Index